Restoring Forests and Trees for Sustainable Development

Restoring Forests and Trees for Sustainable Development

Policies, Practices, Impacts, and Ways Forward

Edited By
PIA KATILA
CAROL J. PIERCE COLFER
WIL DE JONG
GLENN GALLOWAY
PABLO PACHECO
GEORG WINKEL

OXFORD
UNIVERSITY PRESS

OXFORD
UNIVERSITY PRESS

Oxford University Press is a department of the University of Oxford.
It furthers the University's objective of excellence in research, scholarship,
and education by publishing worldwide. Oxford is a registered trade mark of
Oxford University Press in the UK and in certain other countries.

Published in the United States of America by Oxford University Press
198 Madison Avenue, New York, NY 10016, United States of America.

Library of Congress Cataloging-in-Publication Data
Names: Katila, Pia, editor. | Colfer, Carol J. Pierce, editor. | Jong, Wil de, editor. |
Galloway, Glenn, editor. | Pacheco, Pablo, editor. | Winkel, Georg, editor.
Title: Restoring forests and trees for sustainable development :
policies, practices, impacts, and ways forward /
Pia Katila, Carol J. Pierce Colfer, Wil de Jong, Glenn Galloway, Pablo Pacheco, and Georg Winkel
Description: New York, NY : Oxford University Press, [2024] | Includes bibliographical references
Identifiers: LCCN 2024021434 (print) | LCCN 2024021435 (ebook) |
ISBN 9780197683927 (hardback) | ISBN 9780197683934 (epub)
Subjects: LCSH: Forest restoration. | Forest management. | Forest biodiversity conservation.
Classification: LCC SD409 .R493 2024 (print) | LCC SD409 (ebook) |
DDC 634.9/56—dc23/eng/20240607
LC record available at https://lccn.loc.gov/2024021434
LC ebook record available at https://lccn.loc.gov/2024021435

DOI: 10.1093/9780197683958.001.0001

Printed by Integrated Books International, United States of America

Disclaimer

The development of this publication was made possible by the financial support of the Ministry for Foreign Affairs of Finland and Natural Resources Institute Finland, and the in-kind contributions from many universities and research and development organizations, but the views expressed in this book do not necessarily represent those of these organizations.

Contents

Preface

Forests are crucial for human well-being, sustainable development, and ultimately, life on Earth. The crucial ecosystem services that forests provide are widely acknowledged, yet their provision is seriously threatened by continuing deforestation and forest degradation. In response to these detrimental processes, there has been an unprecedented recognition over the past few decades of the urgent need to restore forest ecosystems. This is reflected in various commitments worldwide. Forest restoration is an important element in global environmental agreements and in the United Nation's Sustainable Development Goals. Furthermore, the United Nations declared 2021–2030 the "Decade on Ecosystem Restoration," and international initiatives, such as the global Bonn Challenge and the complementary New York Declaration on Forests, promote the restoration of millions of hectares of degraded forest lands. These are supported by several regional initiatives and numerous pledges made by national governments, nongovernmental and civil society organizations, multilateral organizations, and private sector actors. The importance of ensuring effective ecosystems restoration has also been explicitly stated in the Kunming-Montreal Global Biodiversity Framework.

Forest restoration and tree planting are often promoted as win-win actions that benefit both people and nature and as important solutions to mitigate climate change and enhance biodiversity. Yet, restoration is not undertaken on idle, empty lands; on the contrary, the lands targeted for restoration are almost always used by local or non-local actors for different purposes. This raises a number of questions, for example, what are the goals of restoration, how is restoration understood, what is considered restoration, what approaches and technologies are employed, what lands or forests are restored, how are these areas currently used and by whom, what are the related trade-offs, who makes the relevant decisions, how is restoration financed, what are the benefits and costs of restoration, and who bears the costs and who benefits?

These questions and the complexity of the issues around ecologically, economically, and socially sustainable and just forest restoration and tree planting challenged and motivated the editorial board of the International Union of Forests Research Organizations-Special Project World Forests, Society and Environment (IUFRO-WFSE) to develop this book. The overarching aim is to provide a comprehensive overview of forest restoration drivers, commitments, approaches, and achievements and to present the current state of knowledge and range of ideas regarding interacting social, economic, political, and ecological aspects of restoring forests and trees. The book focuses on the linkages between forest restoration, biodiversity, and ecosystem services and addresses forest restoration from the point of view of socioeconomic development, equity, and governance considerations in the context of shifting societal goals. It provides science-based information for developing sustainable restoration policies and programs, guiding the planning and implementation of restoration efforts in different contexts, and developing pathways and approaches to mitigate the social, cultural, economic, and environmental risks associated with restoration as well as enhance the multiple benefits to be accrued in the short and long run.

WFSE is a global, independent network of scientists and experts. It facilitates collaboration among a large number of scientists and experts from different parts of the world and is coordinated by the Natural Resources Institute Finland (Luke). WFSE supports sustainable forest-related development, livelihoods, and inclusive growth, and the related policy processes. It focuses on topics in the forests, society, and environment interphase that the scientific community recognizes as important, but which do not seem to be receiving adequate attention from the policy community. This is the fifth large volume developed by WFSE. The previous volumes focused on broad topics, such as the Sustainable Development Goals and their impacts on forests and people, forest development discourses, sustainable forest management, and forest-related development.

The development of this book was a collaborative effort of six editors and 62 scientists from research, conservation, and development organizations and universities from around the world. We invited internationally renowned scientists with diverse backgrounds and research interests, representing many disciplines, to lead the development of the different chapters of the book. These lead authors were asked to form and lead international author teams to develop their respective chapters.

The initiation of the process for developing this book collided with the COVID-19 pandemic; as such, the workshops planned for bringing together the editors and lead authors to discuss the development of the different chapters were organized as virtual meetings instead of in-person events. While the virtual events could not fully replace the in-person meetings, the discussions and exchange of ideas and commenting on the chapter drafts in these virtual meetings and via e-mails enabled the editors and lead authors to discuss common issues and support the development of each chapter.

Reflecting the projects and discourses around forest restoration, one of the recurring issues in the discussions among the authors was the different understandings of what is meant by and included in forest restoration. The various views and understandings of restoration as well as what is considered important for "successful" restoration mirrors differences between disciplines. By discussing forest restoration from a multitude of perspectives, this book aims to present a range of understandings and ideas of and approaches to restoration and to highlight the importance of cross-disciplinary approaches in forest-restoration-related research as well as for developing, planning, and implementing forest restoration policies and practices.

Acknowledgments

The development of this book relied on the voluntary in-kind contributions of the large group of authors. Without their dedication and generosity, the development of this book would not have been possible. Many of them contributed to this book on their own time. We sincerely thank all of them for their commitment and outstanding efforts for developing the different chapters of the book. We also thank the authors' and editors' host organizations for supporting part of the authors' work.

For ascertaining the scientific quality of the book, Chapters 2–11 were reviewed by the internationally renowned scientists and experts listed below. Their suggestions and comments were very important for improving the quality and relevance of these chapters. We are grateful for the time they devoted to reviewing the chapters and their crucial contributions for the development of this book.

Reviewers:

Päivi Abernathy
Cristina Adams
Raf Aerts
Stibniati Atmadja
Brett Bennett
Marco Boscolo
Susan Cook-Patton
Helen Ding
Matthew Fagan
Harry Fischer
Robert Fisher
Forrest Fleischman
Manuel Guariguata
Reem Hajjar
Joy E. Hecht
Karen Holl
Anne Larson
Adrian Martin
Adrian Newton
Cora van Oosten
Lesley Potter
Francis E. (Jack) Putz
Vijay Ramprasad
Edward Webb
Leland Werden

We are grateful for the financial support from the Ministry for Foreign Affairs of Finland that made the development and publishing of this open-access book possible. We also thank the Natural Resources Institute Finland (Luke) for supporting the coordination of the WFSE project, and the Secretariat of IUFRO for support in administrative issues.

The editors,
Pia Katila, Carol J. Pierce Colfer, Wil de Jong, Glenn Galloway, Pablo Pacheco, Georg Winkel

Abbreviations and acronyms

ACM	Adaptive Collaborative Management
ADB	Asian Development Bank
AFoCO	Asian Forest Cooperation Organization
AFR100	African Forest Landscape Restoration Initiative
AGB	Aboveground Biomass
ALS	Airborne Laser Scanning
ANR	Assisted Natural Regeneration
APFC	Asia-Pacific Forestry Commission
APFNet	Asia-Pacific Network for Sustainable Forest Management and Rehabilitation
BC	Bonn Challenge
BCR	Benefit-Cost Ratios
BEF	Biodiversity Ecosystem Functioning
BGB	Belowground Biomass
CBA	Cost-Benefit Analysis
CBD/UNCBD	United Nations Convention on Biological Diversity
CICES	Common International Classification of Ecosystem Services
CDM	Clean Development Mechanism
CIFOR	Center for International Forestry Research
COP	Conference of the Parties
CSR	Corporate Social Responsibility
ECCA	Europe, the Caucasus and Central Asia
ECOWAS	Economic Community of West African States
EJ	Environmental Justice
EPS	Extracellular Polymeric Substance
ES	Ecosystem Services
EU	European Union
FAO	Food and Agricultural Organization of the United Nations
FLEGT	Forest Law Enforcement, Governance, and Trade
FLR	Forest Landscape Restoration
FMNR	Farmer Managed Natural Regeneration
FPIC	Free, Prior, and Informed Consent
FRA	Forest Resources Assessment
FTE	Full-Time Equivalent
GCF	Green Climate Fund
GIS	Geographic Information System
GEF	Global Environment Facility
GGW	African Great Green Wall for the Sahara and the Sahel Initiative

GPFLR	Global Partnership on Forest Landscape Restoration
IADB	Inter-American Development Bank
IAM	Integrated Assessment Model
ICRAF	World Agroforestry Centre
IFAS	The International Fund for Saving the Aral Sea
IPBES	Intergovernmental Science-Policy Platform on Biodiversity and Ecosystem Services
IPCC	Intergovernmental Panel on Climate Change
IPLCs	Indigenous Peoples and Local Communities
ITTO	International Tropical Timber Organization
IUCN	International Union for Conservation of Nature
IUFRO	International Union of Forest Research Organizations
LDN	Land Degradation Neutrality
MCA	Multi-Criteria Analysis
MEA	Millennium Ecosystem Assessment
MSP	Mechanical Site Preparation
NBS	Nature-Based Solution
NCP	Nature's Contributions to People
NCS	Natural Climate Solutions
NDC	Nationally Determined Contribution
NGO	Nongovernmental Organization
NGP	National Greening Program
NPV	Net Present Value
NTFP	Non-Timber Forest Product
NWFP	Non-Wood Forest Product
NYDF	New York Declaration on Forests
OECD	Organisation for Economic Co-operation and Development
PES	Payment for Ecosystem Services
PGPB	Plant Growth Promoting Bacteria
PGR	Plant Growth Regulator
PWS	Payment for Watershed Services
REDD +	Reducing Emissions from Deforestation and Forest Degradation, and fostering conservation, sustainable management of forests and enhancement of forest carbon stocks
ROAM	Restoration Opportunities Assessment Methodology
ROI	Return on Investment
RRI	Rights and Resources Initiative
SAARC	South Asian Association for Regional Cooperation
SDG	Sustainable Development Goal
SER	Society for Ecological Restoration
SOM	Soil Organic Matter
SSP	Shared Socioeconomic Pathway
TEV	Total Economic Value
UN	United Nations

UNCCD	United Nations Convention to Combat Desertification
UNDP	United Nations Development Programme
UNECE	United Nations Economic Commission for Europe
UNEP	United Nations Environment Programme
UNFCCC	United Nations Framework Convention on Climate Change
WB	World Bank
WEF	World Economic Forum
WRI	World Resources Institute
WWF	World Wide Fund for Nature

Contributors

Tania Ammour, Senior Advisor to the Regional Direction of the International Union for Conservation of Nature (IUCN), Regional Office for Mexico, Central America and the Caribbean.

Nathaniel Anderson, Research Forester at United States Forest Service, Rocky Mountain Research Station.

Craig Beatty, Lead Specialist Forests Strategy and Research at World Wildlife Fund (WWF) Washington, DC.

Christophe Besacier, Senior Forestry Officer and Coordinator of the Forest and Landscape Restoration Mechanism (FLRM) at Food and Agriculture Organization of the United Nations (FAO).

Jürgen Blaser, Professor for International Forestry at Bern University of Applied Sciences, School of Agricultural, Forest, and Food Sciences.

Mareike Blum, Postdoctoral Researcher at the Mercator Research Institute on Global Commons and Climate Change (MCC) in Berlin.

Frans Bongers, Professor of Tropical Forest Ecology at Wageningen University & Research, Department of Environmental Sciences, Forest Ecology and Forest Management Group.

Pedro Brancalion, Professor of Tropical Forestry at Department of Forest Sciences, Luiz de Queiroz College of Agriculture, University of São Paulo and partner at Re.green, a company dedicated to large-scale tropical forest restoration.

David Brand, Executive Chairman at New Forests, Pty, Ltd.

Maria Brockhaus, Professor of International Forest Policy at the University of Helsinki.

Rens Brouwer, Ph.D. Candidate at Wageningen University & Research, Department of Environmental Sciences, Forest Ecology and Forest Management Group.

Susan Chomba, Director of Vital Landscapes for Africa at the World Resources Institute (WRI).

Carol J. Pierce Colfer, Senior Associate at Center for International Forestry Research-World Agroforestry (CIFOR-ICRAF).

Wil de Jong, Professor at the Chair of Silviculture, Faculty of Environment and Natural Resources, Freiburg University; Adjunct Professor at Renmin University of China, School of Agricultural Economics and Rural Development; Professor Emeritus at Kyoto University.

Ida N. Djenontin, Assistant Professor at Penn State University, Department of Geography and Visiting Fellow at London School of Economics and Political Science, Grantham Research Institute on Climate Change and the Environment.

Houria Djoudi, Senior Scientist at Center for International Forestry Research-World Agroforestry (CIFOR-ICRAF), Sustainable Landscapes and Livelihoods Team.

Adeline Dontenville, Project Manager at European Forest Institute (EFI), EU REDD Facility.

R. Kasten Dumroese, Senior Scientist/National Nursery Specialist at United States Forest Service, Rocky Mountain Research Station and Director, Western Center for Native Plant Conservation and Restoration Science.

Stephen Elliott, Associate Professor at Chiang Mai University, Biology Department, Co-founder and Research Director of Forest Restoration Research Unit.

Mélanie Feurer, Research Associate at Bern University of Applied Sciences, School of Agricultural, Forest, and Food Sciences.

Glenn Galloway, Director of the Master of Sustainable Development Practice Program at University of Florida Center for Latin American Studies and the Center for African Studies.

Joannès Guillemot, Researcher at CIRAD Joint Research Unit Eco&Sols and Visiting Professor at the University of São Paulo.

Sven Günter, Head of Working Area Forestry Worldwide at Thünen Institute of Forestry.

Alejandra Hernández Guzmán, Lecturer at Wageningen University & Research, Department of Environmental Sciences, Forest Ecology and Forest Management Group.

Alan Heinze, Researcher at Center for Research and Assistance in Technology and Design of the State of Jalisco (CIATEJ).

Vladan Ivetić, Professor at University of Belgrade, Faculty of Forestry.

Orjan Jonsson, Forestry Officer at Food and Agriculture Organization of the United Nations (FAO), Forest Resources Assessment Team.

Mawa Karambiri, Policy and Technical Engagement Specialist for the Sahel at Center for International Forestry Research-World Agroforestry (CIFOR-ICRAF), Ouagadougou, Burkina Faso.

Pia Katila, Senior Scientist at Natural Resources Institute Finland (LUKE) and Coordinator of the International Union of Forest Research Organizations' Special Project World Forests, Society and Environment (IUFRO-WFSE).

Rodney J. Keenan, Honorary Professor at University of Melbourne, School of Agriculture, Food and Ecosystem Sciences, Faculty of Science.

Watit Khokthong, Lecturer and Research Advisor at Chiang Mai University, Department of Biology and Forest Restoration Research Unit (FORRU), Faculty of Science, Chiang Mai University.

Michael Kleine, Deputy Executive Director at International Union of Forest Research Organizations (IUFRO) and Coordinator of the Special Programme for Development of Capacities (IUFRO -SPDC).

Daniela Kleinschmit, Professor of Forest and Environmental Policy at University of Freiburg.

Agata Konczal, Researcher at Wageningen University & Research, Department of Environmental Sciences, Forest and Nature Conservation Policy Group.

Koen Kramer, Director of Science at Land Life Company and Professor at Wageningen University & Research, Department of Environmental Sciences, Forest Ecology and Forest Management Group.

Markus Kröger, Professor at University of Helsinki, Department of Social Sciences, Global Development Studies.

Daniel Kübler, Scientist at Thünen Institute of Forestry, Working Area Forestry Worldwide.

Mait Lang, Associate Professor of Remote Forest Monitoring and Head of Department at Tartu Observatory, Department of Remote Monitoring.

Anne M. Larson, Principal Scientist and Team Leader at Center for International Forestry Research-World Agroforestry (CIFOR-ICRAF), Governance, Equity, and Wellbeing Team.

Magnus Löf, Professor and Head of Department at Swedish University of Agricultural Sciences, Southern Swedish Forest Research Centre.

Madelon Lohbeck, Assistant Professor at Wageningen University & Research, Department of Environmental Sciences, Forest Ecology and Forest Management Group.

Bas Louman, Programme Coordinator and Country Advisor Viet Nam at Tropenbos International.

Palle Madsen, Principal, Lead Forest Restoration and Practice Research at InNovaSilva ApS.

Jelena Markovic, Research Associate at Bern University of Applied Sciences, School of Agricultural, Forest, and Food Sciences.

Hemant Ojha, Associate Professor at University of Canberra and Australian National University and Principal Advisor at Institute for Study and Development Worldwide (IFSD).

Symphorien Ongolo, Researcher at the French National Institute for Sustainable Development (IRD–France), Joint Research Unit SENS and Research Associate at University of Göttingen, Department of Forest and Nature Conservation Policy.

Pablo Pacheco, Global Forests Lead Scientist at World Wide Fund for Nature (WWF) and Senior Associate at Center for International Forestry Research-World Agroforestry (CIFOR-ICRAF).

Jwalin Patel, Associate at IDinsight, New Delhi.

Marielos Peña-Claros, Professor at Wageningen University & Research, Department of Environmental Sciences, Forest Ecology and Forest Management Group.

Cindy Prescott, Professor at University of British Columbia, Department of Forest and Conservation Sciences.

Sabaheta Ramcilovik-Suominen, Associate Research Professor and Academy of Finland Research Fellow at Natural Resources Institute Finland (LUKE).

Sabine Reinecke, Research Associate at University of Freiburg, Chair of Silviculture.

Benjamin de Ridder, Forestry Officer at Food and Agriculture Organization of the United Nations (FAO), Forest and Landscape Restoration Mechanism (FLRM).

Cesar Sabogal, Independent, former Forestry Officer for Forest Management at Food and Agriculture Organization of the United Nations (FAO).

Douglas Sheil, Professor at Wageningen University & Research, Department of Environmental Sciences, Forest Ecology and Forest Management Group.

Kenichi Shono, Forestry Officer at the Food and Agriculture Organization of the United Nations (FAO), Forest Biodiversity and Restoration Team.

Silvio Simonit, Environmental Policy and Climate Change Specialist at Food and Agriculture Organization of the United Nations (FAO) Partnership and Liaison Office in Mexico.

Fergus Sinclair, Chief Scientist at Center for International Forest Research-World Agroforestry (CIFOR-ICRAF).

Almeida Sitoe, Professor at Eduardo Mondlane University, Faculty of Agronomy and Forestry.

John Stanturf, Visiting Professor at Estonian University of Life Sciences, Institute of Forestry and Engineering; Chair of Forest and Land Management and Wood Processing Technologies; and Senior Researcher, Forest Restoration Specialist at InNovaSilva ApS.

Takuya Takahashi, Professor at the University of Shiga Prefecture, School of Environmental Science.

Yitagesu Tekle Tegegne, Senior Researcher at European Forest Institute (EFI), Governance Programme; Coordinator of the Circular Bioeconomy Alliance; and Leader of EFI Global Forest Governance Team.

David Andrew Wardell, Principal Scientist at Center for International Forestry Research-World Agroforestry (CIFOR-ICRAF), Value Chains, Finance, and Investment Team.

Georg Winkel, Professor at Wageningen University & Research, Department of Environmental Sciences, Forest and Nature Conservation Policy Group.

Luo Xi, Program Officer at Asia-Pacific Network for Sustainable Forest Management and Rehabilitation (APFNet).

Timothy Young, Professor at University of Tennessee, Department of Forestry, Wildlife and Fisheries, Center for Renewable Carbon.

Rene Zamora Cristales, Director of Initiative 20x20 Secretariat and Senior Manager of Restoration Policy at World Resources Institute (WRI).

Pieter Zuidema, Professor, Tropical Forest Ecology at Wageningen University & Research, Department of Environmental Sciences, Forest Ecology and Forest Management Group.

1
Introduction
Restoring forests and trees for sustainable development

Pia Katila, Carol J. Pierce Colfer, Wil de Jong, Glenn Galloway, Pablo Pacheco, and Georg Winkel

1.1 The aim of this book

The 1987 Report of the World Commission on Environment and Development: Our Common Future defines sustainable development as "development that meets the needs of the present without compromising the ability of future generations to meet their own needs" (UN 1987, 43). This is the most often cited definition of sustainable development. It comprises three interdependent and mutually reinforcing pillars that need to be balanced—economic development, social development, and environmental protection (UN 2002, 2). This principle is reiterated in Agenda 2030 and its 17 sustainable development goals (SDGs), adopted in 2015 by the United Nations (UN) member states. Agenda 2030 states that the goals "are integrated and indivisible and balance the three dimensions of sustainable development: the economic, social, and environmental" (UN 2015a, 1). It further affirms that "social and economic development depends on the sustainable management of our planet's natural resources" (UN 2015a, 9). The main principles of Agenda 2030 center on People, Planet, Prosperity, Peace, and Partnerships, emphasizing human well-being, especially ending poverty and hunger, encouraging just and inclusive societies, prosperity for all, and the need to sustainably manage natural resources and protect the planet from degradation. It envisages a world "in which humanity lives in harmony with nature" (UN 2015a, 4).

Forests and trees provide important ecosystem services that contribute to all aspects of human well-being and are fundamental for global sustainability and sustainable development. The provision of these services is threatened by continuing forest loss and forest degradation and associated biodiversity loss. Deforestation and land degradation together with climate change have been recognized as constituting a major threat to the provision of crucial ecosystem services, thus jeopardizing human welfare (IPBES 2018; WEF 2021). Scientists have shown that we have already crossed some of the boundaries that define the safe "operating environment" for human existence, particularly those for climate change and biodiversity (Steffen et al. 2015).

As a response to these detrimental processes, in recent decades there has been an unprecedented increase in the acknowledgment of the urgent need to restore forest ecosystems along with increased political commitments toward this aim (Humphreys et al. 2019). While there is a long history of reforestation and tree planting for various purposes in different parts of the world, such as for timber production, fuelwood, or

Pia Katila, Carol J. Pierce Colfer, Wil de Jong, Glenn Galloway, Pablo Pacheco, and Georg Winkel, *Introduction*
In: *Restoring Forests and Trees for Sustainable Development*. Edited by: Pia Katila, Carol J. Pierce Colfer, Wil de Jong, Glenn Galloway, Pablo Pacheco, and Georg Winkel, Oxford University Press.
DOI: 10.1093/9780197683958.003.0001

watershed protection (Lamb 2011; 2014), the current attention to forest restoration and tree planting is in theory being undertaken to provide much broader social and ecological benefits and to support many facets of sustainable development. Forests and their conservation, restoration, and management are specifically included in SDG 15 Life on Land, but they are importantly linked to almost all other SDGs (Katila et al. 2019).

There is a large and growing body of literature that focuses on forest restoration, much of which concentrates on ecological and technical aspects. While this knowledge is crucial for restoration efforts, there is increasing recognition of the importance of the social, institutional, economic, cultural, and political factors that influence restoration goals and outcomes; how, when and where restoration takes place; as well as the impacts that restoration efforts have on livelihoods in the locations and landscapes where implementation occurs. Yet, there is a lack of comprehensive understanding of the social, economic, political, and environmental contexts of restoration activities and the related trade-offs and synergies; there is a need for interdisciplinary research that draws on the wide range of experiences gained in different socioeconomic and ecological contexts (Chazdon et al. 2017; Elias et al. 2021; Hagen et al. 2013; Löfqvist et al. 2023; Mansourian and Parrotta 2018; Perring et al. 2015; Wortley et al. 2013).

Advances in ecological science, silvicultural management, and gene technology, as well as technological developments such as the internet, mobile phones, and drones are rapidly changing restoration practices and opening up new arenas for collaboration with local communities. Furthermore, developments in GIS and satellite technology provide better data for analyzing land and forest cover and for estimating the needs and potentials for forest restoration at different scales from local to national to global. The accumulation of scientific information and technological developments calls for a systematic assessment and evaluation of the current state of knowledge, the possibilities and constraints for its utilization, and identification of knowledge gaps.

Scientists have often seen their task as providing the answers to be taken up "as is." It is increasingly being recognized, however, that technologies will be accepted, adapted, or rejected depending on the differing interests, knowledge, capabilities, and goals of local stakeholders. Moreover, increasing awareness of the value of indigenous knowledge is pushing us toward more collaborative and responsive approaches. There is a new openness to holistic and adaptive approaches that strive to understand and integrate indigenous knowledge and to tailor technology that is responsive and appropriate to particular social and ecological contexts.

As support for forest restoration gains momentum, there is a growing recognition among scientists and technicians that greater attention must be devoted to the socioecological contexts where restoration is implemented and especially to issues of power and politics, which crucially shape restoration goals, policies, and approaches and the related outcomes. How restoration is understood, which targets are chosen, and the related aims, measures, and policies implemented can lead to very different outcomes: social and ecological trajectories; implications for biodiversity, ecosystem services, and livelihoods involving potential synergies and trade-offs.

This book contributes to finding and developing nature- and people-based solutions to the crucial environmental challenges that threaten sustainable development and human well-being. It supports the development of sustainable restoration policies and programs, the planning and implementation of restoration efforts in different contexts, and the choosing and development of pathways and approaches to mitigate the social, cultural, economic, and environmental risks associated with restoration.

The chapters of this book provide a comprehensive overview of the current state of knowledge regarding the interacting social, economic, political, and ecological aspects of restoring forests and trees, bringing together the latest scientific information and technical innovations. Each chapter of the book focuses on a different aspect of forest restoration. Chapter 2 explores the history of forest restoration, providing a review of historical cases and assessing their outcomes and legacies for contemporary forest restoration. The next two chapters explore the governance of (forest) landscape restoration: Chapter 3 analyzes the institutions, interests, and ideas underlying the (de)legitimizing narratives of restoration efforts and driving related policy and practice; Chapter 4 explores issues of justice in restoration governance, drawing on environmental and decolonial justice perspectives and presenting ways of enhancing justice in international interventions. Chapter 5 explores how restoration is proposed and implemented across selected world regions, related achievements and challenges, and how various contextual conditions influence these processes. The following three chapters examine the services forests provide, including benefits, synergies, and trade-offs: Chapter 6 does so with respect to climate change mitigation and adaptation; Chapter 7, biodiversity and ecosystem services; and Chapter 8, local development and livelihoods. Chapter 9 assesses the costs and benefits—economic, financial, social, and environmental—of forest restoration, drawing on available studies and data across different restoration options. Chapter 10 considers the financial, ecological, political, and social feasibility of forest restoration targets. Chapter 11 examines technological advances in the field of forest restoration, providing a survey of innovative techniques that are useful across wide ecological and socioeconomic settings for forest stand initiation. The final chapter synthesizes the findings of the previous chapters and discusses their implications for the current debates on forest restoration. It summarizes the major challenges of forest restoration and suggests how they might be overcome, reflecting on the best ways to translate the global restoration project into feasible pathways of doing restoration in a just and sustainable manner.

A comprehensive review of the existing literature serves to identify important information needs and knowledge gaps in current scientific and practical knowledge that have a bearing on the outcomes of ecologically, economically, and socially sustainable restoration.

1.2 Restoration as a nature- and people-based solution to global challenges

The critical global challenges that threaten human welfare and its crucial foundation—nature—range from biodiversity loss and climate change to interrelated problems of poverty, hunger, inequality, and global health risks, such as the COVID-19 pandemic. The common denominator of these challenges is that they are largely caused by human (in)action severely altering natural ecosystems and leading to resource degradation, pollution, and climate warming (Díaz et al. 2019). Importantly, different populations suffer the consequences of such (in)actions—some conducted in situ, some elsewhere—to differing degrees. Echoing the results of the analyses of planetary boundaries (Steffen et al. 2015), the World Economic Forum's Global Risks Report 2021 (WEF 2021) ranked human-induced environmental damage, extreme climate, climate action failure, and biodiversity loss among the most severe long-term risks.

The central role nature plays in addressing global challenges is increasingly being recognized; nature features more and more in the approaches to address and find solutions to different social, economic, and ecological problems and challenges. Various terms are used to refer to such approaches: ecosystem approaches, ecosystem-based adaptation and mitigation, green infrastructure, and natural climate solutions, among others. They all focus on the role of ecosystems, ecosystem functions, and ecosystem services for addressing the global challenges and finding sustainable solutions to support the social, ecological, and economic aspects of sustainable development and to address climate change. Nature-based solutions (NBS) is an umbrella term that covers a wide range of these approaches (Cohen-Shacham et al. 2019). In recent years NBS has been adopted in international environmental policy discourses (e.g., IPCC, UNFCCC, IPBES) (Chausson et al. 2020) and in the European Union's new perspective for building innovative synergies between nature, society, and the economy (Faivre et al. 2017). These approaches have been criticized for emphasizing ecological issues and processes and failing to adequately include local actors and to consider the needs, capacities, and knowledge of local people who use and manage the ecosystems (Fleischman et al. 2020; Stavroula Melanidis and Hagerman 2022).

The International Union for Conservation of Nature (IUCN 2016, 166) defines NBS as "actions to protect, sustainably manage, and restore natural or modified ecosystems, that address societal challenges effectively and adaptively, simultaneously providing human well-being and biodiversity benefits" aiming "to support the achievement of society's development goals and safeguard human well-being in ways that reflect cultural and societal values and enhance the resilience of ecosystems, their capacity for renewal and the provision of services; Nature-based Solutions are designed to address major societal challenges, such as food security, climate change, water security, human health, disaster risk, social and economic development" (IUCN 2016, 167). The European Commission refers to NBS as "solutions that are inspired and supported by nature, which are cost-effective, simultaneously provide environmental, social and economic benefits and help build resilience. Such solutions bring more, and more diverse, nature and natural features and processes into cities, landscapes and seascapes, through locally adapted, resource-efficient and systemic interventions."[1] As these definitions indicate, forest and forest landscape restoration are important elements of NBS; they are crucial for the global efforts to advance sustainable development and to achieve the goals of the Paris Agreement on Climate Change, Agenda 2030, and the 17 SDGs.

One common feature of these approaches is their emphasis on uncertainty and diversity within the interconnected and mutually influencing systems into which restoration efforts must be inserted. The diversity of contexts, hardly debatable, leads to the conclusion that restoration solutions must be tailored to individual cases and circumstances; local communities must be involved in adaptive processes that incorporate attention to their own sociocultural and livelihood realities as well as to external influences and changing circumstances (Colfer et al. 2022; Colfer and Prabhu 2023). Just as the natural contexts must be attended to, so must the human.

Restoration of forests and tree planting are undertaken for many reasons, for example, to increase food and water security, provide renewable energy, boost rural employment, and help protect biodiversity. The contributions of restoration to addressing sustainability challenges may *in theory* be realized through three different but closely interlinked pathways: 1) through the process of restoration and tree planting, e.g., implementing restoration can provide job opportunities for local people, increase incomes, and contribute to increased capacities in restoration-related issues; 2) through the improved or restored forest/trees and the ecosystem services they provide, e.g., planted trees and restored forests improve water quality and increase the availability of non-timber forest products (NTFPs), wood fuels, and timber, possibly also providing opportunities for increased incomes, payments for ecosystem services, and improved food security; and 3) through changes in institutions, policies, and decision-making that govern land and resource use and management, and the sharing of costs and benefits (Adams et al. 2016; Erbaugh and Oldekop 2018; Wilson and Cagalanan 2016). The evidence on the outcomes, effectiveness, and equitability of restoration efforts along these pathways vary greatly, emphasizing the importance of the socioeconomic context in the locations and landscapes where restoration is undertaken (Löfqvist et al. 2023).

1.3 Forest restoration in global and regional agreements and commitments

Forest restoration is a complex undertaking that is closely linked to the global environmental agreements and related goals (Jauffret et al. 2020). Under the Convention for Biological Diversity (CBD), the post-2020 global biodiversity framework adopted in December 2022 aims at bringing 30% "of degraded terrestrial, inland waters, and coastal and marine ecosystems under effective restoration" (CBD 2022, 9). The prevention of forest and tree loss and the restoration or rehabilitation of degraded areas feature prominently in the efforts to combat desertification and land degradation under the UN Convention to Combat Desertification (UNCCD), established in 1994, and are seen as essential elements in countries' efforts toward land degradation neutrality (UNCCD 2019). The Paris Agreement of the UNFCCC (United Nations Framework Convention on Climate Change) makes an explicit reference to the need to "take action to conserve and enhance . . . sinks and reservoirs of greenhouse gases . . . including forests" (UN 2015b, Article 5, 6) and encourages countries to implement and support reducing emissions from deforestation and forest degradation and the role of conservation, sustainable management of forests, and enhancement of forest carbon stocks in developing countries (REDD+) (UN 2015b). In the UNFCCC's 26th Conference of the Parties 2021, over 140 countries reaffirmed their commitments to the global UN treaties, specifically to "sustainable land use, and to the conservation, protection, sustainable management and restoration of forests, and other terrestrial ecosystems" (Glasgow Leaders' Declaration on Forests and Land Use 2021).

In these global agreements, the role of forest restoration in addressing challenges has not been as clearly articulated for economic and social challenges as it has for environmental ones. While Agenda 2030 recognizes that the SDGs balance the economic, social, and environmental dimensions of sustainable development and that they are integrated and indivisible, the different goals and targets are presented in a siloed fashion without systematically considering complex linkages among them (Boas et al. 2016) or across the economic, social, and environmental spheres. Goal 15 focuses on the protection, restoration, and sustainable use of terrestrial ecosystems with Target 15.2 calling for restoration of degraded forests and a substantial increase in afforestation and reforestation; but this target is not integrated with the other SDGs, and their targets, and the efforts toward reaching them (Sayer et al. 2019).

In alignment with the global environmental agreements and the SDGs, the global voluntary restoration initiatives—the Bonn Challenge and the UN Decade on Ecosystem Restoration (2021–2030)—aim at restoring millions of hectares of degraded ecosystems and deforested landscapes through voluntary actions spearheaded by countries, organizations, and private entities. The underlying discourse of these initiatives integrates the different goals of the more siloed global agreements and emphasizes the potential contributions of restoration to support ecological integrity, address climate change, and enhance human well-being.

The Bonn Challenge seeks to facilitate the restoration of 350 Mha of degraded forest lands by 2030. Complementary international declarations, the New York Declaration on Forests (adopted 2014) and the Glasgow Leaders' Declaration (adopted in connection with the UNFCCC COP 26 in 2021), support this ambitious goal and have been endorsed by approximately 200 entities, including national and subnational governments, multinational companies, indigenous groups, and nongovernment organizations (Forest Declaration Platform 2023). Supporting the UN Decade on Ecosystem Restoration, the World Economic Forum's initiative engages the private sector to conserve, restore, and grow 1 trillion trees by 2030 (WEF 2020).

Regional initiatives supporting the Bonn Challenge include Initiative 20x20 in Latin America and the Caribbean, which aimed at protecting and restoring 20 Mha of forests, farms, pasture, and other landscapes by 2020, and the African Forest Landscape Restoration Initiative, which aims to restore 100 Mha in Africa by 2030 (Humphreys et al. 2019). Another regional initiative, the European Union-funded Regreening Africa project, aims to restore 1 Mha in Ethiopia, Ghana, Kenya, Mali, Niger, Rwanda, Senegal, and Somalia[2] (see Chapters 3 and 4). Furthermore, the Asia-Pacific Economic Cooperation (APEC) countries reached their 2007 Forest Cover Goal and reforested more than the targeted 20 Mha by 2020 (APFNet 2021).

Worldwide, in relation to the above-mentioned initiatives, multilateral organizations, national governments, nongovernmental and civil society organizations, and private sector actors have pledged and initiated restoration efforts that aim at restoring millions of hectares of deforested and degraded lands (Chazdon et al. 2017; IUCN 2015). It is crucial to consider how these aspirational goals and commitments are translated to national level targets and further to local or project-level objectives, and how adaptive and responsive these are to local conditions and restoration potential and the related technical, economic, and social opportunities and constraints. Attention to restoration governance at all scales is essential: how restoration is understood, how and by whom restoration objectives are decided upon and what outcomes are desired, how restoration is implemented, and who benefits and who bears the costs of restoration are all issues that have important implications for restoration outcomes and the linkages between restoration and global challenges at different spatial and temporal scales (Brancalion and Holl 2020; Chazdon et al. 2021). Options for building on local knowledge and restoration capacities and combining them with scientific and technical knowledge have been under-recognized and underutilized (Lake et al. 2018).

Governance is an important cross-cutting theme that relates to all aspects of restoration addressed in the different chapters of this book. While important governance issues are mentioned and discussed across chapters, two chapters focus especially on restoration governance. Chapter 3 focuses on forest landscape restoration (FLR) governance and power relations at the international level. It concentrates on the organizations and governance processes relevant for international FLR governance by analyzing the different institutional logics that drive FLR. The analyses are structured around institutions, interests, and ideas. Chapter 4 discusses land management and restoration-related equity and justice issues by using a decolonial environmental justice framework, with a special focus on indigenous and local communities.

1.4 Forest ecosystem services and drivers of forest loss

Following the Common International Classification of Ecosystem Services (CICES), the ecosystem services that forests provide can be grouped into three main categories: provisioning, regulation and maintenance, and cultural services (Haines-Young and Potschin 2014). *Provisioning services* include outputs and processes related to nutrition, material, and energy (such as cultivated crops, timber, NTFPs, water for drinking and other purposes, genetic materials, and biomass-based energy). *Regulation and maintenance services* include: mediation of waste, toxins, and other environmental contaminants (including filtration/sequestration/storage/accumulation by forest ecosystems); and mediation of flows, such as hydrological cycle and water flow maintenance, flood and storm protection, and maintenance of physical, chemical, and biological conditions (including water conditions, pest and disease control, and climate regulation). *Cultural services* relate to physical, intellectual, and spiritual interactions with environmental settings (including educational, aesthetic, symbolic, existence, sacred, and religious aspects) as well as the physical use of the environment, e.g., for recreation and tourism. Cultural meanings and services are often closely linked with the provisioning functions of forests.

All three types of services are intrinsically connected to the ecosystem functions, processes, and structures that generate them; thus, the capacity of the ecosystem to supply these services depends directly on the condition and biodiversity of the ecosystem in question (Haines-Young and Potschin 2014). Biological diversity is a critical foundation for the continued supply of ecosystem services; it can directly influence the availability of different services, such as the presence of viable populations of edible and medicinal plants or pollinating insects. The relationship between biodiversity and ecosystem service provision is highly complex and depends on the service in question, but generally the relationship is found to be positive (Harrison et al. 2014; Thompson et al. 2012). Most of the global terrestrial biodiversity can be found in forests, especially in tropical forests (FAO and UNEP 2020; Parrotta et al. 2012). The linkages between restoration, biodiversity, and ecosystem services are further discussed in Chapter 7. The role of restoration in mitigating climate change and adapting to it is discussed in Chapter 6.

Ecosystem functions, processes, and structures are often far more intimately connected with local communities' knowledge and environmental management than is typically recognized or acknowledged. Such management can involve swidden cultivation that grants long enough fallow periods for soils to recover and plant communities to reseed and regrow; management actions that prevent devastating fires due to excessive fuel loads, or that control the extent of purposely set fires; care and weeding of valuable wildlings of use to people; or replanting of desired species.

Globally, land-use change is the most important driver of forest and biodiversity loss (FAO 2018a; FAO and UNEP 2020; IPBES 2019). Deforestation and land and forest degradation together with climate change constitute a major threat to biodiversity and the provision of crucial ecosystem services, thus jeopardizing sustainable development and human welfare and increasing risks for current and future generations. As an example, recent research suggests that the conversion of natural habitats increases the risk of zoonotic diseases in humans, such as the global COVID-19 pandemic (Gibb et al. 2020). "Land-use change is a globally significant driver of pandemics and caused the emergence of more than 30% of new diseases reported since 1960" (IPBES 2020, 3). The role of site-specific population increase in such patterns can no longer be ignored.

According to the Global Forest Resources Assessment 2020 (FAO 2020), forests cover 4.06 billion ha, 31% of the global land area. Although the trend in the annual rate of deforestation[3] has been decreasing since the 1990s, during 1990–2020 forest loss totaled an estimated 420 Mha, of which 90% took place in the tropics. Annual forest loss decreased from 15.8 Mha per year in 1990–2000, to 15.1 Mha per year in 2000–2010, 11.8 Mha per year in 2010–2015, to 10.2 Mha per year in 2015–2020. Since 1990 the global primary forest[4] area has decreased by over 80 Mha; currently about one-third of global forests are primary forests.

Deforestation and forest degradation threaten biodiversity and forest ecosystems' capacity to provide ecosystem services. Degradation affects different ecosystem services in different ways, even increasing the provision of some while decreasing the provision of others. The perceptions about what is considered a preferred landcover, or a degraded condition, can also vary according to land use and cultivation systems in use (Dove 2019). Furthermore, effects of degradation and the related changes in ecosystem services are felt differently by different social groups; those who are more vulnerable and most dependent on natural resources are likely to be hurt the most (Potts et al. 2018).

It is estimated about one-quarter of the world's land area is degraded (FAO 2011). There is no commonly agreed upon definition of forest degradation, but in general terms it refers to "changes in forest structure, dynamics, and functions resulting mostly from human-induced causes relative to a preferred condition" (Thompson et al. 2013). While reliable global data on degradation does not exist, a significant area of tropical forest is estimated to be degraded (FAO 2020). A recent study shows that the extent and rate of forest degradation in the Brazilian Amazon was comparable to or even surpassed the deforestation rate during 1992–2014 (Matricardi et al. 2020).

Estimates of the degraded or deforested forest lands (or landscapes) with theoretical biophysical potential for restoration range from 0.9 (Bastin et al. 2019) to 2 billion ha (Minnemeyer et al. 2011), yet the areas where restoration is actually a viable option are likely to be much lower due to financial, operational, political, and social constraints (Zeng et al. 2020). Chapter 10 explores the feasibility of the current commitments considering ecological and technical feasibility, potential public and private finance, alternative land uses, products and markets, and social and cultural considerations. Chapter 9 further analyzes the economic aspects of restoration by focusing on the costs and benefits across multiple restoration options. As already alluded to, the social feasibility of forest restoration is intrinsically connected to restoration governance at and across different scales. The participation of local actors, their tenure rights, equity considerations, and the distribution of restoration costs and benefits are crucial issues. Local-level restoration governance and equity are discussed in more depth in Chapter 4, and livelihood related issues in Chapter 8.

Forest restoration aims, at least partly, at reversing ongoing global deforestation and forest degradation. In this context it is pertinent to look at the spatial and temporal linkages between deforestation and degradation and forest restoration. The analyses of the direct causes of deforestation paint a rather uniform picture: agricultural expansion, either for cropland or for livestock grazing, is the primary cause of forest conversion in all regions except Europe (FAO 2021; Pacheco et al. 2021). According to the FAO Remote Sensing Survey, in 2000–2018 more than 50% of global deforestation was caused by cropland expansion, and about 40% was the result of the increase in livestock grazing (FAO 2021).

An earlier analysis of 46 subtropical and tropical countries using FRA 2010 data concludes that agriculture contributed to about 80% of deforestation in 2000–2010. Commercial agriculture was the most important proximate cause for deforestation in these countries, accounting for 40%, followed by subsistence agriculture accounting for 33%. Clear differences were observed between different tropical regions. Commercial agriculture was a more important driver in Latin America (68%) than in Asia and Africa (about 35%), while the role of subsistence agriculture was similar across all regions, accounting for 27–40% of deforestation (Hosonuma et al. 2012). Pacheco et al. (2021) reached similar conclusions by using data compiled from different data sets for 2004–2017 on 24 deforestation fronts in the tropics and subtropics, where most forest loss is concentrated. Expansion of commercial agriculture is found to play a large role in Asia, while small-scale agriculture persists as the main driver of deforestation in Africa. Pacheco et al. (2021) highlight the role of tree plantations,[5] land speculation, infrastructure, and mining as drivers of deforestation in some locations.

Hosonuma et al. (2012) analyze forest degradation drivers and find that the main driver in Asia and Latin America is logging and timber extraction (70%), while in Africa it is fuelwood collection and charcoal production (48%). Across these regions, timber extraction and logging account for 52% of forest degradation, fuelwood collection and charcoal production 31%, uncontrolled fire 9%, and livestock grazing 7%.

The regional differences in the drivers of deforestation and degradation, responses to control them, and related outcomes are further analyzed in Chapter 5.

From the above it can be concluded that a large share of the areas deforested in recent decades are currently used for agriculture. Restoring these areas would mean reforesting and afforesting these lands or increasing the number of trees on agricultural land through the expansion of different agroforestry and silvopastoral systems. Addressing forest degradation would require natural or assisted regeneration of the

Table 1.1 Forest area change in 20 tropical forest countries 2010–2020 and their restoration commitments

20 tropical forest countries	Forest area 2020; Mha	Forest area change 2010–2020; Mha	Plantations change 2010–2020; Mha	Naturally regenerating forests change 2010–2020 Mha	Restoration commitment Mha
Brazil	496.6	−15.0	3.9	−18.9	22.0**
Dem. Rep. of Congo	126.2	−11.0	0.0	−11.0	8.0*
Indonesia	92.1	−7.5	0.3	−7.9	–
Peru	72.3	−1.7	0.1	−1.8	3.2**
India	72.2	2.7	0.5	2.2	26.0***
Angola	66.6	−5.6	−0.1	−5.5	–
Mexico	65.7	−1.3	0.0	−1.285	8.5**
Colombia	59.1	−1.7	0.0	−1.7	1.0**
Bolivia	50.8	−2.3	0.0	−2.3	–
Venezuela	46.2	−1.3	0.4	−1.6	–
Tanzania	45.8	−4.2	0.0	−4.2	5.2*
Zambia	44.8	−1.9	−0.0	−1.9	2.0*
Mozambique	36.7	−2.2	0.0	−2.3	1.0*
Papua New Guinea	35.9	−0.3	0.0	−0.3	–
Myanmar	28.5	−2.9	0.1	−3.0	–
Gabon	23.5	−0.1	0.0	−0.1	–
Central African Rep.	22.3	−0.3	0.0	−0.3	3.5*
Republic of the Congo	22.0	−0.1	0.0	−0.1	2.0*
Nigeria	21.6	−1.6	−0.0	−1.6	4.0*
Cameroon	20.3	−0.6	0.0	−0.6	12.0*

Sources: Forest area data from FAO 2020. Restoration commitments as of 1 August 2023 from *African Forest Landscape Restoration Initiative, https://afr100.org/countries; **Initiative 20x20, https://initiative20x20.org/regions-countries; ***Bonn Challenge, https://www.bonnchallenge.org/pledges?field_related_regions_target_id = 9.

degraded forest areas. It is important to keep in mind that these different restoration approaches and how they are managed have important implications for biodiversity, leading to the provision of different ecosystem services. For example, information on the restoration plans under the Bonn Challenge or in national schemes from 24 countries covering about 75% of the total pledged area indicates that a large part of the commitments (45%) would be reached by monoculture tree planting. While in most countries this means increasing the area of timber plantations, in several countries a larger area is planned to be planted for non-wood purposes such as oil palm, rubber, cocoa, and cashew (Lewis et al. 2019). The conflicts of such restoration efforts with local and indigenous peoples' livelihoods (agriculture) dramatizes the most "wicked problem" of restoration.

Table 1.1 presents the forest area changes in 20 large tropical forest countries in 2010–2020. Of these countries, forest area increased only in India, where both planted forest area and area under natural regeneration increased. In all other countries, forest area continued to decrease. The last column shows the countries' commitments under the Bonn Challenge and related regional initiatives for the area to be restored by 2030. For most countries, achieving these targets would mean halting deforestation and for some, considerable increase in forest cover within the next seven years. Yet, comparing forest area trends in these countries over the last decade to the pledges raises feasibility questions. These considerations are explored further in the book. In the first instance, the early historical restoration efforts, the drivers of forest restoration, and the types of forests that have resulted from previous restoration efforts are analyzed in Chapter 2. Then, using national and regional data, Chapter 5 analyzes the regional and national variations in restoration goals, approaches, outcomes, and challenges. Finally, Chapter 10 focuses on the feasibility of restoration, bringing together ecological, technical, financial social, and political perspectives.

1.5 Restoration goals, modalities, potential benefits, synergies, and trade-offs

1.5.1 Restoration modalities

In general terms, restoration refers to "any intentional activity that initiates or accelerates the recovery of an ecosystem from a degraded state" (IPBES 2018, 666). In relation to forests and trees, restoration includes a diversity of understandings and modalities. These differ in respect to the original state and status of the land, goals of restoration in terms of the desired outcomes, and the related biodiversity and ecosystem services. They also differ with respect to the actors and to the restoration measures, i.e., what actions are undertaken, how, where, and by whom (Mansourian 2018).

One differentiating aspect is the intensity of measures taken, which can range from ceasing the sources of manmade disturbance and protecting the site, allowing it to recover through *natural regeneration*, to measures involving assisting or accelerating recovery, e.g., seeding or planting and subsequent silvicultural practices. With respect to the intensity of measures taken and labor, other inputs, and finance needed, the different modalities form a continuum of interventions (Chazdon et al. 2021).

Reclamation refers to measures aimed at returning vegetation on a severely degraded and often bare land where vegetation has been destroyed, e.g., by mining (Stanturf et al. 2014).

Another important differentiation is based on the classification or use of the land before restoration. *Reforestation* refers to returning trees through planting and/or deliberate seeding to a site that is classified as forest, while *afforestation* refers to establishing trees on a site that was not previously, or not for a long time, used or classified as forest and implies a land use change from non-forest to forest (FAO 2018b).

Restoration efforts can also be differentiated according to the desired outcomes. The objectives of restoration efforts range from ecological to social, economic, and cultural, often including a combination of objectives with varying priorities.

Ecological restoration refers to "the process of assisting the recovery of an ecosystem that has been degraded, damaged, or destroyed" (SER 2004, 3). It seeks to partially or fully restore the native ecosystem and "aims to move a degraded ecosystem to a trajectory of recovery that allows adaptation to local and global changes, as well as persistence and evolution of its component species" (Gann et al. 2019, 7), but not necessarily to its historic state as it is recognized that the biotic and environmental conditions might have changed, e.g., due to climate change (ibid.). People's preferences for the landscape may also have changed over time.

Ecosystem restoration, is often used interchangeably with ecological restoration. In connection to the UN Decade of Forest Restoration, ecosystem restoration is described as a means of "assisting in the recovery of ecosystems that have been degraded or destroyed, as well as conserving the ecosystems that are still intact."[6] However, one differentiating principle between these concepts is that "ecological restoration addresses biodiversity conservation and ecological integrity, whereas some approaches to ecosystem restoration may focus solely on the delivery of ecosystem services" (Gann et al. 2019, 35). When understood this way, ecosystem restoration comes closer to the concept of *rehabilitation*, which can be understood as "reinstating a level of ecosystem functioning for renewed and ongoing provision of ecosystem services potentially derived from nonnative ecosystems as well" (Gann et al. 2019, 7).

Furthermore, the size of the area targeted for restoration can range from small areas of less than one hectare, such as area enclosures in Kenya (Mureithi et al. 2016), to large-scale restoration programs that target millions of hectares, such as national programs in China (Yin and Yin 2010).

Tree planting has historically been an important strategy for afforestation and reforestation. It includes the establishment of *large-scale plantations* as well as *smallholder and community tree planting* (De Jong et al. 2021). The latter often form part of agroforestry and silvopastoral systems, i.e., the integration of trees and woody shrubs in crop and livestock production systems (Miller et al. 2019). Large-scale plantations have mostly been established for the production of timber or wood for pulp and paper, while smallholder and community plantations have also targeted livelihood improvements. The integration of trees on farms and in agroforestry systems and practices can improve productivity and increase the resilience of smallholder production systems against climate change-related risks (Lasco et al. 2014).

The move toward landscape-scale natural resource management within recent decades is also reflected in restoration efforts. Landscapes encompass forest and non-forest ecosystems and different land uses. *Forest landscape restoration* (FLR) can include different combinations of the restoration modalities mentioned above as well as agroforestry and silvopastoral practices and plantation establishment. FLR is a "process that aims to regain ecological functionality and enhance human well-being in deforested and/or degraded landscapes . . . [It is] a means of regaining, improving, and maintaining vital ecological and social functions, in the long-term leading to more resilient and sustainable landscapes" (Besseau et al. 2018, 18). The principles of FLR include a focus on landscapes, restoring ecological functionality and aiming to create multiple benefits; these principles embrace a wide range of strategies—involving stakeholders, maintaining and enhancing natural ecosystems within landscapes, and enhancing resilience through adaptive management (Besseau et al. 2018). FLR reflects the increased use of adaptive and collaborative approaches in natural resource management in recent decades (Butler and Schultz 2019; Sayer and Boedhihartono 2018).

The choice of restoration modality is often limited by the state of the degraded area. For instance, in heavily degraded areas natural regeneration might not be possible, but trees can still be established through planting. Other limiting factors are the biophysical conditions, such as precipitation, soil, and altitude as well as the availability of labor, different technologies, and related capacities. These issues are further described in Chapter 11. However, the objectives of restoration are equally important in guiding which modality or combination of modalities to use. It is crucial to consider who defines the objectives and decides upon restoration modalities as well as who will bear the costs and who will reap the benefits (Chapter 4).

1.5.2 Restoration goals, potential benefits, synergies, and trade-offs

Motivations for tree planting and forest restoration are manifold, ranging from mitigating climate change to increasing timber production to improving local livelihoods. During the past decade there is an increased focus on the potential of forest restoration to address global environmental challenges, especially climate change and biodiversity loss. Concerns for the continuous provision of wood for industrial use and energy purposes, which historically have motivated tree planting and plantation development (Chapter 2), are accentuated by the projections showing increasing demand for wood products by a growing and more affluent global population. In addition, the need to replace fossil fuel-intensive energy sources and construction materials with renewable ones for climate change mitigation is driving wood demand. Recently, FLR has been widely endorsed as a people-centered approach for restoring degraded forest and agricultural landscapes and as an integrated solution for addressing ecological and social challenges, strengthening the inclusion of smallholders and local communities in restoration, and improving local livelihoods. The motivations, logics, and power relations that drive forest restoration at the global scale are discussed in Chapter 3 and the related impacts, outcomes, and challenges at the local scale in Chapters 4 and 8.

The outcomes of restoration efforts are strongly influenced by a range of interacting socioeconomic, political, and cultural factors. In addition, the various technical approaches to restoration will lead to ecologically different forest outcomes (secondary forests, plantations, agroforestry, or silvopastoral systems) and will thus provide different sets of ecosystem services and different levels of biodiversity (Wilson et al. 2017; Chapter 7). This will further define how restoration contributes to addressing global sustainability challenges and whether it will contribute to local livelihoods and support transition to adaptive and sustainable societies. For example, large-scale plantations by outside actors, local smallholder tree-planting activities in communal lands or smallholder farms, forest protection, and agroforestry practices can result in very different outcomes. An important related issue for the potential outcomes is what lands or land uses are targeted, i.e., whether restoration focuses on deforested or degraded forest land or landscapes, and what type of uses characterize these lands. Landscape approaches to restoration often include the restoration of different natural ecosystems as well as areas with crops and agroforests (Chazdon and Uriarte 2016).

The potential of forest restoration and tree planting to address local development and livelihoods is often discussed on the basis of the benefits that (existing) forests and trees provide; the actual benefits from restoration and related trade-offs have not been systematically studied (cf. Adams et al. 2016, Erbaugh and Oldekop 2018). On average, 22% of total household income in developing countries originates from forests (Angelsen et al. 2014). Reflecting the important role of forests and trees in rural livelihoods emphasized in the discourse around forest restoration and tree planting, these activities are often seen to hold considerable potential for livelihood improvement and income generation. This is accentuated by the close relationship between forests and poverty. A large part of the rural poor, more than 1.6 billion people, depend on forests for at least a portion of their livelihoods (WB 2004). According to FAO (2018a), an estimated 250 million people live below the extreme poverty line (USD 1.25/day) in and around tropical forests and savannas; the vast majority are in Africa (63%) and Asia (34%). This is not unrelated to the high birth rates that continue to trouble forest frontiers more generally (Sellers 2017). An estimated 60 million indigenous people depend almost fully on forests (WB 2004) and, in developing countries, about 1.2 billion people rely on agroforestry systems (WB 2004). Given that forests thrive under conditions of low population density, these statistics are troubling. According to FAO (2022), in 72 countries and territories that reported agroforestry, it covers in total 45.4 Mha. Erbaugh et al. (2020) estimate that close to 300 million people live on tropical forest restoration opportunity land. The linkages between restoration and local livelihoods are further discussed in Chapter 8 and the related governance and equity issues in Chapter 4.

The discourse around restoration emphasizes the important potential restoration has for climate change mitigation. Forests are the largest terrestrial carbon sink, and their restoration offers considerable potential to capture atmospheric CO_2 and thus mitigate climate change (Payn et al. 2015). Increasing forests and tree cover is estimated to hold vast potential for mitigating climate change by storing up to "an equivalent of 25% of the current atmospheric carbon pool" (Bastin et al. 2019). This would require a 25% increase in forested area (Bastin et al. 2019). However, the distribution and especially the scale of climate change mitigation opportunities, and the associated approaches, provided by forest restoration are still debated. Tropical areas offer the largest potential for afforestation, but there the trade-offs between afforestation and food production will likely be highest (Doelman et al. 2019). Chapter 6 focuses on restoration and climate change mitigation and adaptation. It analyzes and discusses different climate change mitigation potential estimates and forest restoration strategies as well as linkages to and implications for other land use-based mitigation strategies.

Important synergies among environmental objectives are possible. Strassburg et al. (2020) show that restoring 15% of converted lands worldwide could lead to avoiding 60% of the expected species extinctions while simultaneously sequestering vast quantities of carbon dioxide. They emphasize the importance of addressing the global challenges of biodiversity loss and climate change in a synergistic fashion (Strassburg et al. 2020). Which restoration modalities are used is crucial: large-scale plantations for bioenergy or timber production can increase carbon sequestration, at least in the short term, but they can be detrimental for both biodiversity and water availability and can also have negative impacts on local livelihoods and cultures.

It is important to recognize that the positive synergies between restoration/tree planting and specific sustainability challenges can obscure important trade-offs that result from the change in land use and land use practices. Trade-offs are likely to occur between environmental and social benefits (Uriarte and Chazdon 2016), such as between biodiversity and increased timber production for industrial use or between cultural services for local communities and timber production (Brancalion and Holl 2020; Bullock et al. 2011). Moreover, there are important differences in how the impacts of these trade-offs are distributed among different actors and over time (Byg et al. 2017). These are further discussed in Chapters 4 and 8. The trade-offs between different ecosystem services are discussed in Chapter 7.

1.6 Concepts used in this book

Unless otherwise specified, the term "forests" is used in this book as a general term and shorthand for forests and other land uses in which trees are an important component. The term "forest restoration" is understood broadly, referring to different approaches to bring back forests and trees to where they have existed before, including natural regeneration, reforestation, tree planting, plantation establishment, agroforestry systems, and FLR.

It is important to highlight the crucial differences between these approaches. These are discussed from different perspectives in the chapters to focus attention on the complexity and risks of restoration, while also pointing to measures and considerations to enhance its effectiveness. Each chapter uses the concepts that best serve its purpose, reflecting its perspective and scope.

Notes

1. https://ec.europa.eu/info/research-and-innovation/research-area/environment/nature-based-solutions_en
2. https://regreeningafrica.org/
3. FAO defines deforestation as the conversion of forest to other land uses (FAO 2020).
4. Defined as naturally regenerated forests of native tree species where there are no clearly visible indications of human activity and the ecological processes are not significantly disturbed (FAO 2020).
5. In Pacheco et al. (2021), conversion of natural forest to tree plantations was considered deforestation. In contrast, the FAO definition of deforestation is the conversion of forest to another land use or the permanent reduction of the tree canopy cover below the minimum 10% threshold. The FAO definition of forests includes rubberwood, cork oak, and Christmas tree plantations (http://www.fao.org/3/ap862e/ap862e.pdf).
6. https://www.decadeonrestoration.org/what-ecosystem-restoration

References

Adams, C., S.T. Rodrigues, M. Calmon, and C. Kumar. 2016. "Impacts of Large-Scale Forest Restoration on Socioeconomic Status and Local Livelihoods: What We Know and Do Not Know." *Biotropica* 48 (6): pp. 731–744.

Amezaga, J., J. Bathurst, A. Iroumé, et al. 2019. "SDG 6: Clean Water And Sanitation – Forest-Related Targets and Their Impacts on Forests and People." In *Sustainable Development Goals: Their Impacts on Forests and People*, edited by P. Katila, C.J.P. Colfer, W. de Jong, G. Galloway, P. Pacheco, and G. Winkel, pp. 178–205. Cambridge: Cambridge University Press.

Angelsen, A., P. Jagger, R. Babigumira, et al. 2014. "Environmental Income and Rural Livelihoods: Global-Comparative Analysis." *World Development* 64: pp. S12–S28. DOI:10.1016/j.worlddev.2014.03.006.

APFNet (Asia-Pacific Network for Sustainable Forest Management and Rehabilitation). 2021. *Achieving the APEC 2020 Forest Cover Goal.* A synthesis of economy reports. https://www.apfnet.cn/uploads/media/211025/1-211025113A4.pdf.

Bastin, J.-F., Y. Finegold, C. Garcia, et al. 2019. "The Global Tree Restoration Potential." *Science* 365 (6448): pp. 76–79. DOI:10.1126/science.aax0848.

Besseau, P., S. Graham, and T. Christophersen (eds.). 2018. *Restoring Forests and Landscapes: The Key to a Sustainable Future.* Vienna: Global Partnership on Forest and Landscape Restoration. https://www.forestlandscaperestoration.org/images/gpflr_final%2027aug.pdf.

Boas, I., F. Biermann, and N. Kanie. 2016. "Cross-Sectoral Strategies in Global Sustainability Governance: Towards a Nexus Approach." *International Environmental Agreements* 16: pp. 449–464. DOI:10.1007/s10784-016-9321-1.

Brancalion, P.H.S., and K.D. Holl. 2020. "Guidance for Successful Tree Planting Initiatives." *Journal of Applied Ecology* 57 (12): pp. 2349–2361. DOI:10.1111/1365-2664.13725.

Bullock, J.M., J. Aronson, A.C. Newton, R.F. Pywell, and J.-M. Rey-Benayas. 2011. "Restoration of Ecosystem Services and Biodiversity: Conflicts and Opportunities." *Trends in Ecology and Evolution* 26 (10): pp. 541–549. DOI:10.1016/j.tree.2011.06.011.

Butler, W.H., and C.A. Schultz (eds.). 2019. *A New Era for Collaborative Forest Management: Policy and Practice Insights from the Collaborative Forest Landscape Restoration Program.* London: Earthscan/Routledge.

Byg, A., P. Novo, M. Dinato, et al. 2017. "Trees, Soils, and Warthogs – Distribution of Services and Disservices from Reforestation Areas in Southern Ethiopia." *Forest Policy and Economics* 84: pp. 112–119.

CBD. 2022. *Kunming-Montreal Global biodiversity framework. Draft decision submitted by the President.* Conference of the parties to the convention on biological diversity Fifteenth meeting—Part II Montreal, Canada, 7–19 December 2022. CBD/COP/15/L.25. https://www.cbd.int/doc/c/e6d3/cd1d/daf663719a03902a9b116c34/cop-15-l-25-en.pdf.

Chausson, A., B. Turner, D. Seddon, et al. 2020. "Mapping the Effectiveness of Nature-Based Solutions for Climate Change Adaptation." *Glob Change Biol.* 26: pp. 6134–6155. DOI:10.1111/gcb.15310.

Chazdon, R.L., P.H.S. Brancalion, D. Lamb, L. Laestadius, M. Calmon, and C. Kumar. 2017. "A Policy-Driven Knowledge Agenda for Global Forest and Landscape Restoration." *Conservation Letters* 10 (1): pp. 125–132.

Chazdon, R.L., D.A. Falk, L.B. Banin, et al. 2021. "The Intervention Continuum in Restoration Ecology: Rethinking the Active–Passive Dichotomy." *Restoration Ecology* 21 August 2021. e13535. DOI:10.1111/rec.13535.

Chazdon, R.L., and M. Uriarte. 2016. "Natural Regeneration in the Context of Large-Scale Forest and Landscape Restoration in the Tropics." *Biotropica* 48 (6): pp. 709–715.

Chazdon, R.L., S.J. Wilson, E. Brondizio, M.R. Guariguata, and J. Herbohn. 2021. "Key Challenges for Governing Forest and Landscape Restoration across Different Contexts." *Land Use Policy* 104: 104854. DOI:10.1016/j.landusepol.2020.104854.

Cohen-Shacham, E., A. Andrade, J. Dalton, et al. 2019. "Core Principles for Successfully Implementing and Upscaling Nature-Based Solutions." *Environmental Science and Policy* 98: pp. 20–29. DOI:10.1016/j.envsci.2019.04.014.

Colfer, C.J.P. and R. Prabhu (eds.). 2023. *Responding to Environmental Issues through Adaptive Collaborative Management: From Forest Communities to Global Actors.* London: Earthscan.

Colfer, C.J.P., R. Prabhu, and A. Larson (eds.). 2022. *Adaptive Collaborative Management of Forest Landscapes: Villagers, Bureaucrats, and Civil Society.* London: Earthscan.

De Jong, W., J. Liu, and H. Long. 2021. "The Forest Restoration Frontier." *Ambio* 50 (12): pp. 2224–2237. DOI:10.1007/s13280-021-01614-x.

Díaz, S., J. Settele, E.S. Brondízio, et al. 2019. "Pervasive Human-Driven Decline of Life on Earth Points to the Need for Transformative Change." *Science* 366 (6471). DOI:10.1126/science.aax3100.

Doelman, J.C., E. Stehfest, D.P. van Vuuren, et al. 2019. "Afforestation for Climate Change Mitigation: Potentials, Risks and Trade-offs." *Global Change Biology* 26: 1576–1591. DOI:10.1111/gcb.14887.

Dove, M.R. 2019. "Climate Change and the Politics and Science of Traditional Grassland Management." In *Grasslands and Climate Change,* edited by D. J. Gibson, and J. A. Newman, pp. 276–292. Cambridge: Cambridge University Press.

Elias, M., D. Joshi, and R. Meinzen-Dick. 2021. "Restoration for Whom, by Whom? A Feminist Political Ecology of Restoration." *Ecological Restoration* 39 (1–2): pp. 3–15.

Erbaugh, J.T., and A. Oldekop. 2018. "Forest Landscape Restoration for Livelihoods and Well-Being." *Current Opinion in Environmental Sustainability* 32: 76–83. DOI:10.1016/j.cosust.2018.05.007.

Erbaugh J.T., N. Pradhan, J. Adams, et al. 2020 "Global Forest Restoration and the Importance of Prioritizing Local Communities." *Nature Ecology and Evolution* 4: pp. 1472–1476. DOI:10.1038/s41559-020-01282-2.

Faivre, N., M. Fritz, T. Freitas, B. Boissezon, and S. Vandewoestijne. 2017. "Nature-Based Solutions in the EU: Innovating with Nature to Address Social, Economic and Environmental Challenges." *Environmental Research* 159: pp. 509–518. DOI:10.1016/j.envres.2017.08.032.

FAO. 2011. *The State of the World's Land and Water Resources for Food and Agriculture (SOLAW)—Managing Systems at Risk.* London: Earthscan.

FAO. 2018a. *The State of the World's Forests 2018: Forest Pathways to Sustainable Development.* Rome: FAO.

FAO. 2018b. *Terms and Definitions. FRA 2020.* Forest Resources Assessment Working Paper 188. Rome: FAO. https://www.fao.org/3/I8661EN/i8661en.pdf.

FAO. 2020. *Global Forest Resources Assessment 2020—Main Report.* Rome: FAO.

FAO. 2021. *FAO Remote Sensing Survey Reveals Tropical Rainforests under Pressure as Agricultural Expansion Drives Global Deforestation.* https://www.fao.org/3/cb7449en/cb7449en.pdf.

FAO. 2022. *The State of the World's Forests 2022. Forest Pathways for Green Recovery and Building Inclusive, Resilient and Sustainable Economies.* Rome: FAO. DOI:10.4060/cb9360en.

FAO and UNEP. 2020. *The State of the World's Forests 2020. Forests, Biodiversity and People.* Rome. DOI:10.4060/ca8642en.

Fleischman, F., S. Basant, A. Chhatre, et al. 2020. "Pitfalls of Tree Planting Show Why We Need People-Centered Natural Climate Solutions." *BioScience* 70 (11): pp: 947–950. DOI:10.1093/biosci/biaa094.

Forest Declaration Platform. 2023. Who Are the NYDF Endorsers? https://forestdeclaration.org/about.

Gann, G.D., T. McDonald, B. Walder, et al. 2019. "International Principles and Standards for the Practice of Ecological Restoration." Second edition. *Restoration Ecology* 27 (S1–S46). DOI:10.1111/rec.13035.

Gibb, R., D.W. Redding, K.Q. Chin, et al. 2020. "Zoonotic Host Diversity Increases in Human-Dominated Ecosystems." *Nature* 584: pp. 398–402. DOI:10.1038/s41586-020-2562-8.

Glasgow Leaders' Declaration on Forests and Land Use. 2021. https://ukcop26.org/glasgow-leaders-declaration-on-forests-and-land-use/.

Hagen, D., K. Svavarsdottir, C. Nilsson, et al. 2013. "Ecological and Social Dimensions of Ecosystem Restoration in the Nordic Countries." *Ecology and Society* 18 (4): p. 34. DOI:10.5751/ES-05891-180434.

Haines-Young, R., and M. Potschin. 2014. "Typology/Classification of Ecosystem Services." OpenNESS Synthesis Paper No 2. https://oppla.eu/sites/default/files/uploads/spclassificationofecosystemservices.pdf.

Harrison, P.A., P.M. Berry, G. Simpson, et al. 2014. "Linkages between Biodiversity Attributes and Ecosystem Services: A Systematic Review." *Ecosystem Services* 9: pp. 191–203. DOI:10.1016/j.ecoser.2014.05.006.

Holl, K.D., and T.M. Aide. 2011. "When and Where to Actively Restore Ecosystems?" *Forest Ecology and Management* 261 (10): pp. 1558–1563. DOI:10.1016/j.foreco.2010.07.004.

Hosonuma, N., M. Herold, V. De Sy, et al. 2012. "An Assessment of Deforestation and Forest Degradation Drivers in Developing Countries." *Environ. Res. Lett.* 7 (4): p. 044009. DOI:10.1088/1748-9326/7/4/044009.

Humphreys, D., B. Singer, K. McGinley, et al. 2019. "SDG 17: Partnerships for the Goals: Focus on Forest Finance and Partnerships." In *Sustainable Development Goals: Their Impacts on Forests and People,* edited by P. Katila, C.J.P. Colfer, W. de Jong, G. Galloway, P. Pacheco, and G. Winkel, pp. 541–576. Cambridge: Cambridge University Press.

IPBES. 2018. *The IPBES Assessment Report on Land Degradation and Restoration,* edited by L. Montanarella, R. Scholes, and A. Brainich. Bonn: IPBES Secretariat.

IPBES. 2019. *Summary for Policymakers of the Global Assessment Report on Biodiversity and Ecosystem Services of the Intergovernmental Science-Policy Platform on Biodiversity and Ecosystem Services.* Bonn: IPBES Secretariat.

IPBES. 2020. *Workshop Report on Biodiversity and Pandemics of the Intergovernmental Platform on Biodiversity and Ecosystem Services.* Bonn: IPBES Secretariat. DOI:10.5281/zenodo.4147317.

IUCN. 2015. *Leaders Pledge to Restore Additional 18 Million Hectares of Critical Landscapes as Part of Global Target.* https://www.iucn.org/content/leaders-pledge-restore-additional-18-million-hectares-critical-landscapes-part-global-target.

IUCN. 2016. IUCN *Resolutions, Recommendations and other Decisions. World Conservation Congress Honolulu, Hawai'i, United States of America 6–10 September 2016.* Gland: IUCN. https://portals.iucn.org/library/sites/library/files/documents/IUCN-WCC-6th-005.pdf.

Jauffret, S., B. Bodin, A. Vidal, et al. 2020. "Forest and Landscape Restoration: Enhancing Synergies between the Rio Conventions." *Unasylva* 71 (252): pp. 92–101.

Katila, P., C.J.P. Colfer, W. de Jong, G. Galloway, P. Pacheco, and G. Winkel (eds.). 2019. *Sustainable Development Goals: Their Impacts on Forests and People.* Cambridge: Cambridge University Press.

Lake, F.K., C.P. Giardina, J. Parrotta, and I. Davidson-Hunt. 2018. "Considering Diverse Knowledge Systems in Forest Landscape Restoration." In *Forest Landscape Restoration. Integrated Approaches to Support Effective Implementation,* edited by S. Mansourian, and J. Parrotta, pp. 37–46. Abingdon-on-Thames: Routledge.

Lamb, D. 2011. *Regreening the Bare Hills. Tropical Forest Restoration in the Asia-Pacific Region.* Dordrecht: Springer.

Lamb, D. 2014. *Large-Scale Forest Restoration.* New York: Routledge.

Lasco, R.D., R.J.P. Delfino, and M. L. O. Espaldon. 2014. "Agroforestry Systems: Helping Smallholders Adapt to Climate Risks While Mitigating Climate Change." *WIREs Climate Change* 5: pp. 825–833. DOI:10.1002/wcc.301.

Lewis, S.L., C. E., Wheeler, E.T. Mitchard, and A. Koch. 2019. "Regenerate Natural Forests to Store Carbon." *Nature* 568: pp. 25–28. https://www.nature.com/articles/d41586-019-01026-8.

Löfqvist, S., F. Kleinschroth, A. Bey, et al. 2023. "How Social Considerations Improve the Equity and Effectiveness of Ecosystem Restoration." *BioScience* 73: pp. 134–148. DOI:10.1093/biosci/biac099.

Mansourian, S. 2018. "In the Eye of the Beholder: Reconciling Interpretations of Forest Landscape restoration." *Land Degradation & Development*. 29: pp. 2888–2898.

Mansourian, S., and J. Parrotta (eds.). 2018. *Forest Landscape Restoration. Integrated Approaches to Support Effective Implementation*. London: Routledge.

Matricardi, E.A.T., D.L Skole, O.B. Costa, M.A Pedlowski, J.H. Samek, and E. Miguel. 2020. "Long-Term Forest Degradation Surpasses Deforestation in the Brazilian Amazon." *Science* 369 (6509): pp. 1378–1382. DOI:10.1126/science.abb3021.

Miller, D.C., P.J. Ordoñez, S.E. Brown, et al. 2019. "The Impacts of Agroforestry on Agricultural Productivity, Ecosystem Services, and Human Well-Being in Low-and Middle-Income Countries: An Evidence and Gap Map." *Campbell Systematic Reviews* 16: p. e1066. https://doi.org/10.1002/cl2.1066.

Minnemeyer, S., L. Laestadius, N. Sizer, C. Saint-Laurent, and P. Potapov. 2011. *Atlas of Forest and Landscape Restoration Opportunities*. Washington: World Resources Institute. https://www.wri.org/data/atlas-forest-and-landscape-restoration-opportunities.

Mureithi, S.M., A. Verdoodt, J.T. Njoka, C.K.K. Gachene, and E. Van Ranst. 2016. "Benefits Derived from Rehabilitating a Degraded Semi-Arid Rangeland in Communal Enclosures, Kenya." *Land Degradation & Development* 27: pp. 1853–1862. DOI:10.1002/ldr.2341.

Pacheco, P., K. Mo, N. Dudley, et al. 2021. *Deforestation Fronts: Drivers and Responses in a Changing World*. Gland: WWF.

Parrotta, J.A., C. Wildburger, and S. Mansourian (eds.). 2012. *Understanding Relationships between Biodiversity, Carbon, Forests and People: The Key to Achieving REDD+ Objectives. A Global Assessment Report. Prepared by the Global Forest Expert Panel on Biodiversity, Forest Management, and REDD+*. IUFRO World Series Volume 31. Vienna: IUFRO.

Payn, T., J-M. Carnus, P. Freer-Smith, et al. 2015. "Changes in Planted Forests and Future Global Implications." *Forest Ecology and Management* 352: pp. 57–67.

Perring, M.P., R.J. Standish, J.N. Price, et al. 2015. "Advances in Restoration Ecology: Rising to the Challenges of the Coming Decades." *Ecosphere* 6 (8): p. 131. DOI:10.1890/ES15-00121.1.

Potts, M.D., T. Holland, B.F.N. Erasmus, et al. 2018. "Land Degradation and Restoration Associated with Changes in Ecosystem Services and Functions, and Human Well-Being and Good Quality of Life." In *The IPBES Assessment Report on Land Degradation and Restoration*, edited by L. Montanarella, R. Scholes, and A. Brainich, pp. 341–432. Bonn: IPBES Secretariat.

Sayer, J., and K.A. Boedhihartono. 2018. Integrated Landscape Approaches to Forest Restoration. In *Forest Landscape Restoration,* edited by S. Mansourian, and J. Parrotta, pp. 83–99. London: Routledge.

Sayer, J., D. Sheil, G. Galloway, et al. 2019. "SDG 15: Life on Land – The Central Role of Forests in Sustainable Development." In *Sustainable Development Goals: Their Impacts on Forests and People*, edited by P. Katila, C.J.P. Colfer, W. de Jong, G. Galloway, P. Pacheco, and G. Winkel, pp. 482–509. Cambridge: Cambridge University Press.

Sellers, S. 2017. "Family Planning and Deforestation: Evidence from the Ecuadorian Amazon." *Popul Environ* 38 (4): pp. 424–447. DOI:10.1007%2Fs11111-017-0275-1.

SER (Society for Ecological Restoration International Science & Policy Working Group). 2004. *The SER International Primer on Ecological Restoration*. Tucson: Society for

Ecological Restoration International. https://cdn.ymaws.com/www.ser.org/resource/resmgr/custompages/publications/ser_publications/ser_primer.pdf.

Stanturf, J.A., P.J. Palik, and R.K. Dumroese. 2014. "Contemporary Forest Restoration: A Review Emphasizing Function." *Forest Ecology and Management* 331 (1): pp. 292–323. DOI:10.1016/j.foreco.2014.07.029.

Stavroula Melanidis, M., and S. Hagerman. 2022. "Competing Narratives of Nature-Based Solutions: Leveraging the Power of Nature or Dangerous Distraction?" *Environmental Science & Policy* 132: pp. 273–281. DOI:10.1016/j.envsci.2022.02.028.

Steffen, W., K. Richardson, J. Rockström, et al. 2015. "Planetary Boundaries: Guiding Human Development on a Changing Planet." *Science* 347: pp. 736–746. DOI:10.1126/science.1259855.

Strassburg, B.B.N., A. Iribarrem, H.L. Beyer, et al. 2020. "Global Priority Areas for Ecosystem Restoration." *Nature* 586: pp. 724–729. DOI:10.1038/s41586-020-2784-9.

Thompson, I.D., J. Ferreira, T. Gardner, et al. 2012. "Forest Biodiversity, Carbon and Other Ecosystem Services: Relationships and Impacts of Deforestation and Forest Degradation." In *Understanding Relationships between Biodiversity, Carbon, Forests and People: The Key to Achieving REDD+ Objectives. A Global Assessment Report,* edited by J.A. Parrotta, C. Wildburger, and S. Mansourian, pp. 22–49. IUFRO World Series Volume 31. Vienna: IUFRO.

Thompson, I.D., M.R. Guariguata, K. Okabe, et al. 2013. "An Operational Framework for Defining and Monitoring Forest Degradation." *Ecology and Society* 18 (2): p. 20.

UN. 1987. *Report of the World Commission on Environment and Development: Our Common Future.* New York: UN.

UN. 2002. *Report of the World Summit on Sustainable Development.* Johannesburg: UN.

UN. 2015a. *Resolution Adopted by the General Assembly on 25 September 2015. Transforming Our World: The 2030 Agenda for Sustainable Development.* A/70/L.1. https://documents.un.org/doc/undoc/gen/n15/291/89/pdf/n1529189.pdf?token=X37UyenSTz8Eb3axcb&fe=true.

UN 2015b. *Paris Agreement.* United Nations, *Treaty Series*, vol. 3156, p. 79. https://treaties.un.org/doc/Treaties/2016/02/20160215%2006-03%20PM/Ch_XXVII-7-d.pdf.

UNCCD 2019. *Forests and Trees: At the Heart of Land Degradation Neutrality.* Bonn: United Nations Convention to Combat Desertification. https://www.unccd.int/sites/default/files/documents/2019-08/Forest%20brochure%20-%20web.pdf.

Uriarte, M., and R.L. Chazdon. 2016. "Incorporating Natural Regeneration in Forest Landscape Restoration in Tropical Regions: Synthesis and Key Research Gaps." *Biotropica* 48 (6): pp. 915–924.

WB. 2004. *Sustaining Forests: A Development Strategy.* Washington: World Bank.

WEF. 2020. *One Trillion Trees—World Economic Forum Launches Plan to Help Nature and the Climate.* https://www.weforum.org/agenda/2020/01/one-trillion-trees-world-economic-forum-launches-plan-to-help-nature-and-the-climate/.

WEF. 2021. *The Global Risks Report 2021.* Geneva: World Economic Forum.

Wilson, S.J., and D. Cagalanan. 2016. "Governing Restoration: Strategies, Adaptations and Innovations for Tomorrow's Forest Landscapes." *World Development Perspectives* 4: pp. 11–15.

Wilson, S.J., J. Schelhas, R. Grau, A. Sofía Nanni, and S. Sloan. 2017. "Forest Ecosystem-Service Transitions: The Ecological Dimensions of the Forest Transition." *Ecology and Society* 22 (4): p. 38. DOI:10.5751/ES-09615-220438.

Wortley, L., J.-M. Hero, and M. Howes. 2013. "Evaluating Ecological Restoration Success: A Review of the Literature." *Restoration Ecology* 21 (5): pp. 537–543.

Zeng, Y., T.V. Sarira, L. R. Carrasco, et al. 2020. "Economic and Social Constraints on Reforestation for Climate Mitigation in Southeast Asia." *Nature Climate Change* 10 (9): pp. 842–844.

Yin, R., and Q. Yin. 2010. "China's Primary Programs of Terrestrial Ecosystem Restoration: Initiation, Implementation, and Challenges." *Environmental Management* 45: pp. 429–441. DOI:10.1007/s00267-009-9373-x.

2

Forest flickers of history. Early modern woodland restoration and how it shapes postmodern options

Wil de Jong, Takuya Takahashi, Georg Winkel, Agata Konczal, Nathaniel Anderson, and David Andrew Wardell

2.1 Introduction

Forest restoration, here understood as bringing back forests where once they stood, has gained major attention in multiple international forums (Erbaugh et al. 2020). The area of restored forests has increased dramatically since the mid-20th century. For instance, the global area of planted forests has increased from 65 Mha around the early 1960s to close to 300 Mha in 2020 (de Jong et al. 2021; FAO 2020).

Forest restoration has been taking place in many countries for much longer. Well-known examples are from European countries, many of which were able to reverse a progressive decline in their national forest cover into a sustained forest cover increase (Mather 2001), much of which continues today (FAO 2020). Other, less well-known cases are of forest restoration undertaken by European colonial powers in countries like India, Indonesia, Myanmar, and the Philippines. Poorly researched cases include Japan, mainland southeast Asia, and China. Historical forest restoration has left little long-lasting imprints in most of these examples.

Given the prominence of forest restoration on the global policy agenda, looking back at past forest restoration experiences, such as those listed above, is relevant. For several countries, forest restoration in the early modern period and the first half of the 20th century resulted in significant contributions to their national forest estates which, in the Global Forest Resource Assessment statistics, qualify as planted forests (de Jong et al. 2021). Past forest restoration experiences can provide insights into the drivers that resulted in forest restoration and into how knowledge, interests, and politics shape forest restoration trajectories and outcomes. Understanding these past experiences might also provide insights into the environmental and social impacts of forest restoration. We argue that these past experiences might inform modern-day forest restoration decision-making and implementation.

In this chapter we understand forest restoration to be efforts to restore a tree cover that by prevalent definitions can be qualified as a forest. The return of a tree cover in a certain area of land can be the result of planting or seeding tree species but can also result from natural regeneration. The tree-planting efforts of interest are those in which a substantial, contiguous area of land is planted with trees with the intention to create a tree cover that meets a definition of forest—either the Food and Agriculture

Wil de Jong, Takuya Takahashi, Georg Winkel, Agata Konczal, Nathaniel Anderson, and David Andrew Wardell, *Forest flickers of history. Early modern woodland restoration and how it shapes postmodern options* In: *Restoring Forests and Trees for Sustainable Development.* Edited by: Pia Katila, Carol J. Pierce Colfer, Wil de Jong, Glenn Galloway, Pablo Pacheco, and Georg Winkel, Oxford University Press. DOI: 10.1093/9780197683958.003.0002

Organization of the United Nations (FAO) definition or a national definition. Natural regeneration can be the result of active avoidance or abandonment of activities that inhibit spontaneous tree growth. This might be because land use interests have changed or because land protecting measures have been introduced.

Our chapter reviews historical cases of forest restoration to answer the following questions: 1. How common and widespread was forest restoration in the past? 2. How was forest restoration undertaken and implemented, by whom, and for what purpose? 3. What were the environmental consequences? 4. Who benefitted and who suffered from forest restoration and in which sense? 5. What can be recognized as a historical forest restoration legacy? 6. What lessons can be drawn from past experiences that are relevant for modern-day forest restoration?

The chapter is structured as follows. Section 2.2 develops an analytical frame, including an outline of a protocol to collect information and data from the selected cases of early modern forest restoration. Section 2.3 provides the results of the research on those cases. Section 2.4 discusses the evidence presented in Section 2.3 and provides answers to the formulated questions of the previous paragraph. Section 2.5 concludes.

2.2 Analytical framework, cases, and research protocol

We limit the historical scope of our analysis of forest restoration up to the mid-20th century. We consider later forest restoration efforts to be part of modern-day forest restoration practices (de Jong et al. 2021). Such later practices took place in markedly different settings. By the mid-20th century many former territories held as colonies had become independent countries. This ended forest restoration implemented by colonial governments in the now-independent countries. The end of World War II (WWII) saw the foundation of FAO, which has been an important facilitator of the modern global forest restoration trajectory. It also marked the beginning of international development cooperation in which forest restoration efforts play an important role.

The lower historical boundary of the analysis is harder to define. Ancient structured and organized efforts to restore forests are known from India and China, among others (Evans 2009; Rahman 1993). These have, to our knowledge, left no significant imprint on contemporary forests estates anywhere. In Section 2.3, we briefly review references in the literature to ancient forest restoration but do not examine these examples in-depth. For cases of early modern forest restoration, however, not having much imprint on contemporary forest landscapes does not necessarily exclude them from the analysis. Examples from Asia, including mainland Southeast Asia (Peluso 1994), and China (Saito 2009) represent serious forest restoration efforts that resulted in restored forests; forests that were subsequently converted back to agricultural land. These cases constitute remarkable forest restoration efforts, and their analyses contribute to providing answers to the research questions formulated in Section 2.1.

We selected relevant cases to be included in this study. We formulated a list of questions to be posed for each of the cases researched for this chapter (Table 2.1). These include the magnitude of early modern forest restoration (question 1), which suggests the relative importance that was assigned to the problem of disappearing forests and the need to bring them back. Past forest restoration, as is today, was undertaken

Table 2.1 Questions for cases of early modern forest restoration

Questions paper tries to answer	Questions posed to the cases
1. How common and widespread was forest restoration in the past?	When did forest restoration begin? When did the forest restoration end? How large was the targeted area?
2. How was forest restoration undertaken and implemented, by whom, and for what purpose?	How was forest restoration undertaken? Who implemented forest restoration? What were the arrangements, including policies and legislation, incentives and agreements, and compensations? For what reasons was restoration considered necessary?
3. Who benefitted and who suffered from forest restoration and in what sense?	Who benefitted from restoration, in the short term and long term, and what were those benefits? Who was disadvantaged because of forest restoration and what were the losses?
4. What were the environmental consequences of forest restoration?	What were the short-term and long-term positive and negative outcomes for regulating ecosystem services and biodiversity? Who were the beneficiaries of positive outcomes for regulating ecosystem services and biodiversity? Who was negatively affected by restoration resulting in ecosystem disservices or biodiversity loss?
5. What can be recognized as a historical forest restoration legacy?	Based on analysis of answers to questions 1–4
6. What lessons can be drawn from past experiences that are relevant for modern-day forest restoration?	Based on analysis of answers to questions 1–4

for various reasons, by different actors, under different arrangements, and with the support of variable policies and legislation (question 2). These are important dimensions, not least because it is a relevant question to what extent they influenced forest restoration in subsequent time periods. The same point can be made for questions 3 and 4, which inquire about the political economic reality of forest restoration. Forest restoration is intended to generate benefits, provisioning and regulating ecosystem services; but restoration choices likely limit the benefits for some while enhancing the benefits for others. Questions about biodiversity gains and losses, and the risk the latter might pose for humanity, adds a contemporary evaluation criterion. These considerations might or might not have been relevant at the time a given forest restoration was initiated.

We relied on our own knowledge to identify relevant past forest restoration events (e.g., de Jong et al. 2021), but also undertook a literature search for cases of which we were unaware. Once we had identified cases of interest to be included in our analysis, we expanded the literature search to find answers to the questions in the second column of Table 2.1. We used Scopus and Google Scholar as the main databases to find information. We also undertook less structured searches in Google to identify references or documents not found in the academic databases.

2.3 Cases of early modern forest restoration

2.3.1 Japan: From the Shogunate era into westernization

Early modern period (1603–1867)

Since the beginning of the Edo Shogunate in 1603, Japan was ruled by a Shogun and major feudal lords. Each of the 250 independent domains was governed by a feudal lord, but the Shogun did hold power over them (Totman 1998). Society was dependent on agriculture and gradually developed countrywide commercial activities. The population almost doubled during the 17th century and then stagnated until the mid-19th century (Kito 2000). In the 16th and 17th centuries the rampant use of high-quality timber for constructing castles, other large buildings, and castle towns led to the wide-scale depletion of forests.

The Shogunate and many major feudal lords established Lords' Forests (*ohayashi*), which they used to secure timber, especially for construction. These were in the form of reserved mountains (*tomeyama*) and reserved trees (*tomeki*), which could not be harvested even when they were growing on farmers' housing sites. For example, in the middle of the 17th century, two farmers in the Owari domain were executed because they peeled off the bark of some of those reserved trees (TIHF 2012). Starting in the late 17th century, rotation management was practiced in some of the Lords' Forests to ensure sustainable harvesting (TIHF 2012). As these regulations alone could not sustain resources, more active afforestation measures were undertaken since the 18th century. Several feudal lords launched profit-sharing and afforestation schemes with farmers (TIHF 2012).

As the Shogunate and feudal lords were dependent on rice tributes as land taxes from farmers, they were eager to protect rice crop farming. Due to the expansion of agricultural lands and population growth, forest lands were heavily exploited for extracting fodder, fertilizer, and fuel during the 16th and 17th centuries. This caused erosion, flooding, and landslides, which negatively affected rice crops. The Shogunate issued several ordinances to prevent this, for instance, the Ordinance for the Mountains and Rivers in Several Countries (*shokoku sansen okite*), which stipulated that farmers plant seedlings on denuded sites along rivers to stop erosion (TIHF 2012; Totman 1998). The Shogunate dispatched inspectors to check whether farmers observed these ordinances. Additionally, the Shogunate and major feudal lords designated protection forests to protect water sources (TIHF 2015). The protection forests were also set up to protect agricultural lands on seashores from sand, sea wind, or high tide (TIHF 2015).

In the 17th century, entrepreneurial plantation forestry was practiced by Japanese farmers and merchants in farmers' forests, for instance, in Yoshino, Yamaguni, Owase, Ohme, and Nishikawa (TIHF 2012). In the early modern era, as the transportation of harvested timber was very costly, only areas with transportation opportunities, that is, proximity to waterways or to large cities, could afford to establish entrepreneurial forestry. Agricultural books (*nohsho*) that included technical knowledge on forest plantations were published in large numbers.

In summary, during this period, despite the doubling of the population and little reliance on fossil fuels, Japan sustained its forests. Saito (2009) estimates that while the population doubled during the 17th century, the total area of woodlands decreased from 27 to 25.5 Mha in the same period. Political regime policies and entrepreneurial plantation forestry are believed to have played important roles. The benefits of planted forests were accrued to the Shogunate, major feudal lords, and residents in general. Shogunate and feudal lords secured timber resources to meet their own needs, while planted forests helped protect agricultural production, from which common people benefitted. Nevertheless, residents had to endure restrictions on the use of forests. The Edo period political regime continued until the Meiji Restoration in 1868, the beginning of a national state and the modernization of Japan. Proto-industrial activities were also observed during the later Edo period and developed further during the Meiji period. The knowledge about forests gained during the Edo period was transferred to the next era. In 1879, the succeeding government began collecting documents about policies during this era to consult with them when drafting forest legislation (Totman 1998). During the late 19th century, many forest owners and the government referred to technological and institutional knowledge about entrepreneurial forestry in Yoshino (Fujita 1995).

The Meiji and later years (1867–1920)

The period between the Meiji Restoration in 1868 and the 1920s saw the introduction of Western science, technology, and institutions. A capitalistic economy was established in Japan at the turn of the century, yet traditional and agricultural components of society were still prominent in the 1920s. The Meiji government established national forests and imperial forests, which were transformed mainly from Lords' Forests (Handa 1990). The differentiation between national forests and imperial forests, and nonnational forests owned by local residents created many disputes because forest lands assigned to both categories had been used by the same forest users during previous periods (Totman 2007). Newly introduced knowledge, mainly from Germany, was applied to the management of national and imperial forests. Responding to the nationwide flooding in previous years, the government prepared its first Forest Act in 1897, which designated protection forests in national and nonnational forests (Handa 1990).

The new government abolished forest regulations enacted by the Shogunate and feudal lords, and this was followed by an intensification of forest use by farmers, which again led to erosion and floods (Takahashi 1888). Local leaders, farmers, merchants, and local bureaucrats responded to this by organizing residents and regulating forest use and also by planting trees. Local branches of the Meiji government, the prefectures, supported such efforts by providing knowledge and materials. Prefectural forest councils, the members of which were appointed by governors who were appointed by the central government, designated protection forests (Furuido 2018; Shiga Prefectural Local Forest Council 1898–1938).

An increase in Japan's population from 32.3 million in 1850 to 43.8 million in 1900 accompanying the growth of capitalistic industries did not lead to a drastic decrease in Japan's forests. The total area of woodland was 25.5 Mha in 1850 and 24.3 Mha in 1900 (Saito 2009). The benefits from forests accrued mainly to elite groups of the Meiji society, including aristocrats, bureaucrats, and the owners of agricultural and forest lands. Farmers, who made up a majority in society and many of whom owned land, gained from forest regulating services.

As these examples show, throughout the early modern to modern era, efforts to restore Japan's forest and to assure a stable forest cover were linked to agricultural production and disaster prevention and other major societal interests. Despite fluctuations in forest cover as a result of periods of increased deforestation, the evidence suggests that Japan never experienced drastic and long-term deforestation (Saito 2009). The country counteracted deforestation trends with policies and forestry activities, maintaining about 70% of its land area under forest cover until today (Totman 1998).

2.3.2 Germany: From resolving the "Holznot" to multifunctional forests

The history of German forest restoration starts at the beginning of the 19th century. Connected to the establishment of forest science, "modern" forest laws, and forestry administrations, these restoration efforts transformed large parts of the German landscape. Impacts went well beyond the German border: a core part of "German forestry" influenced many other countries' forest management, ideologically and technically (Scott 1998).

The turn of the 19th century was a decisive period in German history. The expansion of post-revolutionary France led to the French occupation of the politically fragmented German states. Political restoration followed the defeat of Napoleon, which consolidated particularly Prussian state power in Germany. It reduced the power of the church and other religious institutions, and of local noble families, who previously ruled much of the fragmented territory. The political restoration also served as the catalyst for the advancement of a centralized "modern" nation-state in Germany, established in 1871, and for economic liberalism, resulting in the expansion of an industrial sector in the country.

Around 1800, the German countryside was dominated by small-scale agriculture. Pre-historic forests that once covered most of the country had long been pushed back to less fertile soils, nonproductive mountains, wetlands, and a few aristocratic hunting grounds. The remaining forest area, around 20% of the country, was used by the agrarian society as woodland pasture, for shifting cultivation especially in mountainous regions, the feeding of pigs in oak forests, and the use of the upper organic soils for fertilization and feed. Large areas of the territory, especially on sandy soils in the north and east of the country, were used for fuel wood production using coppicing. Small numbers of trees were cut to be used for construction where mature forests remained near settlements. The landscape was intensively used to supply food and feed for an increasing population and domestic animals, resulting in integrated forest and agricultural uses on poor soils.

At the dawn of German industrialization and urbanization, wood was a critical resource for energy and construction materials. This resulted in intense expert debates during the 18th century of a perceived or expected wood shortage (Holznot) that might constrain industrial and urban development. This debate paved the way for forest policy change in the 19th century (Radkau 1986; Schott 1997).

During the first half of the 19th century, the socioeconomic land-use situation began to change. Increasing agricultural productivity and trade decreased the importance and profitability of extensive agroforestry land use on marginal sites. This resulted in rural impoverishment and strong migration to industrializing urban centers, including overseas, such as the United States and Brazil. Immigration occurred particularly from rural areas with high proportions of marginal lands. The combination of these socioeconomic changes, the establishment of forest science as a discipline, and the new state-centered planning of the economic transition toward industrialization lay the ground of the new forest restoration projects by the German states and, after 1871, the German Empire.

Forest restoration followed two closely interrelated approaches. First, on heathlands and other marginal agricultural lands, large-scale reforestation occurred. In many cases, fast-growing conifer trees, such as European spruce (*Picea abies*) and Scotch pine (*Pinus sylvestris*) were planted in pure stands. These species were seen as advantageous, as they can produce industrial and construction wood in a short time and are well adapted to poor soils and open lands. Second, the transformation of already extant forests into high forests for timber production began. Silvicultural practices changed drastically, for instance from different types of coppice forestry or shifting cultivation to high-production forests. This resulted in a separation between agriculture and forestry. Forestry shifted to primarily wood production, and agriculture was pushed out of forest areas.

The new forest restoration project was accompanied by changes in forest legislation as new forest laws were enacted. This new forestry regime came up against resistance from the rural population. For example, in the Western German High Eifel (Hocheifel) region around 1800 about 45% of the land area was heathland where sheep pasturing and extensive shifting cultivation were practiced. When the Prussian state began large-scale afforestation in this region, local farmers objected to the conversion of heathlands into coniferous forests. In some cases, the military was called upon to protect the newly planted trees. Forest restoration that was started in the early 19th century continued throughout the 20th century. Less than 1% of the heathland was left in the Hocheifel by the end of the 20th century. Today these remnant heathlands are under nature conservation status for their exceptional biodiversity and scenic and recreational value.

In 1800, Germany's forests made up 20% of the country's land area, while today it is 31%. The expansion concentrated in the country's mountainous areas, but also in the north and northeast in regions with poor, sandy soils. The forest restoration project drastically altered the tree-species composition of the forests: from forests with largely native broadleaf trees to a high proportion of coniferous forests with trees native to the country but often planted outside of their natural habitats (e.g., in low elevations instead of mountain areas).

German forest restoration concurred with a major transition from an agrarian to an industrial society. Economic life changed from local markets to national and international commodity markets; a ruling of scattered territories by the church, noble families, and free cities was replaced by a centralized state with a professional bureaucracy; and a reliance on local knowledge was replaced by a turning to science and technology for answers and solutions. Strong debates over a shortage of wood as energy for a growing society were decisive for legitimizing forest restoration (Radkau 1986). Afforestation (*Waldvermehrung*), sustainable timber management (*forstliche Nachhaltigkeit*) and protection of the forest area from conversion into other land uses (*Walderhaltung*) became the three central goals of German forestry and were institutionalized in forest laws. Also key was the institutionalization and development of autonomous, hierarchically organized state forest services that had the mandate to implement the three goals (Winkel 2007). Ambitious natural science research programs began during this time, aimed at demonstrating the positive effects of forests for society (Pistorius et al. 2012). Biodiversity impacts and carbon storage were yet to be of concern. However, the soil and water protection functions of forests were emphasized in expert and policy debates, and they served to legitimize forest protection and restoration.

Retrospectively, the pre-restored forest landscape, with a huge variety of often small-scale agricultural and agroforestry uses, is now perceived to be higher in biodiversity than the large-scale forest areas planted with a handful of species. The German forest restoration project and the abolition of forestry in agriculture led to an improvement in soil productivity, increased forest growth, and boosted carbon storage. It appears that the synergy and trade-off patterns of the 19th century German forest restoration project are similar to those now debated for forest restoration in the tropics.

2.3.3 Poland: Restoring forests for restoring the nation

Between 1795 and 1918, the north and west of Poland was part of the Prussian Kingdom, the south was part of the Austrian Habsburg Monarchy, and the east was part of the Russian Empire. During these years, forest restoration on Polish lands took place under the influence of legislation according to the forestry model of the state that controlled the respective area (Broda 2006a). In the Austrian and Prussian partitions, liberal economic theory and land and forest ownership in private hands provided the state with greater income. Large-scale sale of state property and forests had taken place early during this time. In the middle of the 19th century, a reverse process began in the Prussian partition. Centralization, nationalization, and strengthening of the role of the Prussian state led to the enlargement of state forests (Scott 1998). In the Austrian Polish territory afforestation took place on wasteland and land unsuitable for agriculture. Agricultural land was purchased from local landowners and subjected to afforestation.

Afforestation in the north and west of Poland, part of the Prussian Kingdom, was mostly pine monocultures managed according to the prevailing German forest science principles (Scott 1998). Clear-cutting and the removal of valuable trees from mature forests, as well as ongoing deforestation, took place in the Austrian partition, while restoration in private forests was rare (Broda 2006a). Forest restoration began in the Russian partition after 1852, but despite these efforts, forest cover declined until 1907 (Figure 2.1).

In 1919, just after World War I (WWI), about 240,000 ha of deforested land previously held by the partitioning states became Polish state forests. Clear-cutting during the war amounted to 213,615 ha, and burned forests, destroyed young stands, and areas with depleted soils amounted to 26,500 ha (Jaworski 2006). During the 1919–1922 period, the forest area in the Polish territory continued to decline despite restoration efforts, as a result of policies to hand out land to smallholders and landless peasants. During the later interwar period, however, the total Polish forest area increased from 19.6% in 1922 to 22.2% in 1937 (Broda 2006b).

Between 1918 and 1939 the share of state-owned forests increased by 7% because of afforestation by the State Forests Service of some 700 000 ha. The State Forest Service implemented afforestation on privately owned land that was exchanged for arable land or purchased. In September 1936, a new Presidential Decree stipulated that the total forest area owned and managed by the State Forests Service could not decrease. Consequently, the decree required the State Forests Service to acquire privately owned forests, as well as non-forest land whose restoration was in the interest of the Polish state (Broda 2006b). Despite these regulations, during the interwar period private forest owners, who owned and managed 66% of the total forest area, reforested 180 000 ha of land (Broda 2006b).

Restoration efforts by the State Forest Service were hindered by lack of access to seeds and seedlings. In the 1920s the negative effects of nonlocal provenances were already visible in Polish forests. The policy then became the use of forest seeds harvested in Polish forests only (Maciejowski 1935) and establishing forest nurseries

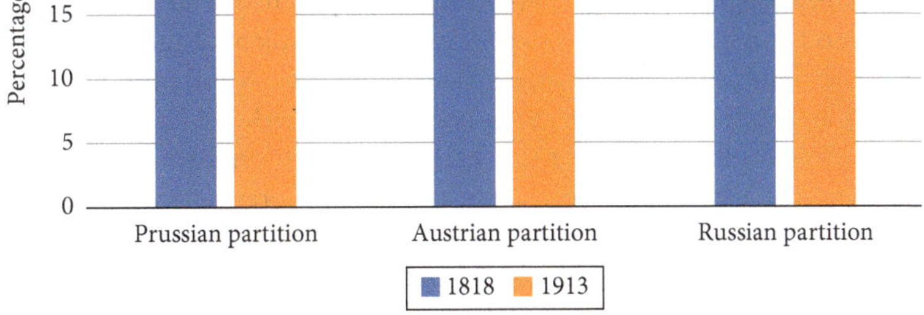

Figure 2.1 Forest cover change in Poland between 1815 and 1913 in the different partitions (Source: Miklaszewski 1928)

(Jaworski 2006). Between 1919 and 1929, 90–95% of reforestation was done with pine (Broda 2000). In 1925 an outbreak of pine beauty (*Panolis flammea*) destroyed significant areas of forest stands in northern Poland and led to salvage logging of stands of mostly nonlocal provenances, called Bavarian pine. These were replanted with domestic provenances. Between 1919 and 1939 in the area managed by the State Forest Service, more forest was restored than was clear-cut.

The necessity for forest restoration stimulated the development of forest science in Poland. The topics of restoration and regeneration were vividly debated among forest scientists during the entire interwar period. The possibility of natural and seminatural regeneration was explored and, when possible, recommended, e.g., to allow for the natural regeneration of fir and beech (Jaworski 2006). As a result, restoration shifted focus to improving the health and quality of existing stands. The Polish State Forests Service prioritized changing stands' species composition to species suitable for habitats in which they were planted with the use of local provenances. The principles of forestry copied from the German model and German foresters, following the German-Austrian forestry traditions, were reconsidered by recognizing the unique Polish forest context, for instance, in newly opened forestry technical and high schools. Polish forestry and the need for forest restoration became deeply linked with the "reborn Polish nation" ideology. Knowledge about Polish forests and their specificity was tightly linked with love for the country (Konczal 2020).

2.3.4 United States: Eastern reforestation following westward expansion

Widespread deforestation occurred throughout the Eastern United States from 1750 through to the later part of the 19th century. Of the estimated 340 Mha of forest land east of the Great Plains prior to European settlement, only 190 Mha remained by 1920, with 56 Mha estimated to be original forest not degraded by grazing, logging, burning, or other human disturbance (Williams 1989). As the population grew, eastern forests were cleared for settlement and agriculture, including crops and grazing. As the country industrialized, it had a voracious appetite for timber and fuelwood for wood products and energy. Fuelwood and charcoal demand for cooking and heating grew to include fuel for furnaces, boilers, steam engines, and eventually electric power generation. Wood demand grew rapidly for structural lumber and building materials, shipbuilding, railroad ties, mining trusses, and many other uses, including wood and resin-derived chemicals like methyl alcohol, tannic acid, and turpentine. By the late 1800s, paper made industrially from wood pulp was common; and by 1900, large paper companies were active in the northeast (Irland 1999).

Patterns of deforestation during this period were not uniform but varied across the region in time and space according to a variety of forest characteristics and socioeconomic drivers. Near the densely populated northeast coastal cities and adjacent countryside, settled and agricultural lands dominated. For example, agricultural land use peaked in the state of Massachusetts between 1830 and 1885 when tilled land, hay, and pasture were the dominant land cover; forest was reduced to 20–40% of land cover, depending on the subregion (Hall et al. 2002). Farther north, in the more sparsely populated forests of interior Maine, timber harvesting was widespread without concurrent land clearing for agriculture. The coastal and settled areas of the south followed similar patterns as the northeast, but commercial harvesting in the interior southern states intensified later, around the turn of the 20th century, following deforestation in the northeast and north-central states near the Great Lakes (Walker 1999). This industry was often characterized by temporary boom–bust operations, leaving abandoned mills and infrastructure in their wake. Much of the industrial logging in both the north and south was done without consideration of regeneration. By the late 19th century, many of the states that are now covered in relatively well-stocked forests had larger proportions of poorly stocked timberlands and agricultural land than mature forest.

It was not always clear to landowners at the time that cut-over timberlands would be able to regrow merchantable trees again (Walker 1991); but thanks to favorable soils and climate, much of the clear-cut industrial timberland regenerated naturally through seed and sprouting, though the stocking, composition, and condition of the forest was often different from the original forest. For example, forests in the Allegheny region of Pennsylvania, heavily dominated by hemlock, birch, beech, and maple in 1800, came to be dominated by cherry, maple, and oak by the mid-20th century (Irland 1999). In the northeast, agricultural production fell dramatically after 1890, and much of the farmland that had been in crops and pasture was no longer used for agriculture. After peaking around 1880, a long period of agricultural "abandonment" followed (Hart 1968). Reforestation of abandoned farmland often occurred through natural regeneration in the absence of tilling, grazing and burning, with some establishment of plantation forests, especially after 1900. A similar trend occurred later but was less pronounced in the southeast, where farmland area peaked in 1920 and then declined sharply when croplands were abandoned, following the disintegration of the prevailing sharecropping system (Waisanen and Bliss 2002). In the southeast, reforestation was aided by the conversion of some abandoned agricultural lands to plantation pine forests and by the recognition that degraded, cut-over industrial timberlands could be rehabilitated to pine forest through good management that included seeding, planting, fire exclusion, and protection from wild boars (Hart 1968; Williams 1989).

The reforestation of the eastern United States occurred primarily on private land under generally well-established individual property rights for the time. Currently, of the estimated 134 Mha of forest land across all northeast and southern states, approximately 84% is privately owned by individuals, families, and corporations. However, this does not mean that public policies at the national, state, and local levels did not play a role in reforestation.

A national forest agency was established in 1876, which became known as the United States Forest Service in 1905. This designated a federal agency specifically devoted to the conservation, management, and use of the nation's forests for the public good. The rise of scientific forestry in the United States was further codified by the proliferation of university forestry programs devoted to training professional foresters, including early programs in North Carolina, New York, Connecticut, and Georgia that were established around the turn of the 20th century. The Smith-Lever Act (1914) established partnerships between agricultural colleges and the United

States Development Agency to support agricultural extension work, including farmer and landowner education in forestry. In the south, the establishment and expansion of the National Forest System and the rehabilitation of cut-over pine land by professional foresters was instrumental in their recovery (Walker 1999).

Among the robust forest policy activity of the time, several policies are of note. The Weeks Act (1911) authorized the federal government to purchase private land to protect watersheds and their streams and waterways from sedimentation and other effects of deforestation and degradation. This act was instrumental in expanding the National Forest System in the east. It was subsequently updated by the Clarke–McNary Act (1924), which expanded land buying authority and allowed the Department of Agriculture to work cooperatively with the states to improve forest protection, especially to protect water supplies and control wildfire, as well as to secure the continuous production of timber. Legislation also played a critical role in establishing nurseries to produce seedlings, developing state forest agencies, and assisting private landowners in reforestation, especially by providing seedlings and direct technical assistance through agricultural extension and other means. In some cases, the government provided direct incentives to convert farmland to conservation use, such as in the Soil Bank Act (1956). Some eastern private lands continue to be harvested in exploitative ways that degrade their condition, especially when subdivided and fragmented for real estate development (Germain et al. 2007).

Early 20th century US forest restoration is widely hailed as a success story. However, land clearing, logging practices, and diseases significantly altered the composition of eastern forests, as did the intentional and unintentional introduction of non-native invasive species. Most notable are the near extermination of mature American chestnut (*Castanea dentata*), which was once widespread and commercially valuable, due to introduced chestnut blight (*Cryphonectria parasitica*), and the severe loss of longleaf pine (*Pinus palustris*) habitat in the south, where cutover stands transitioned to other species, often to low-grade, poor-quality deciduous forests (Walker 1999). The loss of biodiversity in the case of longleaf pine was severe. Furthermore, it is disputed that reforestation in the east necessarily resulted in net gain in forest cover nationally (Ramankutty et al. 2010). As western forests opened to harvesting, many associated environmental problems were exported from east to west. More recently, a reversal of forest transition appears to be occurring in some eastern states from deforestation associated with residential and commercial development (Jeon et al. 2014).

The eastern transition from agricultural land use and exploitative harvesting back to forest was facilitated by a combination of broad socioeconomic forces and societal changes occurring at a time when the government and society more broadly were beginning to perceive and understand the importance of forests, not just for wood supply but also for water and other values (Godfrey 2012). Scientific forestry was emerging in the United States as a professional discipline, and universities were implementing vigorous agricultural extension in cooperation with state and federal government agencies to guide the transition. Certainly, farmers and agricultural producers in the east were hit hard by these changes. However, the benefits of reforestation that accrued proximately to private landowners were ultimately shared more widely among a population that was increasingly concerned with forest conservation for water, recreation, wildlife, aesthetics, and a broader pallet of ecosystem services.

2.3.5 Colonial forest restoration

An important contribution of early modern forest restoration can be attributed to efforts by colonial regimes. The countries that held colonies changed in the course of history. Prominent colonial regimes included Belgium, Germany, Great Britain, France, Italy, Japan, Netherlands, Spain, Portugal, and the United States. Territories held under colonial control changed equally over time. A recognized phenomenon is what Siiskonen (2015) refers to as the scramble for colonial territories toward the end of the 19th century, a period when European countries competed for the control of ever-larger areas of the southern hemisphere, especially in Africa and Asia.

Colonial governments had much interest in forests in their colonial territories. Not only were forests an important natural resource that could be exploited for profit, which colonial powers expected to obtain from their colonies, but also because it was widely believed among colonial land specialists and related academic circles that deforestation contributed to climate change, which negatively affected other land use sectors. This "desiccation debate" was held among British, Dutch, German, and French colonial land specialists alike. Desiccation is a decrease in rainfall, a lowering of the water table, and a drying up of surface waters. Deforestation caused by local agricultural extension, grazing, and fires was held to be a main cause of desiccation (Siiskonen 2015). The appropriation of customary lands, however, was also shaped by the articulation of national timber (or wood fuel) and global environmental crisis narratives. Locals were persistently framed as profligate land and resource users. They encountered forestry arguments unfolding in different places and at different times (Barton 2002), comprising three main elements: the appropriation of lands to create national networks of forest reserves, the establishment of forestry departments to oversee the introduction of "scientific forestry" principles and multifaceted efforts to regulate and control bushfires, and the production and marketing of wood fuels and other forest products.

Several commentators hold that the desiccation discourse served to restrain forest and land use of the indigenous population in some colonial territories (e.g., Endfield and Nash 2002). It was rejected by the famous forester Dietrich Brandis in 1911, following his experience in British India (Potter 2003). But the belief in the validity of desiccation theory greatly influenced conservation policies in some colonial territories and left its imprint on forest restoration plans (Grove 1995; 1997).

Colonial commercial plantations

The largest area of planted forests in colonial territories was in commercial plantations. Only a handful of species were planted. The timber species most in demand internationally was teak, used for shipbuilding. It was already being produced in Java by local rulers at the time when first Portuguese and later Dutch administrations took control of much of what is now Indonesia. The Dutch colonial government engaged German foresters, as did several other colonial governments in India and African countries (Siiskonen 2015), to take care of the teak forests in the early 19th century. British colonial authorities also engaged French foresters trained in the French Royal Forestry School established in Nancy in 1824, preceding the British equivalent by a century.

All teak forests in Java were planted forests, and much of it was planted following a "taungya" arrangement. Under this arrangement, which was implemented variously across regions in Asia, local farmers were engaged to plant teak and were allowed to interplant food crops until the growing trees inhibited food production. Taungya was also applied with other commercial species. An estimated 300,000 ha of teak forest existed in Java in 1950, 5 years after the country gained independence. Great Britain continued to exploit natural teak forests, especially in Burma, until the mid-20th century.

Teak was planted by the British Forest Service in India starting in 1842 and in Burma in 1856 (Evans 2009). It is estimated that, at the beginning of the 20th century, India had some 50,000 ha of planted teak (Bennet 2014). Commercial tree planting in India was not limited to teak. Eucalyptus and acacia species were planted in the Nilgiri plateau, and eucalyptus species in the Mysore region. These plantations were intended to meet regional firewood and other wood demands (Bennet 2014) and to yield Mysore gum (Evans 2009). The total area of planted forests on the Nilgiri plateau, however, appears to have been not much more than 500 ha (Bennet 2014).

The planting of teak, eucalyptus, and acacia was followed by tree planting in other areas of southern India, Australia, and the Cape Colony. Teak began to be planted outside Asia, for instance, in what is now Nigeria (Horne 1966) and Ghana (Kadambi 1972). British foresters also turned to pine species, which began to be extensively planted in South Africa, Australia, and New Zealand. Evans (2009), for instance, reports that 180,000 ha of plantations of various pine species existed in South Africa and 9,800 ha of planted pine forests in Queensland, Australia.

Forest restoration and the desiccation conundrum

Colonial administrations were also interested in other forest commodities, such as shea nuts in West Africa (Wardell et al. 2021). As seen above, forest restoration was also considered necessary to address the perceived negative environmental impacts of local land use. Davis (2016) undertakes an extensive review of the prominence of the desiccation discourse. It was debated extensively among foresters and related professionals with responsibilities or interest in colonial territories. The discourse led to calls for policies to implement reforestation in many instances, including the planting of two protective tree belts across West Africa following the Anglo-French Forestry Mission between December 1936 and February 1937. These plans were abandoned due to WWII but have since become the focus of the contemporary Great Green Wall program (Wardell 2020).

How much the desiccation debate resulted in forest restoration is difficult to know. Multiple authors point to the limited impact of the desiccation debate on forests and land use (Davis 2016; Fairhead and Leach 2000a; Siiskonen 2015). This limitation resulted from the difficulties in implementing policies that restricted land use practices perceived to degrade forests and in implementing other conservation measures and forest restoration activities. (Fairhead and Leach 2000a). Siiskonen (2015), for instance, mentions plans by the German administration to establish thousands of hectares of date palm (*Phoenix dactylifera*) along the Swakop River in the Namibia desert; but only 50 ha were planted. Similarly in 1934, the French colonial administration of Guinea planned to "rewood the degraded Niger upstream catchment areas," but these plans "foundered in the face of limited budget allocation" (Fairhead and Leach 2000a) and the greater interest in developing the (agricultural) Office du Niger. In French Algeria, a League for the Reforestation of Algeria was established in 1882, which gained broad membership among French and a few Algerian settlers. In 1912, a dedicated reforestation service was founded. Much of the efforts to achieve lofty objectives to restore, or rather to transform, Algerian rural areas to an ideal imagined forested landscape (Ford 2008) focused, as elsewhere, on restraining the land use practices of local land users. During the second half of the 19th century evidence is available of the planting of only 5,400 ha (Merdas et al. 2017), when the country had 3.2 Mha of forestland at that time (Ford 2008).

In Namibia, there are reports of forest restoration by local farmers under the colonial regime. Settlers extensively planted fruit trees near the Angola-Namibian border, on land they had occupied since the 1920s. Kreike (2009) refers to this as a "fruit tree frontier." The tree vegetation included planted fruit trees but also spontaneous regenerating fruit trees and other tree species. Similar cases of forest restoration among rural communities in remote regions who engaged in low technology, low-input land use are reported for South America, Africa, and Asia (Fairhead and Leach 2000b; Padoch and Peters 1993; Posey 1985). Commercial interest in trees resulted in their protection and reproduction in swidden fallows, but also more actively tended forest gardens. These are practices that predate the modern forest restoration era. There is some evidence that these practices were influenced by the land use administrations of colonial powers. The example given by Kreike (2009) suggests that colonial land administration influenced land occupation, vegetation conversion, and the subsequent fruit tree frontier formation. Another example is the adaptation of swidden fallow restoration among rural farmers in Southeast Asia. Rubber trees introduced into the region by British and Dutch colonial officials began to be planted by local farmers in swidden fallow fields. This increased the proportion of managed fallows in swidden fallow agriculture landscapes (de Jong 2001; Dove 2011).

In summary, the colonial administrations had much interest in forests in the foreign territories they controlled. In several colonial territories, particularly after budget restrictions were introduced in the early 1900s, timber was a key commodity that helped pay for colonial administration. This is one of the reasons that the colonial forestry ideas were exported from India and Burma to many other anglophone colonies. Forests constituted a valuable economic resource but were also recognized for regulating ecosystem services. It was argued that deforestation would cause wide-scale desiccation problems that would negatively affect agricultural commodity production, one of the primary incomes derived by colonial regimes. Actual planting of forests, however, was primarily done with species that had commercial value, such as teak, pine, acacia, and eucalyptus, and a few less common species. This also happened in rural areas outside the view of colonial governments. It appears that forest restoration to address the desiccation problem happened only occasionally. The benefits of planted forests accrued primarily to the colonial regimes. Both the land grabbing for the planting of commercial trees and the desiccation narrative concurred with policies and measures meant to constrain local land use practices that resulted in the removal of tree vegetation—land use practices that were largely evaluated negatively by colonial (and also some postcolonial) professionals (e.g., Whittlesey 1939).

One can draw parallels between colonial forest restoration discourse and practice and its contemporary counterpart (Davis and Robbins 2018). Colonial regimes were widely concerned about the negative impacts of deforestation and hailed forest restoration as an important measure to remediate those impacts. Not enough actual restoration appears to have taken place, however, to achieve environmental benefits. Most forest restoration in colonial territories pursued by colonial administrations was commercial tree planting. Such commercial tree planting, the limited efforts to restore forests for environmental benefits, and the prevailing narratives to justify both were used to condemn and constrain local land use practices. Meanwhile, local land use and sometimes forest restoration was adapted to the colonial forestry regimes.

2.4 Discussion

We structure the discussion of the evidence provided in Section 2.3 by returning to the questions posed in Section 2.1. Responding to question one, how common and widespread was forest restoration in the past, the evidence indicates that forest restoration did happen in early modern times and even before. The determination of how widespread such pre or early modern restoration was, or to extract learning from any other relevant characteristics of these initiatives, would require an in-depth research of historical sources, which is beyond the scope of this chapter.

The examples reviewed in this chapter, and other cases reported in the literature, illustrate that forest restoration happened in multiple locations around the world. In Japan early modern forest restoration began in the 17th century, both to grow timber and to combat erosion that negatively affected rice production. Through its restoration efforts, Japan managed to prevent a serious decline of its forest area and maintain a relatively stable forest cover since the 17th century. In Europe, forest restoration varied considerably by country. Several European countries experienced a first wave of forest recovery during the 14th century as a result of a decline in population, caused by bubonic plague epidemics. However, in many cases this recovery was not long-lasting. Early restoration took place in India and Egypt, for instance, and evidence exists of pre-modern tree planting in Roman Italy, and by cloisters in lands they controlled (Farrell et al. 2000). Mather and Needle (1998), Rudel et al. (2005), and Meyfroidt and Lambin (2011) suggest historical moments when European countries experienced their forest transition inflection point, after which forest restoration took place. While forest restoration did occur before these moments, it did not exceed deforestation, especially when considered over a longer period.

Meyfroidt and Lambin (2011) place the beginning of "modern" forest recovery in European countries in the 17th century for Scotland and the UK, in the early 19th century for Czechia, Denmark, France, and Hungary, and in the second half of the 19th century for Austria, Germany, Poland, Portugal, Slovenia, Sweden, and Switzerland. In Bulgaria, European Russia, Ireland, Italy, Ukraine, and the United Kingdom, forest cover began to increase only in the early 20th century (Meyfroidt and Lambin 2011). Of the countries outside Europe, only Japan, the United States, and New Zealand are listed as having experienced forest cover increase before the 1950s.

The countries listed above are the only ones that experienced forest transition before the mid-20th century. This indicates that forest transition is by no means equivalent to forest restoration taking place, as the cases of colonial forest restoration discussed in the previous section show. Forest restoration did take place before the mid-20th century in multiple countries in Europe as well as in tropical regions (de Jong 2010; Meyfroidt and Lambin 2011). But the evidence suggests that, at a global scale, forest restoration practices accelerated in the second half of the 19th century, and even more so since the first half of the 20th century.

The second question, how was forest restoration undertaken and implemented, by whom, and for what purpose, also requires nuanced answers. The evidence in this chapter and from other cases (Chokkalingam et al. 2005; Songster 2003; Wardell 2020) indicates that forest restoration was pursued to generate a range of ecosystem services. The primary purpose of tree planting was often wood or timber production; in fewer cases, it was for other products that qualify as provisioning ecosystem services. However, the purpose of some of the early forest restoration was also to generate environmental benefits rather than producing consumables or commodities. Or, in terms of the ecosystem services classification, an important part of pre-1950s forest restoration was aimed at generating regulating ecosystem services. The reforestation in the vicinity of Nuernberg, Germany was to stop erosion from sandy hills. Standing out is the desiccation debate among forestry experts who commented on or had forestry responsibilities related to colonial forestry administration or exploitation and resulting efforts to restore forestlands (Davis 2016; Fairhead and Leach 2000a; Grove 1994; Siiskonen 2015).

The evidence gathered in Section 2.3 suggests that government actors played an important role in promoting or facilitating forest restoration in all the locations reviewed. This was the case in Japan, Germany, Poland, the United States, and in most colonial forest restoration cases, but also, for instance, in post-imperial China (Songster 2003) and the Philippines (Chokkalingam et al. 2007). Government agencies often engaged with people who lived in the locations where forest restoration took place. Arrangements to undertake restoration included hiring people (less common during the early years) and formal contracts with companies or entrepreneurs. In the case of Japan, enacting regulations that required local residents to plant trees was also included. Arrangements such as taungya, which was used in several locations in Southeast Asia, can be viewed as benefit-sharing arrangements between government actors and local residents, although more benefits accrued to the forest departments than to local people (Peluso 1994).

The desiccation debate assumed a direct link between deforestation and negative environmental consequences, including lowered water flow and rainfall patterns. This debate eventually died out, not least because these causal linkages were questioned (Davis 2016; Fairhead and Leach 2000a). It is noteworthy that in recent years the argument linking forests and rain patterns has returned in full force (Ellison et al. 2017; Sheil 2009). One can also argue that today's climate change and the role assigned to forests in solving emerging challenges mirrors the desiccation debate, but in a postcolonial, globalized world. Only now, the former colonial powers lack the legitimacy to prevent tropical deforestation for the sake of protecting their agricultural interests at home or in tropical countries, to stop the negative effects that deforestation has on rainfall patterns, or to reduce the contribution of deforestation to CO_2 emissions. Conservation measures and interventions have globalized, and forest protection and restoration are considered important measures to address the global climate crisis. Narratives are devised to suggest that forest residents will equally benefit from these measures, mirroring arguments made as the desiccation debate unfolded (de Jong et al. 2021).

The third question relates to who benefitted from forest restoration and in what sense, and who suffered detrimental impacts where and when it happened. This question refers to the consequences for the environment of early modern and pre-contemporary forest restoration. In the early industrialized countries of Europe, Japan, the United States, and New Zealand, reforestation concurred with a trend away from deforestation, that is, it is linked to forest transition. This was not the case in many colonially occupied countries: they depended on natural resource use and also often maintained considerable areas of natural forests, even though they experienced net deforestation. In these countries, forest restoration and forest transition are less closely linked (de Jong 2010), as areas restored to forests again succumbed to deforestation forces. However, some restored forests escaped such a fate, for instance teak forests planted in the 18th century or earlier in Indonesia. In other places, such as Japan, past forest restoration aided in assuring a stable forest cover that has not varied much over centuries.

The legacy of past forest restoration has manifested itself in other ways. Foresters recorded their experiences and shared their knowledge and ideas, which were passed on and influenced the administration and practice for many years to come, even after countries that experienced colonial dominance became independent (e.g., de Jong et al. 2021; Potter 2003).

The cases reviewed here demonstrate that restored forests have not always resulted in biodiversity conservation nor the creation of complex and diverse ecosystems. This is particularly the case where forest restoration was undertaken for wood or timber production, in which case the "new forests" were commonly single-species forests with limited biodiversity and ecological diversity and a limited capacity to provide regulating ecosystem services. As observed in Section 2.3.2, arguments are made that restored forests from the past replaced more diverse agriculture-forestry-cultural landscapes, and that biodiversity and ecological diversity in the new forests declined. In some places, this has resulted in calls to reverse past forest restoration and restore these early modern agriculture-forestry-cultural landscapes. More prominently, it has led to the idea to transform plantations into more biodiverse, multiple tree-species forests.

As is the case today, who benefits from forest restoration depends very much on where it is done, for what purpose, and how it is implemented. Where the state facilitated or coerced forest restoration, the benefits accrued primarily to (colonial) governments or its associates, mostly entrepreneurs affiliated with the governments. Where forest restoration was successful, these forests generated regulating ecosystem services that benefitted local and nonlocal people as well as provisioning ecosystem services that benefitted forest users.

The forest restoration histories, however, have a well-known darker side. Feudal lords in Japan imposed restrictions on forest uses on local communities to further their own interests at the expense of those communities. A common phenomenon in multiple locations was the eviction of rural residents (and/or their agricultural practices) from land that was to be subjected to forest restoration or the imposition of the use of lands for restoration against the preference of original users of those lands.

The answer to the final two questions, what can be recognized as a historical forest restoration legacy and what lessons can be drawn from past experiences that are relevant for modern-day forest restoration, can be summarized as follows. The possibility and option to plant trees has likely been known to people since agriculture developed. Where societies or communities benefitted from trees and forests but cleared forests for land use, the possibility and option to restore forests was recognized early on and did happen in pre-modern times. When societies began to experience negative consequences of land use conversion for agriculture and its impact on forests, declining availability of forests products, and increased erosion and flooding linked to deforestation, efforts began to bring back forest cover. This happened in different places independently—in Japan, Indonesia, and various countries in Europe, possibly even in various tropical regions (e.g., Balée 1994).

What is now generally understood as forest restoration started when societies had organized as states or along similar society structures. Forest restoration was proposed, organized, and coordinated by actors with authority. However, this did not always result in long-term impact on the ground. It needed to be pushed through conflicting interests, resulting in a top-down perspective that has a legacy in today's forest restoration thinking. Not uncommon, the actors with authority imposed forest restoration to address the interests of particular groups of society. This happened in Japan, where feudal lords sought to ensure the supply of rice they taxed from their subordinates; in colonies concerned by the negative impacts of deforestation on commercial agriculture or estate production; and, in Germany, where a shortage of wood supply was seen as a threat for the success of the industrialization project. One can argue that this is a second enduring legacy of the world's forest restoration tradition: one that is becoming especially evident in recent years when the global community, including world leaders participating in the UNFCCC COP26, are calling for a stop to deforestation and to implement forest restoration to solve the looming global climate crisis. Current forest restoration plans are expected to happen in multiple locations in the world to benefit the global community. Whether local residents will enjoy benefits from forest restoration remains to be seen, while they may or may not experience negative impacts from climate change (de Jong et al. 2021; see also Chapter 10).

Another long-lasting legacy of early modern and pre-contemporary forest restoration relates to the global exchange in forest restoration ideas, knowledge, and practices and how this has influenced contemporary forest restoration thinking. When forest restoration became part of land use practice, policy, and administration in multiple locations in the world, it became heavily influenced by "scientific forestry," as happened to forestry in general. Scientific forestry is claimed to have emerged in Europe, especially Germany and France (Farrell et al. 2000), but it subsequently influenced forestry ideas not only in European colonies but also in many other countries, where forestry and forest agencies became part of the public natural resource administrations. In examples known to us, countries have set up their own national forest restoration programs and related research and experimentation institutions and programs

(Chokkalingam et al. 2007; Jong 2010). These generally follow a tradition that can be linked to 19th-century forest restoration ideas and practices, which themselves are directly linked to the global dispersal of the said European scientific forestry tradition from that same era.

If one reads documentation and academic studies on forests and forestry in the past, the dominant focus is on how societies evolved, how forests provide benefits and opportunities, and how this, largely negatively, impacted forests themselves. Most analyses of society–forest interactions make only passing reference to forest restoration, if any. The evidence synthesized here suggests a lot more has happened that fits under the term forest restoration, too little of which is known or has been explored. This is a gap in our knowledge and understanding that needs to be remedied. This points to the need to undertake more research on past forest restoration practice and experiences.

2.5 Conclusion

Forest restoration has been around for a long time. It gained prominence in the 19th and early 20th centuries, a prominence that has not diminished, although many of its ideas have since changed profoundly (de Jong et al. 2021). Early modern forest restoration heritage endures, as do the forests that persist to this day, as a legacy that is morally judged differently by different actors and that is connected to varying contexts and governance traditions, including problematic ones, such as colonial states and governments. It can be reasonably expected that this might be similar when contemporary forest restoration is analyzed in the future. To better capture the lessons from past forest restoration it is advisable to invest in historical forest restoration research.

Acknowledgments

The authors thank Carol J. Pierce Colfer and Glenn Galloway for providing comments on early versions of this chapter. We also thank two anonymous reviewers for reading and commenting on the chapter and making helpful suggestions.

References

Balée, W. 1994. *Footprints of the Forest: Ka'apor Ethnobotany—the Historical Ecology of Plant Utilization by an Amazonian People.* New York: Columbia University Press.

Barton, G.A. 2002. *Empire Forestry and the Origins of Environmentalism.* Cambridge: Cambridge University Press.

Bennet, B. 2014. "The Origins of Timber Plantations in India." *The Agricultural History Review* 62 (1): pp. 98–118.

Broda, J. 2000. *History of Forestry in Poland.* Poznan: AR in Poznan Press (in Polish).

Broda, J. 2006a. "Forest Heritage after the Period of Partitions and Post-War Conditions." In *The History of Polish Forests and Forestry 1924–2004. The Interwar Period,* edited by A. Szujecki, pp. 27–54. Warsaw: The State Forests Information Centre (in Polish).

Broda, J. 2006b. "Forest Cover, Type of Ownership and Administrative Division of State Forests." In *The History of Polish Forests and Forestry 1924–2004. The Interwar Period*, edited by A. Szujecki, pp. 63–78. Warsaw: The State Forests Information Centre (in Polish).

Chokkalingam, U., A.P. Carandang, J.M. Pulhin, R.D. Lasco, R.J.J. Peras, and T. Toma. 2007. *One Century of Forest Rehabilitation in the Philippines. Approaches, Outcomes and Lessons.* Bogor: CIFOR.

Chokkalingam, U., C. Sabogal, E. Almeida, et al. 2005. "Local Participation, Livelihood Needs, and Institutional Arrangements: Three Keys to Sustainable Rehabilitation of Degraded Tropical Forest Lands." In *Forest Restoration in Landscapes: Beyond Planting Trees*, edited by S. Mansourian, D. Vallauri, and N. Dudley, pp. 405–414. New York: Springer.

Davis, D.K. 2016. *The Arid Lands: History, Knowledge, Power.* London: MIT Press.

Davis, D.K., and P. Robbins. 2018. "Ecologies of the Colonial Present: Pathological Forestry from the Taux de Boisement to Civilised Plantations." *Environment and Planning E: Nature & Space* 1 (4): pp. 447–469. DOI:10.1177/2514848618812029.

De Jong, W. 2001. "When New Technologies Meet Traditional Forest Management: The Impact of Rubber Cultivation on the Forest Landscape in Borneo." In *The Impact of Technological Change in Agricultural Change on Deforestation*, edited by A. Angelsen and D. Kaimowitz, pp. 367–381. Wallington: CABI International.

De Jong, W. 2010. "Forest Rehabilitation and Its Implication for Forest Transition Theory." *Biotropica* 42 (1): pp. 3–9. DOI:10.1111/j.1744-7429.2009.00568.x.

De Jong, W., J. Liu, and H. Long. 2021. "The Forest Restoration Frontier." *Ambio* 50: pp. 2224–2237.

Dove, M. 2011. *The Banana Tree at the Gate: A History of Marginal People and Global Markets in Borneo.* New Haven: Yale University Press.

Ellison, D., C.E. Morris, B. Locatelli, et al. 2017. "Trees, Forests and Water: Cool Insights for a Hot World." *Global Environmental Change* 43: pp. 51–61.

Endfield, G., and D.J. Nash. 2002. "Drought, Desiccation and Discourse: Missionary Correspondence and Nineteenth-Century Climate Change in Central Southern Africa." *Geographical Journal* 168 (1): pp. 33–47.

Erbaugh, J.T., N. Pradhan, J. Adams, et al. 2020. "Global Forest Restoration and the Importance of Prioritizing Local Communities." *Nature Ecology and Evolution* 4: pp. 1472–1476.

Evans, J. 2009. *Planted Forests: Uses, Impacts and Sustainability.* Wallingford: CABI.

Fairhead, J., and M. Leach. 2000a. "Desiccation and Domination: Science and Struggles Over Environment and Development in Colonial Guinea." *The Journal of African History* 41 (1): pp. 35–54.

Fairhead, J., and M. Leach. 2000b. *Misreading the African Landscape; Society and Ecology in a Forest-Savanna Mosaic.* Cambridge: Cambridge University Press.

FAO. 2020. *Global Forest Resource Assessment 2020.* Rome: FAO.

Farrell, E.P., E. Fuerer, D. Ryan, et al. 2000. "European Forest Ecosystems: Building the Future on the Legacy of the Past." *Forest Ecology and Management* 132: pp. 5–20.

Ford, C. 2008. "Reforestation, Landscape Conservation, and the Anxieties of Empire in French Colonial Algeria." *The American Historical Review* 113 (2): pp. 341–362.

Fujita, Y. 1995. *Establishment of Plantation Forestry in Japan.* Tokyo: Kokin Shoten (in Japanese).

Furuido, H. 2018. "Significance of Hoanrin (Protective Forest) as Measures in Forest Policies: Implication for Land-Use Management (Land, Society and the Forest)." *Forest Economy* 70 (12): pp. 10–17 (in Japanese).

Germain R., N. Anderson, and E. Bevilacqua. 2007. "The Effects of Forest Land Parcelization and Ownership Transfers on Nonindustrial Private Forest Land Forest Stocking in New York." *Journal of Forestry* 105 (8): pp. 403–408.

Godfrey, A. 2012. *From Prairies to Peaks: A History of the Rocky Mountain Region of the U.S. Forest Service, 1905–2012.* Washington: U.S. Department of Agriculture, Forest Service and Department of the Interior, National Park Service.

Grove, R.H. 1994. "A Historical Review of Early Institutional and Conservationist Responses of Fears of Artificially-Induced Global Climate Change: The Deforestation-Desiccation Discourse, 1500–1860." *Chemosphere* 29 (5): pp. 1001–1013.

Grove, R.H. 1995. *Green Imperialism: Colonial Expansion, Tropical Island Edens and the Origins of Environmentalism 1600–1860.* Cambridge: Cambridge University Press.

Grove, R.H. 1997. *Ecology, Climate and Empire: Colonialism and Global Environmental History, 1400–1940.* Cambridge: White Horse Press.

Hall B., G. Motzkin, D.R. Foster, M. Syfert, and J. Burk. 2002. "Three Hundred Years of Forest and Land-Use Change in Massachusetts, USA." *Journal of Biogeography* 29: pp. 1319–1335.

Handa, R. (ed.). 1990. *Forest Policy (Rinseigaku).* Tokyo: Bun'eido (in Japanese).

Hart, J.F. 1968. "Loss and Abandonment of Cleared Farm Land in the Eastern United States." *Annals of the Association of American Geographers* 58 (3): pp. 417–440.

Horne, J.E.M. 1966. "Teak in Nigeria." *Nigerian Information Bulletin (New Series)* 16: p. 38.

Irland, L.C. 1999. *The Northeast's Changing Forests.* Cambridge: Harvard University Press.

Jaworski, A. 2006. "Development of Forest Management." In *The History of Polish Forests and Forestry 1924–2004. The Interwar Period,* edited by A. Szujecki, pp. 143–181.Warsaw: The State Forests Information Centre (in Polish).

Jeon, S. B., P. Olofsson, and C.E. Woodcock. 2014. "Land Use Change in New England: A Reversal of the Forest Transition." *Journal of Land Use Science* 9 (1): pp. 105–130.

Kadambi, K. 1972. *Silviculture and Management of Teak.* Bulletin No. 24. Nacogdoches: School of Forestry, Stephen F. Austin State University.

Kito, H. 2000. *Japanese History Based on Population Data.* Tokyo: Kodansha (in Japanese).

Konczal, A.A. 2020. "Why Can a Forest Not Be Private? A Post-Socialist Perspective on Polish Forestry Paradigms – An Anthropological Contribution." *Forest Policy and Economics* 117: pp. 1–11.

Kreike, E. 2009. *Deforestation and Reforestation in Namibia: The Global Consequences of Local Contradictions.* Princeton: Markus Wiener Publishers.

Maciejowski, K. 1935. "Silviculture in State Forest Enterprises." *Polish Forest* 15 (12): pp. 347–358 (in Polish).

Mather, A.S. 2001. "The Transition from Deforestation to Reforestation in Europe." In *Agricultural Technologies and Tropical Deforestation,* edited by A. Angelsen, and D. Kaimowitz, pp. 35–52. Wallingford: CABI.

Mather, A.S., and C.L. Needle. 1998. "The Forest Transition: A Theoretical Basis." *Area* 30 (2): pp. 117–112.

Merdas S., T. Mostephaoui, and M. Belhamara. 2017. "Reforestation in Algeria: History, Current Practice and Future Perspectives." *Reforesta* 3: pp. 116–124.

Meyfroidt, P., and E.F. Lambin. 2011. "Global Forest Transition: Prospects for an End to Deforestation." *Annual Review of Environmental Resources* 36: pp. 343–371.

Miklaszewski, J. 1928. *State Forests in Poland. Vol 1.* Warsaw: Trade Union of Foresters in the Republic of Poland (in Polish), (in Polish).

Padoch, C., and C. Peters. 1993. "Managed Forest Gardens in West Kalimantan, Indonesia." In *Perspectives on Biodiversity: Case Studies of Genetic Resource Conservation and Development,* edited by C.S. Potter, J.I. Cohen, and D. Janczewski, pp. 167–176. Washington: AAAS.

Peluso, N. L. 1994. *Rich Forests, Poor People.* Berkeley: University of California Press.

Pistorius, T., H. Schaich, G. Winkel, et al. 2012. "Lessons for REDDplus: A Comparative Analysis of the German Discourse on Forest Functions and the Global Ecosystem Services Debate." *Forest Policy and Economics* 18: pp. 4–12.

Posey, D.A. 1985. "Indigenous Management of Tropical Forest Ecosystems: The Case of the Kayapó Indians of the Brazilian Amazon." *Agroforestry Systems* 3 (2): pp. 139–158.

Potter, L. 2003. "Forests Versus Agriculture. Colonial Forest Services, Environmental Ideas and the Regulation of Land Use Change in South-East Asia." In *The Political Ecology of Tropical Forests in Southeast Asia, Historical Perspectives,* edited by L.T. Po, W. de Jong, and K.-I. Abe, pp. 29–71. Kyoto: Kyoto University Press.

Radkau, J. 1986. "Zur angeblichen Energiekrise des 18. Jahrhunderts: Revisionistische Betrachtungen über die 'Holznot'." *Vierteljahrschrift für Sozial-und Wirtschaftsgeschichte* 73 (H.1): pp. 1–37.

Rahman, L.M. 1993. "History of Forest Conservation in Indo Bangladesh Aranya." *A Forestry Periodical Society for the Bangladesh Civil Service (Forest)* 2 (2): pp. 21–24.

Ramankutty, N., E. Heller, and J. Rhemtulla. 2010. "Prevailing Myths about Agricultural Abandonment and Forest Regrowth in the United States." *Annals of the Association of American Geographers* 100 (3): pp. 502–512.

Rudel, K.R., O.T. Coomes, E. Moran, et al. 2005. "Forest Transitions: Towards a Global Understanding of Land Use Change." *Global Environmental Change* 15 (1): pp. 23–31.

Saito, O. 2009. "Forest History and the Great Divergence: China, Japan, and the West Compared." *Journal of Global History* 4: pp. 379–404.

Schott, D. (ed.). 1997. *Energie und Stadt in Europa: von der vorindustriellen" Holznot" bis zur Ölkrise der 1970er Jahre.* Stuttgart: Franz Steiner Verlag.

Scott, J.C. 1998. *Seeing Like a State. How Certain Schemes to Improve the Human Condition Have Failed.* New Haven: Yale University Press.

Sheil, D. 2009. "How Forests Attract Rain: An Examination of a New Hypothesis." *BioScience* 59: pp. 341–347.

Shiga Prefectural Local Forest Council. 1898–1938. *Minutes of Shiga Prefectural Local Forest Council* [in Japanese].

Siiskonen, H.O. 2015. "The Concept of Climate Improvement: Colonialism and Environment in German South West Africa." *Environment and History* 21 (2): pp. 281–302.

Songster, E. 2003. "Cultivating the Nation in Fujian's Forests: Forest Policies and Afforestation Efforts in China, 1911–1937." *Environmental History* 8 (3): pp. 452–473.

Takahashi, T. 1888. *Concerns about Forests.* https://dl.ndl.go.jp/info:ndljp/pid/802306?itemId=info%3Andljp%2Fpid%2F802306&__lang=en, (in Japanese).

TIHF (Tokugawa Institute for the History of Forestry). 2012. *Edo Studies on Forests.* Tokyo: Tokyodo (in Japanese).

TIHF (Tokugawa Institute for the History of Forestry). 2015. *Edo Studies on Forests 2.* Tokyo: Tokyodo (in Japanese).

Totman, C. 1998. *The Green Archipelago: Forestry in Preindustrial Japan.* Berkeley: University of California Press.

Totman, C. 2007. *Japan's Imperial Forest, Goryōrin, 1889–1945.* Kent: Global Oriental.

Waisanen P.J., and N.B. Bliss. 2002. "Changes in Population and Agricultural Land in Conterminous United States Counties, 1790 to 1997." *Global Biogeochemical Cycles* 16 (4): pp. 84-1–84-19.

Walker, L.C. 1991. *The Southern Forest: A Chronicle.* Austin: University of Texas Press.

Walker, L.C. 1999. *The North American Forests: Geography, Ecology and Silviculture.* New York: CRC Press.

Wardell, D.A. 2020. "Groundnuts and Headwaters Protection Reserves. Tensions in Colonial Forest Policy and Practice in the Northern Territories of the Gold Coast." In *Commonwealth Forest & Environmental History. Empire Forests and Colonial Environments in Africa, the Caribbean, South Asia and New Zealand*, edited by V. Damodaran, and R. D'Souza, pp. 357–401. Delhi: Primus Books.

Wardell, D.A., M. Elias, M. Zida, et al. 2021. "Shea (*Vitellaria paradoxa* C.F. Gaertn.) – A Peripheral Empire Commodity in French West Africa, 1894–1960." *International Forestry Review* 23 (4): pp. 511–533.

Whittlesey, D. 1939. "Shifting Cultivation." *Economic Geography* 13 (1): pp. 35–52.

Williams, M. 1989. *Americans and Their Forests: A Historical Geography.* New York: Cambridge University Press.

Winkel, G. 2007. *Nature Protection Policy in the German Forest: Appraisals, Analyses and the Development of a Story Line.* Freiburg: Freiburger Schriften zur Forst- und Umweltpolitik 13 (in German).

3

Forest (landscape) restoration governance

Institutions, interests, ideas, and their interlinked logics

Daniela Kleinschmit, Mareike Blum, Maria Brockhaus, Mawa Karambiri, Markus Kröger, Sabaheta Ramcilovic-Suominen, Sabine Reinecke, and Symphorien Ongolo

3.1 Introduction

Forest landscape restoration (FLR) has gained increased international attention in the political sphere. It resonates with several United Nation's (UN) Sustainable Development Goals (SDGs), as restoring land is expected to help tackle climate change, support biodiversity, and ensure human well-being (UNCCD 2018). In the decades leading up to the SDGs, numerous initiatives evolved focusing on or related to FLR, promoting FLR as a solution to these challenges, including regulating the hydrological cycle, protecting soil, as well as improving livelihoods of the poor. These initiatives have emerged at diverse levels from the local to the global, the latter prominently represented by the Bonn Challenge and the New York Declaration on Forests (NYDF). A full range of actors, comprising public and private actors as well as civil society, are involved in propagating, supporting, and implementing FLR.

Current FLR concepts and initiatives seek to advance century-long experience with site-level forest restoration efforts to larger landscapes or even global scales. This global FLR agenda is mirrored in a global proliferation of restoration investments and interventions and a firm belief that it is "possible to enhance the overall benefits to people and biodiversity at that scale" (Mansourian 2005, 11).

However, critical voices are questioning the win-win narratives of FLR (and other global initiatives), pointing to potential losses and losers and new conflicts of interest (Mansourian 2021; Seymour 2020). Within the forest sector, early critics concentrated on technical problems of implementing FLR such as possible unintended ecological effects (e.g., Thomas et al. 2015). Lately, concerns are growing over the social and economic implications of when FLR is translated into mere tree-planting schemes, including flaws in the governance of FLR (Pritchard 2021). One issue is the asymmetries in land access and control over resources that can lead to the exclusion of local communities, pastoralists, and minorities, specifically women (e.g., Chazdon et al. 2021; Elmhirst et al. 2017; Turner et al. 2021). Existing studies often concentrate on case studies at the local level (e.g., Baynes et al. 2019). Nonetheless, they suggest paying greater attention to how politics and often asymmetric power relations drive restoration (Elias et al. 2021; Pritchard 2021).

This chapter aims to respond to a clear gap in knowledge relating to FLR and to broaden the perspective beyond single case studies and local-level analysis. It

Daniela Kleinschmit, Mareike Blum, Maria Brockhaus, Mawa Karambiri, Markus Kröger, Sabaheta Ramcilovic-Suominen, Sabine Reinecke, and Symphorien Ongolo, *Forest (landscape) restoration governance* In: *Restoring Forests and Trees for Sustainable Development.* Edited by: Pia Katila, Carol J. Pierce Colfer, Wil de Jong, Glenn Galloway, Pablo Pacheco, and Georg Winkel, Oxford University Press. © Oxford University Press 2024. DOI: 10.1093/9780197683958.003.0003 contributes to the FLR debate by critically reflecting upon and discussing the international governance of FLR, understanding governance as "the formal and informal bundles of rules, roles and relationships that define and regulate the social practices of state and nonstate actors in international affairs" (Slaughter et al. 1998, 371). While this analysis focuses on international governance, we draw on examples from local and regional levels to reflect the practical and social implications and interlinkages among different levels of governance. We take a historically informed perspective and review the different rationales at play that underlie FLR and other tree-planting schemes. The chapter starts from the assumption, following Nicholls and Huybrechts (2016), that FLR as a political arena reflects power asymmetries rather than a balance of interests; it is characterized by distinct power relations and is dominated by particular rationales legitimized through specific narratives.

We analyze FLR governance and explore polymorphous power relations in respective governance arrangements. We do so with a tentative, interpretative framework drawing on literature about policy change, which highlights three overlapping and often competing dimensions: institutions, interests, and ideas (e.g., Brockhaus and Angelsen 2012; Campbell 1998; Esping-Andersen 1996; Hall and Thelen 2008; Hay 2004; Korpi and Palme 2003; Schmidt 2008; Scott 2008). With our analytical framework, we pursue the following research questions: How do institutions, interests, ideas, and their interlinkages constitute different FLR logics? Which legitimation narratives underpin each respective logic? Where are the silences in the specific narratives? Figure 3.1 provides an overview of our theoretical dimensions and conceptual lens.

Institutions, interests, and ideas serve as analytical lenses when identifying and characterizing three key logics driving FLR today, and its predecessors over time. Each logic reflects different interests, underpinned by distinct beliefs and narratives that legitimize FLR actions, while being embedded in institutional structures that are characterized by path dependencies and stickiness (Backstrand and Lovbrand 2006; Delabre et al. 2020; Hajer and Versteeg 2005; Leipold et al. 2019; Schmidt 2008). We argue that these logics are shaping current FLR processes and outcomes; hence we place them at the center of our analysis. To analyze the

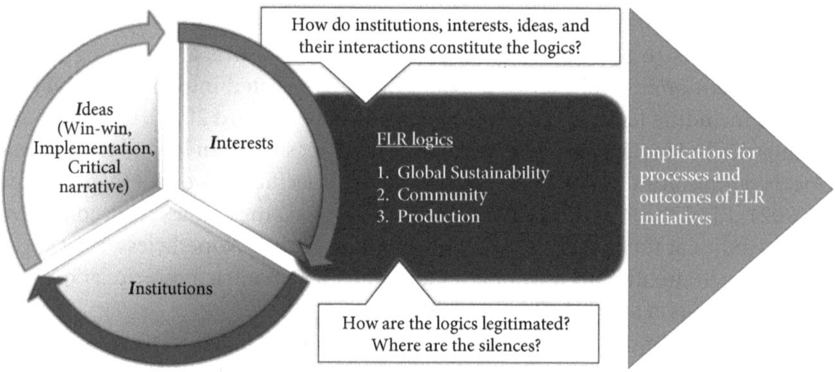

Figure 3.1 Analytical framework to unpack FLR logics

different ideas and meanings of FLR, we build on three narratives that implicitly define and legitimize the FLR logics: win-win, socio-technological implementation, and the critical FLR narrative.

In the following sections, we describe the scope and methodology of our analysis (Section 3.2), which includes a conceptualization of institutions, interests, and ideas and a particular focus on three different FLR legitimation narratives (win-win, implementation, critique). We then apply our analytical framework to examine the three FLR logics we have identified (Section 3.3). The results will be discussed in Section 3.4 and Section 3.5 concludes the chapter.

3.2 Scope and methodology

The substantive scope and main data sources of this assessment include relevant studies carried out within different academic disciplines including economics, ecology, political science, and sociology. In this chapter, we focus on international FLR governance, while empirical insights are derived from multiple geographical areas.

The chapter provides an in-depth 3I analysis (institutions, interests, ideas: the latter includes three legitimation narratives) of different FLR logics: (i) sustainability, (ii) community, and (iii) production. Logics may overlap and refer to multiple interests at the same time. The logics also differ from each other by highlighting different discursive elements and/or remaining silent regarding others. We identify these logics through a review of historic and contemporary scientific literature and technical reports that refer to and rationalize FLR and tree-planting schemes and that provide insights into related policies and politics. We then analyze these three FLR logics by tracing their respective institutions, interests, and ideas based on existing knowledge and literature on past and current developments related to the governance of tree planting and forest restoration. As a final step, we critically discuss the logics and the underlying power relations by asking how these logics are constituted and legitimized with different narratives.

3.2.1 Institutions, interests, and ideas in FLR

Institutions in international governance refer to the norms that define and regulate appropriate behavior. Institutions, appearing as instruments, statements, or other forms, are a central component of social structure and may vary in the degree of institutionalization (Bernstein 2000). Institutions are not "neutral" rules out there, but the outcomes of political processes in which different beliefs and discourses have been competing. We argue that institutions, interests, and ideas are intimately interlinked (cf. Schmidt 2008). While existing institutions are the (sometimes unintended) consequence of earlier institutional decisions, these typically reflect the (privileged) preferences of the most powerful or dominant actors. Likewise, existing institutions shape the preferences and power relations of actors in a given governance setting.

Interests describe the preferences of actors; depending on the distribution of resources and power in a governance setting, certain actors may be more capable of attaining and exercising their interests than others. Within international governance processes, private and other non-state actors may gain authority alongside governmental authorities (Büthe 2004; Odoom 2016), adding to the power of states and intergovernmental/international organizations (Dobbin, Simmons, and Garrett 2007; Kahler and Lake 2003).

Ideas (including beliefs and discourses) affect policy and practice, as they frame the problem and present a limited set of choices of what is "reasonable" or what is put forward as "feasible," for example, in a context of forest and climate change (Brockhaus and Angelsen 2012). With the argumentative turn in environmental governance and the study of public policy and planning (Hajer 1995), the role of discursive legitimation gained momentum. (De)Legitimation refers to the process of making particular governance goals, procedures, or outcomes (less) more acceptable or plausible to society or selected audiences (Blum 2020; Steffek 2009).

3.2.2 (De)Legitimation narratives in FLR

Legitimation (or delegitimation) narratives are a set of ideas that assign (or take away) relevance and legitimacy to specific governance features, scales, and subject positions, while neglecting others (Bäckstrand and Lövbrand 2007, 2016; Blum 2020). At the same time, (de)legitimation narratives support certain interests and thus empower and disempower specific FLR actors. Elements in these (de)legitimation narratives can be linked to narratives and discourses already established in the wider governance literature (Bäckstrand and Lövbrand 2006, 2007, 2016), in particular, win-win promises in ecological modernization discourses, managerial and science/expert emphases in green governmentality, and critical contestations of these two narratives in civic environmentalism and climate justice. In order to unpack the power relations of how FLR is justified and positioned, we briefly outline here three FLR (de)legitimation narratives that are connected to broader meta-discourses (Bäckstrand and Lövbrand 2006) and that will be used to critically reflect tradeoffs and goal conflicts within the dimension of ideas.

FLR win-win narrative

Win-win narratives of FLR highlight the multiple benefits that may be simultaneously achieved within one shared landscape, or within a mosaic of landscapes. Often, a monetary compensation logic underpins the win-win narrative, pursuing promises of ecosystem service payments for farmers or carbon credits for the private sector. In the context of the Reforestation of the Brazilian Atlantic forest both "natural regeneration" and "Brazilian livelihoods" are presented in one such win-win scenario (Silva 2021). This narrative nourishes ideas that there is a place to satisfy everyone's needs in a perfectly integrated landscape (Reinecke and Blum 2018). This conceptualization of FLR as a win-win situation resonates well with arguments of ecological modernization, which emphasize that environmental governance arrangements are a positive-sum game and that sustainable development, economic growth, and climate protection can be achieved simultaneously (Bäckstrand and Lövbrand 2006). Yet the possible tradeoffs and political conflicts over access and rights to land are not considered. Risks at the local level are perceived as manageable, and an optimistic view prevails that they can be addressed by well-designed institutions.

The win-win narrative holds that successful FLR projects are required to follow good forest governance principles agreed upon at the global level "considering tenure, policies, and institutions [as well as] transparency, accountability, and public participation" (Mansourian 2016, 271). FLR efforts are expected to use integrative techniques of monitoring, capacity building, and benefit sharing (Pistorius and Kiff 2017; Sayer et al. 2013), or to adopt a landscape approach, which can help to avoid conflicts and contestation. The win-win narrative suggests that different, even competing, interests need to and can be aligned across a landscape. Collaborative multi-stakeholder processes allow the engagement of relevant stakeholders. The optimistic faith in efficient multi-stakeholder governance, however, may naively underestimate preexisting power structures in which these processes are embedded (Faysse 2006; Karambiri and Brockhaus 2019).

FLR implementation narrative

The FLR implementation narrative is embedded in green governmentality and ecological modernization discourses on forest governance (Bäckstrand and Lövbrand 2006). It acknowledges tradeoffs that may exist between climate mitigation, adaptation, rural development, economic growth, and biodiversity conservation goals. The main challenge under this narrative is the implementation of FLR, which is, however, considered manageable. Framed as a genuine problem of implementation, the narrative focuses on the political, financial, and technical interventions that serve as key tools for fostering implementation (Mansourian et al. 2020; Reinecke and Blum 2018). Central actors are the state, the private sector, and the research and academic sector; one of the main concerns is related to the ability of establishing an institutional setting that is able to change incentive structures and attract the private sector (Reinecke and Blum 2018) (see example in Box 3.1).

FLR critical narrative

The critical FLR narrative goes further than the concerned variation of the implementation narrative. The critical narrative considers FLR as an expression of neocolonialism. It argues that in FLR the claims tying exploitative tree plantation activities to some sort of restoration narrative in forestry can be traced back to colonial times in the 19th century (Kröger 2013a; Murray 1992; Smiet 1990). Rather than referring to a genuinely technical implementation challenge, the critical narrative rejects and delegitimizes FLR projects that are not truly participatory and community centered. In essence, FLR is fundamentally challenged and perceived as not much more than "old wine in new bottles," which will not contribute to a desirable transformation in the land use sector, including changes away from unsustainable business-as-usual practices. It is considered problematic that the reforestation of "degraded" or "non-productive" forest landscapes consists of planting trees to establish forest plantations, while ecosystems and informal land uses are neglected (Aerts and Honnay 2011; Barr and Sayer 2012; Lukas and Peluso 2020).

Critical research on FLR focusing on political and social questions is only beginning to emerge, together with first practical implementation experiences (Fleischman et al. 2020; Reinecke and Blum 2018). A review of this literature suggests that certain issues and concerns known from environmental policy and governance initiatives preceding FLR are in fact resurfacing in the debates on FLR (for an overview see Brockington and Igoe 2006; Delabre et al. 2020; Mansourian 2016; McDermott and Ituerte-Lima 2016). These overlapping issues and concerns include: (i) the focus on carbon rather than on the entire socio-ecological system (Lewis et al. 2019), (ii) the risk of green grabbing, where, for instance, rural people are resettled or denied access to make space for industrial restoration projects (Erbaugh et al. 2020), (iii) the prioritization of approaches oriented to productivity and top-down governance (Fleischman et al. 2020; Reinecke and Blum 2018), (iv) the anthropocentric focus on wilderness prevailing in the "Global North" and misunderstandings of ecosystems (Bond et al. 2019; Vetter 2020), (v) the lack of acknowledgment of local communities and traditional knowledge (Erbaugh et al. 2020; Fleischman et al. 2020), and finally, (vi) the lack of consideration of justice and legitimacy (Fleischman et al. 2020).

3.3 Three logics of FLR

Building on the reviewed literature, we shed light on three different logics that—as we argue—currently dominate FLR: the global sustainability logic, the community logic, and the production logic. We analyze how institutions, interests, and ideas constitute these logics and—as in relation to the ideas dimension—we reflect on how they are discursively legitimated via different narratives as identified in our analytical framework. Figure 3.2 gives an overview of the investigated logics and introduces specific

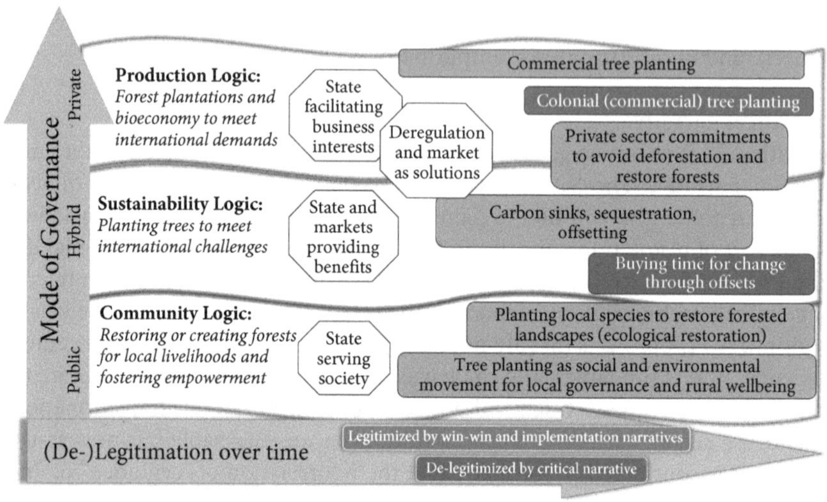

Figure 3.2 Logics and discursive elements promoting tree planting in the Global South over time and under different modes of governance

justifications and objectives related to these. The figure also highlights the evolution of these logics over time and implications for governance by state, markets, and society as well as core elements of the legitimation narratives.

3.3.1 The global sustainability logic: Planting trees to meet international challenges

Afforestation, reforestation, and forest restoration have been popular responses to the global environmental crisis for decades; tree planting as panacea features prominently on international and domestic policy agendas (Bäckstrand and Lövbrand 2006).

Institutions

In 1987 the Brundtland Commission published the "Report of the World Commission on Environment and Development: Our Common Future,"[1] marking the first major global initiative to promote restoration of forest landscapes. The report recommended a set of international policy measures for sustainable land use, including restoration of degraded areas. Building on this report, the 1992 Rio de Janeiro Earth Summit gave rise to three international conventions: Convention on Biological Diversity (CBD),[2] United Nations Convention to Combat Desertification (UNCCD),[3] and United Nations Framework Convention on Climate Change (UNFCCC).[4] While the three sister conventions are not focused on forests and their restoration as such, they offer an international framework of global norms for forest restoration and forest plantations.

Within this historical context, and due to its embeddedness in the institutional environment of UN conventions, the Bonn Challenge can be described as a landmark for FLR (Laestadius et al. 2015). It builds on the science-driven process in the early years of the millennium to jointly develop and pursue a concept for FLR (for an overview see Mansourian 2021). The Bonn Challenge is a policy-driven, multi-stakeholder platform promoting FLR in global environmental governance. It was launched by the Government of Germany in 2011, led by the Ministry of Environment and the International Union for Conservation of Nature (IUCN) with support from the Global Partnership on Forest Landscape Restoration (GPFLR). The Bonn Challenge aims to restore 150 Mha of the world's deforested and degraded land by 2020. Building on the NYDF, its target was extended to 350 Mha to be restored by 2030 (Mansourian 2021).

These processes paved the way toward the UN's Declaration of a "Decade on Ecosystem Restoration 2021–2030" (UN Decade), as adopted by the UN General Assembly in March 2020. Supported by more than 70 nations, several international organizations (including United Nations Environment Programme (UNEP), Food and Agriculture Organization of the UN (FAO), and IUCN, and other global initiatives), the UN Decade shares the same FLR goals as the Bonn Challenge and NYDF. The UN Decade aims to facilitate a broad public appreciation for healthy ecosystems to strengthen measures designed to restore degraded ecosystems while contributing to the goals of the three UN Conventions (UNFCCC, CBD, and UNCCD) as well as to the SDGs.

FLR is embedded in the paradigm of sustainable development, which is formalized via the UN's SDGs. Specifically, FLR is meant to support the SDG target 15.3, which calls on countries to achieve land degradation neutrality (LDN) by 2030. This goal is supported by the UNCCD, which integrated and adopted LDN at its 12th Conference of the Parties in October 2015 (UNCCD 2015a, decision 3). Moreover, FLR is closely linked to SDG 2—no hunger, improving food security, and promoting sustainable agriculture—as it builds on the integrated landscape approach, which promotes the sustainable use of both forest and agricultural land (FAO 2022a; IUCN 2015). An important part of the global sustainability logic is the vision of improving food security by enhancing land productivity, creating regional value chains, and improving local food supplies.

Interests

The global FLR initiatives draw on and are supported by a broad set of multinational organizations that promote sustainable development, however with different agendas: for example, UN Environment focuses on nature and climate; the FAO, sustainable agriculture and conservation; the World Bank (WB), economic development. All of these organizations are involved directly or indirectly in the Bonn Challenge or its regional spinoffs, the UN Decade, the NYDF, etc. Following the working logics and interests of these organizations, FLR is meant to contribute in multiple ways to the achievement of the SDGs and the three major UN conventions. Restoring forests and landscapes is a practical way to increase food and water security, sequester carbon, enhance adaptability and resilience to climate change, and minimize the risks associated with conflicts over natural resources and large-scale migration.

With IUCN at the heart of its organization, the Bonn Challenge is furthermore nurtured by strong conservation ideas. Based on the concept of "ecosystem restoration," as supported in particular by the Society for Ecological Restoration (SER), planting trees should recover native ecosystems and prioritize nature-based solutions (Ecosystem Restoration Thematic Group 2021; SER International Science & Policy Working Group 2004). The Bonn Challenge explicitly references Aichi Biodiversity Target 15 of the CBD—to restore 15% of all degraded land by 2020—reaffirming the ecological focus of the convention.

The need to protect or restore forests has been increasingly linked to the global agenda to address climate change. As a consequence, the Bonn Challenge and many FLR projects are closely tied to national strategies of climate action on deforestation. Forests play a critical role in UN framework agreements, for example, the Paris Agreement under UNFCCC, and are addressed in political instruments like afforestation and reforestation projects under the Clean Development Mechanism or Reducing Emissions from Deforestation and Forest Degradation (REDD+, where restoration may fall under "+" activities or the second D). The NYDF, agreed at the 2014 UN Climate Summit, has been endorsed by more than 50 governments, more than 50 companies, and more than 50 societal groups. It takes up the aims and implementation of the Bonn Challenge and links them with the climate logic of carbon mitigation. With the declaration the engagement of the private sector becomes a strong motive supported by the involved national governments, nourishing the idea that global sustainability goals, such as ecological restoration or rural development, can go hand in hand with business development and exploring new markets.

Ideas

The global sustainability logic shares key arguments and framings with the "Win-Win" narrative, with one "win" referring to people and their livelihoods, the second win referring to nature and climate. The promise of simultaneously enabling rural development for local people, mitigating climate change, and improving ecosystem health makes FLR an attractive idea for several UN communities and funding organizations. From the perspective of the critical narrative, this neglects possible tradeoffs or goal conflicts (Griscom et al. 2017). FLR links to other global logics such as REDD + or tree plantations for global carbon offsetting. The global logic to effectively mitigate carbon emissions by planting trees has often led to a decrease of biodiversity due to industrial tree plantations (Overbeek et al. 2012). Goal conflicts are suspected to arise, for example, between climate change, biodiversity benefits, and other needs—specifically those of the current users of "degraded" or "abandoned" land (Mansourian 2021). In the NYDF context, the aim of mitigating carbon through ecological restoration prominently ties the FLR agenda to ideas of marketability and economic development (i.e., carbon credit benefits). In this sense, a logic that highlights carbon mitigation may be in greater alignment with traditional aims pursued in timber or biomass markets than with the ecological restoration and biodiversity motives.

Finally, there is a risk that promises of planting trees tomorrow might legitimize deforestation today or distract from efforts to keep forests and trees standing (Seymour 2020; Verchot et al. 2010).

3.3.2 The community logic: Restoring or creating forests for local livelihoods and fostering empowerment

Since the 1970s global forest governance has embraced community-based forestry as a new paradigm, with tree planting being encouraged on private and community lands to advance participation and democracy and to serve environmental and economic aims (Gilmour 2016)—although with limited success (Karambiri and Brockhaus 2019). Yet, planting trees is also part of a (postcolonial) struggle for place and political belonging, with the Green Belt movement led by Wangari Maathai as an example (Hunt 2014).

Institutions

The norms of the community logic are linked to values such as justice for, participation with, and empowerment of local actors, values also prominently pronounced in FLR policy arenas. Erbaugh et al. (2020) find that 294.5 million people live on tropical forest lands considered suitable for restoration. Findings from the field show strong hopes that agroecological agroforestry (Ollinaho and Kröger 2021) may help alleviate poverty and promote (rural) development, ensuring food and energy security for local and forest-dependent people (e.g., Reed et al. 2017; Reij et al. 2009). In this logic, local communities are to play a central role as key stakeholders and beneficiaries in building up FLR institutions. The logic builds on existing local institutions in implementing FLR planning, activities, and monitoring. This FLR logic requires that local communities are assigned corresponding rights in decision-making, multi-stakeholder participation, and project implementation (Reinecke and Blum 2018). Regarding inclusiveness and empowerment goals, this logic aspires to involve more women in FLR initiatives as a way to enhance gender-balanced participation in rural development and local democracy (Adams et al. 2016; Erbaugh and Oldekop 2018; Sijapati Basnett et al. 2017).

Interests

The community logic represents the interests of different types of nongovernmental organizations (NGOs), funders, and community actors to gain development-related project funding by involving and supporting local actors. Local associations in partnership with transnational conservation NGOs form the key actor coalitions that drive extensive livelihood-focused, forest restoration initiatives in the Global South. Up to 2.5 billion people in smallholder agriculture directly benefit from forest resources and services, with almost one billion engaging in the collection of fuelwood from forests (FAO and UNEP 2020). In some cases, planting trees has even turned into a social movement and is narratively promoted as an expression of reclaiming (local) rights over land. The Green Belt Movement initiated in 2006 by Wangari Muta Maathai as Trillion Trees partnership is a prominent example of such local–international collaboration with a strong focus on and purpose grounded in livelihood and empowerment motives. This illustrates that local NGOs often play a crucial role in facilitating the community logic and are pursuing altruistic and strategic interests simultaneously. Box 3.1 provides more information about the Great Green Wall for the Sahara and the Sahel initiative. Such initiatives are also an expression of criticism against FLR processes that do not take people and their needs and rights seriously but are just "repeopling" supposedly bottom-up initiatives that are in fact mainly driven by public organizations or corporate interests. Sporadically, coalitions have formed between local associations, transnational conservation NGOs, and opinion leaders, supported by private firms as part of their corporate social and/or environmental responsibility programs (Van Oosten et al. 2014). Bilateral and multilateral development cooperation have added to such multi-stakeholder coalitions by providing the financial or environmental/development projects' infrastructure.

A central motive in the community logic is rural development. Land tenure, particularly its insecurity, seems to mark a major concern for proponents of this rationale. Tenure, and hence the question of who controls the land, defines who will make decisions about how the land is used. Most of the lands used by local communities, especially in Sub-Saharan Africa, are managed under customary or other informal rights regimes, which often lack formal recognition by the state. Another related disincentive for communities to plant trees or restore forest ecosystems is that rights to land and rights to trees are often legally separated (Chomba et al. 2020; Warnholtz et al. 2017). Yet, interest in tenure and rights seems to have shifted from a community focus toward a concern over tenure security for those investing in nature-based solutions, as a recent analysis of REDD + governance found (McDermott et al. 2022).

Box 3.1 The Great Green Wall for the Sahara and the Sahel initiative

Following the 1970s drought in the Sahel, the African Great Green Wall (GGW) for the Sahara and the Sahel initiative was launched by the presidents of the Community of Sahel-Saharan States, during their 7th session in 2005 in Ouagadougou, and backed by the African Union in 2007. The GGW aims to plant a tree wall 8000 km long, 15 km wide across 11 countries,[5] in the Sahel from Senegal in the west to Djibouti in the east to combat desertification. The Pan-African Agency of the GGW (PAAGGW) including representatives of national governments, international organizations, private sector, and civil society was established in 2010 to oversee and coordinate the initiative's implementation. The UNCCD, the European Union, WB, IUCN, FAO, and the Global Environment Facility (GEF) are among the key international investors in the GGW.[6]

The core idea of the GGW to halt desertification, that is, the spread of the Sahara Desert, took root in French and British colonial legacies. They put forward an environmental narrative that blames indigenous people, their lifestyles (e.g., pastoralism) and livelihoods (e.g., shifting cultivation) for desertification through overgrazing, deforestation, and land degradation.[7] The GGW vision does not challenge these claims but rather echoes them, as pointed out by Bellefontaine et al. (2011), including the assumption of idle lands in the Sahel, while the region hosts mosaic landscapes with integrated agro-sylvo pastoral schemes governed by diverse traditions.

Over time, shifts in the vision and narratives surrounding the GGW have occurred. Desertification is no longer considered an isolated environmental mismanagement by the Sahel dwellers, but rather, is seen as embedded in the global climate crisis. This new narrative is that countries who are historically responsible for carbon emissions should help the most vulnerable countries adapt to adverse effects of climate change (Behnke and Mortimore 2016). Additionally, countries (and non-state actors) perceive GGW as a globalized opportunity to meet their ambitions and nationally determined contributions (NDCs) with regard to the Paris Agreement boosted by the diverse forest landscape restoration initiatives. Moreover, the GGW vision has broadened from tree-belt planting to mosaic landscape restoration and integrated ecosystem management with high ambitions.[8] Thus, the new vision integrates a diversity of land use and production systems, natural assisted regeneration, and sustainable land and water conservation to improve people's livelihoods.[9] It also has turned into an emblematic resource: for example, during the One Planet summit in January 2021, the GGW was referred to as the first flagship of the UN Decade on Ecosystem Restoration, and its projects were praised as nature-based solutions that the UNEP, the Global Environment Facility, and other donors support.[10]

Furthermore, mixed development aid, and market-driven governance and funding structures are emerging, justified by the lack of performance of the GGW so far, with estimates of 4%[11] to 15%[12] of targets achieved since its inception. Nongovernmental and development agencies are mobilizing funding to scale up restoration in drylands to support the GGW, for example, with the project

"Closing the gaps in the Great Green Wall."[13] The private sector investment in the GGW is still minor, but there is growing interest and efforts to bring them onboard as they are said to be critical for the achievement of the GGW targets.[14]

These developments call for more serious efforts to investigate—and eliminate—the colonial echoes in narratives that legitimize claims over land and resources; otherwise the GGW risks simply reproducing inequalities deeply engrained in the land sector in the Global South.

However, it must be acknowledged that success stories of local empowerment have been observed, for example, in the case of Nepal (Laudari et al. 2022).

Ideas

In the community logic, FLR is justified with empowering local people via land tenure and rural development and links to the win-win logic: reforestation (first win) and empowering local people (second win). At the same time, tenure rights can also be framed as a legal "implementation challenge," which neglects the political struggles. Setting this focus, governments are often presented as legitimate actors and key agents in FLR processes, reaffirming a top-down rationale of implementation rather than enabling and fostering social movements as FLR "from below." Within this dominant focus on the state, local governments and communities serve rather as beneficiaries and recipients of "proper" political processes. However, numerous actors, including policy actors themselves, share doubts about central governments' capacities and political will to successfully change land-use practices and power imbalances (Laudari et al. 2022).

3.3.3 The production logic: Forest plantations and bioeconomy

Planting trees and restoring forests to ensure the provision of goods and services to the benefit of those owning or claiming the land has a long tradition: planting trees in swidden forest gardens or ladangs managed by local communities, planting and restoring claimed land as hunting grounds by bishops and kings, and establishing large-scale tree plantations by church, (colonial) states, or corporations (Bryant 1998; Sargent and Bass 1992). Ongoing primary forest loss coupled with increasing demand for timber and biomass have served as justification for the establishment of large-scale plantations and restoration initiatives (Sargent and Bass 1992; Solberg et al. 2014), including a more recent focus on tree plantations for purposes of carbon sequestration.

Institutions

The production rationale as it unfolds in FLR shows strong parallels to the way the concept of bioeconomy has gained traction internationally in the institutional context of forestry over the last two decades (Pülzl et al. 2014). Under the production logic, dominant actors and existing practices of large-scale tree planting and management shape institutional arrangements for FLR (or bioeconomy respectively), by connecting to environmental concerns and objectives of FLR while fostering timber production by expanding forest plantations on degraded lands. Plantations are the most popular restoration plan of the Bonn Challenge and other schemes, "planting vast monocultures of trees as profitable enterprises," in countries such as Brazil, China, Indonesia, and others (Lewis et al. 2019, 26). For instance, India has restored 9.8 Mha since 2011 and has committed to restoring 21 Mha of degraded land by 2030; largely this area results from tree plantations and agroforestry (Borah et al. 2018). While some claim the restoration objectives have been largely reached (FAO 2022b), others point to social and ecological problems with these large-scale restoration schemes (He 2014; McElwee and Nghi 2021; Vetter 2020). Lamb et al. (2012, 15) provide an explanation for why plantation monocultures persist: it is the dominant restoration activity, as it is the "simplest type of new forest." It can be managed on short rotations and, as such, is perceived by landowners as more profitable and easier to manage. Even though there is an expectation that these forests are managed differently than traditional plantations, with greater attention to biodiversity and other goods and services at later stages, this approach still seems to be in conflict with global interests in FLR, where FLR claims over degraded land have mainly been triggered, or at least justified and legitimized, by environmental concerns.

The idea of using forest plantations as a restoration strategy has a long history (Wenhua 2004) well before the Bonn Challenge. In this tradition, restoration has a timber and biomass production focus to derive economic benefits through plantations (Laestadius et al. 2015). Planting new trees was a necessity deriving from the overexploitation and radical degradation of forest land in different regions of the world, driven by exactly the same mercantile use of forest resources and emerging global commodity markets that built the core motive for later on restoring those productive capacities. In Box 3.2 in this chapter, a critical perspective on "productive" forest industry FLR projects in the Global South provides some examples. In Finland, these dynamics arose with the rise of tar capitalism, a process that devastated large forest areas to produce export tar from pine trees; this process lasted until the early 20th century, replacing many common swidden forestlands with privatized land holdings devoid of forests (Toivanen and Kröger 2019). Elsewhere in Europe, such dynamics were driven by industrial use and commodity markets, including agriculture (Mather 2001). A similar situation occurred in imperial China (Marks 1996). A global rush for forests in the tropics occurred to satisfy the growing demand of industrialized and growing economies. Colonial powers intensively exploited forestlands in the Global South, often expropriating lands from local populations and exporting the goods to developing European economies, for example, the British Empire establishing teak plantations in Burma (Bryant 1998).

Interests

Historically, this logic of mercantile use of forest was jointly implemented by colonial administrations and a network of pro-colonial private companies (Hardin and Bahuchet 2011; Peluso and Vandergeest 2001). Additionally, local chiefs were used as the extended hand of colonial power (e.g., Western Africa (Amanor 1999)). At the time of independence for most developing countries (between 1950 and 1980), the

Box 3.2 A critical perspective on "productive" forest industry FLR projects in the Global South

Tree plantations have been framed as being part of FLR initiatives, especially by large tree-plantation, expanding pulp, charcoal, and carbon credit corporations (Kröger 2014). These corporations are based not only in Sweden, Finland, Canada, the USA, Japan, and Singapore but also, and increasingly, include the core regions and key sectorial corporations within Global South countries like Suzano from Brazil, and Chilean pulp corporations (Kröger 2012; Kröger 2013a). These schemes, which replace pampa landscapes with eucalyptus monocultures, for example in Uruguay, have thus been called forestry extractivism (Kröger and Ehrnström-Fuentes 2021), a subcategory of agrarian extractivism, rather than a form of forest restoration or maintenance.

The tree-planting industries emphasize that they are also setting aside or even restoring some degraded forests, for example, in the Atlantic rainforest areas of Brazil. Critics point out, however, that the corporations themselves did massive deforestation in these rainforests, especially before 1993 (Kröger 2013b) and have not recomposed these areas with native forests as ordained by courts (Kröger 2013b). Such tree plantation conflicts also include cases from Africa, especially in energy-wood and carbon credit tree plantation expansions, where locals that have been dispossessed have occupied large areas, similar to the landless and indigenous movements' thousands-of-hectares plantation occupations in Brazil's Bahia and Espirito Santo states (Gerber 2011; Overbeek et al. 2012).

Development cooperation funds from Nordic countries are involved in these conflicts through their funding of tree-planting schemes, especially in Africa (Bergius et al. 2018; Lyons and Westoby 2014). Local people have expressed frustration from being excluded from these enclave investment areas, seeing local food prices rise, and water levels decrease, among other grievances (Kröger 2014).

The dynamics of tree plantation expansion have been somewhat different in China, where smallholders and villages have been able to secure comparatively more control over the process. Mega-pulp investment projects have been remolded into other kinds of investments, for example in the Guangxi province, which witnessed heavy conflicts around Stora Enso (Xu 2018). Currently, there is massive expansion of all kinds of tree plantations, not just eucalyptus, in China. Digitalization, like the mobile app Ant Forest with over 500 million users, suggests there is an ongoing financialization and datafication of tree-planting schemes, which is becoming more deeply entrenched in the carbon-neutrality aims of countries and corporations.

Growing awareness that forest restoration schemes can have possible negative socio-ecological impacts, e.g., on biodiversity or subsistence of communities, has led to suggestions for more careful considerations of local knowledge and experiences. These sounder schemes should strive toward agroforestry to restore entire deforested, pasture-ravaged areas (Ollinaho and Kröger 2021). For instance, in some areas where the several landless movements in Brazil's Bahia state have managed to win through their struggles with pulp corporations such agroforestry schemes are currently expanding, partly assisted by technical experts funded by corporations through state-mediation of the conflicts. However, sometimes such schemes are not recognized or supported by international forest governance, due to a narrow (corporate-influenced) definition of deforestation and forest restoration.

Many tree-planting areas are currently experiencing rising conflicts around land ownership and control. These result from the 1970s–1980s authoritarian regimes giving leeway or ordering illegal land grabbers to lay ground for establishing plantations for corporations, by grabbing indigenous lands, e.g., in Bahia pulp companies claim to have land ownership without formal land ownership titles. Yet, they are still expanding and driving indigenous groups to the margins. In particular in contexts like in Latin America the de jure and de facto respect for ethno-territorial rights depends on the given government and larger governance setting, including constitutions and regionally dominant political economic groups (Kröger and Lalander 2016).

These problems of expanding tree plantations are not just limited to the tropics but are also visible in North America, such as in British Colombia (Simard 2021), and in Europe, such as Finland and Sweden, where natural, seminatural and even the most species-rich forests are being removed under the guise of a bioeconomy discourse (Kröger 2016; Kröger and Raitio 2017). These forest removals, followed by monoculture planting of spruce mostly, are being increasingly contested in social and conventional media, as well as in physical protests; for example, in 2021 different environmental NGO activists started to block and denounce the cutting of the last remaining forest patches in Finland.

dominant forest ownership held by colonial powers was transferred to the newly established postcolonial state. While the new states started establishing state-owned logging firms in post-socialist countries to pursue timber extraction, other public and private companies in the tropics developed fully commercial forest plantations (rubber, cocoa, teak, eucalyptus, palm oil, etc.), often using the narrative of reforestation or restoring degraded forest lands (Feintrenie 2014; Laungaramsri 2012). In the 1980s several critiques were raised by different social movements about the establishment of such (large-tree) plantations, for example, in India, Indonesia, Chile, and Vietnam, where campaigns by environmental NGOs and rural people tried to prevent the clearing of natural forests for tree plantations. Despite contestations, claims of positive implications for large-scale tree plantations—for example, in terms of labor effects or local income generation—persist (Malkamäki et al. 2018; Nolte and Ostermeier 2017), and are repeated within the FLR context, even though contested too (Elias et al. 2021; McElwee and Nghi 2021).

Following the logic of forest product markets, international organizations with a "business-friendly" agenda still support forest plantations for FLR, especially in the Global South, for example, the International Tropical Timber Organization (ITTO) and the WB (ITTO 2022; WB 2017). Actors supporting this logic argue that green supply chains are essential for realizing a bioeconomy approach for FLR, highlighting the importance of FLR and its linkages to value chains, livelihoods, markets, investments, and financing mechanisms (ITTO 2022). This market-focused logic is also supported by the United Nations Forum on Forests, the United Nations Strategic Plan for Forest 2017–2030, and the Collaborative Partnership on Forests (UNEP and FAO 2022). All of the above-named organizations do in fact acknowledge that conventional plantations are unlikely to provide multi-functional and integrated outputs as aimed for by the FLR concept.

Ideas
Critics warn that large-scale plantations (of exotic tree species) may fall short of other expectations tied to FLR, for example, concerning carbon sequestration and/or biodiversity conservation (Bastin et al. 2019; Brancalion et al. 2019). In response, technical discussions are ongoing about how plantations may be modified to deliver additional goods and services without significantly compromising the primary function of economically viable roundwood production.

The logic resonates strongly with the win-win narrative, as it emphasizes productivism and seems to ignore goal conflicts (Kleinschmit et al. 2014; Kröger and Raitio 2017; Ramcilovic-Suominen and Pülzl 2018). Moreover, the core ideas of bioeconomy, to rely on bio-based resources and to replace fossil-based products with renewable products, are linked to the production logic of FLR. Critics argue that the bioeconomy concept is being (mis)used to relegitimize "policy goals from previous industrial forestry eras," safeguarding and prioritizing timber production over ecological concerns (Kröger and Raitio 2017, 6). These interlinkages of bioeconomy and FLR assumptions and ideas can be observed in Kyrgyzstan, for example. In the country's Green Economy Program 2020–2023, reforestation and fast-growing tree plantations are among the core points to foster its bioeconomy (Partnership for Action on Green Economy 2019). We argue that FLR's productivity logic shares problematic ideas with the bioeconomy approach as both increase the economic pressure on forests and are likely to cause new conflicts, which are so far neglected in the debate.

3.4 Discussing the dominant logics of FLR

The aim of this chapter is to better understand and critically reflect on the international governance of FLR by unpacking logics driving restoration efforts and the underlying power relations embedded in the specific institutions, interests, and ideas of each of these logics. We ask how these logics are constituted and legitimized with different narrative, some more powerful than others. Our analysis illustrates that in multilevel governance settings policy levels may not always be perfectly separable; significant power asymmetries exist within and between levels, which lends discursive power to different narratives. Acknowledging the limitations of categories as simplifications and generalizations of practically complex and even interwoven interpretations of FLR, three key logics are identified and analyzed: (i) the global sustainability logic: planting trees to meet international challenges, (ii) the community logic: restoring or creating forests for local livelihoods and fostering empowerment, and (iii) the production logic: forest plantations and bioeconomy. These logics are characterized by different levels of popularity and predominance. In Figure 3.2 these logics are visualized, embedded in their mode of governance and providing an idea about the time of existence.

The sustainability-focused logic of planting trees is currently very prominent in the international FLR governance arena. With the rise of initiatives like the Bonn Challenge and the NYDF, which are connected to UN-led programs related to the SDGs and the UN Decade, the focus is primarily on regaining ecological functions of forests including biodiversity conservation and carbon capture. The dominance of this logic rests in the support of powerful international and transnational actors, like UN organizations and governments, but also actors from academia, civil society, or the private sector supporting the implementation of pledges through technical or financial resources. The logic tends to follow the assumption that planting more trees is good in principle; yet it uses problematic concepts such as "abandoned land" or "terre vacant," with the majority of such "idle or degraded" lands being in the Global South. Such framing reinforces formalized and "legal" approaches to land use, with claims of such lands by the state, justified by the notion that the land is not used otherwise. In many contexts, we see the problem that FLR then risks neglecting informal or customary land use practices such as swidden and mobile husbandry, firewood collection, or subsistence farming on supposedly empty land. Moreover, there are also potential biases in producing and relying on global mapping of restoration potentials in arid land (see Fagan 2020). Recently, calls for a rights-based approach to FLR have become louder, questioning the simplistic assumptions about abandoned landscapes with little value (Mansourian 2016; Van Oosten and Merten 2021).

We argue that the global sustainability logic naively underestimates persistent tradeoffs between climate goals, local food production, and biodiversity conservation, as well as ignores important considerations about who will benefit from FLR and how FLR needs to be implemented accordingly, e.g., bottom-up vs. top-down governance. Based on our analysis, we see the need to move away from "business as usual" approaches in forest (restoration) projects, which tend to reproduce existing power relations and neglect tradeoffs and potential conflicts. The currently dominant FLR logics are likely to produce many "win-lose" or "lose-lose" projects if historical power structures are not recognized and addressed. In order to create space for transformative change away from unsustainable tree planting and to overcome the domination of a "Western Gaze" to policy, valuing, and thinking, different logics need to emerge (Ramcilovic-Suominen 2022). Alternative approaches to study and inspire sustainability governance and FLR include decoloniality (Escobar 1995; Mignolo 2011; Rodríguez and Inturias 2018; see also Chapter 4), just socio-ecological transformations (Martin et al. 2020; Menton et al. 2020), convivial conservation (Büscher and Fletcher 2019), and local democracy (Ribot 2006; Walters et al. 2021). These conceptualizations all highlight the importance of local actors and social movements in bringing about desired changes for people and the environment.

The second logic of FLR—the community logic—highlights local livelihood and empowerment and has been promoted in the Global South, for example, as community-based forestry by development agencies since the 1970s. Likewise, it may be seen as a reactive logic, responding to the tradeoffs and conflicts arising from the practices pursued under the two other logics. This is particularly the case when community-based forestry is applied as a top-down approach. In contrast, FLR as a bottom-up approach is driven by social, often ecofeminist movements. While the overall community logic and related narratives have gained popularity over time, the more radical strand related to social movements is represented by rather marginalized actor groups. However, attention within the FLR arena is increasing, as the tenure rights of marginalized groups, especially indigenous peoples, youth, and women, have been acknowledged as relevant for sustainable (forest) land and resource use. The actors behind the wider logic and its respective stakes include not only local actors, but also international and local movements, NGOs, development agencies, and social scientists, among others. While this logic underpins singular flagship projects like the Green Belt Movement or Farmer Managed Natural Regeneration, considerations of livelihoods and rights have begun to inspire FLR institutions that adhere to SDGs and an ecological logic, contributing to the appearance of global FLR programs offering a compelling vision to unite the global community in its aim to regain ecological integrity and enhance human well-being. Furthermore, triggered by fundamental critiques in science and civil society, and fostered by stakeholder engagement and consultation, the integration of local people to effectively implement FLR has raised awareness.

The third logic, production and bioeconomy, reflects the mainstream global FLR movement that presents productive plantations as an opportunity for the restoration of multiple landscape functions (Sayer and Elliot 2005). Creating regional value chains should, according to this logic, directly benefit local people and enable an economy based on renewable materials. While we consider value chains and local livelihoods as crucial, the question of who benefits most from these new value chains demands to be addressed. This productivity-focused logic follows the same historically dominant and persistent exploitative economic logic behind global timber production and neocolonial trade relationships among world regions.

We recognize that the FLR concept is evolving. It has structures in place that in principle should allow for learning, e.g., large global exchange events such as the Global Landscapes Forum. However, the question remains: who is given a voice in such events—and given by whom? Learning requires information exchange beyond echo chambers (Malkamäki et al. 2021); what can be considered still a largely peripheral debate on tenure rights is expected to gain more prominence and to be taken on board in FLR agendas. Yet, market interests should not be underestimated in this debate: the rise of the issue of tenure on the FLR agenda is not only advocated by NGOs and social movements, but also actively driven by private sector interests in safeguarding their investments. Whether concerns like tenure rights will fundamentally change FLR practices or will only create mostly symbolic changes remains an empirical question yet to be answered.

For FLR to move away from business-as-usual rationales of exploiting forests, forest resources, and people, and in order to overcome the pitfalls that seem inherent to its current design and implementation, decolonizing the mindsets and ideas shaping FLR governance requires much more space on the global FLR agenda. This chapter's critical review of motivations and expectations in FLR and an acknowledgment of the tradeoffs and potential losers of international FLR policymaking is only a first step in this direction.

3.5 Conclusions

Three key distinct logics that, to varying degrees, drive FLR policy and practice today have been identified in this chapter: (i) the global sustainability logic: planting trees to meet international challenges, (ii) the community logic: restoring or creating forests for local livelihoods and fostering empowerment, and (iii) the production logic: forest plantations and bioeconomy. Currently FLR follows two dominant logics, focusing on global sustainability and production arguments. The former underestimates persistent tradeoffs between different goals and ignores the essential question on who will benefit from FLR. In contrast, the production logic follows a historically dominant and persistent exploitative economic logic behind global timber production and neocolonial trade relationships. Currently the FLR concept is further evolving and learning opportunities are in place recognizing trade-offs and conflicts. However, if FLR follows the dominant logics, it might produce as well in the future many "win-lose" and/or "lose-lose" situations. The question of inherent historical power structure in FLR design and implementation and the question of who benefits are not addressed.

Acknowledgments

Sabaheta Ramcilovic-Suominen gratefully acknowledges the funding by the Academy of Finland (grant number: 332353) that supported her work on this book chapter.

Notes

1. https://www.unccd.int/issues/land-and-sustainable-development-goals.
2. https://www.cbd.int/
3. https://www.unccd.int/
4. https://unfccc.int/
5. The 11 countries of the GGW are Senegal, Burkina Faso, Mali, Mauritania, Niger, Nigeria, Tchad, Sudan, Ethiopia, Eritrea, and Djibouti.
6. https://www.unccd.int/publications/great-green-wall-implementation-status-and-way-ahead-2030
7. https://theconversation.com/africas-got-plans-for-a-great-green-wall-why-the-idea-needs-a-rethink-78627
8. https://www.oneplanetsummit.fr/en/coalitions-82/great-green-wall-accelerator-193
9. https://www.unccd.int/publications/great-green-wall-implementation-status-and-way-ahead-
10. https://www.unep.org/news-and-stories/story/good-news-africas-great-green-wall
11. https://earth.org/great-green-wall-in-africa-4-complete/
12. https://www.greatgreenwall.org/about-great-green-wall
13. https://www.iucn.org/theme/ecosystem-management/our-work/global-drylands-initiative/gdi-projects/closing-gaps-great-green-wall-linking-sectors-and-stakeholders-increased-synergy-and-scaling

14. https://www.iucn.org/theme/ecosystem-management/our-work/global-drylands-initiative/gdi-projects/closing-gaps-great-green-wall-linking-sectors-and-stakeholders-increased-synergy-and-scaling

References

Adams, C., S.T. Rodrigues, M. Calmon, and C. Kumar. 2016. "Impacts of Large-Scale Forest Restoration on Socioeconomic Status and Local Livelihoods: What We Know and Do Not Know." *Biotropica* 48 (6): pp. 731–744.

Aerts, R., and O. Honnay. 2011. "Forest Restoration, Biodiversity and Ecosystem Functioning." *BMC Ecology* 11 (1): p. 29.

Amanor, K.S. 1999. *Global Restructuring and Land Rights in Ghana: Forest Food Chains, Timber, and Rural Livelihoods*. Uppsala: Scandinavian Institute of African Studies.

Bäckstrand, K., and E. Lövbrand. 2006. "Planting Trees to Mitigate Climate Change: Contested Discourses of Ecological Modernization, Green Governmentality and Civic Environmentalism." *Global Environmental Politics* 6: pp. 50–75.

Bäckstrand, K., and E. Lövbrand. 2007. "Climate Governance Beyond 2012." In *The Social Construction of Climate Change: Power, Knowledge, Norms, Discourses*, edited by M.E. Pettenger, pp. 123–147. Burlington: Ashgate Publishing.

Bäckstrand, K., and E. Lövbrand. 2016. "The Road to Paris: Contending Climate Governance Discourses in the Post-Copenhagen Era." *Journal of Environmental Policy & Planning* 21 (5): pp. 519–532. DOI:10.1080/1523908X.2016.1150777.

Barr, C.M., and J.A. Sayer. 2012. "The Political Economy of Reforestation and Forest Restoration in Asia-Pacific: Critical Issues for REDD+." *Biological Conservation* 154: pp. 9–19.

Bastin, J.F., Y. Finegold, C. Garcia, et al. 2019. "The Global Tree Restoration Potential." *Science* 365 (6448): pp. 76–79.

Baynes J., J. Herbohn, N. Gregorio, W. Unsworth, and É.H. Tremblay. 2019. "Equity for Women and Marginalized Groups in Patriarchal Societies during Forest Landscape Restoration: The Controlling Influence of Tradition and Culture." *Environmental Conservation* 46: pp. 241–246.

Behnke, R., and M. Mortimore. 2016. *The End of Desertification? Disputing Environmental Change in the Drylands*. Berlin: Springer Earth System Sciences.

Bellefontaine, R., M. Bernoux, B. Bonnet, et al. 2011. *The African Great Green Wall Project. What Advice Can Scientists Provide?*, edited by I. Amsallem and S. Jauffret. CSFD Topical issue. Montpellier: Agropolis International. https://agritrop.cirad.fr/567880/1/document_567880.pdf.

Bergius, M., T.A. Benjaminsen, and M. Widgren. 2018. "Green Economy, Scandinavian Investments and Agricultural Modernization in Tanzania." *Journal of Peasant Studies* 45 (4): pp. 825–852.

Bernstein, S. 2000. "Ideas, Social Structure and the Compromise of Liberal Environmentalism." *European Journal of International Relations* 6 (4): pp. 464–512.

Blum, M. 2020. "The Discursive Legitimation of Market-driven Climate Governance after the Paris Agreement: The Case of the Global Carbon Offset Markets." Thesis submitted, Universität Freiburg.

Bond, W.J., N. Stevens, G.F. Midgley, and C.E. Lehmann. 2019. "The Trouble with Trees: Afforestation Plans for Africa." *Trends in Ecology & Evolution* 34 (11): pp. 963–965.

Bonn Challenge. 2018. https://www.bonnchallenge.org/.

Borah, B., A. Bhattacharjee, and N.M. Ishwar. 2018. *Bonn Challenge and India: Progress on Restoration Efforts Across States and Landscapes*. New Delhi: IUCN and Ministry of Environment, Forest and Climate Change, Government of India.

Brancalion, P.H.S., A. Niamir, E. Broadbent, et al. 2019. "Global restoration opportunities in tropical rainforest landscapes." *Science Advances* 5 (7). DOI:10.1126/sciadv.aav3223.

Brockhaus, M., and A. Angelsen. 2012. "Seeing REDD+ through 4Is: A Political Economy Framework." In *Analysing REDD+: Challenges and Choices*, edited by A. Angelsen, M. Brockhaus, W.D. Sunderlin, and L. Verchot, pp. 15–30. Bogor: CIFOR.

Brockington, D., and J. Igoe. 2006. "Eviction for Conservation: A Global Overview." *Conservation and Society* 4 (3): pp. 424–470.

Bryant, R.L. 1998. "Power, Knowledge and Political Ecology in the Third World: A Review." *Progress in Physical Geography* 22 (1): pp. 79–94.

Büscher, B., and R. Fletcher. 2019. "Source: Towards Convivial Conservation." *Conservation and Society* 17 (3): pp. 283–296.

Büthe, T. 2004. "Governance Through Private Authority: Non-State Actors in World Politics." Review Article, *Journal of International Affairs* 58 (1): pp. 81–90.

Campbell, J.L. 1998. "Institutional Analysis and the Role of Ideas in Political Economy." *Theory and Society* 27 (3): pp. 377–409.

Chazdon, R.L., S.J. Wilson, E. Brondizio, M.R. Guariguata, and J. Herbohn. 2021. "Key Challenges for Governing Forest and Landscape Restoration across Different Contexts." *Land Use Policy* 104: pp. 104854. DOI:10.1016/j.landusepol.2020.104854.

Chomba, S., F. Sinclair, P. Savadogo, M. Bourne, and M. Lohbeck. 2020. "Opportunities and Constraints for Using Farmer Managed Natural Regeneration for Land Restoration in Sub-Saharan Africa." *Frontiers in Forests and Global Change* 3 (3): pp. 571–679.

Delabre, I., E. Boyd, M. Brockhaus, M., et al. 2020. "Unearthing the Myths of Global Sustainable Forest Governance." *Global Sustainability* 3: pp. 1–10.

Dobbin, F., B. Simmons, and G. Garrett. 2007. "The Global Diffusion of Public Policies: Social Construction, Coercion, Competition, or Learning?" *Annual Review of Sociology* 33: pp. 449–472.

Ecosystem Restoration Thematic Group. 2021. "Third Global Forum on Ecological Restoration." https://cdn.ymaws.com/www.ser.org/resource/resmgr/publications/global_forum_preliminary_rep.pdf.

Elias, M., D. Joshi, and R. Meinzen-Dick. 2021. "Restoration for Whom, by Whom? A Feminist Political Ecology of Restoration." *Ecological Restoration* 39 (1–2): pp. 3–15.

Elmhirst, R., B. Sijapati Basnet, M. Sicawati, and D. Ekowati. 2017. "Gender Issues in Large Scale Land Acquisition: Insights from Oil Palm in Indonesia." https://rightsandresources.org/wp-content/uploads/2017/05/Gender-Issues-in-Large-Scale-Land-Acquisition-Insights-from-Oil-Palm-in-Indonesia_April-2017_CIFOR.pdf.

Erbaugh, J.T., and J.A. Oldekop. 2018. "Forest Landscape Restoration for Livelihoods and Well-being." *Current Opinion in Environmental Sustainability* 32: pp. 76–83.

Erbaugh, J.T., N. Pradhan, J. Adams, et al. 2020. "Global Forest Restoration and the Importance of Prioritizing Local Communities." *Nature Ecology & Evolution* 4 (11): pp. 1472–1476.

Escobar, A. 1995. *Encountering Development: The Making and Unmaking of the Third World.* Princeton: Princeton University Press.

Esping-Andersen, G. 1996. *Welfare States in Transition: National Adaptations in Global Economies.* Thousand Oaks: Sage.

Fagan, M.E. 2020. "A Lesson Unlearned? Underestimating Tree Cover in Drylands Biases Global Restoration Maps." *Global Change Biology* 26 (9): pp. 4679–4690.

FAO. 2022a. *The Forest and Landscape Restoration Mechanism.* https://www.fao.org/in-action/forest-landscape-restoration-mechanism/background/vision/en/.

FAO. 2022b. *The State of the World's Forests 2022. Forest Pathways for Green Recovery and Building Inclusive, Resilient and Sustainable Economies.* Rome: FAO. DOI:10.4060/cb9360en.

FAO and UNEP. 2020. *The State of the World's Forests 2020. Forests, Biodiversity and People.* Rome. DOI:10.4060/ca8642en.

Faysse, N. 2006. "Troubles on the Way: An Analysis of the Challenges Faced by Multi-Stakeholder Platforms." *Natural Resources Forum* 30 (3): pp. 219–229. DOI:10.1111/j.1477-8947.2006.00112.x.

Feintrenie, L. 2014. "Agro-Industrial Plantations in Central Africa, Risks And Opportunities." *Biodiversity and Conservation* 23 (6): pp. 1577–1589.

Fleischman, F., S. Basant, A. Chhatre, et al. 2020. "Pitfalls of Tree Planting Show Why We Need People-Centered Natural Climate Solutions." *BioScience* 70 (11): pp. 1–4.

Gerber, J.F. 2011. "Conflicts over Industrial Tree Plantations in the South: Who, How and Why?" *Global Environmental Change* 21 (1): pp. 165–176.

Gilmour, D. 2016. "Forty Years of Community-Based Forestry: A Review of Its Extent and Effectiveness." FAO Forestry Paper 176. Rome: FAO.

Griscom, B.W., J. Adams, P.W. Ellis, et al. 2017. "Natural Climate Solutions." *Proceedings of the National Academy of Sciences* 114: 11645–11650.

Hajer, M. 1995. *The Politics of Environmental Discourse.* New York: Oxford University Press.

Hajer, M., and W. Versteeg. 2005. "A Decade of Discourse Analysis of Environmental Politics: Achievements, Challenges, Perspectives." *Journal of Environmental Policy & Planning* 7 (3): pp. 175–184.

Hall, P.A., and K. Thelen. 2008. "Institutional Change in Varieties of Capitalism." *Socio-Economic Review* 7 (1): pp. 7–34.

Hardin, R., and S. Bahuchet. 2011. "Concessionary Politics: Property, Patronage, and Political Rivalry in Central African Forest Management." *Current Anthropology* 52 (S3): pp. 113–125.

Hay, C. 2004. "Ideas, Interests and Institutions in the Comparative Political Economy of Great Transformations." *Review of International Political Economy* 11 (1): pp. 204–226.

He, J. 2014. "Governing Forest Restoration: Local Case Studies of Sloping Land Conversion Program in Southwest China." *Forest Policy and Economics* 46: pp. 30–38.

Hunt, K.P. 2014. "It's More Than Planting Trees, It's Planting Ideas: Ecofeminist Praxis in the Green Belt Movement". *Southern Communication Journal* 79 (3): pp. 235–249.

ITTO. 2022. *Forest Landscape Restoration.* https://www.itto.int/sustainable_forest_management/forest_landscape/.

IUCN. 2015. *Enhancing Food Security through Forest Landscape Restoration: Lessons from Burkina Faso, Brazil, Guatemala, Viet Nam, Ghana, Ethiopia and Philippines.* Gland: IUCN. https://portals.iucn.org/library/node/45774.

Kahler, M., and D.A. Lake. 2003. *Governance in a Global Economy: Political Authority in Transition*. Princeton: Princeton University Press.

Karambiri, M., and M. Brockhaus. 2019. "Leading Rural Land Conflict as Citizens and Leaving It as Denizens: Inside Forest Conservation Politics in Burkina Faso." *Journal of Rural Studies* 65: pp. 22–31.

Kleinschmit, D., B.H. Lindstad, B.J. Thorsen, et al. 2014. "Shades of Green: A Social Scientific View on Bioeconomy in the Forest Sector." *Scandinavian Journal of Forest Research* 29: pp. 402–410.

Korpi, W., and J. Palme. 2003. "New Politics and Class Politics in the Context of Austerity and Globalization: Welfare State Regress in 18 Countries, 1975–95." *American Political Science Review* 97: pp. 425–446.

Kröger, M. 2012. "Neo-mercantilist Capitalism and Post-2008 Cleavages in Economic Decision-making Power in Brazil." *Third World Quarterly* 33 (5): pp. 887–901.

Kröger, M. 2013a. "Globalization as the 'Pulping' of Landscapes: Forestry Capitalism's North-South Territorial Accumulation." *Globalizations* 10 (6): pp. 837–53.

Kröger, M. 2013b. *Contentious Agency and Natural Resource Politics*. London: Routledge.

Kröger, M. 2014. "The Political Economy of Global Tree Plantation Expansion: A Review." *Journal of Peasant Studies* 41 (2): pp. 235–261.

Kröger, M. 2016. "The Political Economy of 'Flex Trees': A Preliminary Analysis." *Journal of Peasant Studies* 43: pp. 886–909.

Kröger, M., and M. Ehrnström-Fuentes. 2021. "Forestry Extractivism in Uruguay." In Agrarian *Extractivism in Latin America*, edited by B.M. McKay, A. Alonso-Fradejas, and A. Ezquerro-Cañete, pp. 186–207. London: Routledge.

Kröger, M., and R. Lalander. 2016. "Ethno-Territorial Rights and the Resource Extraction Boom in Latin America: Do Constitutions Matter?" *Third World Quarterly* 37 (4): pp. 682–702.

Kröger, M., and K. Raitio. 2017. "Finnish Forest Policy in the Era of Bioeconomy: A Pathway to Sustainability?" *Forest Policy and Economics* 77: pp. 6–15.

Laestadius, L., K. Buckingham, S. Maginnis, and C. Saint-Laurent. 2015. "Before Bonn and beyond: The History and Future of Forest Landscape Restoration." *Unasylva* 245 (66): pp. 11–18.

Lamb, D., J. Stanturf, J. and P. Madsen. 2012. "What Is Forest Landscape Restoration?" In *Forest Landscape Restoration: Integrating Natural and Social Sciences*, edited by J. Stanturf, D. Lamb, and P. Madsen, pp. 3–23. Dordrecht: Springer.

Laudari, H.K., K. Aryal, T.Maraseni, et al. 2022. "Sixty-Five Years of Forest Restoration in Nepal: Lessons Learned and Way Forward." *Land Use Policy* 115: pp. 106033. DOI:10.1016/j.landusepol.2022.106033.

Laungaramsri, P. 2012. "Frontier Capitalism and the Expansion of Rubber Plantations in Southern Laos." *Journal of Southeast Asian Studies* 43 (3): pp. 463–477.

Leipold, S., P.H. Feindt, G. Winkel, and R. Keller. 2019. "Discourse Analysis of Environmental Policy Revisited: Traditions, Trends, Perspectives." *Journal Of Environmental Policy & Planning* 21 (5): pp. 445–463. DOI:10.1080/1523908X.2019.1660462.

Lewis, S.L., C.E. Wheeler, E.T.A. Mitchard, and A. Koch. 2019. "Restoring Natural Forests Is the Best Way to Remove Atmospheric Carbon." *Nature* 568: pp. 25–28.

Lukas, M.C., and N.L. Peluso. 2020. "Transforming the Classic Political Forest: Contentious Territories in Java." *Antipode* 52 (4): pp. 971–995.

Lyons, K., and P. Westoby. 2014. "Carbon Colonialism and the New Land Grab: Plantation Forestry in Uganda and Its Livelihood Impacts." *Journal of Rural Studies* 36: pp. 13–21.

Malkamäki, A., D. D'Amato, N.J. Hogarth, et al. 2018. "A Systematic Review of the Socio-Economic Impacts of Large-Scale Tree Plantations, Worldwide." *Global Environmental Change* 53: pp. 90–103

Malkamäki, A., P.M. Wagner, M. Brockhaus, A. Toppinen, and T. Ylä-Anttila. 2021. "On the Acoustics of Policy Learning: Can Co-Participation in Policy Forums Break Up Echo Chambers?" *Policy Studies Journal* 49 (2): pp. 431–456.

Mansourian, S. 2005. "Overview of Forest Restoration Strategies and Terms." In *Forest Restoration in Landscapes: Beyond Planting Trees*, edited by S. Mansourian, D. Vallauri, and N. Dudley, pp. 8–13. New York: Springer.

Mansourian, S. 2016. "Understanding the Relationship between Governance and Forest Landscape Restoration." *Conservation and Society* 14: pp. 267–278.

Mansourian, S. 2021. "Disciplines, Sectors, Motivations and Power Relations in Forest Landscape Restoration." *Ecological Restoration* 39 (1–2): pp. 16–26.

Mansourian, S., J. Parrotta, P. Balaji, et al. 2020. "Putting the Pieces Together: Integration for Forest Landscape Restoration Implementation." *Land Degradation & Development* 31 (4): pp. 419–429.

Marks, R.B. 1996. "Commercialization without Capitalism: Processes of Environmental Change in South China, 1550–1850." *Environmental History* 1 (1): pp. 56–82.

Martin, A., M.T. Armijos, B. Coolsaet, et al. 2020. "Environmental Justice and Transformations to Sustainability, Environment." *Science and Policy for Sustainable Development* 62 (6): pp. 19–30.

Mather, A.S. 2001. "The Transition from Deforestation to Reforestation in Europe." In *Agricultural Technologies and Tropical Deforestation*, edited by A. Angelsen, and D. Kaimowitz, pp. 35–52. Wallingford: CABI in association with Centre for International Forestry Research.

McDermott, C., B. Vira, J. Walcott, et al. 2022. "The Evolving Governance of REDD+ ." In: *Forests, Climate, Biodiversity and People*, edited by J. Parotta, S. Mansourian, C. Wildburger, and N. Grima, pp. 21–60. IUFRO World Series Vol. 40. Vienna: IUFRO.

McDermott, C.L., and C. Ituarte-Lima. 2016. "Safeguarding What and for Whom? The Role of Institutional Fit in Shaping REDD+ in Mexico." *Ecology and Society* 21 (1): p. 9.

McElwee, P., and T.H. Nghi. 2021. "Assessing the Social Benefits of Tree Planting by Smallholders in Vietnam: Lessons for Large-Scale Reforestation Programs." *Ecological Restoration* 39 (1–2): pp. 52–63.

Menton, M., C. Larrea, S. Latorre, et al. 2020. "Environmental Justice and the SDGs: From Synergies to Gaps and Contradictions." *Sustainability Science* 15: pp. 1621–1636.

Mignolo, W.D. 2011. *The Darker Side of Western Modernity: Global Futures, Decolonial Options*. Durham: Duke University Press.

Murray, M.J. 1992. "'White Gold' or 'White Blood'? The Rubber Plantations of Colonial Indochina, 1910-40." *The Journal of Peasant Studies* 19 (3–4): pp. 41–67.

Nicholls, A., and B. Huybrechts. 2016. "Sustaining Inter-organizational Relationships Across Institutional Logics and Power Asymmetries: The Case of Fair Trade." *Journal of Business Ethics* 135: pp. 699–714.

Nolte, K., and M. Ostermeier. 2017. "Labour Market Effects of Large-Scale Agricultural Investment: Conceptual Considerations and Estimated Employment Effects." *World Development* 98: pp. 430–446.

Odoom, D. 2016. "Conceptualizing Decentralization: Reflecting on some Existing Literature." *Applied Research Journal* 1 (4): pp. 21–38.

Ollinaho, O., and M. Kröger. 2021. "Agroforestry Transitions: The Good, the Bad and the Ugly." *Journal of Rural Studies* 82: pp. 210–221.

Overbeek, W., M. Kröger, and J.F. Gerber. 2012. "An Overview of Industrial Tree Plantations in the Global South: Conflicts, Trends and Resistance Struggles." *Environmental Justice Organizations Liabilities and Trade Report.* http://www.ejolt.org/2012/06/an-overview-of-industrial-tree-plantations-in-the-global-south-conflicts-trends-and-resistance-struggles/

Partnership for Action on Green Economy. 2019. "Green Economy Week 2019: Green Economy-Strong Regions-Sustainable Development of the Country." https://www.un-page.org/files/public/concept_note_-_ge_week_-_kr.pdf.

Peluso, N.L., and P. Vandergeest. 2001. "Genealogies of the Political Forest and Customary Rights in Indonesia, Malaysia, and Thailand." *The Journal of Asian Studies* 60 (3): pp. 761–812.

Pistorius, T., and L. Kiff. 2017. *From a Biodiversity Perspective: Risks, Trade-Offs, and International Guidance for Forest Landscape Restoration.* Freiburg: UNIQUE.

Pritchard, R. 2021. "Politics, Power and Planting Trees." *Nature Sustainability* 4 (11): pp. 932–932.

Pülzl, H., D. Kleinschmit, and B. Arts. 2014. "Bioeconomy—An Emerging Meta-Discourse Affecting Forest Discourses?" *Scandinavian Journal of Forest Research* 29: pp. 386–393.

Ramcilovic-Suominen, S. 2022. "Envisioning Just Transformations in and Beyond the EU Bioeconomy: Inspirations from Decolonial Environmental Justice and Degrowth." *Sustainable Science:* pp. 1–16. DOI:10.1007/s11625-022-01091-5.

Ramcilovic-Suominen, S., and H. Pülzl. 2018. "Sustainable Development – a 'Selling Point' of the Emerging EU Bioeconomy Policy Framework?" *Journal of Cleaner Production* 172 (5): pp. 4170–4180.

Reed, J., J. van Vianen, J. Barlow, and T. Sunderland. 2017. "Have Integrated Landscape Approaches Reconciled Societal and Environmental Issues in the Tropics?" *Land Use Policy* 63: pp. 481–492.

Reij, C., G. Tappan, and M. Smale. 2009. "Agroenvironmental Transformation in the Sahel. Another Kind of 'Green Revolution.'" *IFPRI Discussion Paper* 00914. Washington: The International Food Policy Research Institute.

Reinecke, S., and M. Blum. 2018. "Discourses across Scales on Forest Landscape Restoration." *Sustainability* 10 (3): p. 613.

Ribot, J.C. 2006. "Choose Democracy: Environmentalists' Socio-Political Responsibility." *Global Environmental Change* 16 (2): 115–119. DOI:10.1016/J.GLOENVCHA.2006.01.004.

Rodríguez, I., and M.L. Inturias. 2018. "Conflict Transformation in Indigenous Peoples' Territories: Doing Environmental Justice with a 'Decolonial Turn.'" *Development Studies Research* 5 (1): 90–105.

Sargent, C., and S. Bass. 1992. *Plantation Politics: Forest Plantations in Development.* London: Earthscan. DOI:10.4324/9781315066882.

Sayer, J., and C. Elliot. 2005. "The Role of Commercial Plantations in Forest Landscape Restoration." In: *Forest Restoration in Landscapes - Beyond Planting Trees*, edited by S. Mansourian, D. Vallauri, and N. Dudley, 379–384. New York: Springer.

Sayer, J., T. Sunderland, J. Ghazoul, et al. 2013. "Ten Principles for a Landscape Approach to Reconciling Agriculture, Conservation, and Other Competing Land Uses." *Proceedings of the National Academy of Sciences of the United States of America* 110: 8349–8356.

Schmidt, V. 2008. "Discursive Institutionalism: The Explanatory Power of Ideas and Discourse." *Annual Review of Political Science* 11: pp. 303–326.

Scott, W.R. 2008. "Approaching Adulthood: The Maturing of Institutional Theory." *Theory and Society* 37 (427): pp. 427–442.

SER International Science & Policy Working Group. 2004. The SER International Primer on Ecological Restoration. https://cdn.ymaws.com/www.ser.org/resource/resmgr/custompages/publications/ser_publications/ser_primer.pdf.

Seymour, F. 2020. "Seeing the Forests as Well as the (Trillion) Trees in Corporate Climate Strategies." *One Earth* 2 (5): pp. 390–393.

Sijapati Basnett, B., M. Elias, M. Ihalainen, and A.M. Paez Valencia. 2017. "Gender Matters in Forest Landscape Restoration: A Framework for Design and Evaluation." *CIFOR Brief* 6685.

Silva, D. 2021. Brazil's Atlantic Forests Are Naturally Regenerating Much Faster than Expected. https://forestsnews.cifor.org/74340/brazils-atlantic-forests-are-naturally-regenerating-much-faster-than-expected?fnl=en.

Simard, S. 2021. *Finding the Mother Tree: Uncovering the Wisdom and Intelligence of the Forest.* London: Penguin Books.

Slaughter, A-M., A. Tulumello, and S. Wood. 1998. "International Law and International Relations Theory: A New Generation of Interdisciplinary Scholarship." *The American Journal of International Law* 92: pp. 367–397.

Smiet, A.C. 1990. "Forest Ecology on Java: Conversion and Usage in a Historical Perspective." *Journal of Tropical Forest Science* 2 (4): pp. 286–302.

Solberg, B., L. Hetemäki, M.I. Kallio, A. Moiseyev, and H.K. Sjølie. 2014. Impacts of Forest Bioenergy and Policies on the Forest Sector Markets in Europe—What Do We Know? Technical Report 89. Joensuu: European Forest Institute.

Steffek, J. 2009. "Discursive Legitimation in Environmental Governance." *Forest Policy and Economics* 11 (5–6): pp. 313–318.

Thomas, E., R. Jalonen, J. Loo, J., and M. Bozzano. 2015. "Avoiding Failure in Forest Restoration: The Importance of Genetically Diverse and Site-Matched Germplasm." *Unasilva* 245 (66): pp. 29–36.

Turner, M.D., T. Carney, L. Lawler, et al. 2021. "Environmental Rehabilitation and the Vulnerability of the Poor: The Case of the Great Green Wall." *Land Use Policy* 111: p. 105750.

UNEP and FAO. 2022. "About the UN Decade." https://www.decadeonrestoration.org/about-un-decade.

United Nations Convention to Combat Desertification (UNCCD). "Land and Sustainable Development Goals." https://www.unccd.int/issues/land-and-sustainable-development-goals. Accessed April 4, 2024.

Van Oosten, C., P. Gunarso, I. Koesoetjahjo, and F. Wiersum. 2014. "Governing Forest Landscape Restoration: Cases from Indonesia." *Forests* 5 (6): pp. 1143–1162.

Van Oosten, C., and K. Merten. 2021. *Securing Rights in Landscapes: Towards a Rights Based Landscape Approach.* Report WCDI-21-135. Wageningen: Wageningen Centre for Development Innovation, Wageningen University & Research.

Verchot, L.V., E. Petkova, K. Obidzinski, et al. 2010. *Reducing Forestry Emissions in Indonesia.* Bogor: CIFOR.

Vetter, S. 2020. "With Power Comes Responsibility—A Rangelands Perspective on Forest Landscape Restoration." *Frontiers in Sustainable Food Systems* 4: 549483.

Walters, G., M. Baruah, M. Karambiri, et al. 2021. "The Power of Choice: How Institutional Selection Influences Restoration Success in Africa." *Land Use Policy* 104: p.104090.

Warnholtz, S., Gerardo, Fernández, M., Smyle, J., and J. Springer. 2017. *Securing Forest Tenure Rights for Rural Development: Lessons from Six Countries in Latin America.* Washington: PROFOR. https://www.profor.info/sites/profor.info/files/PROFOR_SecuringForestTenureRights_0.pdf.

Wenhua, L. 2004. "Degradation and Restoration of Forest Ecosystems in China." *Forest Ecology and Management* 201 (1): 33–41.

World Bank (WB). 2017. Forests/Landscapes. https://www.worldbank.org/en/results/2017/12/01/forestslandscapes.

Xu, Y. 2018. "Political Economy of Land Grabbing inside China Involving Foreign Investors." *Third World Quarterly* 39 (11): 2069–2084.

4
Decolonial environmental justice in landscape restoration

Sabaheta Ramcilovic-Suominen, Susan Chomba, Anne M. Larson, and Fergus Sinclair

4.1 Introduction

This chapter focuses on possible ways of enhancing justice in international landscape restoration interventions. The main contribution of this chapter is the application of a decolonial environmental justice (DEJ) framework to landscape restoration. The framework enables us to identify the major barriers to justice in large-scale, international restoration projects and to suggest strategies to strengthen non-imposing, de-/anti-colonial approaches. The key argument is that restoration projects, no matter the intentions, will not succeed if they ignore or violate local notions of justice, knowledge systems, as well as institutional and belief systems. The DEJ framework highlights the importance of epistemic justice, self-determination, and self-governance, among others, leading to a set of recommendations that are different from, yet related to other relevant perspectives (e.g., Osborne et al. 2021).

The chapter proceeds as follows. Section 4.1 provides a general introduction to the concept of landscape restoration, followed by an outline of some of the major actors and logics of international political initiatives for landscape restoration. Section 4.2 provides a brief history of community engagement in nature conservation and, more recently, in landscape restoration, emphasizing the importance of engaging local communities in such initiatives. Section 4.3 proposes an environmental justice (EJ) framework from a decolonial perspective, and Section 4.4 identifies some of the major barriers to effective, just, and equitable restoration. Finally Section 4.5, drawing on these empirical and theoretical insights, highlights the key conditions to enable a more just, equitable, and locally empowering landscape restoration, providing some examples of existing efforts, and Section 4.6 concludes the chapter.

A landscape comprises a mosaic of different land uses (e.g., forests, agroforests, farmlands, mangroves, wetlands, grasslands), goes beyond a human/nonhuman dichotomy, and focuses on more-than-human beings (Abram 1997; McGregor et al. 2020). Drawing on Chazdon et al. (2017, 126) we define landscape restoration as "the process of regenerating or restoring socioecological systems that have lost some, or all of its functions and values, which may include agricultural productivity, watershed functions, biodiversity as well as cultural diversity and cultural values." Despite this broad definition, we limit our analytical focus to landscape restoration initiatives that promote tree planting or natural regeneration for the purpose of landscape restoration. Second, given our focus on justice and local communities, we place particular

Sabaheta Ramcilovic-Suominen, Susan Chomba, Anne M. Larson, and Fergus Sinclair, *Decolonial environmental justice in landscape restoration* In: *Restoring Forests and Trees for Sustainable Development.* Edited by: Pia Katila, Carol J. Pierce Colfer, Wil de Jong, Glenn Galloway, Pablo Pacheco, and Georg Winkel, Oxford University Press.
DOI: 10.1093/9780197683958.003.0004

attention on landscape restoration initiatives that affect, involve, and directly or indirectly concern local peoples and communities.

An estimated 294.5 million people live on land that has potential for landscape restoration in the Global South (Erbaugh et al. 2020). These millions of mainly smallholder farmers, as well as Indigenous Peoples, make daily decisions that shape landscapes. Global reviews of restoration potential indicate that at least 74% of this potential area for restoration falls under "mosaic restoration," that is, areas with varied population densities where forests are combined with smallholder agriculture, settlements, and other land uses (Erbaugh et al. 2020; Minnemeyer et al. 2011). These landscapes are comprised of secondary forests, farmlands, grazing lands, water bodies, urban areas, and other land uses (Minnemeyer et al. 2011). Promoting landscape restoration in such conditions requires an in-depth understanding of historical, political, and justice-related conceptualizations and considerations across various scales and societal groups, and by extension, across different ontologies and epistemologies, including but not limited to indigenous ones (Winter 2021).

Globally, political momentum for landscape restoration is growing, as demonstrated by the recent launch of the United Nations Decade on Ecosystem Restoration (2021–2030), the Bonn Challenge—i.e., the commitment to restore 350 Mha by 2030—and the New York Declaration on Forests. These have in turn spurred regional and national commitments, such as the African Forest Landscape Restoration Initiative (AFR100)—to restore 100 Mha in Africa by 2030—and the 20x20 Initiative—to restore 20 Mha of degraded land in Latin America and the Caribbean by 2020—among others. These political declarations on restoration, for the most part, emphasize the importance of planting trees as conceived by international and state actors, donor agencies, and the scientific elite (i.e., for climate change mitigation and biodiversity protection), whereas local communities' roles, knowledge systems, cultural and political agency, as well as vulnerabilities remain peripheral (Erbaugh et al. 2020; Fleischman et al. 2020). Examples include the World Economic Forum's one trillion[1] trees campaign, the Trillion Trees Initiative,[2] the Ethiopian government pledge to plant five billion[3] trees under the Green Legacy, and most recently, the "30x30 target" under the Convention on Biological Diversity, aiming to protect at least 30% of the Earth's surface by 2030 (Waldron et al. 2020)[4] (for an overview of international forest and landscape restoration initiatives, see Chapter 3).

Even if initiatives include general statements regarding the importance of communities or community benefits, most international and state-orchestrated tree planting programs adopt a large-scale approach to planting trees, climate mitigation, and biodiversity protection with little attention to social and cultural aspects (Chapter 3). Under these conditions, important questions remain unaddressed, such as: Why and for whom are the trees being planted? What types of trees—e.g., native or exotic—support which functions and purposes? Where are they being planted and who owns the land? Who sacrifices for landscape restoration and how are the benefits distributed? What are the (un)intended consequences and how are they being addressed? (Elias et al. 2021; Fleischman et al. 2020; Holl and Brancalion 2020). Large-scale tree planting initiatives tend to either establish monoculture plantations or use a few exotic species with little local socioecological significance; they often fail to include local communities in their planning and decision-making processes, or to prioritize local livelihoods (Erbaugh et al. 2020; Holl and Brancalion 2020; McElwee and Nghi 2021). Consequences include aggravating biodiversity loss (e.g., through monocultures), reducing water availability for farming, increasing competition with crops for light, nutrients, or water (e.g., by introducing fast-growing trees in agricultural lands), and other negative effects on ecosystem services (Elias et al. 2021; Holl and Brancalion 2020).

Seeing trees and other vegetation solely as "natural climate solution" (Griscom et al. 2017) risks overestimating their potential for solving a climate crisis that is rooted in the historical use of fossil fuels (Bastin et al. 2019), shifting attention away from the need to curb emissions by industrialized countries (Holl and Brancalion 2020). It perpetuates a simplistic notion of trees as carbon sinks instead of as part of complex ecosystems that store as well as release carbon (Fleischman et al. 2020). It also ignores that these ecosystems provide food, water, fiber, and income, which are foundational for rural livelihoods in the Global South (Erbaugh et al. 2020). In addition, seeing nature in this way represents a form of epistemic injustice, violating many local and Indigenous Peoples' relational onto-epistemologies (Gebara 2020; McGregor 2014; Whyte 2017; Winter 2021). This raises critical questions about who sets the agenda for restoration, through which governing institutions (traditional, state), with what priorities (carbon sequestration, biodiversity conservation, local livelihoods, customary practices), and with what "acceptable" trade-offs.

The current landscape of restoration initiatives is rooted in similar institutional settings, driven by international actors and interests prioritizing conservation and climate (Hope 2021; Kashwan 2017; Kashwan et al. 2021). Thus, the potential negative effects of externally designed restoration projects on local people (and vice versa) are often overlooked or even seen as a necessary trade-off. A blanket goal of "protecting" at least 30% of land and ocean for "nature" (Waldron et al. 2020) is a good example of a socially and culturally blind approach that risks socioecological violence and injustice. Prioritizing nature conservation over the people living there fails to account for potentially significant losses for local peoples—in terms of cultural identities, heritages, territories, livelihoods, and access to farmlands, grazing lands, or fisheries—and may violate their right to prior consultation and consent. Such initiatives often lead to local sacrifices, placing the burden and consequences of restoration on local actors in favor of ecological values as defined by the global community, the prime beneficiary (Agrawal et al. 2021).

Restoration initiatives that take this approach are highly unlikely to succeed as they risk being met with both active and passive local resistance. In India[5] for example, indigenous communities actively resisted the government's efforts to "restore" their land with teak plantations, which they deemed to be of little ecological or cultural significance (Kashwan 2017). In Kenya, farmers resisted the Plantation Establishment and Livelihood Improvement Scheme with deliberate but secretive killing of seedlings and young trees. Other forms of passive resistance, what Scott (1985) refers to as "weapons of the weak," include sabotage, feigned ignorance, and use of arson. Similarly, fortress conservation models, which are characterized by "protected area management" through the forceful exclusion and/or eviction of local communities from protected areas in order to achieve biodiversity or climate change goals, have faced outright defiance and resistance, including protests. Examples include conservation and protected areas in Tanzania (Brockington 2002; Kijazi 2015) and elsewhere (Büscher et al. 2012).

Addressing questions of where, on whose lands, and what species are planted requires looking not only at species diversity and suitability for different ecological and climatic conditions (soil, precipitation, and altitude, at present and with future changing climate), but also their suitability for local sociocultural, sociopolitical and socioeconomic conditions. Asking why and for whom trees are being planted means dealing with the diversity of interests, ideas, concerns, logics, and ideologies of different groups defined by wealth, age, gender, ethnicity, caste, autochthony, migration, and other social differentiations within a community (Larson et al. 2019). It also implies taking into account the complexities associated with the local institutions that define land rights and tree ownership, the power relations between different groups in the community, and the imbalances in access and use rights (Colfer et al. 2021; Monterroso et al. 2019; Ribot and Peluso 2003). If these aspects are not addressed, tree planting and restoration can exacerbate inequalities, violence, and injustices, especially when benefits to land restoration are skewed in favor of socioeconomically dominant groups through elite capture, thereby entrenching systemic injustices.

4.2 A historical account of community engagement

We understand "community" as a group that is diverse with varying degrees of heterogeneity across cultures, ethnicities, socioeconomic statuses, and politics. Communities include people not only of different genders, ages, castes, races, wealth, ethnicities, religions, etc., but also with different interests, ideas, worldviews, and ideologies. All of these factors can affect their means, powers, and abilities to influence and shape the world around them, including decisions to manage, govern, and use the natural environment (Agrawal and Gibson 1999; Ribot and Peluso 2003). Within a community, different people are likely to be affected in different ways by land degradation. For example, rural women who collect firewood and water are disproportionately affected by the scarcity of trees used for fuel and the loss of local water sources, as they have to travel longer distances in search of these resources (Nchanji et al. 2021). Youth are disproportionately affected by the loss of local jobs and income sources. Indigenous peoples are disproportionately affected when protected area policies prohibit their access to places they consider their home (Arsenault et al. 2019; Dawson et al. 2021; Whyte 2014).

The composition of rural communities is also relevant and needs to be taken into consideration. While many Indigenous Peoples and traditional communities have lived in the same place for generations, others were displaced during the colonial and postcolonial era, through enslavement as well as assimilation policies, and have made their home in new locations and ecosystems. For instance, in Latin America highland Indigenous Peoples (e.g., from the Andes) have moved into lowland tropical forests, while different native tribes from across what is now known as the United States of America were placed in Native reservations. Nonindigenous farmers may have less of a collective identity and have often moved into new ecosystems, especially rainforests, spontaneously or through planned colonization (Larson 2010) and "trans-migration" policies (Elmhirst 1999).

Finally, power asymmetries between different segments of communities should be noted, as they influence inclusion and exclusion in restoration initiatives and other forest and nature protection projects (Nchanji et al. 2021; Ramcilovic-Suominen and Kotilainen 2020). It is common for international environmental policy initiatives, including those of landscape restoration, to be designed and implemented with little consideration of this heterogeneity, or local social and cultural complexity, including but not limited to power dynamics; this ends up altering the social and political fabric, governing structures, and decision-making processes (Hope 2021; Kashwan et al. 2021; Mansourian 2021).

Restoring nature and landscapes has been an integral part of local community interactions with ecosystems for millennia (Dawson et al. 2021; McGregor et al. 2020). Community engagement in formal land restoration projects is also not new. Early schemes for granting tenure over forest lands to local communities, such as those in India and Nepal beginning in the 1970s, were as much about granting prescribed management responsibilities and labor requirements as they were about rights (Larson and Pulhin 2012; Saxena 1997). In these cases, tenure and rights over land were often granted for degraded areas with the responsibility to restore the land through reforestation (Saxena 1997). These were some of the first (formal) community-based "restoration" policies.

A wave of initiatives promoting participation in natural resource management and conservation swept most of the Global South in the 1980s and 1990s in forestry, wildlife, marine ecosystems, and related fields (Blaikie 2006; Ribot and Larson 2013). While these approaches took a variety of forms in different countries and contexts, their aim was to involve local communities (in the broad sense of the term) in decision-making, implementation, and benefit sharing around natural resource management. Nevertheless, the issues being addressed, from the definition of the problems to the solutions and implementation goals, were often externally defined. "Participation" in many of these activities focused on how to reach, inform, or train local people, and primarily how to obtain their consent for implementing the various interventions for sustainable resource or forest management. For these reasons, and notwithstanding some initiatives that better aligned program and local objectives, participation of this kind has been referred to as the "tyranny of participation" (Kothari 2001) and "symbolic violence" rather than participation that is empowering (Ojha et al. 2009). Over the years there has been a growing awareness of the importance of more substantive and meaningful participation (e.g., Colfer 2004; Colfer et al. 2021).

Around the same time, in the 1980s and 1990s, decentralized forest management was also emerging, but was more politically motivated, aimed at expanding democracy, accountability, and responsiveness in natural resource management and reversing central governments' capture of resources that occurred during colonial and postcolonial regimes. Colonial and postcolonial state management in much of the Global South is characterized by state or public ownership of land, forests (sometimes even individual trees on farms), wildlife, and fisheries, denying local communities access to resources (Blaikie 2006; Ribot et al. 2006; Ribot and Larson 2013). Nevertheless, many central governments saw decentralization as an opportunity to devolve responsibilities to subnational governments and local communities—but seldom authority, resource rights, or legal tenure (Ribot et al. 2006). The key lessons from this wave of decentralizations that are pertinent for restoration are government tendency toward maintaining control, local community need for continued bargaining over rights (Larson and Soto 2008), and responsibilization of local people (Mustalahti and Agrawal 2020), which often results in more responsibilities and less rights (Ramcilovic-Suominen and Mustalahti 2022).

By the 2000s, international attention to "rights-based approaches" was growing. This is evident in progress on indigenous land tenure rights through International Labor Organization Convention 169 (passed in 1989) on indigenous and tribal peoples and, more recently, by the 2007 United Nations Declaration on the Rights of Indigenous Peoples. Nevertheless, like decentralization reforms, many forest (and forest land) tenure reforms aimed at Indigenous Peoples and other customary communities, especially those outside of Latin America, prioritized granting responsibilities to communities rather than substantive rights (Larson and Pulhin 2012). More recent trends, however, show an increase in more substantive "ownership rights" being granted to Indigenous Peoples from 2013 to 2017 in comparison with the previous 5-year period (RRI 2018). The Rights and Resources Initiative (RRI) classifies ownership as legal recognition for an unlimited duration of rights of access, withdrawal, management, exclusion, due process, and due compensation. In practice, however, there is wide variation in local ideas about territory and nature; even the idea of "ownership" itself can be an imposition (Arsenault et al. 2019; Gebara 2020; McGregor 2014; McGregor et al. 2020; Temper 2019). Importantly, states claim legal authority of over two-thirds of the global forest area, much of which is also claimed by Indigenous Peoples and local communities (RRI 2018).

The concept of land restoration therefore unfolds against the backdrop of previous efforts to prescribe roles and responsibilities to communities and indigenous groups, and the ongoing challenges and power dynamics in granting tenure rights and in decision-making over nature.

4.3 Environmental justice and decolonial environmental justice

In this section we outline a three-dimensional EJ framework, followed by a short introduction to decoloniality and DEJ, summarized in the Table 4.1. The three-dimensional EJ framework emphasizes participation, distribution, and recognition as the three key justice dimensions, distinguishing between procedural, distributional, and recognitional justice (Fraser 2009; Schlosberg 2007; Walker 2012). *Procedural justice* concerns participation—who participates in decision-making processes and project implementation—as well as representation—who is represented in the decisions and who is not (Fraser 2009). This also relates to the question of how nominally participating actors *represent* the interests and concerns of the entire constituency or concerned group (Fraser 2009; Sikor 2013), which is particularly challenging for multi-stakeholder processes. *Distributional justice* relates to the distribution of burdens, costs, harms, or risks (e.g., pollution, waste, exposure to toxins, etc.) and of benefits and opportunities (e.g., access to green spaces, clean air, water, economic

Table 4.1 Operationalizing and questioning the three dimensions of environmental justice

Types of justice	Environmental justice (EJ)	Decolonial EJ (DEJ)
Procedural	Participation/representation	Self-governance
Recognitional	Recognition	Self-recognition, self-determination
Distributional	Fair distribution of costs and benefits (of nature)	Questions the duality of humans and nature, and other binaries

opportunity, etc.) (Schlosberg 2007). Finally, *recognitional justice* emphasizes the importance of recognizing people's identities and cultures and highlights the harms of mis-recognition and mistreatment due to identity and belonging. Recognition in decision-making processes and their outcomes therefore relates to or enables both procedural and distributional justice (Fraser 2009; Schlosberg 2007).

As a critique to EJ, DEJ draws on the theory of decoloniality. To outline this theory we draw on decolonial scholars, including Fanon (1967), Mignolo (2008), Maldonado-Torres (2007), and Quijano (2000). Decoloniality is most widely defined as a critique of modernity. To decolonize implies to undo various harms of colonialism and coloniality, together with undoing or abolishing the very structures that help reproduce them (Fanon 1963, 1967). It also means to reverse the various forms of violence and oppression from colonialism and coloniality that persist today, including extractivism and cultural domination. Discussing decoloniality, therefore, brings to focus the concept of coloniality. Coloniality as a concept implies that the colonial logics, mindsets, and economic and power structures did not stop with the end of formal colonization, which refers to the moments in history when former colonies, at least most of them, gained independence. Hence, while political colonization ended, coloniality—that is, the colonial mindsets, logics, metaphysics, values, ontologies, and relationships with the rest of nature, and with previously colonized peoples—continues to exist to the present day and is the very foundation of our global political and economic system. Coloniality is reproduced by and manifests itself through various phenomena, including global capitalism, patriarchy, continued imperialism, modernity as we know it; it is also very visible in global relations and environmental governance (Hickel et al. 2021; Sultana 2022). Various forms of coloniality are discussed in the literature, even if all of them cannot be disentangled from one another, including coloniality of mind (epistemology), being (ontology), and power (politics) (Escobar 2007; Fanon 1963, 1967). By extension then, we talk about decoloniality and the need to decolonize these same entities—mind, being, power—which implies decolonizing the political, the epistemic, and the ontological spheres (Escobar 2007; Fanon 1963,1967; Mignolo 2007; 2008; Quijano 2000). Decoloniality of mind (Fanon 1967) and imaginary (i.e., cultures, meanings, significations, categories, worldviews, etc.) (Latouche 2015) refers to decolonizing the epistemic and ontological spheres. Decoloniality of the material or political implies undoing concrete political, cultural, and economic colonial structures and harms. This includes, for instance, land restitution and autonomous representation (Tuck and Yang 2012).

Drawing on decolonial and postcolonial theories, several scholars (Alvarez and Coolseat 2018; Rodríguez and Inturias 2018; Temper 2019) have critiqued the EJ theory, including the three-dimensional EJ framework, on various grounds. One of the main critiques is that the EJ theory is based on Western or Eurocentric (we use these terms interchangeably) epistemology and ontology, which decolonial scholars argue (re)produces onto-epistemic injustices and marginalizations as they delegitimize and/or render invisible other ways of knowing and other knowledge systems (Escobar 1998; Mignolo 2008; Quijano 2000; de Sousa Santos 2008; Whyte 2017; 2020). A second major critique concerns political authority, which decolonial scholars argue is reserved only for state authority in the EJ framework. This does not mean that non-state actors (e.g., nongovernmental organizations, local people, or businesses) do not participate and/or are not represented or recognized, but that the means and terms of participation, representation, and recognition are defined by the state (Pellow 2018). Scholars argue that neither participation nor recognition go far enough to ensure the self-governing authority of informal (i.e., non-state) traditional institutions, and their legal systems and structures (Alvarez and Coolseat 2018), which represents a form of political denial or oppression (Blaser et al. 2010; Rodríguez 2020; Temper 2019). The problem occurs when external notions of justice are applied in spaces with different ontologies, epistemologies, and legal and political structures, leading to epistemic injustice and political denial, or to what Alvarez and Coolseat (2018, 6) call "coloniality of justice."

DEJ has thus emerged as a critique and response to the three-dimensional EJ framework. Drawing on Alvarez and Coolseat (2018) as well as Temper (2019), we describe DEJ in direct relation to the three-dimensional EJ framework. Concerning *distributional justice*, DEJ suggests the need to rethink the very idea of objectifying and distributing nature, and with that the need to rethink the Eurocentric human–nature duality. Scholars have also highlighted the contradiction in the idea of equal distribution of harm as a precondition for justice: not only does equal exposure to pollution (for example) not serve justice, but also, it diverts attention from the need to address the sources of the harm (pollution) itself (Alvarez and Coolseat 2018). The critique is of an onto-epistemological nature, as the idea of distributional justice rests upon the Eurocentric view of a human–nature duality and assumes that the (rest of) nature is a resource that can be owned and distributed among humans. Nature in this view is objectified and seen as separate from humans, which violates many Indigenous People's relational views on nature—for example, ones of responsibility, kinship, trust, consent, and/or reciprocity (McGregor 2014; Whyte 2017). DEJ suggests the need to rethink, transcend, and deconstruct the human–nature dualism (see also Gebara 2020). This position is gaining prominence, also among non-decolonial and nonindigenous scholars of environmental policy (e.g., Biermann 2020). DEJ argues, rather than discussing how to distribute burdens and opportunities, there is a need to take a step back and focus on the questions of epistemology and ontology when theorizing justice, with an aim to ensure that people draw on their own worldviews.

Concerning *recognition*, while of course recognizing other cultures and identities is very important, DEJ scholars argue that recognition is limited, does not lead to self-determination, and can even be counterproductive (Alvvarez and Coolseat 2018). Self-determination, also referred to as self-recognition (Coulthard 2014), can be defined as the right to identify and conceptualize one's own desires, struggles, needs, and interests on one's own and outside of the dominant conceptualizations, categories, and knowledge and legal systems. Self-determination is not possible within the limits of the dominant epistemic, ontological, and legal domains; seeking to achieve it, as recognitional justice suggests, can have counterproductive results by unintentionally further marginalizing already marginalized positions, cultures, and struggles for independence and autonomy (Coulthard 2014). This can happen when dominant groups—elites, academics—even those with the best of intentions, are unaware of local conceptualizations, categories, cosmologies, knowledge, and legal systems. While self-determination is closely related to epistemic or cognitive justice (Rodriguez 2020), it is also a matter of reimagining, reclaiming, and reaffirming one's own ideas of the self through cultural practices, spirituality, politics of difference, and politics of alternatives (Coulthard 2014; Temper 2019).

Finally, concerning *procedural justice*, the DEJ perspective argues that participation and representation only ensure that decision-making processes conform with the dominating actors' "rules of the game." DEJ argues what is needed is the recognition of self-governance authority, implying that an authority other than state (i.e., indigenous and other traditional authorities) would have the right to make decisions about how to govern their territories. To give an example, various forestry and natural resource protection projects are implemented in contexts of legal pluralism, where both formal (or state) and informal (or customary) governing structures coexist (Lund 2011). External and top-down restoration initiatives typically work closely with state actors, who naturally promote and make use of state laws and often ignore customary governing structures, rendering them irrelevant or, worse still, illegal. For instance, Ramcilovic-Suominen et al. (2021) find that Hmong people in Laos do not always wish to participate in forestry projects, which they feel ignore their worldviews and traditional institutional structures.

We argue that DEJ allows for ontologically and epistemologically different ideas and logics, different political, legal, and governing structures to coexist along with the dominant Eurocentric ideas and dominant state-centric authority and institutions (Alvarez and Coolseat 2018; Rodriguez 2020; Rodriguez and Inturias 2018). It is more sensitive and responsive to the cultural and legal plurality and therefore more applicable to culturally and legally diverse or pluralistic societies, which is often where landscape restoration is de-facto implemented. It is also better equipped to recognize intended and unintended mechanisms of subordination and marginalization through the imposition of Eurocentric worldviews and therefore more applicable to non-Western and postcolonial societies and cultures (Alvarez and Coolseat 2018; Temper 2019). For these reasons, we find the DEJ better positioned to support the cause of advancing justice; we also find it useful to apply this framework in the context of international environmental policy and governance, including forest and landscape restoration. Nonetheless, we do not argue that it is a substitution for the three-dimensional EJ framework in all cases, nor do we argue for its universal applicability.

4.4 Major barriers to just and effective landscape restoration

Four major barriers to just and effective international landscape restoration are derived from an analysis of landscape restoration using a DEJ perspective. The analysis focuses on tensions between global and local levels; it does not look at barriers within the local level, such as the domination and marginalization of certain societal groups or inequalities within a community regarding gender, class, or age.

The first barrier for just and effective restoration in international landscape restoration initiatives concerns *prioritizing global over local knowledge systems, logics, and politics.* Nature and restoration mean different things to different peoples, societies, and cultures, as well as to different actors within the groups. These meanings depend not only on their immediate concerns and interests, but also on their ideas and worldviews regarding their place and relation with the rest of nature (Arsenault et al. 2019; Gebara 2020). In Eurocentric philosophies of life and modern ecological sciences, such as restoration ecology, the need for restoration is commonly explained in two ways: (i) the need for "returning to some original pristine state of nature" (Bradshaw 1996; Bradshaw and Chadwick 1980), or (ii) for "regaining ecological functionality to provide ecosystem services and enhance human well-being across degraded landscapes" (Chazdon et al. 2017). The former is problematic as it separates people and ecosystems, even if they have coexisted and co-shaped each other for millennia. The latter can be described as a utilitarian and anthropocentric approach, which prioritizes human well-being in ecosystem restoration. Neither places significant, if any, attention on relational ontology that promotes coexistence and reciprocal relations between humans and other species and forms of life (Winter 2021).

In contrast to the perspectives prominent in modern Eurocentric ecological sciences, for many indigenous and traditional communities restoration is not separate from their knowledge systems, worldviews, experiences, and relations with nature. In other words, it is integrated into their day-to-day interactions, their livelihoods and cultural practices, rather than a separate activity that they need to suddenly attend to. There is strong evidence that many such communities, guided by their traditional knowledge and practices, have maintained socioecological vitality and resilience—that is, the vitality and resilience of their ecosystems as well as their cultures, well-being, and local economies (Dawson et al. 2021; Fleischman et al. 2020; Kashwan et al. 2021; McGregor 2014; McGregor et al. 2020; Sinclair and Joshi 2000). It is thus crucial for restoration efforts not to separate or exclude local people from their territories and practices; rather, they need to empower local communities, counter the processes that undermine their culture, knowledge, and traditional practices; and to partner with them.

Global restoration initiatives need to recognize local communities' roles, agency, and knowledge systems and to draw on these when designing and implementing landscape restoration efforts. Examples of traditional/indigenous knowledge that have been applied in restoration include the use of the fast-growing tree Balsa (*Ochroma pyramidale*)—to accelerate succession toward mature forests and to control invasive species—by the Lacandon Maya of Southern Mexico (Douterlungne et al. 2010). Farmer Managed Natural Regeneration, a widespread approach in the West African Sahel, is built around traditional management of trees in African savannas, including selecting suitable species that provide beneficial relationships with crops and livestock and tree management practices, such as thinning, coppicing, pruning, etc. Coupled with livestock management and other enabling policy interventions, their approach presents greater promise as a restoration practice in arid and semi-arid areas compared to tree planting (Chomba et al. 2020).

The second barrier concerns the *tendency to target small-scale forest use and livelihood activities rather than large-scale degradation drivers.* The drivers of land degradation are multiple and interconnected. Initiatives for conservation, or landscape restoration, ignore large-scale systematic and for-profit drivers while disproportionately focusing on small-scale drivers (see Lambin et al. 2001). They ignore the key drivers of deforestation and land degradation, which may include national policy (e.g., resettling villagers from highlands to lowlands, fostering large-scale land investments and driving forest degradation and insecure land tenure). As local people resist policies and projects with harmful effects, this resistance in turn leads to attacks and violence against them, with over 1,733 documented deaths of land and environmental defenders, between 2012 and 2021 (Global Witness 2022).

Different drivers originate at different scales, from global to local, and have varied temporal and spatial dimensions; undesired effects take time to emerge and often emerge in places far from the original action or initiative (Ingalls et al. 2018). Local communities often have a good understanding of the relevant drivers pertinent to their areas, including at times the external ones (Crossland et al. 2018). Global restoration initiatives need to recognize this and may consider incorporating local communities in identifying and mapping the drivers to be tackled rather than bringing a predefined list of problems/drivers and solutions. The global drivers, for example, global demand for wood or other deforestation-causing commodities (soy, coffee, cocoa, palm oil, etc.), manifest in increased logging and corporate seizures of land, of which communities are painfully aware, though the complexity of these indirect drivers may be harder to discern. The large-scale, market- and policy-triggered drivers are surely harder to tackle and are often ignored, in part due to the vested interests of powerful political and corporate actors. One must also consider the externally funded project's tight time schedules (Ramcilovic-Suominen 2019), one important aspect that must change for better long-term success.

The third barrier for just and effective landscape restoration relates to *the risk of offshoring burdens of global landscape restoration to local communities.* Studies show that the benefits of restoring degraded lands outweigh the costs (Nkonya et al. 2016; Waldron et al. 2020). While we do not want to argue whether this is always the case, the question is: benefits and costs for whom? Waldron et al. (2020) claim that the benefits of expanding protected areas, focusing only on forests and mangroves, would be between 170–534 billion USD per year by 2050, mainly from avoided flooding, climate change, soil loss, and coastal storm damage. This estimate is based on the assumption that these restoration efforts will be successful, which is by no means guaranteed. The question that receives less attention is who bears the burdens and losses for generating this sum, and how is this sum shared. More importantly, what is at stake for local communities is far more than the money—their cultures, territories, heritages, and ways of life (Whyte 2014; 2020) cannot be compensated for.

It is unrealistic to expect local communities to participate in and sustain restoration efforts by contributing labor, time, and land to a restoration process that has been designed mostly "from outside," with a view of benefits that may differ from local perspectives. However, different groups and individuals within a local community expect different kinds of benefits and therefore what constitutes a benefit or a cost should be defined locally. For example, pastoralists and herders may desire feed and fodder for their livestock, while crop farmers may want to curb soil erosion and increase crop productivity; both may be in conflict initially with the goal of an international project developer to increase the climate mitigation potential of the landscape. Many rural people have immediate livelihood needs and cannot always wait (only) for long-term benefits. As demonstrated by various examples of restoration practices that are codesigned with local people—such as watershed restoration that applies cross-jurisdictional planning with communities in Latin America and Asia (Armitage 2005; Sabatier et al. 2005), or restoration codesigned with local women and in line with local traditions (Mukasa et al. 2022)—global restoration efforts must go beyond short-term project payments to farmers to conserve or plant trees. The bottom line is that global restoration practices need to ensure that local people are not the ones to sacrifice their livelihoods for ecological restoration while someone else defines the objectives and rules of the game. This does not mean that international actors have no role to play but that their current role and space of influence needs to be discussed and redistributed downward.

Finally, the fourth barrier to just landscape restoration is that the *state actors and structures are perceived as the only legitimate source of authority, while customary institutions are ignored.* Typically, international environmental policy and governance initiatives are negotiated, implemented, and enforced by relying on the nation-state as the main source of authority and legitimacy. This does not necessarily mean that the non-state actors do not participate but that participation and representation are conceptualized within the dominant state political authority and structures. Assigning the political authority to the state only is embedded within the foundation of international environmental governance, where representative rather than direct democracy is a rule of the game; therefore, only states have voting rights. Adhering to this principle, however, becomes an issue given legal pluralism (Tamanaha 2007) and parallel authorities in postcolonial countries, which leads to domination and denial of indigenous and other traditional sources of political authority (Lund 2011; Ubink 2011). Bias against customary and collective systems has led many countries to prioritize individual land rights, sometimes leading to the breakdown of the local and customary institutions that support sustainability. Such biases continue even in countries that have passed laws recognizing community rights, where states still fail to provide enabling conditions to exercise those rights (Aggarwal et al. 2021). Procedures for rights recognition tend to be cumbersome, confusing, and expensive (Notess et al. 2020).

In many countries, customary and statutory rules and authorities coexist. As a result, in the case of Mali, Niger, and Senegal, for example, the communities' rights to land were separated from their rights to trees, and the rights to trees from the rights to tree products, such as fruits, leaves, and nuts (McLain et al. 2018). The 2017 Malian Agricultural Land Policy recognizes customary rights to extract resources on agricultural land, but the rules of exploitation are contradicted by the state's forest policy.

This legal plurality, where statutory and customary laws coexist, often include overlaps and contradictions, making it difficult to clarify the rights, roles, and responsibilities of the local communities in land restoration. Ensuring user rights to trees is possible, however, as has been seen in Niger for example. Gradually granting greater user rights to those who plant, tend, and work toward regenerating the land is becoming more common as well.

4.5 How to avoid reproducing past injustices

While we do not deconstruct landscape restoration as a neocolonial project (as is often done for nature conservation initiatives (Büscher et al. 2012; Hope 2021; Kashwan et al. 2021), we do, however, argue that landscape restoration as promoted in the international restoration policy initiatives alludes to a neocolonial, top-down, and "interventionalist" project (Chapter 3). These initiatives are in fact embedded within the institutional and organizational settings that uphold neocolonial approaches to nature conservation (Hope 2021; Kashwan et al. 2021), hence the risk of reproducing injustices and violence is great. As we adopt a DEJ perspective, it would be counterintuitive to suggest a blueprint for external actors to follow in designing landscape restoration projects. Besides, decolonial and locally just restoration is embedded within diverse and varied contexts, with particular landscapes and particular everyday practices, interactions, and relationalities within these landscapes; it therefore should not be designed as a "one-size-fits-all" solution. Rather, it is important to consider how to engage in these diverse contexts, what Sarmiento Barletti et al. (2020, 15) refer to as "designing for engagement." What we offer is a set of questions for practitioners to reflect upon, when designing and implementing landscape restoration initiatives, and for scholars to reflect on when analyzing landscape restoration from a decolonial perspective. We also propose key conditions for avoiding reproducing injustices and neocolonial approaches in landscape restoration policy and practice.

Applying a decolonial lens to global environmental policy more broadly (Ramcilovic-Suominen 2022; Rutazibwa 2018) to global environmental policy, specifically landscape restoration, allows us to identify the following questions for reflection and analytical inquiry:

i. *Colonial mindsets, logics, and structures*: To what extent does the landscape restoration initiative question and attempt to unsettle or dismantle the mindsets, mandates, logics, structures, and binaries (e.g., developed/undeveloped, superior/inferior) established by, rooted in, and instituted within imperial and colonial structures?
ii. *Geography*: Who decides on the location of a landscape restoration initiative and to what extent do the deciding actors depend directly and immediately on the land for their subsistence and livelihoods?
iii. *Motivation and key objectives*: Who defines and decides upon the key motivations and objectives of a landscape restoration initiative: global actors, domestic state actors, different groups of local people?

iv. *Interests, concerns, and needs*: Whose interests, concerns, and needs does the initiative serve and whose does it neglect or violate: those of global actors, local communities, dominant groups, or those of the marginalized groups?
v. *Costs, benefits, and trade-offs*: Who defines what is considered a cost and what is a benefit (e.g., goods, services, products, and/or end results) in a landscape restoration initiative, and who decides how these are provided or denied across scales and among societal groups?
vi. *Modes of operation*: To what extent are the modes of operation and the actual restoration process (e.g., planting species, restoration techniques) defined by those who are to be directly affected by the initiative?
vii. *Institutions and knowledge systems*: To what extent are the institutional structures and knowledge systems that are used to design, negotiate, and implement the restoration process distinct from the customary knowledge systems and institutional structures?

Drawing from the DEJ perspective (Section 4.3), and the identified major barriers for justice (Section 4.4), we propose three conditions as key for ensuring justice and local empowerment in landscape restoration. This does not mean that these are the only conditions worth ensuring in landscape restoration, nor does it mean that following these conditions will resolve all the challenges, needs, and goals in landscape restoration. It means that they are important for promoting and facilitating what we argue to be the key precondition for a decolonially just landscape restoration. Our suggestions advance and relate to the various justice dimensions, including procedure and recognition, but go beyond to highlight the importance of epistemic justice, self-determination struggles (identity, culture, territorial rights), and self-governance of traditional governing authority (application of traditional legal and institutional structures). These are exemplified in existing efforts and initiatives, with which some of the authors are directly familiar. Often, however, the existing examples do not reflect the issue in question (e.g., locally defined benefits and trade-offs in Subsection 4.5.2), and in some cases we lack examples from practice altogether (e.g., for landscape restoration that nurtures self-determination and self-governance).

4.5.1 Landscape restoration is embedded in local knowledge systems and agency

Due to the evidence that Indigenous Peoples and their knowledge systems play a key role as stewards of the environment (Corrigan et al. 2018; Dawson et al. 2021; Kashwan et al. 2021), there is a renewed interest in traditional and indigenous knowledges, which certainly is important and something to celebrate. While there is a lot of potential in incorporating traditional forms of knowledge (Arsenault et al. 2019), the interaction and cocreation between local and external/scientific actors and knowledges needs to be done with a fair amount of care, mindful of risks such as indigenous knowledge appropriation and commodification (Arsenault et al. 2019; McGregor 2014; Whyte 2017).

In this context it should be kept in mind that local knowledge is embedded and cannot be disentangled from local people's daily interactions with their environment or from their legal systems. Hence, the cultural, epistemic, and political (i.e., cultural livelihood practices and relations with the environment and the governing structures) are closely intertwined. Epistemic justice and the right to rely on their knowledges is associated with local political empowerment, control over territory, and the ability to influence key decisions (Blaser et al. 2010; Whyte 2014). It relates to local political agency, concerns, and interests.

Co-defining the meanings, objectives, underlying motivations, and logics together with local communities at the very beginning of the landscape restoration process ensures that these initiatives are in line with the traditional knowledge rooted in the given place and the people of that place. Grounding an initiative in traditional and/or indigenous knowledges from the beginning should also ensure that such knowledge systems have equal footing with the more dominant ones, such as scientific knowledge. This does not imply that the latter is less relevant but simply different. Incorporating traditional knowledge is often done at a later stage of a policy or project initiative, where it is approached as "local wisdom" that requires scientific validation, hence rendering it as "less than" scientific knowledge (Winter 2021). We want to highlight that epistemic justice implies epistemic plurality, the coexistence of different epistemic and ontological systems rather than hierarchies between them. Box 4.1 presents examples of existing tools and approaches for the integration of indigenous and local knowledge systems in restoration practices.

Apart from the concerns related to the coexistence of local/global, internal/external, one should keep in mind that the members of the same community may have different concerns and interests, despite a shared onto-epistemological orientation. Local communities may be diverse and heterogenous, consisting of different ethnicities, languages, and socioeconomic classes (Agrawal and Gibson 1999). Those with more/less power and resources (i.e., "local elites"), but also those of a certain

Box 4.1 Examples of existing tools and approaches for the integration of indigenous and local knowledge systems in restoration practices

Landcare is an example of a community-based and -led approach to sustainable landscape management (Catacutan et al. 2009, 2015). Landcare is an established movement in the Philippines, South Africa, and Uganda. Landcare groups undertake community vision mapping to define how they would like to see their productive and restored landscape. In this way the community defines their own landscape vision and can determine what actions need to take place to achieve this vision using their local capacities and help from external actors.

Several other tools and approaches have been developed to support community participatory mapping and option identification and cocreation. For example, (i) combination of local knowledge with structured stakeholder engagement for inclusively designing a more diverse set of agroforestry options in the Democratic Republic of Congo (Dumont et al. 2019) and (ii) participatory mapping of degraded lands through community perception or integrating geographic information system (GIS) maps (Petri et al. 2019).

Box 4.2 Examples of existing tools for integrating gender in restoration initiatives

Tools that have been developed by scientists to capture and address gender differences in development/scientific interventions, particularly in agriculture include:

i. the Gender and Inclusion Toolbox: Participatory Research in Climate Change and Agriculture,[6]
ii. the Gender and Climate Change Research in Agriculture and Food Security for Rural Development training guide,[7] and
iii. the Gender Box for addressing gender bias in forest management.[8]

For multi-stakeholder forums and processes in landscape management, Getting it Right[9] supports participation for women and Indigenous Peoples.

While these tools demonstrate the recognition and will to address gender disparities in decision-making about and access to resources and benefits, they are often externally defined, technocratically framed, and administered as a "protocol" that is likely to make significant changes to socioculturally held norms and practices at the local level. Other tools, such as those for inclusive landscape management[10] and for gender and social inclusion[11]—aimed in part to be self-administered for reflection in multi-stakeholder processes or partnerships—have been codeveloped between researchers and these diverse stakeholder groups, including with indigenous women in Peru, to analyze their participation in the governance of their communities and territories.

gender or ethnicity, will have more/less ability to exercise their agency (Lund and Saito-Jensen 2013). Hence, attention needs to be paid to power relations, identities, interests, and levels of marginalization (Tole 2010). For example, in communities where social and cultural norms prescribe lower priority to women, or to marginalized ethnic groups (in the cases of ethnically mixed communities), it can be challenging to ensure that the interests and preferences of the marginalized groups are considered (Agarwal 2001). Similarly, women and migrant communities tend to have more insecure land and resource tenure rights, which reduces their ability to benefit from restoration (and other) decisions (Jhaveri 2021; Larson et al. 2019). The same is observed in terms of different age groups, ethnicities, religions, sexual orientations, and the like. Reconciliation within the same community, where needed, and ensuring that the voices and concerns of traditionally marginalized groups are heard while recognizing and respecting the local norms is a challenge to be navigated (Evans et al. 2021). Some existing efforts and examples are given in Box 4.2.

4.5.2 Benefits and trade-offs are locally defined and decided upon

Putting to use local peoples' knowledges and their cultural and political agency is important, not only in terms of procedural and epistemic justice but also in terms of distributional effects, when deciding who benefits from an initiative and in what ways. Climate change is certainly a key concern, and should be treated as such, but using it to legitimise various trade-offs in relation to local livelihoods and ways of life, as is commonly done at the international policy level, is missusing the emergency to justify violence (Cramb et al. 2009; Ramcilovic-Suominen and Nathan 2020). Besides, local knowledge systems and ways of life promote rather than hinder regenerative practices that are *also* climate friendly (Dawson et al. 2021; Kashwan et al. 2021). We recommend supporting such regenerative and place-based practices rather than importing external ideas of "climate-smart" technologies, which, once again, empower external actors.

Regarding benefits and trade-offs, externally defined benefits, such as monetary compensations for performance, hardly represent a sufficient motivation for local and Indigenous Peoples to join an initiative, as Dawson et al.'s (2021) synthesis of literature shows. They are also oriented to the short-term motivations and can disrupt local livelihoods (Ramcilovic-Suominen and Kotilainen 2020). Defining benefits and trade-offs locally, however, does not imply that everyone will benefit in the same manner; internal deliberations and negotiations are power-laden processes that may not lead to an ideal outcome or a solution that satisfies all. Often the limitations of groups are structurally embedded in laws and cultural norms and practices. Nonetheless, local deliberation over benefits and sacrifices builds local ownership and commitment to an initiative; it is an important element that adds continuity even when things do not go as planned.

Benefits versus trade-offs in landscape restoration will vary from one sociopolitical, cultural, and ecological space to another. Nevertheless, the existing literature suggests they are often related to at least one of the following issues: (i) degradation drivers and (ii) land tenure rights. Section 4.4 identifies the tendency of international policy mechanisms to target small-scale rather than large-scale degradation drivers and livelihood activities, such as traditional shifting cultivation (Cramb et al. 2009), while ignoring the larger and more profitable agricultural expansions (Ramcilovic-Suominen and Nathan 2020). Targeting such drivers not only delegitimizes and denies local knowledge but also diminishes important sources of local livelihoods (Dressler et al. 2016). Local community mapping of degradation drivers and cocreation of mechanisms for addressing these can be a first step in the right direction. Tools such as participatory mapping of land degradation (Box 4.2) can be applied to provide an opportunity for communities to indicate where degradation is occurring and to come up with suitable restoration options, where feasible.

Many countries and initiatives have undertaken efforts to secure community tenure rights (RRI 2018), but progress has been disappointing (Aggarwal et al. 2021). There is nothing simple about the process. Challenges include competition over land, bias against customary and collective tenure rights, technical and capacity deficits, among many others (see Larson and Springer 2016). Competition over access to and control over land refers to multiple demands on the same land, from both outside and inside communities, which can lead to elite capture. Even if communities have secure rights, it does not mean everyone within the community does; women, for example, are often also subject to local community and household rules and norms that restrict access to land and resources; even where national laws protect women's land rights, these are often poorly implemented (Jhaveri 2021; Monterroso et al. 2019). Increasingly, countries are embracing joint land ownership and titling in the name of husband and wife through land certification that provides legal rights to both spouses; examples include Rwanda and Ethiopia.

While tenure remains an important challenge under the current international governance regime, when examined through a DEJ lens, the notion of tenure and property may invoke epistemic injustices, imposing a Western conception of ownership (McGregor et al. 2020; Temper 2019). This relates to conflicts of worldviews or ontologies (Escobar 1998; 2007) and it is against relational ontology where we are a form of nature other forms of nature, such as trees and land with abundant life endowed into it cannot be owned by us (Winter 2021). From a DEJ perspective, the priority should therefore be a better alignment with and consideration of the local human–nature relations and the traditional legal systems in which such relations are engraved. Depending on the local imaginaries of justice, this may expand to include justice for nonhuman species (Tschakert et al. 2021; Winter 2021).

4.5.3 Self-determination and self-governance are promoted

Self-determination and self-governing authority as defined in Section 4.4 cannot be realized under the externally designed international and state policies, no matter how well intended they might be (Escobar 1998; Mignolo 2008; Quijano 2000; Whyte 2017, 2020). Self-determination and self-governance should primarily be approached as a political right, which is obviously tightly linked to culture and tradition. Advancing self-determination and self-governing authority would require, for example, that the traditional indigenous and/or customary authorities and their ruling structures and legal practices be strengthened (provided they are in place) or established (if not) as a valid form of legal authority under international governance regimes. This calls for nothing short of decolonization of and in international governance, which is essentially an institution rooted in colonial mindsets and power hierarchy (Ziai 2012, 2016), and for repatriation of indigenous lands (Tuck and Yang 2012). These are also necessary preconditions for self-determination of Indigenous Peoples and self-governance of indigenous lands is on the agenda, which we believe is a "must" if decoloniality is to be more than a metaphor.

Such progressive ideas significantly expand the narrow and more common approaches where local people are recognized as "equal" and called to participate within externally derived policy initiatives and structures. Keeping the historical context in mind, it is important to remember that many colonized people in settler-colonial countries (e.g., the USA, Canada, and Australia) and in postcolonial countries in Africa and Asia, for example, are likley to still perceive these structures as ones of oppression and remnants of colonial rule (Blaser et al. 2010; Pellow 2018; Whyte 2014, 2020).

While this broad and highly political agenda may appear distant from policy initiatives such as landscape restoration, it is the foundation for devising decolonial and just landscape restoration. For example, it suggests that relying solely on statutory legal traditions when it comes to issues such as rights, benefit sharing, income, or management activities is highly problematic. We return to property and tenure as an example of this. State policies and legal frameworks define legal ownership over forests and other nature entities, framing land and life on it as objects deprived of agency, rights, and roles. Following this logic, the individuals will care better for such objects if they have property rights over these—an approach promoted widely and globally that has led a number of countries to clarify rights to various "commodities," including carbon in forests (Streck 2021). Although they may be well intended, in many cases and places the imposition of property rights over nature is a legacy of colonialism and an act of domination of state legal structures over indigenous ones (Rodriguez and Inturias 2018; Temper 2019). Moreover, in situations where tenure is the only formally recognized way to ensure long-term rights and access to land and other "resources," local and indigenous communities opt for it and, in some cases, abuse their indigenous status to appropriate land (Astuti and McGregor 2017). As the evidence grows of the importance of Indigenous and/or traditional and customary institutions and arrangements for biodiversity and climate (Corrigan et al. 2018), notions of indigenous knowledge and even epistemic justice are increasingly recognized and called for (IPBES 2019). Yet they remain at the margins of concrete policy initiatives, including landscape restoration.

4.6 Conclusion

In this chapter we define landscape restoration rather broadly to include regeneration and restoration of various socioecological system functions and values, which may include agricultural productivity, watershed functions, and biodiversity as well as cultural diversity and cultural values. Our focus is limited to landscape restoration initiatives that propose tree planting for landscape restoration purposes.

Using the DEJ perspective, we identify some common barriers to, and propose several enabling factors for, locally just landscape restoration. DEJ challenges the universality of justice principles, including participation, distribution, and recognition, highlighting the need to deconstruct and go beyond these concepts and ideas. It highlights the importance of epistemic justice, self-determination struggles, and self-governance, of going beyond the dominant categories and conceptualizations, knowledges, and legal systems. These are preconditions for justice that does not oppress (Alvarez and Coolseat 2018).

Given our foci on landscape restoration processes that affect, involve, or in any other way concern local communities, as well as our DEJ perspective, the identified barriers for justice emphasize exclusively the tensions between global and local levels rather than local-level challenges (e.g., including domination and marginalization of certain societal groups, gender bias, class, and other inequalities within a community), the existence of which we acknowledge. Similarly, the DEJ perspective arguing for the decoloniality of knowledge and of political authority, for the promotion of self-determination and self-governance, and for an open exploration about benefits and sacrifices is better positioned to emphasize the potentials of locally defined problems and solutions over globally defined ones. DEJ emphasizes the importance of seeking to establish and, where needed, regenerate and reclaim marginalized identities, cultures, knowledges, and authorities. Acknowledging that these approaches are not without their own hurdles, the authors believe that they are necessary for challenging cultural and political domination in international development and sustainability efforts (Ziai 2016). Despite the usefulness of this framework for landscape restoration policies, we recognize that our analysis does not directly address the important aspects of future generations and justice for other-than-human species. Such concerns are crucial in terms of landscape restoration, and we call for future studies to expand our analytical scope to include justice for future humans, and to go beyond human dimensions in their analyses of justice (Tschakert et al. 2021). Indigenous justice framings are particularly useful in this endeavor (McGregor et al. 2020; Whyte 2020; Winter 2021).

This chapter sheds light on options and possibilities for what can be done to shift current landscape restoration projects toward more locally owned and just initiatives that enhance local empowerment, epistemic justice, and local people's self-determination and self-governance. In this chapter, we include reference to some currently existing tools and approaches that are used to promote equitable landscape restoration, including integrating local knowledge, community visioning and mapping of degradation drivers, and a gender-balanced approach to restoration. While these examples are important steps forward, they also show that there is much left to do for a just, decolonial landscape restoration that empowers local actors to exercise their agency and governing structures.

Acknowledgments

Sabaheta Ramcilovic-Suominen gratefully acknowledges the funding from the Academy of Finland (grant number: 332353) that supported her work on this book chapter.

Notes

1. https://www.1t.org
2. https://trilliontrees.org/home
3. https://www.africanews.com/2020/05/13/green-legacy-initiative-ethiopia-targets-5-billion-trees-this-year/
4. The '30x30' initiative has faced significant resistance: https://politicalecologynetwork.org/2021/02/04/200-ngos-and-experts-warn-against-un-plan-to-protect-30-percent-of-the-planet-by-2030/; https://openlettertowaldronetal.wordpress.com
5. https://undark.org/2020/01/06/india-indigenous-trees/
6. https://cgspace.cgiar.org/bitstream/handle/10568/45955/CCAFS_Gender_Toolbox.pdf?sequence=7
7. https://www.fao.org/3/md280e/md280e00.htm
8. https://www.cifor.org/publications/pdf_files/OccPapers/OP-82.pdf
9. https://www.cifor.org/knowledge/publication/7973/
10. https://www.cifor.org/toolboxes/tools-for-managing-landscapes-inclusively/
11. https://www.cifor-icraf.org/news/feature/gender-and-social-inclusion/

References

Abram, D. 1997. *Spell of the Sensuous: Perceptual Language in a More-Than-Human World.* New York: Vintage Books.

Agarwal, B. 2001. "Participatory Exclusions, Community Forestry, and Gender: An Analysis for South Asia and a Conceptual Framework." *World Development* 29 (10): pp. 1623–1648.

Aggarwal, S., A.M. Larson, C. McDermott, P. Katila, and L. Giessen. 2021. "Tenure Reform for Better Forestry: An Unfinished Policy Agenda." *Forest Policy and Economics* 124: p. 102376.

Agrawal, A., K. Bawa, D. Brockington, et al. 2021. "An Open Letter to the Lead Authors of 'Protecting 30% of the Planet for Nature: Costs, Benefits and Implications.'" Published by Open Letter to Waldron. https://www.resilience.org/stories/2021-01-12/an-open-letter-to-the-lead-authors-of-protecting-30-of-the-planet-for-nature-costs-benefits-and-implications/.

Agrawal, A., and C.C. Gibson. 1999. "Enchantment and Disenchantment: The Role of Community in Natural Resource Conservation." *World Development* 27 (4): pp. 629–649.

Alvarez, L., and B. Coolsaet. 2018. "Decolonizing Environmental Justice Studies: Perspective, Capitalism Nature Socialism." *Capitalism, Nature, Socialism* 31 (2): pp. 50–69. DOI:10.1080/10455752.2018.1558272.

Armitage, D. 2005. "Adaptive Capacity and Community-Based Natural Resource Management." *Environmental Management* 35 (6): pp. 703–715.

Arsenault, R., C. Bourassa, S. Diver, D. McGregor, and A. Witham. 2019. "Including Indigenous Knowledge Systems in Environmental Assessments: Restructuring the Process." *Global Environmental Politics* 19 (3): pp. 120–132. DOI:10.1162/glep_a_00519.

Astuti, R., and A. McGregor. 2017. "Indigenous Land Claims or Green Grabs? Inclusions and Exclusions within Forest Carbon Politics in Indonesia." *The Journal of Peasant Studies* 44 (2): pp. 445–466. DOI:10.1080/03066150.2016.1197908.

Bastin, J. F., Y. Finegold, C. Garcia, et al. 2019. "The Global Tree Restoration Potential." *Science* 365 (6448): pp. 76–79.

Biermann, F. 2020. "The Future of 'Environmental' Policy in the Anthropocene: Time for a Paradigm Shift." *Environmental Politics* 32 (1–2): pp. 61–80. DOI:10.1080/09644016.2020.1846958.

Blaikie, P. 2006. "Is Small Really Beautiful? Community-Based Natural Resource Management in Malawi and Botswana." *World Development* 34 (11): pp. 1942–1957.

Blaser, M., R. De Costa, D. McGregor, and W. D. Coleman (eds.). 2010. *Indigenous Peoples and Autonomy, Insights for a Global Age.* Vancouver: UBC Press.

Bradshaw, A. D. 1996. "Underlying Principles of Restoration." *Canadian Journal of Fisheries and Aquatic Sciences* 53 (S1): pp. 3–9.

Bradshaw, A. D., and M. J. Chadwick. 1980. *The Restoration of Land: the Ecology and Reclamation of Derelict and Degraded Land.* Berkeley: University of California Press.

Brockington, D. 2002. *Fortress Conservation: The Preservation of Mkmazi Game Reserve, Tanzania.* Indiana: Indiana University Press.

Büscher, B., S. Sullivan, K. Neves, J. Igoe, and D. Brockington, D. 2012. "Towards a Synthesized Critique of Neoliberal Biodiversity Conservation." *Capitalism Nature Socialism* 23 (2): pp. 4–30.

Catacutan, C., C. Muller, M. Johnson, and D. Garrity. 2015. "Landcare: A Landscape Approach at Scale." In *Climate-Smart Landscapes: Multifunctionality in Practice*, edited by P. van Noordwijk, M. Minang, O. Freeman, C. Mbow, J. de Leeuw, and D. Catacutan, pp. 151–161. Nairobi: World Agroforestry Centre.

Catacutan, D., C. Neely, M. Johnson, H. Poussard, and R. Youl. 2009. *Landcare: Local Action – Global Progress*. Nairobi: World Agroforestry Centre.

Chazdon, R. L., P. H. Brancalion, D. Lamb, L. Laestadius, M. Calmon, M., and C. Kumar. 2017. "A Policy-Driven Knowledge Agenda for Global Forest and Landscape Restoration." *Conservation Letters* 10 (1): pp. 125–132.

Chomba, S., F. Sinclair, P. Savadogo, M. Bourne, and M. Lohbeck. 2020. "Opportunities and Constraints for Using Farmer Managed Natural Regeneration for Land Restoration in Sub-Saharan Africa." *Frontiers in Forests and Global Change* 3: p. 122.

Colfer, C. J. P. (ed.). 2004. *The Equitable Forest Diversity, Community, and Resource Management*. New York: Routledge.

Colfer, C. J. P., R. Prabhu, and A. Larson (eds.). 2021. *Adaptive Collaborative Management in Forest Landscapes: Villagers, Bureaucrats and Civil Society*. London: Earthscan/Routledge.

Corrigan, C., H. Bingham, S. Yichuan, et al. 2018. "Quantifying the Contribution to Biodiversity Conservation of Protected Areas Governed by Indigenous Peoples and Local Communities." *Biological Conservation* 227: pp. 403–412. DOI:10.1016/j.biocon.2018.09.007.

Coulthard, G. S. 2014. *Red Skin, White Masks: Rejecting the Colonial Politics of Recognition*. Minneapolis: University of Minnesota Press.

Cramb, R. A., C. J. Pierce Colfer, W. Dressler, et al. 2009. "Swidden Transformations and Rural Livelihoods in Southeast Asia." *Human Ecology* 37 (3): pp. 323–346.

Crossland, M., L. Winowiecki, T. Pagella, K. Hadgu, and F. Sinclair. 2018. "Implications of Variation in Local Perception of Degradation and Restoration Processes for Implementing Land Degradation Neutrality." *Environmental Development* 28: pp. 42–54.

Dawson, N. M., B. Coolsaet, E. J. Sterling, et al. 2021. "The Role of Indigenous Peoples and Local Communities in Effective and Equitable Conservation." *Ecology and Society* 26 (3): p.19. DOI:10.5751/ES-12625-260319.

de Sousa Santos, B. 2008. *Another Knowledge is Possible: Beyond Northern Epistemologies*. London: Verso.

Douterlungne, D., S. I. Levy-Tacher, D. J. Golicher, and F. R. Dañobeytia. 2010. "Applying Indigenous Knowledge to the Restoration of Degraded Tropical Rain Forest Clearings Dominated by Bracken Fern." *Restoration Ecology* 18 (3): pp. 322–329.

Dressler, W. H, D. Wilson, J. Clendenning, et al. 2016. "The Impact of Swidden Decline on Livelihoods and Ecosystem Services in Southeast Asia: A Review of the Evidence from 1990 to 2015." *Ambio* 46: pp. 291–310.

Dumont, E. S., S. Bonhomme, T. F. Pagella, and F. L. Sinclair. 2019. "Structured Stakeholder Engagement Leads to Development of More Diverse and Inclusive Agroforestry Options." *Experimental Agriculture* 55 (S1): pp. 252–274.

Elias, M., D. Joshi, and R. Meinzen-Dick. 2021. "Restoration for Whom, by Whom? A Feminist Political Ecology of Restoration." *Ecological Restoration* 39 (1–2): pp. 3–15

Elmhirst, R. 1999. "Space, Identity Politics and Resource Control in Indonesia's Transmigration Programme." *Political Geography* 18: pp. 813–835. DOI:10.1016/S0962-6298(99)00029-3.

Erbaugh, J. T., N. Pradhan, J. Adams, et al. 2020. "Global Forest Restoration and the Importance of Prioritizing Local Communities." *Nature Ecology & Evolution* 4 (11): pp.1472–1476.

Escobar, A. 1998. "Whose Knowledge, Whose Nature? Biodiversity, Conservation, and the Political Ecology of Social Movements." *Journal of Political Ecology* 5: pp. 53–82.

Escobar, A. 2007. "Worlds and Knowledges Otherwise 1: The Latin American Modernity/Coloniality Research Program." *Cultural Studies* 21 (2–3): pp. 179–210.

Evans, K., I. Monterroso, D. B. Ombogoh, et al. 2021. *Getting It Right, a Guide to Improve Inclusion in Multistakeholder Forums.* Bogor: CIFOR. https://www.cifor.org/knowledge/publication/7973/.

Fanon, F. 1963. *The Wretched of the Earth.* Translated by C. Farrington. New York: Grove Weidenfeld. https://abahlali.org/wp-content/uploads/2011/04/Frantz-Fanon-The-Wretched-of-the-Earth-1965.pdf.

Fanon, F. 1967. *Black Skin, White Masks.* New York: Grove Press.

Fleischman, F., S. Basant, A. Chhatre, et al. 2020. "Pitfalls of Tree Planting Show Why We Need People-Centered Natural Climate Solutions." *BioScience* XX: pp. 1–4. DOI:10.1093/biosci/biaa094.

Fraser, N. 2009. *Scales of Justice: Reimagining Political Space in a Globalizing World.* New York: Columbia University Press.

Gebara, M.F. 2020. "Thinking Beyond the Human-Nature Dichotomy." In *Frontiers of Development in the Amazon: Riches, Risks and Resistances,* edited by A. A. R. Ioris, R. R. Ioris, and S. Shubin, pp. 233–257. Lanham: Lexington Books.

Global Witness. 2022. Decade of Defiance. Ten Years of Reporting Land and Environmental Activism Worldwide. https://www.globalwitness.org/en/campaigns/environmental-activists/decade-defiance/#decade-killings-globally.

Griscom, B. W., J. Adams, P. W. Ellis, et al. 2017. "Natural Climate Solutions." *Proceedings of the National Academy of Sciences* 114 (44): pp. 11645–11650.

Hickel, J., D. Sullivan, and H. Zoomkawala. 2021. "Plunder in the Post-Colonial Era: Quantifying Drain from the Global South through Unequal Exchange, 1960–2018." *New Political Economy* 26 (6): pp. 1030–1047. DOI:10.1080/13563467.2021.1899153.

Holl, K. D., and P. H. Brancalion. 2020. "Tree Planting is not a Simple Solution." *Science* 368 (6491): pp. 580–581.

Hope, J. 2021. "Conservation in the Pluriverse: Anti-Capitalist Struggle, Knowledge from Resistance and the 'Repoliticisation of Nature' in the TIPNIS, Bolivia." *Geoforum* 124: pp. 217–225. DOI:10.1016/j.geoforum.2021.04.006.

Ingalls, M. L., P. Meyfroidt, P. Xuan To, M. Kenney-Lazar, and M. Epprecht. 2018. "The Transboundary Displacement of Deforestation under REDD+: Problematic Intersections between the Trade of Forest-Risk Commodities and Land Grabbing in the Mekong region." *Global Environmental Change* 50: pp. 255–267. DOI:10.1016/j.gloenvcha.2018.04.003.

IPBES. 2019. *Summary for Policymakers of the Global Assessment Report on Biodiversity and Ecosystem Services of the Intergovernmental Science-Policy Platform on Biodiversity and Ecosystem Services.* Bonn: IPBES Secretariat. DOI:10.5281/zenodo.3553579.

Jhaveri, N. 2021. *Forest Tenure Pathways to Gender Equality: A Practitioner's Guide.* Bogor: CIFOR. https://www.cifor.org/knowledge/publication/7909.

Kashwan, P. 2017. *Democracy in the Woods: Environmental Conservation and Social Justice in India, Tanzania, and Mexico.* New York: Oxford University Press.

Kashwan, P., R. V. Duffy, F. Massé, A. P. Asiyanbi, and E. Marijnen. 2021. "From Racialized Neocolonial Global Conservation to an Inclusive and Regenerative Conservation." *Environment: Science and Policy for Sustainable Development* 63 (4): pp. 4–19. DOI:10.1080/00139157.2021.1924574.

Kijazi, M. 2015. "Climate Emergency, Carbon Capture and Coercive Conservation on Mt. Kilimanjaro." In *Carbon Conflicts and Forest Landscapes in Africa*, edited by M. Leach and I. Scoones, pp. 78–98. London: Routledge.

Kothari, U. 2001. "Power, Knowledge and Social Control in Participatory Development." In *Participation: The New Tyranny?* edited by B. Cooke, and U. Kothari, pp. 139–152. London: Zed Books.

Lambin, E. F., B. L. Turner, H. J. Geist, et al. 2001. "The Causes of Land-Use and Land-Cover Change: Moving beyond the Myths." *Global Environmental Change* 11 (4): pp. 261–269. DOI:10.1016/S0959-3780(01)00007-3.

Larson, A. M. 2010. "The 'Demonization' of Rainforest Migrants, or: What Conservation Means to Poor Colonist Farmers." In *Beyond the Biophysical: Knowledge, Culture and Politics in Agriculture and Natural Resource Management*, edited by L. German, J. Ramisch, and R. Verma, pp. 49–71. Dordrecht: Springer.

Larson A. M., I. Monterroso, and P. Cantuarias. 2019. "Gender and the Formalization of Native Communities in the Peruvian Amazon." InfoBrief no. 238. Bogor: CIFOR.

Larson, A. M., and J. Pulhin. 2012. "Enhancing Forest Tenure Reforms through More Responsive Regulations." *Conservation and Society* 10 (2): pp. 103–113.

Larson, A. M., and F. Soto. 2008. "Decentralization of Natural Resource Governance Regimes." *Annual Review of Environment and Resources* 33: pp. 213–239.

Larson, A. M., and J. Springer. 2016. "Recognition and Respect for Legitimate Tenure Rights." Natural Resource Governance Series. IUCN Working Paper. https://www.iucn.org/sites/default/files/2022-07/tenure_rights_final.pdf.

Latouche, S. 2015. "Decolonization of Imaginary." In *Degrowth: A Vocabulary for a New Era*, edited by G. D'Alisa, F. Demaria, and G. Kallis, pp. 11–7120. Abingdon: Routledge.

Lund, C. 2011. "Fragmented Sovereignty: Land Reform and Dispossession in Laos." *Journal of Peasant Studies* 38 (4): pp. 885–905.

Lund, J. F., and Saito-Jensen, M. 2013. "Revisiting the Issue of Elite Capture of Participatory Initiatives." *World Development* 46: pp. 104–112.

Maldonado-Torres, N. 2007. "On the Coloniality of Being." *Cultural Studies* 21 (2–3): pp. 240–270. DOI:10.1080/09502380601162548.

Mansourian, S. 2021. "Disciplines, Sectors, Motivations and Power Relations in Forest Landscape Restoration." *Ecological Restoration* 39 (1–2): pp. 16–26.

McElwee, P., and T. H. Nghi. 2021. "Assessing the Social Benefits of Tree Planting by Smallholders in Vietnam: Lessons for Large-Scale Reforestation Programs." *Ecological Restoration* 39 (1–2): pp. 52–63.

McGregor, D. 2014. "Traditional Knowledge and Water Governance: The Ethic of Responsibility." *AlterNative: An International Journal of Indigenous Peoples* 10 (5): pp. 493–507. DOI:10.1177/117718011401000505.

McGregor, D., M. Sritharan, and S. Whitaker. 2020. "Indigenous Environmental Justice and Sustainability." *Current Opinion in Environmental Sustainability* 43: pp. 35–40. DOI:10.1016/j.cosust.2020.01.007.

McLain, R., S. Lawry, M. R. Guariguata, and J. Reed. 2018. "Toward a Tenure-Responsive Approach to Forest Landscape Restoration: A Proposed Tenure Diagnostic for Assessing Restoration Opportunities." *Land Use Policy* 104: 103748. DOI:10.1016/j.landusepol.2018.11.053.

Mignolo, W. 2007. "Introduction. Coloniality of Power and De-Colonial Thinking." *Cultural Studies* 21 (2–3): pp. 155–167.

Mignolo, W. 2008. "Epistemic Disobedience and the Decolonial Option: A Manifesto." *Subaltern Studies: An Interdisciplinary Study of Media and Communication* 26 (7–8): pp. 159–181. DOI:10.1177/0263276409349275.

Minnemeyer, S., L. Laestadius, N. Sizer, C. Saint-Laurent, and P. Potapov. 2011. "A World of Opportunity for Forest and Landscape Restoration". The Global Partnership on Forest Landscape Restoration. http://pdf.wri.org/world_of_opportunity_brochure_2011-09.pdf.

Monterroso, I., A. M. Larson, E. Mwangi, N. Liswanti, and Z. Cruz-Burga. 2019. *Mobilizing Change for Women within Collective Tenure Regimes.* Washington: Resource Equity. https://landwise-production.s3.amazonaws.com/2022/03/CIFOR_Mobilizing-Change-Women-Within-Collective-Tenure-Regimes_2019_Final-1.pdf

Mukasa, C., A. Tibazalika, A. Banana, E. Mwangi, and T. Mutimukuru-Maravanyika. 2022. "Strengthening Women's Tenure Rights and Participation in Community Forestry in Central Uganda." In *Adaptive Collaborative Management of Forest Landscapes: Villagers, Bureaucrats and Civil Society*, edited by C. J. P. Colfer, R. Prabhu, and A. Larson, pp. 106–132. London: Earthscan/Routledge.

Mustalahti, I., and A. Agrawal. 2020. "Research Trends: Responsibilization in Natural Resource Governance." *Forest Policy and Economics* 121: p. 102308. DOI:10.1016/j.forpol.2020.102308.

Nchanji, Y. K., S. Ramcilovic-Suominen, and J. Kotilainen. 2021. "Power Imbalances, Social Inequalities and Gender Roles as Barriers to True Participation in National Park Management: The Case of Korup National Park, Cameroon." *Forest Policy and Economics* 130: p.102527. DOI:10.1016/j.forpol.2021.102527.

Nkonya, E., W. Anderson, E. Kato, et al. 2016. "Global Cost of Land Degradation." In *Economics of Land Degradation and Improvement–A Global Assessment for Sustainable Development,* edited by E. Nkonya, A. Mirzabaev, and J. von Braun, pp. 117–165. Springer Open.

Notess, L., P. Veit, I. Monterroso, et al. 2020. "Community Land Formalization and Company Land Acquisition Procedures: A Review of 33 Procedures in 15 Countries." *Land Use Policy* 110: 104461. DOI:10.1016/j.landusepol.2020.104461.

Ojha, H. R., J. Cameron, C. Kumar. 2009. "Deliberation or Symbolic Violence? The Governance of Community Forestry in Nepal." *Forest Policy and Economics* 11: pp. 365–374. DOI:10.1016/j.forpol.2008.11.003.

Osborne, T., S. Brock, R. Chazdon, et al. 2021. "The Political Ecology Playbook for Ecosystem Restoration: Principles for Effective, Equitable, and Transformative Landscapes." *Global Environmental Change* 70: 102320. DOI:10.1016/j.gloenvcha.2021.102320.

Pellow, D. 2018. "Political Prisoners and Environmental Justice." *Capitalism Nature Socialism* 29 (4): pp. 1–20. DOI:10.1080/10455752.2018.1530835.

Petri, M., R. Biancalani, and L. Lindeque. 2019. "Guidelines for the National Assessment and Mapping of Land Degradation and Conservation." Rome: FAO.

Quijano, A. 2000. "Coloniality of Power and Eurocentrism in Latin America." *International Sociology* 15: pp. 215–232.

Ramcilovic-Suominen, S. 2019. "REDD+ as a Tool for State Territorialization: Managing Forests and People in Laos." *Journal of Political Ecology* 26 (1): pp. 263–281. DOI:10.2458/v26i1.23357.

Ramcilovic-Suominen, S. 2022. "Envisioning Just Transformations in and beyond the EU Bioeconomy: Inspirations from Decolonial Environmental Justice and Degrowth." *Sustainability Science* 18: pp. 707–722. DOI:10.1007/s11625-022-01091-5.

Ramcilovic-Suominen, S., S. Carodenuto, C. McDermott, and J. Hiedanpää. 2021. "Environmental Justice and REDD+ Safeguards in Laos: Lessons from an Authoritarian Political Regime." *Ambio* 50 (12): pp. 2256–2271. DOI:10.1007/s13280-021-01618-7.

Ramcilovic-Suominen, S., and J. Kotilainen. 2020. "Power Relations in Community Resilience and Politics of Shifting Cultivation in Laos." *Forest Policy and Economics* 115: pp. 102–159. DOI:10.1016/j.forpol.2020.102159.

Ramcilovic-Suominen, S., and Mustalahti, I. 2022. "Village Forestry under Donor-Driven Forestry Interventions in Laos." In *Routledge Handbook of Community Forestry*, edited by J. Bulkan, J. Palmer, A. M. Larson, and M. Hobley, pp. 434–448. London: Routledge. DOI:10.4324/9780367488710-33.

Ramcilovic-Suominen, S., and I. Nathan. 2020. "REDD+ Policy Translation and Storylines in Laos." *Journal of Political Ecology* 27 (1): pp. 436–455. DOI:10.2458/v27i1.23188.

Ribot, J. C., A. Agrawal, and A. M. Larson. 2006. "Recentralizing while Decentralizing: How National Governments Reappropriate Forest Resources." *World Development* 34 (11): pp. 1864–1886.

Ribot, J. C., and A. M. Larson (eds.). 2013. *Democratic Decentralization through a Natural Resource Lens: Cases from Africa, Asia and Latin America.* London: Routledge.

Ribot, J. C., and N. L. Peluso. 2003. "A Theory of Access." *Rural Sociology* 68 (2): pp. 153–181.

Rodríguez, I. 2020. "Latin American Decolonial Environmental Justice Approach." In *Environmental Justice: Key Issues*, edited by B. Coolsaet, pp. 78–93. New York: Routledge.

Rodríguez, I., and M. L. Inturias. 2018. "Conflict Transformation in Indigenous Peoples' Territories: Doing Environmental Justice with a 'Decolonial Turn.'" *Development Studies Research* 5 (1): pp. 90–105. DOI:10.1080/21665095.2018.1486220.

RRI. 2018. *At a Crossroads. Consequential Trends in Recognition of Community-Based Forest Tenure from 2002–2017.* Washington: Rights and Resources Initiative.

Rutazibwa, O. U. 2018. "On Babies and Bathwater. Decolonizing International Development Studies." In *Decolonization and Feminisms in Global Teaching and Learning*, edited by S. de Jong, R. Icaza, and O.U. Rutazibwa, pp. 158–180. New York: Routledge.

Sabatier, P. A., W. Focht, M. Lubell, Z. Trachtenberg, and A. Vedlitz (eds.). 2005. *Swimming Upstream: Collaborative Approaches to Watershed Management*. Cambridge: MIT Press.

Sarmiento Barletti, J.P., A.M. Larson, C. Hewlett, and D. Delgado. 2020. "Designing for Engagement: A Realist Synthesis Review of How Context Affects the Outcomes of Multi-Stakeholder Forums on Land Use and/or Land-Use Change." *World Development* 127: 104753. DOI:10.1016/j.worlddev.2019.104753

Saxena, N. C. 1997. *The Saga of Participatory Forest Management in India*. Bogor: CIFOR. www.cifor.org/publications/pdf_files/Books/SP-Saga.pdf.

Schlosberg, D. 2007. *Defining Environmental Justice: Theories, Movements and Nature*. Oxford: Oxford University Press.

Scott, J. C. 1985. *Weapons of the Weak: Everyday Forms of Peasant Resistance*. Yale: Yale University Press.

Sikor, T. (ed.). 2013. *The Justices and Injustices of Ecosystem Services*. London: Earthscan.

Sinclair, F. L., and L. Joshi. 2000. "Taking Local Knowledge about Trees Seriously." In *Forestry, Forest Users and Research: New Ways of Learning*, edited by A. Lawrence, pp. 45–61. ETFRN Series No. 1. Vienna: European Tropical Forest Research Network.

Streck, C. 2021. "REDD+ and Leakage: Debunking Myths and Promoting Integrated Solutions." *Climate Policy* 21 (6): pp. 843–852. DOI:10.1080/14693062.2021.1920363.

Sultana, F. 2022. "The Unbearable Heaviness of Climate Coloniality." *Political Geography* 99: 102638. DOI:10.1016/j.polgeo.2022.102638.

Tamanaha, B.Z. 2007. "Understanding Legal Pluralism: Past to Present, Local to Global." *Sydney Law Review* 29. St. John's Legal Studies Research Paper No. 07-0080. https://ssrn.com/abstract=1010105.

Temper, L. 2019. "Blocking Pipelines, Unsettling Environmental Justice: From Rights of Nature to Responsibility to Territory." *Local Environment* 24 (2): pp. 94–112. DOI:10.1080/13549839.2018.1536698.

Tole, L. 2010. "Reforms from the Ground Up: A Review of Community-Based Forest Management in Tropical Developing Countries." *Environmental Management* 45 (6): pp. 1312–1331.

Tschakert, P., D. Schlosberg, D. Celermajer, et al. 2021. "Multispecies Justice: Climate-Just Futures with, for and beyond Humans." *Wiley Interdisciplinary Reviews: Climate Change* 12 (2): e699.

Tuck, E., and K. W. Yang. 2012. "Decolonization Is Not a Metaphor." *Decolonization: Indigeneity, Education & Society* 1 (1): pp. 1–40.

Ubink, J. 2011. "Stating the Customary: An Innovative Approach to the Locally Legitimate Recording of Customary Law in Namibia." *Traditional justice: Practitioners' perspectives Working Paper Series*. Paper no. 8. Rome: International Development Law Organization.

Waldron, A., V. Adams, J. Allan, et al. 2020. "Protecting 30% of the Planet for Nature: Costs, Benefits and Economic Implications." Working paper. http://pure.iiasa.ac.at/id/eprint/16560/.

Walker, G. 2012. *Environmental Justice: Concepts, Evidence and Politics.* London: Routledge.

Whyte, K. P. 2014. "Indigenous Women, Climate Change Impacts, and Collective Action." *Hypatia* 29 (3): pp. 599–616. DOI:10.1111/hypa.12089.

Whyte, K. P. 2017. "What Do Indigenous Knowledges Do for Indigenous Peoples?" Forthcoming in *Keepers of the Green World: Traditional Ecological Knowledge and Sustainability*, edited by M. K. Nelson, and D. Shilling, pp. 57–83. https://ssrn.com/abstract=2612715.

Whyte, K. P. 2020. "Indigenous Environmental Justice: Anti-Colonial Action through Kinship." In *Environmental Justice Key Issues*, edited by B. Coolsaet, pp. 266–278. London: Routledge. DOI:10.4324/9780429029585-23.

Winter, C. J. 2021. *Subjects of Intergenerational Justice: Indigenous Philosophy, the Environment and Relationships.* London: Routledge. DOI:10.4324/9781003097457.

Ziai, A. 2012. "Postcolonial Perspectives on 'Development.'" ZEF Working Paper Series, No. 103. University of Bonn, Center for Development Research (ZEF).

Ziai, A. 2016. *Development Discourse and Global History: From Colonialism to the Sustainable Development Goals.* New York: Routledge.

5
Regional variation in forest landscape restoration

César Sabogal, Jürgen Blaser, Kenichi Shono, Yitagesu Tekle Tegegne, Rene Zamora Cristales, Almeida Sitoe, Tania Ammour, Adeline Dontenville, Mélanie Feurer, Jelena Markovic, Christophe Besacier, Silvio Simonit, Benjamin de Ridder, and Orjan Jonsson

5.1 Introduction

5.1.1 Background

Land degradation[1] affects an estimated 23% of the world's terrestrial area and is increasing at an annual rate of 5–10 Mha (Stavi and Lal 2014). Currently, land degradation "through human activities is negatively impacting the well-being of at least 3.2 billion people, pushing the planet toward a sixth mass extinction and costing more than 10% of the annual global gross domestic product in loss of biodiversity and ecosystem services" (IPBES 2018, p. xx). A study by Laestadius et al. (2012) suggests that 27% or 1.5 Mha of forest landscapes at the time of study are to some degree degraded, and as much as 52% or 2.8 Mha could be considered fragmented, leaving only 21% (1,112 Mha) as intact forest (Caspari et al. 2013). In a more recent study, Bastin et al. (2019) estimate that 1.7–1.8 billion ha of potential forest land (defined as >10% tree cover) exists in areas that were previously degraded, dominated by sparse vegetation, grasslands, and degraded bare soils. Agriculture is responsible for converting about 30% of forests worldwide (Ramankutty et al. 2018) and is considered the main driver of deforestation: some estimates suggest that globally, between 2000 and 2010 around 80% of deforestation may be attributed to conversion into agriculture and grazing lands (Kissinger et al. 2012; Ramankutty et al. 2018. See Curtis et al. 2018 for more recent estimates).

The high rates of forest loss and forest land degradation[2] have prompted responses favoring forest ecosystem[3]/land restoration[4] at the global, regional, and national levels. Over the past decade, restoring deforested and degraded forest landscapes has become an urgent policy priority (Mansourian et al. 2017; Mansuy et al. 2020).

Forest landscape restoration (FLR) was originally defined as "a process that aims to regain ecological integrity and enhance human well-being in deforested or degraded forest landscapes" (Maginnis and Jackson 2002). The term has broadened in scope beyond forestry and forest ecosystems to take a more holistic and inclusive overview to address the drivers of deforestation and land degradation, to restore ecological functions, and to promote the sustainable use of land and forest resources

César Sabogal, Jürgen Blaser, Kenichi Shono, Yitagesu Tekle Tegegne, Rene Zamora Cristales, Almeida Sitoe, Tania Ammour, Adeline Dontenville, Mélanie Feurer, Jelena Markovic, Christophe Besacier, Silvio Simonit, Benjamin de Ridder, and Orjan Jonsson, *Regional variation in forest landscape restoration* In: *Restoring Forests and Trees for Sustainable Development*. Edited by: Pia Katila, Carol J. Pierce Colfer, Wil de Jong, Glenn Galloway, Pablo Pacheco, and Georg Winkel, Oxford University Press. © Oxford University Press 2024. DOI: 10.1093/9780197683958.003.0005 with positive outcomes at the landscape scale for livelihoods, ecosystem services, and biodiversity conservation (Baker and Eckerberg 2016; Burton and Macdonald 2011; Chazdon et al. 2020; Gann et al. 2019; Gomes et al. 2020; van Oosten et al. 2014).

FLR is gaining momentum globally and has become an important international policy topic in the environmental sector and a major component of nature-based climate solutions (Chazdon et al. 2020). Multiple initiatives, led by governmental and nongovernmental organizations, have been launched with the goal of achieving large-scale restoration targets (Chazdon et al. 2017), many under the banner of the Bonn Challenge—a global initiative to restore 150 Mha by 2020 and 350 Mha by 2030 (Mansuy et al. 2020). Additionally, the United Nations Decade on Ecosystem Restoration 2021–2030, declared in 2019, aims to scale up the restoration of degraded ecosystems as a proven measure to fight the climate crisis and enhance food security, water supply, and biodiversity.

Hundreds of FLR-based initiatives are currently underway worldwide (Chazdon et al. 2020). Most of the 63 nations, regions and companies who have committed to the Bonn Challenge since 2020, totaling over 173 Mha in restoration pledges, are in Africa, Latin America, and Asia (Stanturf and Mansourian 2020). The jurisdictions that have made these pledges reflect a vast range of geographies, political and socioeconomic contexts, biophysical conditions, resources, capacities, and motivations (Dave et al. 2019).

Commitments at the regional scale have materialized in Africa through the African Forest Landscape Restoration Initiative (AFR100)[5] with country commitments totaling 148 Mha (of which 117 Mha belong to the Bonn Challenge) and in Latin America through Initiative 20 × 20[6] with country commitments reaching 21.7 Mha. In Asia, six countries have made Bonn Challenge commitments for 22.6 Mha, but other assessments of 12 countries have identified a total of 65 Mha with restoration potential (Stanturf and Mansourian 2020). Importantly, the effective implementation of the Bonn Challenge in regional initiatives represents an opportunity to expand FLR efforts and contribute to several nationally determined contributions (NDCs).

Country commitments to the Bonn Challenge have been received with some criticism. Bastin et al. (2019) point out the need for better country-level forest accounting.[7] In their review, Stanturf and Mansourian (2020) conclude that, in practice, FLR initiatives are still in their infancy when it comes to fully adhering to the objectives of the FLR approach, "and many initiatives that are labelled FLR would not qualify under its definition or principles" (referring to the six globally agreed principles of FLR, Besseau et al. 2018).

5.1.2 Objective, scope, and methodology

This chapter explores how restoration is proposed and implemented across selected world regions and how various contextual conditions influence these processes. It will be shown that these conditions result in challenges facing restoration efforts, leading to variable outcomes. Gaining a better understanding of the regional/country differences will provide important insights for planning and implementing national/subnational FLR agendas, taking into consideration preconditions and success factors.

This chapter uses the broader and more inclusive concept of FLR: "a means of regaining, improving and maintaining vital ecological, [economic] and social functions, in the long-term leading to more resilient and sustainable landscapes."[8]

Data sources for our review include published material, webpages, technical reports, and project-based information. We also used the extensive region/country-based knowledge and experience from authors and selected source persons to reliably cover as many possible variables as defined for describing the FLR contexts, implementation tools, and challenges.

Restoration across world regions

Differences in FLR are described at the level of selected regions in tropical, subtropical, and/or temperate climate domains of Latin America, Africa, Asia, and Europe. Eight world regions and 17 representative countries were selected based on our familiarity with these geographies, including access to secondary information and country informants. We defined and grouped a number of variables to describe the relevant contexts of, responses to, and challenges for FLR implementation:

1. **Context:** forest resources, deforestation/degradation drivers and pressures, and restoration rationales.

Extent of total forest area, other wooden area, and planted forest area—using FAO Forest Resources Assessment (FRA) 2020 data[9]

Extent of deforestation and land degradation—accumulated deforested area, annual deforestation rate, accumulated degraded area.

Drivers of deforestation and land degradation—direct and indirect causes.

Responses to halt deforestation and (forest) land degradation—main responses at the regional level.

Restoration commitments, targets, and approaches—official country declarations to restore degraded forests/lands; how targets and approaches are defined.

2. **Tools**: conditions or factors in place for the effective, successful implementation of FLR commitments and targets.

Policies and legal frameworks—strategies, plans, and institutional arrangements.

Institutional capacities and effectiveness—(cross-sectoral) articulation and coordination of policies and government programs to integrate human, technical, and financial resources; planning, operations, supervision, and enforcement.

Economic incentives, financing, and investments—public/community, private sector government, international cooperation.

Public partnerships and negotiation platforms—significant initiatives and alliances; national/subnational dialogue and action.

Capacity building, technical assistance, and extension services—stakeholder efforts to effectively work together (groups and institutions); availability and quality of support.

Technological innovation and research—to add value to forest products and services; existence and capacity of research programs.

Technical interventions—predominant field-level methods/techniques/practices/interventions.

3. **Challenges**: Main societal, ecological, and economic issues and demands that support or impede restoration (e.g., in relation to land availability, land use, environmental impacts, etc.) in the given regions.

Restoration in selected countries

For the in-depth review, the FLR process in each chosen country[10] was characterized using a common frame of variables to capture and compare their differences. In addition to those listed above, the variables are:

- Assessment of restoration needs and opportunities
- Stakeholder participation
- Financing accessibility
- Field-level practices/restoration models implemented
- Monitoring
- Innovative aspects
- Accomplishments
- Lessons learned
- Trends in the conditions for effective FLR implementation

Table 5.1 shows the eight selected world regions and the 17 representative countries.

The chapter presents by section: the regional context in terms of forest resources and changes in land cover, the drivers of these changes, the responses to halt

Table 5.1 Selected world regions and countries used to characterize their FLR process

Latin America		**Sub-Saharan Africa**		**Asia-Europe**	
Central America & Mexico[a] *(8 countries)*	Mexico Guatemala El Salvador	**West Africa**[b] *(15 countries)*	Côte d'Ivoire Ghana	**Southeast Asia**[c]*(11 countries)*	Philippines Cambodia
				South Asia[d] *(8 countries)*	Pakistan
Tropical South America[e] *(6 countries)*	Colombia Peru Brazil	**East Africa**[f] *(14 countries)*	Ethiopia Kenya Mozambique	**Central Asia**[g] *(5 countries)*	Uzbekistan
				Eastern & South-Eastern Europe *(13 countries)*	Ukraine Macedonia

[a] It includes Mexico and the seven Central American countries

[b] It includes the geopolitical states of West Africa. Source: https://en.wikipedia.org/wiki/West_Africa

[c] Source: https://en.wikipedia.org/wiki/Southeast_Asia

[d] Source: https://en.wikipedia.org/wiki/South_Asia

[e] It includes five countries sharing the Amazon Basin: Bolivia, Brazil, Colombia, Ecuador, Venezuela and Peru. Guyana and Suriname were not considered as part of this region.

[f] It includes countries in the Horn of Africa, the Nile Valley, the East African Community and Southeast Africa. Source: https://en.wikipedia.org/wiki/East_Africa#Countries,_capitals_and_largest_cities

[g] Source: https://en.wikipedia.org/wiki/Central_Asia

deforestation and land degradation, and the restoration commitments, targets, and approaches (Section 5.2); the implementation tools for FLR, covering: policies and legal frameworks; institutional capacities and effectiveness; economic incentives and investment environment; public–private partnerships and platforms for dialogue and negotiation; capacity building, technical assistance, and extension services; access to technological innovation and research; and main technical interventions (restoration practices) (Section 5.3); a comparison of the regional variation and progress in FLR implementation (Section 5.4); and, lastly, a review of the main challenges and some key lessons (Section 5.5).

5.2 Implementation context for forest landscape restoration

5.2.1 Forest resources and changes in land cover

The eight regions encompass over 80 countries, mostly from tropical/subtropical climatic domains; countries in Central Asia and Eastern & South-Eastern Europe have predominantly temperate forests. As expected, there are large differences between all regions in the extent of forest resources (including both natural and planted forests) and in the magnitude of deforestation and land and forest degradation (Table 5.2).

There are marked differences across the regions and across countries within specific regions (e.g., West Africa, South Asia). Tropical South America is the region with the largest deforested area in the period 2015–2020 (FAO 2020), but some countries in East Africa and in Southeast Asia show comparatively high deforestation rates in relation to the extent of their natural forests.[11] In general, however, deforestation rates in other tropical regions have tended to decrease or at least stabilize. On the other hand, Central Asia and Eastern & South-Eastern Europe, both with temperate climates, have made gains in forest cover through the expansion of planted forests (a trend reported with FRA data (Payn et al. 2015)).

Estimates of land and forest degradation at the country level are not easily available or comparable, and thus the figures given at the regional level can only serve as reference. Nevertheless, degraded lands and forests are affecting considerable areas, probably representing about one-third of land in Central America & Mexico and up to 90% of forests in West Africa (Blaser et al. 2011).[12]

5.2.2 Drivers of deforestation and land degradation

The expansion of large-scale industrial agriculture—commodity crops such as sugar cane, palm oil, soybean, coffee, cocoa—is a major direct cause of deforestation in Latin America and Asia; in the former, cattle ranching is as well. Clearing for pastures and overgrazing are important drivers in Sub-Saharan Africa and in the dry forests and highlands of Latin America and Asia (Boucher et al. 2011; FAO and RECOFTC 2016; Hosonuma et al. 2012; Kissinger et al. 2012; Pacheco et al. 2021; Rudel 2013;

Table 5.2 Forest resources in the selected eight world regions

Variables	Central America & Mexico	Tropical South America	West Africa	East Africa	Southeast Asia	South Asia	Central Asia	Eastern & South-Eastern Europe
a) Climate domain (Trop = tropical, SubT = subtropical, Temp = temperate)	Trop-SubT	Trop	Trop	Trop-SubT	Trop	Trop-SubT-Temp	Temp	Temp
b) Total forest area, 2020 (Mha)	88.1	771.3	84.1	185.0	207.5	91.8	13.0	44.3
c) Total planted forest, 2020 (Mha)	0.49	14.30	1.65	3.20	17.30	14.17	3.03	7.19
d) Extent of deforestation (Mha)[a]	0.33 (0.47%)	2.5 (0.32%)	0.27[b] (0.32%)	1.24 (0.67%)	0.98 (0.47%)	0.67 (0.73%)	0.01 (0.08%)	0.02 (0.045%)
e) Extent of land and forest degradation[c]	30% of land	230 Mha of land	90% of forests	180 Mha of land	90.7 Mha of forests	> 30% of land just in India	1–8% of forests	unknown

[a] Combined annual amount of deforested area for the period 2015–2020, and % of total forest area.

[b] But high deforestation rates in Togo, Nigeria, Mali, and Benin.

[c] Estimates, in % of total land area or total forest area or in Mha, from varied time periods.

Sloan and Sayer 2015). Mining, infrastructure (roads, reservoirs), urban development (human settlements), human-induced fires and natural disasters are increasingly pressuring forests worldwide. In some tropical regions, illegal logging and the overharvesting of timber and firewood are the main causes of forest degradation. Other drivers include fires, overgrazing, and severe weather events (Boucher et al. 2011; Sloan and Sayer 2015; FAO 2020; Pacheco et al. 2021).

Demand for land is an important driver of change in countries with relatively high population densities, as in Central America, Mexico, and West Africa, whose economies and livelihoods largely depend on the production of agricultural commodities (Redo et al. 2012; UNCCD 2019). Weak governance, forest illegality, and perverse incentives act as important underlying causes of land and forest degradation in Central America & Mexico (Redo et al. 2012; Sesnie et al. 2017). In most tropical South American countries, land tenure conflict is a relevant driver of deforestation, largely due to legal and capacity gaps in addressing land tenure issues (Cabrera et al. 2020; Carvalho et al. 2019; DAR 2015; Fearnside 2017). Underlying causes in this region include poorly defined property rights, limited capacity to comply with norms and apply penalties, market incentives, and changes in the population's attitudes and values (UNEP and ACTO 2009). In West Africa, the lack of alternative livelihoods is considered the major driver, as people depend on forest products for many purposes (Bromhead 2012; Fedele et al. 2021). Indirect drivers in this region include: the lack of forest protection and conservation policies and the limited implementation capacity of existing policies; the limited participation of local communities in forest protection and conservation and the inequitable access to benefits; and the absence of clear and consistent institutions and policies to support tree planting and ecosystem restoration.[13] In East Africa, high population growth (2.65% per year) and high poverty levels (33% of population living on less than USD 1.9/day/person) create a disadvantageous combination that drives deforestation.[14] In Southeast Asia, economic growth has increased demand for forest goods, and agricultural products have become an important indirect driver of deforestation and degradation. Furthermore, poor governance, corruption, tenure uncertainties, and low institutional capacity for planning and enforcement of rules are also critical underlying factors for forest degradation in this region (FAO and RECOFTC 2016).

In South Asia, high population density and low per capita forest area put intense human pressure on forests, resulting in extensive deforestation and degradation. Though industrial and service sector growth is reducing dependence on agriculture across the region, the increasing demand for minerals, energy, and infrastructure drives forest clearance in some locations (FAO 2011, 2019). In Central Asia, forest and land degradation are major problems across the region, where key drivers include fuelwood removal, overgrazing, and uncontrolled logging (FAO and UNECE 2019). This contrasts with Eastern & South-Eastern Europe, where agricultural land abandonment is one of the most important drivers of land use change. Degradation mainly results from earlier intensive use of natural resources; deficiencies in silvicultural, pasture, and agriculture management; and climate change impacts such as droughts, heatwaves, and extreme wildfires (FAO and UNECE 2021).

5.2.3 Responses to halt deforestation and land degradation

Responses to deforestation and degradation vary across countries, ecoregions, and ecosystems. In Central America & Mexico, country strategies focus on improving regulations against environmental degradation, supporting policies and incentives to conserve, sustainably manage, and restore forests (e.g., developing payments for ecosystem services (PES) programs), and improving the governance and community management of landscapes. In tropical South America, initiatives concentrate on the Amazon through government programs supported by civil society organizations and international cooperation (e.g., Reducing Emissions from Deforestation and Forest Degradation (REDD +) initiatives, the European Union's (EU) Forest Law Enforcement, Governance and Trade (FLEGT)). In the Andean ecoregion, where soil and water degradation are critical, responses involve (integrated) watershed management programs, ecological restoration, and tree planting (e.g., in Bolivia, Colombia, Ecuador, and Peru). In the drylands, responses are geared toward preserving biodiversity, promoting avoided deforestation, increasing negative emissions connected to carbon markets, and securing water reservoirs and other key ecosystem services for human well-being.

In response to the dramatic decline of forest cover in West Africa, the 15 member countries of the Economic Community of West African States agreed in 2013 to work together across borders to protect and manage the region's forests and wildlife, adopting the Convergence Plan for the Sustainable Management and Use of Forest Ecosystems in West Africa alongside the Sub-Regional Action Program to Combat Desertification. Nigeria and Ghana have passed through the "transition" of deforestation and are ready to reinvest in landscape restoration (on farmlands) and plantation forestry (on degraded forest reserve land), which should eventually result in a mosaic of land uses, including trees, in the production landscape (Bromhead 2012). In the Sahel, several countries (e.g., Senegal) moved to align forest and woodland policy with ongoing decentralization strategies. Local communities were given tree-harvesting rights, and programs were put in place to support farmer-managed natural regeneration (FMNR) as part of broader food security, agroforestry, and sustainable land management approaches (Bromhead 2012). In East Africa, AFR100 is the largest initiative to tackle the drivers of deforestation and increase tree planting and forest restoration.

Of note for Southeast Asia, in 2011 Indonesia placed a temporary moratorium on forest-clearing for logging or agricultural purposes in primary forest and peatlands, and in the same year the Philippines launched the National Greening Programme (NGP) with the ambitious goal of planting 7.16 Mha of trees on degraded former forestlands by 2028 (Russell 2020). Other efforts have been directed to stamp out illegal logging, particularly with support from the FLEGT program. In parallel, the EU's REDD + Facility is helping partner countries in the region to prevent deforestation by managing land allocation for such purposes as conservation, agriculture, and forestry (Russell 2020).

In South Asia, apart from the South Asian Association for Regional Cooperation (SAARC) Action Plan on Climate Change (2008), which identified the sharing of best practices on sustainable forest management as a key element, there has not been any prominent coordinated effort at the regional level to halt deforestation and land degradation. Each country in the region has different contexts and drivers of forest degradation, to which their respective governments respond, often with the effective participation of local communities—in particular, collaborative community management/social forestry efforts in Nepal and India (Bampton et al. 2007; Misra and Shabbeer 2001). These efforts, together with economic development, have reversed deforestation trends in most countries of the region, although degradation continues as a result of population pressure resulting in increased fuelwood collection, livestock grazing, non-wood forest products (NWFP) collection, and timber harvesting.

In Central Asia, two responses can be highlighted: The International Fund for Saving the Aral Sea (IFAS), established in 1992, aims to improve the socioeconomic and ecological situation in the Aral Sea Basin; and the 2018 Astana Resolution calls for reinforced cooperation on FLR to identify and work to restore degraded lands by 2030 in the frame of the Bonn Challenge/ECCA30 for Europe, the Caucasus and Central Asia (ECCA) countries. In Eastern & South-Eastern Europe, the 2021 United Nations Economic Commission for Europe (UNECE) FLR Eastern & South-Eastern Europe study identified key drivers of forest degradation and the potential for FLR in 17 countries, which gave way to a regional commitment to restore more than 4 Mha of land by 2030 under the Bonn Challenge. The biggest threat is not deforestation but a steady loss of forest quality due to unsustainable forest use, climate change with associated forest fires, pests and diseases, difficult access to forests; and lack of funding.

5.2.4 Restoration commitments, targets, and approaches

All regions are being supported internationally in the development of regional or national FLR agendas. Latin America has had its regional Initiative 20x20 in place since 2014, while over 30 countries in Africa have committed restoration targets to AFR100. In Asia, various organizations are contributing to the development of (sub) regional FLR agendas (the Asia-Pacific Network for Sustainable Forest Management and Rehabilitation (APFNet), the Asian Forest Cooperation Organization (AFoCO), SAARC, IFAS). In 2018, the Asia-Pacific Regional Strategy and Action Plan on Forest and Landscape Restoration was established, primarily covering the member countries of the Asia-Pacific Forestry Commission (APFC). Countries in (Eastern/South-Eastern) Europe, the Caucasus and Central Asia started comparatively later, when they aligned with the Bonn Challenge/ECCA30.[15] There are other international bodies actively supporting countries in the planning, implementation, and monitoring of their FLR commitments. Central America & Mexico, East Africa, Southeast Asia, and South Asia are the most favored regions in terms of international technical and financial support provided by agencies such as the Asian Development Bank (ADB), the Green Climate Fund, the Global Environment Facility (GEF), the Food and Agriculture Organization of the United Nations (FAO), the Inter-American Development Bank (IADB), the International Tropical Timber Organization (ITTO), the International Union for Conservation of Nature (IUCN), the United Nations Development Programme (UNDP), the United Nations Environment Programme (UNEP) and the World Bank (WB).

Some country targets appear to be quite high (e.g., Indonesia with 29.3 Mha, India 21 Mha, Ethiopia 15 Mha, Brazil, 12 Mha, and Mali, 10 Mha). Taking the range of country commitments for restoration between 0.5 and 3 Mha, for a 10-year time horizon (until 2030) it means an annual rate of 50,000 to 300,000 ha to restore. Compared with the annual rate of planted forest area—even if tree planting only represents half or a quarter of the area targeted for restoration—the huge challenge faced by some countries to reach their targets becomes evident: for example, Guatemala averaged 4,024 ha/year of planted forests for 2010–2020 vs. a 1.2 Mha (120,000 ha/year) restoration target; Ethiopia averaged 46,259 ha/year for 2010–2020 vs. a 15 Mha (1.5 Mha/year) restoration target.

Table 5.3 presents an overview of the selected regions' restoration commitments and main technical interventions.

Restoration commitments are not always clearly/explicitly defined, at least in terms of the desired outcomes and the necessary interventions to achieve them. From the available sources, reference is usually made to the restoration approach (e.g., ecological restoration, land rehabilitation) and the technical activities to undertake. The most common ones are the establishment (and/or restocking/management) of tree plantations (in block, as enrichment plantings, shelterbelts, etc.), the promotion of assisted natural regeneration (ANR),[16] agroforestry and silvopastoral systems, area enclosures, fuelwood production, and bioengineering. There are marked differences in the country approaches, however, in the approaches being pursued. The choice of options can depend on the ecosystem to be restored (humid/dry forest, wetlands, mangroves, grassland, etc.), the site condition, past experience, and available resources. Monocultures with exotic species (mainly eucalypts and pines) are often the preferred option for reforestation/afforestation, though native tree/shrub species are specifically mentioned in some countries as the preferred way to restore degraded landscapes.

5.3 Implementation tools for forest landscape restoration

5.3.1 Policies and legal frameworks

Specific policies and legal frameworks promoting FLR implementation are more likely to be found in regions/countries where there has been comparatively more political engagement and international support, as in Latin America. The regional Initiative 20x20 (launched in 2014 under the leadership of the World Resources Institute), plays an important role in creating the political conditions and/or facilitating the technical and financial means to mobilize government commitments and coordinated action for FLR planning and implementation. In Africa, AFR100, started in 2015 as a regional initiative, focuses on communities by supporting FMNR,[17] tree plantings, fuelwood plantations, reforms in the fuelwood and charcoal market (notably in West Africa) (Adeleke 2018; Bromhead 2012), as well as agroforestry systems, productive forest management, and biodiversity conservation (in East Africa) (Chirwa et al. 2015; Lwanga 2003). In both regions there are also important efforts to improve land tenure and access rights.

Table 5.3 Overview of restoration commitments and main technical interventions in the selected regions

Central America & Mexico	Trop. South America	West Africa	East Africa
Commitments			
• 16.5 Mha (7 countries) to the Bonn Challenge (BC) and Initiative 20x20 • Largest commitment: Mexico, 8.5 Mha	• Nearly 20 Mha (5 countries) to the BC and Initiative 20x20 • Largest commitment: Brazil, 12 Mha	• 37 Mha (12 countries) to AFR100 • Largest commitment: Mali, 10 Mha	• 39.3 Mha (9 countries) to AFR100 • Largest commitment: Ethiopia, 15 Mha
Main technical interventions			
• Reforestation (tree plantations with native or exotic species) • Assisted natural regeneration (ANR) • Silvopastures (intensive and trees in line) • Agroforestry systems (permanent crops and trees, e.g., tree shade coffee; and annual crops and trees) • Grassland restoration • Avoided deforestation • Sustainable forest management (both in degraded and secondary forests)	• Reforestation (tree plantations using exotics and native species) • ANR • Agroforestry and silvopastoral systems • Ecological restoration • Activities to prevent deforestation (avoided deforestation)	• Agroforestry practices (native and exotic species for timber, fuelwood or fruit production) • Tree crop plantations (cocoa, coffee, rubber, oil palm, cashew) • FMNR	• Tree plantations (mainly using exotic species) • Natural regeneration • Area exclosures (practiced without additional measures of planting to allow natural regeneration)

Southeast Asia	South Asia	Central Asia	Eastern & Southeast Europe
Commitments			
• No regional targets, only national ones • Largest commitment: Indonesia, 29.3 Mha	• 22.65 Mha (5 countries) to the BC • Largest commitment: India, 21 Mha	• 2.39 Mha (4 countries) to the BC • Largest commitment: Kazakhstan, 1.5 Mha	• More than 4 Mha by 2030 (14 countries) to the BC* • Largest commitment: Ukraine 1.5Mha
Main technical interventions			
• Temporary moratorium on forest-clearing for logging or agricultural purposes in primary forest and peatlands • ANR • Tree plantations (monocultures with exotics, multi-species plantations, including enrichment planting) • Agroforestry (tree home gardens, farm forest) • Ecological restoration (e.g., framework method)	• Tree plantations (native and non-native tree seedlings) • ANR • Agroforestry systems • Bioengineering (such as in stabilizing areas affected by landslides)	• Tree plantations • Shelterbelts of fast-growing species • Windbreaks in agricultural areas • ANR, especially for degraded saxaul forests and for mountain forests) • Improved grazing management	• Forest management (improved silvicultural practices) • Afforestation • Reforestation • ANR

* **The new EU Nature Restoration Law will include commitments to EU member countries in this region** (https://ec.europa.eu/commission/presscorner/detail/en/ip_23_5662).

Source: Authors' compilation based on reports revised up to August 2021, for Eastern and Southeast Europe up to January 2023.

National policies that specifically target FLR can be found in countries that have been implementing restoration/rehabilitation for some time (for decades in the case of the Philippines, Colombia, and Ghana) or that have benefited from several years of international/regional support (e.g., Guatemala, Kenya). FLR strategies most commonly focus on productive restoration,[18] converting degraded lands/forests into productive systems (e.g., Mexico aims to do so with agroforestry and agroecology systems (Gobierno de Mexico 2019); Ethiopia, with small-scale and industrial forest plantations (Lemenih and Kass 2014)). In some countries, restoration policies and the implementation programs or plans are embedded in or linked to national policies/strategies on climate change (e.g., El Salvador, PRISMA 2015). Other countries have set restoration targets as part of their national REDD + strategies (e.g., Côte d'Ivoire articulates a multi-sectoral response to forest degradation and deforestation with emphases on zero-deforestation agriculture, sustainable forest management, and restoration of degraded forests (République de Côte d'Ivoire 2017)).

An important development observed in national strategies or plans is the addressing of tenure issues (for forest land), use rights for forest resources, and community participation in forest restoration and management. This is playing out to a greater degree in Sub-Saharan Africa and Asia (e.g., policy legislative reforms for tree tenure and benefit sharing in Ghana; land reclassification measures in Indonesia and Vietnam; and land registration to enhance tenure security in Ethiopia and the Philippines). In line with these strategies, governments in Southeast Asia are increasingly favoring community-based forestry (e.g., Philippines, Thailand, and Cambodia (FAO and RECOFTC 2016); also observed in South Asia (FAO 2019)).

Central Asia and Eastern & South-Eastern Europe countries are still in an early planning phase of FLR implementation, focusing on technical and financial support from international/regional initiatives and seeking to reinforce regional cooperation and capacity exchange.

Other policies have paved the way for economic compensation tools for forest restoration efforts, mainly in Latin America (e.g., Brazil,[19] Colombia,[20] Ecuador,[21] Guatemala[22]) and Southeast Asia (e.g., Vietnam[23] and the Philippines[24]).

A common policy strategy many countries use is the promotion of extensive (commercial/industrial) tree plantations (e.g., Brazil, Côte d'Ivoire, Ethiopia, Mozambique, the Philippines, and Pakistan) as well as agroforests and tree farms (e.g., Indonesia, Nigeria, and Vietnam).[25] The situation appears to be quite different in Central Asia and Eastern & South-Eastern Europe countries, where restoration is seen as part of standard forest management policies and regulations, including measures for water and soil protection.

5.3.2 Institutional capacities and effectiveness

Most countries reviewed here face challenges in governance structures and capacities, including for decentralized control and decision-making. Institutional ineffectiveness for FLR is a common issue (e.g., Guatemala[26]). This might be in large part because responsibilities for FLR are often split across sectors, which complicates coordination and policy alignment (e.g., Mexico[27] and Mozambique[28]). Moreover, poor institutional coordination across sectors and scales is common (e.g., Brazil[29] and Ethiopia[30]). A particular challenge resulting from poor intersectoral articulation is reflected in incoherence or lack of alignment between environmental and agricultural policies, often generating perverse incentives for deforestation. There is, however, good progress in intersectoral collaboration in some countries such as El Salvador,[31] Guatemala (FLR Roundtable),[32] Kenya (Landscape Restoration Technical Working Group), and the Philippines (National Convergence Initiative).

Furthermore, institutional responsibilities are not always well defined, and donor programs are not always well integrated into national efforts (e.g., Ethiopia). In Central Asia there is a lack of clarity over responsibilities regarding certain land types not covered by corresponding governmental entities.[33] Pakistan, in South Asia, shows a different situation: interprovincial coordination mechanisms at the national level were created by the government-funded "10-Billion-Tree Programme."[34] In Kenya, a county-based FLR Coordination Committee oversees county-level initiatives, which improves monitoring and linkages with the national level.

5.3.3 Economic incentives and investment environment

Most countries in the studied regions do not as yet have direct economic incentives for FLR, though several are in the planning stage. Latin America and Asia seem to be at the forefront, with operational public incentive schemes in some countries. In Central America & Mexico, incentives are mainly in the form of direct payments to landowners willing to protect, sustainably manage, or restore their land. Disbursements are made based on results. The region stands out with examples from Guatemala (with two primary policy instruments: PROBOSQUE, the incentive program for the establishment, recovery, restoration, management, production and protection of forests; and PINPEP,[35] the forestry incentive program for owners of small areas of land with forestry or agroforestry vocation), Mexico (several public programs finance restoration based on agroforestry, e.g., Sembrando Vida[36]), and El Salvador (incentives mainly focused on productive systems to integrate ecosystem service/biodiversity conservation).[37] In tropical South America and Southeast Asia, PES programs/mechanisms have been created and implemented to varying degrees, notably in Ecuador (Socio-Bosque program[38]), Colombia (BanCO2[39]), Brazil (e.g., Acre's State System of Incentives for Environmental Services[40]), Vietnam (Forest PES[41]), Cambodia[42] and the Philippines.[43]

Economic incentives for FLR are not yet directly available in West Africa (though in Ghana, the National Forest Plantation Development Program provides financial assistance to those investing in FLR, especially corporate investors), East Africa, Central Asia, or Eastern & South-Eastern Europe. The Philippines government currently provides diverse socioeconomic incentives, and others could be adapted to support restoration activities; whereas in Pakistan around half of the budgets of the two mega government projects ("10-Billion Tree" and "One-Billion Tree") are to be exclusively spent on FLR through the promotion of ANR and other interventions, and the incentives at the local level include providing seedlings, plantation costs, and compensation for watch wardens for grazing exclusion. Across the regions, government incentives support tree plantation, agroforestry, and silvopastoral systems, mostly through promotional policies, technical assistance, and tax relief (e.g., Mexico, Brazil, Kenya, and Indonesia).

The investment environment for FLR activities is highly variable. Most restoration-like initiatives have traditionally been carried out with public resources, while some countries (particularly in Central Asia and Eastern & South-Eastern Europe) are still striving to create a favorable environment for private investment. In Latin America, largely due to Initiative 20x20, there is a range of private impact investors closely collaborating with governments to invest in FLR.[44] In Central America & Mexico, economic incentive programs are attracting companies and private impact investors. El Salvador has public sector investment for financing restoration and private sector funds, including from cooperatives, coffee growers, organized farmers, and microenterprises (especially from the sugar cane sector), with a focus on restoration of basic grains, agroforestry (e.g., cocoa), and natural ecosystems (mangroves). In Colombia, government efforts are underway to attract investments in the climate-smart transformation of rural landscapes through the development of incentive mechanisms as well as public–private partnerships.[45]

A favorable financing environment for FLR is also present in Southeast Asia. The Philippines' NGP, for example, focuses on tree plantations as a strategy for climate change mitigation, poverty reduction, and opportunities for communities to develop social enterprises; furthermore, the government—with support from the international cooperation community—finances projects to primarily protect critical watersheds and coastal areas.[46] In Sub-Saharan Africa, public investments are often linked to the establishment of commercial forest plantations: in Ethiopia the current investment policy offers tax incentives for establishing large-scale plantations, but corresponding incentives are not extended to private farmers or communities establishing woodlots; in Ghana, FLR investments are primarily linked to REDD +, with a focus on increasing trees on farms and in cocoa landscapes.

The main funding source for some countries comes from international cooperation (e.g., Mozambique's Zambezia Integrated Landscape Management Program, funded by the World Bank,[47] and Uzbekistan's Regional Resilient Landscapes Restoration Program,[48] a joint project the government is developing with the World Bank.

5.3.4 Public–private partnerships and negotiating platforms

Public–private initiatives and partnerships constitute an array of other governance approaches for effective FLR implementation. A promising collaboration of private impact investors with governments is happening in some regions (notably in Latin America). Various international and regional initiatives are facilitating restoration projects, focusing on avoided deforestation, ecological restoration, agroforestry systems, reforestation, and secondary forest restoration as common business models. Several countries are actively engaged in REDD + initiatives as well as FLR-related projects (funded by GEF, GCF, IADB).[49] Plans to carry out large FLR programs in Southeast Asian countries are supported by regional organizations,[50] financial agencies (World Bank, ADB, GEF), and bilateral cooperation. For instance, Indonesia has large government programs for mangroves and peatlands and for restoration concessions. Myanmar has a 10-year government National Reforestation and Rehabilitation Program to improve the economic and environmental conditions of local communities.[51] The Philippines government has partnership arrangements with communities.[52] Vietnam has a pioneer policy on forest PES including watershed protection, landscape beauty protection, and biodiversity conservation for tourism.[53] In South Asia, there is a strong emphasis on and demonstrated success of participatory forestry and industry–farmer partnership arrangements in support of tree growing on farms.[54] In Central Asia, public–private investments are mainly developed for ecotourism and to finance the production of NWFP (nuts, fruits, aromatic plants).[55]

Some countries have developed national (and subnational) platforms for dialogue, negotiation, and action. These platforms have emerged more consistently in Latin America, notably in the case of Guatemala (National Roundtable for FLR[56]), Colombia (Colombian Network of Ecological Restoration[57]), and Brazil (Atlantic Forest Restoration Pact[58]). In West Africa, existing platforms are more limited in scope and range, such as the Cocoa and Forest Initiative of Cote d'Ivoire,[59] a public–private platform to design and monitor the implementation of the cocoa companies' FLR objectives. In East Africa, Kenya has a FLR Technical Working Group (supported by the Kenya Forest Service, Kenya Forestry Research Institute, and the Council of Governors) to coordinate FLR interventions at the decentralized level.[60] In South Asia, Pakistan has provincial and national-level platforms that discuss land and environmental issues, including FLR.[61] In Central Asia there is the Bonn Challenge Regional Ministerial Roundtable for Central Asia Desert Initiative.[62]

5.3.5 Capacity building, technical assistance, and extension services

Across the reviewed regions, capacity building, technical assistance, and/or extension services geared toward effectively implementing FLR commitments at national and subnational levels are often weak or not properly in place. Historically, these three institutional incentives have been directed to foster plantation forestry or forest rehabilitation, with massive investments particularly in parts of Asia and Africa.[63] Some countries have developed institutional and technical capacities to expand degraded (forest) land restoration/rehabilitation. In the Philippines, several government agencies provide training and assistance for FLR, often supported by internationally funded projects.[64] In Kenya, efforts are underway to promote synergies and complementarities among projects/programs/institutions working on capacity building, technical assistance, and extension services for FLR. In the studied tropical countries, quality extension services that target specific beneficiaries—often poor, small-scale farmers and rural communities—are often deficient or even lacking.

5.3.6 Access to technological innovation and research

Regional/country efforts to generate and facilitate access to technological innovation and research in support of FLR is quite varied, reflecting differences in institutional capacity (human and financial resources). In parts of tropical South America, Southeast Asia, and South Asia, there is a critical mass of national governmental and nongovernmental institutions, often supported by international cooperation, actively promoting and working on FLR implementation. Latin American countries (notably Brazil, Mexico, and Colombia) have networks of public and nongovernmental organizations in place, generating and actively disseminating innovations and research results in support of ecological and productive restoration in the most important biomes. The selected Asian regions (in the Philippines, Thailand, and Indonesia in particular) also exhibit a critical mass of institutional work producing valuable knowledge and tools for FLR implementation.

A common research interest among the reviewed countries is defining the best species for the varied sites in need of restoration (e.g., Pakistan, Uzbekistan), and in developing efficient technologies for seed production and nursery establishment and management (e.g., Ethiopia). Some countries have made efforts to explore the potential of lesser used and/or nontraditional species (Ghana), including in secondary forests (Brazil, Guatemala), and of technological innovations to add value to FLR products and services.

5.3.7 Restoration practices

Overall, the preferred restoration intervention has been the establishment of tree plantations (reforestation, afforestation), frequently using monocultures with exotic species (most common in parts of tropical South America, e.g., Brazil and Colombia; also in Southeast Asia, South Asia, and East Africa), in woodlots on small farmers' land for fuelwood or fruit production (West Africa) or to form shelterbelts and windbreaks on agricultural land (Central Asia) (see Table 5.3).

Agroforestry systems are extensively used in all tropical/subtropical regions, with diverse purposes and arrangements: trees associated with cash crops (mainly coffee, cocoa) are popular in the two Latin American regions and in West Africa, while farm forests or home gardens are traditional practices in Asia. Silvopastoral systems are particularly important in Latin America (Colombia, Brazil). ANR is also used in all the regions, notably in West Africa—where FMNR has been applied in extensive areas, e.g., Niger and Mali—Southeast Asia and South Asia—where ANR has been a common rehabilitation practice since the 1990s. Area enclosures to restrict human and animal interference to encourage natural regeneration is another preferred practice (mainly in East Africa).

Some regional approaches include the improvement of forest management practices and rehabilitation measures, particularly targeting community and smallholder forests (West Africa, Central America & Mexico, Central Asia, Eastern & South-Eastern Europe). Finally, measures to avoid deforestation and degradation are highlighted in a few FLR national strategies or plans of forest-rich tropical countries within the framework of national REDD + strategies and as part of adaptation actions to climate change with benefits to local communities (Central America & Mexico, tropical South America, East Africa, and Southeast Asia).

All of the above types of interventions fall within a general approach of favoring productive restoration/rehabilitation, which has been widely encouraged in most of the regions/countries, including as part of business models (with detailed technical and financial information available for investors, as in several Latin American countries). Ecological restoration continues to be a critically important approach in the studied regions but is not explicitly nor sufficiently reflected in the available documentation. The review of past restoration efforts shows the importance in many countries of interventions geared toward restoring original ecosystems/habitats (e.g., Colombia highlands,[65] Thailand with the Framework Method[66]).

5.4 Overview of regional variation and implementation

5.4.1 Common issues and regional differences

The eight reviewed world regions represent contrasting geographical, sociocultural, economic, political, and institutional contexts, but do share some common elements. As expected, the two Latin American regions share much in common, likewise for the African regions and the Asian ones (particularly Southeast Asia and South Asia). All these regions are primarily situated in the tropics and subtropics. The other studied regions, Central Asia and Eastern & South-Eastern Europe, are in the temperate zone and share some similarities.

The main variables where commonalities were found are: (i) institutional capacities and effectiveness for FLR planning and implementation; (ii) capacity building, technical assistance, and extension services; and (iii) the main technical interventions (restoration approaches and options). The major differences were found in: (i) enabling policies and regulations for FLR (in particular with regard to tenure issues); (ii) economic incentives and investment environment; (iii) public–private partnerships and negotiating platforms; and (iv) access to technological innovation and research.

Overall, the similarities and differences listed above are to be understood as factors that contribute to creating favorable conditions for the successful development of FLR processes at the national/subnational levels. In this regard, the reviewed Latin American regions stand out for making relatively good progress in some of the enabling factors, as shown in the previous section. Some countries in both regions have performed better on specific key factors (such as the enactment of policies that are effectively financed and applied), leading to tangible outcomes (area restored, money invested, carbon sequestered, jobs created, etc.). In the case of El Salvador, data from the Second Bonn Challenge progress report (Dave et al. 2019) shows that:

- between 2014 and 2019, a total of 238,950 ha were restored, representing 24% of the national commitment set for 2030;
- up to 2018, approximately 12,235 full-time equivalent (FTE) short-term jobs were provided by FLR implementation, while the management of the restored landscape created 2,715 FTE long-term jobs; between 2018 and 2019,

implementation and maintenance of restoration actions generated an additional 46,108 short-term FTE jobs and 68,984 long-term FTE jobs, respectively;
- from 2014 to 2019 restoration actions, mostly stemming from agroforestry activities, resulted in a total accumulated removal of 1,860,035 tons of CO_2 equivalent;
- between 2014 and 2017, 39% of the restored area was located in key biodiversity areas (a proportion that increased to 54% considering the areas restored between 2018 and 2019).

Unfortunately, comprehensive data on such outcomes are not available for very many countries, indicating the importance of monitoring these initiatives over time.

The FLR context and current progress reported in Southeast Asia and South Asia foster favorable conditions and take into account the important country differences in the policy environments, institutional capacities, and financing accessibility. Some Asian countries have accumulated considerable experience and capacity in dealing with large-scale (government-led) reforestation/rehabilitation projects, creating a comparatively good foundation on which to build their FLR agendas, which—as mentioned in Section 5.3.7—are increasingly promoting the involvement of communities and small-scale landholders (FAO 2019).

In Sub-Saharan Africa, the critical issues at stake for achieving the ambitious restoration commitments will depend on governance arrangements related to effective stakeholder participation, tenure and use rights, and institutional technical and financial capacities at different levels. For East Africa, the analysis points out the comparatively good progress in countries that have created favorable conditions for upscaling FLR targets (e.g., Ethiopia, Kenya). In the case of Ethiopia:

- the growing recognition of the need to restore degraded land (Lemenih and Kassa 2014) triggered by several factors including the scarcity of forest products (mainly woody biomass), has resulted in a strong political will to regreen the country;
- restoration is integrated in key national policies and strategies (the Climate Resilient Green Economy Strategy, the Forest Sector Development Program, and the REDD + Strategy), and the government has established such entities as the Environment Forest and Climate Change Commission and the National Forest Sector Development Programme, encouraging various restoration initiatives (Zeleke and Vidal 2020).

The Central Asia and Eastern & South-Eastern Europe regions are in the early stage of FLR planning and implementation; this is in part explained by their political history (referring to countries from the former Soviet Union) and the nature and main uses of their forest resources, including the fact that planted forests account for a significant proportion of their forests.

5.4.2 Grouping countries according to stages of implementation

From the reviewed information and the knowledge and experience from source persons familiar with the selected countries, we identify distinct situations and degrees of progress with regard to FLR implementation, allowing us to tentatively group the reviewed countries into three broad stages:

- Early stage: FLR planning is in progress; country efforts focus on creating necessary implementation conditions. Countries: Côte d' Ivoire (West Africa), Mozambique (East Africa), Cambodia (Southeast Asia), Uzbekistan (Central Asia), and Ukraine and Macedonia (Eastern & South-Eastern Europe);
- Intermediary stage: FLR implementation is progressing as necessary conditions are increasingly being improved nationwide. Countries: Mexico (Central America & Mexico), Brazil, Colombia, and Peru (tropical South America), Ghana (West Africa), Ethiopia (East Africa), and Pakistan (South Asia);
- Advanced stage: FLR implementation is well in progress after having made good strides with some key conditions, and various facilitating factors (e.g., economic incentives) are well in place, nationwide. Countries: Guatemala and El Salvador (Central America & Mexico), Kenya (East Africa), and the Philippines (Southeast Asia).

This grouping reflects to some extent the large regional differences highlighted in the previous section. The six Latin American countries (Brazil, Colombia, El Salvador, Guatemala, Mexico, and Peru) are comparatively ahead in the implementation of their FLR agendas compared with other regions included in this analysis. The small countries of El Salvador and Guatemala are classified in the advanced stage because the evidence shows they have a number of instrumental conditions and tools for steering FLR processes (e.g., enabling policies, functional stakeholder platforms, public economic incentives, continued capacity building linked with practical-oriented knowledge products and extension services and operating monitoring systems), guided, respectively by a national program and strategy with several years of implementation with notable technical and financial support from international cooperation agencies.

The five African and five Asian countries are spread over the three stages, reflecting differences in their past trajectories: strong government efforts to rehabilitate degraded forest lands and establish industrial forest plantations, generally with strong support from the international cooperation (e.g., the case of the Philippines), governance and institutional structures and capacities, handling of tenure issues, community participation, and conflict negotiation, among other critical aspects for FLR implementation. That Central Asia and Eastern & South-Eastern Europe countries have most recently entered the FLR movement might explain their being in the early stage, though their particular contexts—mostly in the temperate domain and with strong central governments—likely also shapes their FLR processes.

5.5 Main challenges and key lessons

The analyzed regions share challenges in implementing regional/national FLR agendas, resulting from policy, institutional, governance, financing, and/or technical deficiencies. Based on the extensive review and the insights from the authors and source persons consulted, Table 5.4 summarizes the challenges we highlight for each region.

Table 5.4 Some challenges for FLR implementation identified in the reviewed regions

Central America & Mexico	Tropical South America	West Africa	East Africa
• *Financing restoration on the ground*: need to develop carbon markets for small and medium landholders to receive additional income • *Setting the right incentives*: need to align incentives in both the agricultural and environmental sectors and develop coherent policies • *Shifting behaviors*: need for changes in practices for key commodity production such as sugar cane and palm oil • *Investing in research* to improve efficiency and understand the impacts of different restoration practices • *Building up capacity* around FLR at all levels and developing communication strategies • *Empowering a regional movement*: convening governments, regional platforms and national stakeholders to keep the engagement for FLR implementation	• *Integrating policies, institutions* and relevant stakeholders across sectors and jurisdictions to support FLR • *Strengthening current incentives* for native forest protection, forestry and restoration • *Promoting rural extension agents* as stakeholders to support small landholders and local communities in the adoption of sustainable production practices and accessing suitable funding opportunities • *Strengthening security of land tenure* to include various forms of tenure in legal frameworks that promote FLR • *Increasing capacities* (organizational, financial, and technical) that enable FLR implementation	• *Applying a thorough approach in the intensively used drylands* in northern West Africa through agroecological measures, forest rehabilitation and conservation, and *support for diversification and promotion of resilient species* • *Ensuring fair access to land*, particularly wherever animal husbandry comes into play (need of "One Health")[a] • *Making use of opportunities for (cocoa) smallholders to be included in government plans and receive financing* for FLR initiatives in the more humid areas in the southern part of West Africa, where climate change is forcing (cocoa) farmers to shift production into remaining intact forest areas	• *Filling knowledge gaps* on the factors affecting the process (ecological, climatic, biological, social) under different contexts • *Addressing issues with land access and security* in state-owned land by strengthening local communities' effective participation in the process of land allocation (e.g., to commercial tree planting) and fostering positive interactions with government and/or private entrepreneurs • *Promoting sustainable production and value chain development for fuelwood and charcoal industries* • *Promoting forest restoration as an important strategy to increase resilience to extreme climatic events*, particularly droughts

Southeast Asia	South Asia	Central Asia	Eastern & South-Eastern Europe
• *Improving policies related to tenure and access to and use rights* of natural forests • *Strengthening institutions and attention to social issues* required to deal with forest restoration • *Shifting restoration efforts to wider targets/objectives* (such as biodiversity conservation, carbon sequestration, watershed protection, food security and improvement of livelihoods) • *Creating favourable conditions for wider stakeholder participation* • *Diversifying funding* for FLR (currently heavily dependent on public funding)	• *Addressing conflicting interests among stakeholders*, e.g., conservationists vs. private companies vs. local communities • *Addressing costs and demand for labour* as significant barriers • *Addressing high demand for forest products and forest land* driven by high population density • *Reversing environmental degradation* in the drylands • *Improving awareness of and societal demand for environmental services*, given the continuing high human pressure on forest resources • *Strengthening long-term political will* for restoration	• *Addressing conflict with other land uses, including pastures and agriculture*, by shifting the focus of FLR efforts on less densely populated areas or using agroforestry/windbreaks • *Enabling sustainable harvesting practices* in areas where local communities depend on NWFP • *Making FLR economically viable*, and enhancing local community access to reforestation/afforestation and harvesting of forest products	• *Addressing conflict with other sectors* because of contradictory or *poorly aligned policies* and structural administrative & technical problems (different views between state forest and pastureland managers vs municipalities and communities that manage land) • *Addressing land use issues* as a result of infrastructure projects affecting forest, agriculture and grazing • *Encouraging negotiation of divergent views on the use of natural resources* between state agencies, small-scale forest and landowners • *Financing* for FLR projects

[a] *One Health* is a collaborative, multisectoral, and transdisciplinary approach—working at the local, regional, national, and global levels—with the goal of achieving optimal health outcomes recognizing the interconnection between people, animals, plants, and their shared environment. Source: https://www.cdc.gov/onehealth/index.html#:~:text=One%20Health%20is%20a%20collaborative,plants%2C%20and%20their%20shared%20environment.

The *need for policy integration and institutional articulation across sectors is called for in the face of contradictory or poorly aligned policies* related to land use, land access, tenure and resource use rights, and security (Central America & Mexico, tropical South America, East Africa, Southeast Asia, Central Asia, Eastern & South-Eastern Europe), reflecting conflicts and power imbalances among stakeholders and the governments of various countries (e.g., land use issues with infrastructure projects in Eastern & South-Eastern Europe). The *need to align incentives from agricultural and environmental sectors* is particularly highlighted in Central America & Mexico, tropical South America, West Africa and Southeast Asia. As such, we propose that the *application of current incentives should be revised* by increasing public funding for environmental interventions so as to make native forest protection and restoration more attractive than productive interventions (mostly commercial timber plantations).

Governance issues are explicitly identified in some of the regions, particularly in countries where communities and smallholders are key targets of the restoration agendas. In Southeast Asia, this is considered the most critical factor, highlighting the need to *create favorable conditions for wider stakeholder participation*. Other important issues include conflicting stakeholder interests over land use (South Asia, Central Asia) and natural resource use (Eastern & South-Eastern Europe).

Financing is a recurrent challenge, as expected. The institutional, human, and financial capacities for implementing national plans/strategies fall short of the ambitious country commitments. The traditionally strong government role in addressing land and forest degradation, frequently by promoting (extensive) tree planting and soil and water conservation measures, is now faced with increasingly more complex social, environmental, and technical tasks, exacerbated by climate change, rising population pressures, and rural-to-urban migration. In East Africa, extreme climatic events (droughts in particular) are forcing the vulnerable to turn to forests for timber, fuelwood, food, and fodder. High population density in South Asia is driving higher demands for forest products and forest land (for fuelwood, grazing, etc.).

Other challenges distinctively identified and shared in the regions seek to *address weaknesses in capacities* (organizational, technical, administrative, entrepreneurial), extension work, research, technological innovation, and communication. These are part of the set of conditions and tools required for successful FLR implementation.

Other regional challenges highlighted include how restoration targets are defined (Southeast Asia), which approaches to use in diverse landscapes (West Africa), stakeholder needs/demands (West Africa, Central Asia, South Asia), the lack of enough knowledge for restoration (e.g., in view of extreme climatic events (East Africa), and the financial feasibility of restoration (Central Asia, South Asia).

Learning from current and past efforts, connecting the context in which they occur(ed) with the preconditions that influence(d) the processes and the challenges to achieve desired outcomes, can provide valuable insights for decision-makers into the factors or elements contributing to effective/successful FLR implementation on the ground. We highlight the following critical lessons as a way forward:

Align FLR targets with national priorities. All regions are being supported internationally in the development of their (national) FLR agendas, though with varied intensity in terms of technical and financial resources. The rationale for FLR also varies in view of national priorities and how these are set. On the one hand, FLR is politically or technically devised at the country level primarily to address increasing deforestation and land degradation (with all the ecological, economic, and social impacts this brings) and/or climate change (with more frequent natural disasters, fires, landslides, etc.) or to recover land productivity and the supply of forest and agricultural products. The case of El Salvador offers an important lesson in this regard, illustrating that FLR requires starting from the needs of livelihoods and national priorities (food and water security, provision of energy for firewood; and conservation/recovery of biodiversity, jobs, and income) in order to guide the incorporation of trees and forests in different mosaics of land uses, including protected areas and biological corridors. On the other hand, FLR is seen to bring opportunities and potential synergies with other international agendas (e.g., poverty alleviation, zero deforestation, biodiversity conservation, land degradation neutrality, REDD +) and with this comes access to international funding and public and private investments. In Côte d'Ivoire, for instance, one of the main finance opportunities for FLR came from the restoration of cocoa landscapes and the commitments made by the Ivorian government and the cocoa industry to produce and trade deforestation-free cocoa. A message here is for governments to identify opportunities to channel investments and support from local and international companies into FLR priorities.

Ensure coherent policies conducive to investments in FLR. The importance of enabling policies and legal frameworks to promote FLR implementation needs to be stressed as this (pre-) condition plays a key role in attracting public and private investments. A first consideration is to devise restoration policies framed with relevant existing policies, as shown by the experience from Peru (with public restoration policies developed in synergy with other policies such as water resource management, family farming, environmental impact assessment, protected natural areas, environmental education). The case of Guatemala shows that the alignment of agriculture and environmental polices is key in ensuring complementarity and avoiding perverse incentives. The need for integrating the FLR agenda into the national agricultural policy is highlighted in East Africa: a lesson learned in Ethiopia is that the devolution of property rights must be accompanied by other policy changes and reforms in the agricultural sector to serve as a catalyst and provide incentives for participation in FLR initiatives. A lesson from Kenya is that an enabling policy framework is needed also to encourage increased private sector engagement in restoration through corporate social responsibility, PES schemes, and tax incentives linked to sustainability.

Make the FLR process genuinely participatory. A recurrent message from the experiences and lessons learned at the country level is that the inclusive and effective participation of key stakeholders in FLR planning and implementation is critical. For example, in Ghana, key actors participating in restoration processes include district assemblies, traditional authorities, opinion leaders, and local community groups including women and youth. The main lessons coming from the Philippines and Pakistan stress the importance of engaging and empowering communities at the onset of the process. In Mozambique and Ukraine, decentralized participatory approaches that include local communities in implementation are key for FLR initiatives.

Support the generation of short- and long-term economic benefits from community FLR efforts. A stronger case for restoration investment needs to be made to leaders at the country level by determining and pointing out the economic and social returns of FLR approaches in different contextual conditions. Lessons learned point to the need to support community efforts to restore degraded forests/lands with economic incentives targeting short- to long-term benefits. For example, a priority in Kenya's FLR plan is the integration of bio-enterprise development with landscape restoration to provide short-term benefits to the local communities as they await the long-term benefits of restoration (e.g., through the compensation for community labor inputs, sale of NWFP, PES for communal land, and conservation easements). Such efforts point to the importance of promoting restoration activities that generate economic returns with products of high added value.

The effective and successful implementation of the ambitious regional/country commitments and targets calls for more integrated and sustained efforts from key national, regional, and international decision-makers to have the above conditions (and others) strongly in place in the short term. When contrasting historical reforestation/rehabilitation trends with current ambitious targets/commitments, a huge gap becomes evident. Closing this gap requires establishing a comprehensive set of preconditions and tools to initiate, consolidate, and achieve the desired FLR outcomes to the benefit of key stakeholders.

Notes

1. Land degradation may be defined as long-term loss of ecosystem function and productivity caused by disturbances from which land cannot recover unaided (Bai et al. 2008).
2. Degraded forest land is defined by ITTO (2002, 10) as "former forest land severely damaged by the excessive harvesting of wood and/or non-wood forest products, poor management, repeated fire, grazing or other disturbances or land-uses that damage soil and vegetation to a degree that inhibits or severely delays the re-establishment of forest after abandonment."
3. Forest ecosystem restoration "involves returning trees to former forest land and improving the condition of degraded forests. As well as planting native tree species, it can include conserving wild plants and animals and protecting the soils and water sources that are part of the forest ecosystem." Source: https://www.decadeonrestoration.org/types-ecosystem-restoration/forests.
4. Land restoration or rehabilitation is the process of ecological restoration of a site to a natural landscape and habitat, safe for humans, wildlife, and plant communities. Source: https://knowledge.unccd.int/drought-toolbox-action/land-restoration
5. A country-led effort to bring 100 Mha of land in Africa into restoration by 2030. The initiative is a contribution of the African Union to the Bonn Challenge and the Land Degradation Neutrality targets, and the partnership consists of 31 African governments and numerous technical and financial partners. Source: https://afr100.org).
6. Initiative 20x20 includes 18 Latin American and Caribbean countries and three regional programs committed to protecting and restoring more than 52 Mha of degraded land by 2030. Source: https://initiative20x20.org/restoration-projects.
7. The authors compared country-level results to the commitments of 48 countries in the Bonn Challenge, finding that "approximately 10% of countries have committed to restoring an area of land that considerably exceeds the total area that is available for restoration. By contrast, over 43% of the countries have committed to restore an area that is less than 50% of the area available for restoration" (Bastin et al. 2019, 77).

8. We added the economic function to the original definition from the Global Partnership on Forest and Landscape Restoration (Bessau et al. 2018, 18) as we believe it is also important to consider in many landscape restoration efforts.
9. https://doi.org/10.4060/ca9825en
10. Countries were selected in consideration of their efforts in FLR implementation and considering the knowledge and expertise of chapter contributors and country collaborators.
11. SE Asia was one of the global hotspots of forest loss in the period 1990–2000. Since then, however, there has been a reversing trend. In the period 2015–2020, two countries recorded net gains in forest cover: Viet Nam and the Philippines (FAO 2020).
12. Over 90% of the forests in northern West Africa are in a stage of degraded natural or secondary forests, while only about 10% of intact forests remained in 2010. In the humid part of the region, forests are highly fragmented and degraded (Blaser et al. 2011).
13. https://forestsnews.cifor.org/19171/west-africa-adopts-regional-approach-to-manage-forest-ecosystems?fnl=; https://eros.usgs.gov/westafrica/land-cover/deforestation-upper-guinean-forest
14. https://www.worldometers.info/world-population/eastern-africa-population/; https://www.afdb.org/fileadmin/uploads/afdb/Documents/Publications/2019AEO/REO_2019_-_East_Africa_.pdf
15. ECCA30 is a country-led initiative to bring 30 million hectares of degraded and deforested land in Europe, the Caucasus and Central Asia into restoration by 2030. Source: https://www.bonnchallenge.org/sites/default/files/resources/files/%5Bnode%3Anid%5D/ECCA30.pdf.
16. A blend of active planting and passive restoration, where local people intervene to help trees and native vegetation naturally recover by eliminating barriers and threats to their growth, leaning on their knowledge of the land and on ancestral traditions. Source: https://www.wri.org/insights/what-assisted-natural-regeneration-benefits-definition.
17. FMNR is a low-cost land restoration technique used to combat poverty and hunger among poor farmers by increasing food and timber production and resilience to climate extremes. Source: https://www.wvi.org/sites/default/files/2019-12/FMNR%20Publication%203Dec_Online_0.pdf.
18. A means to harmonize decisions and tensions about land use; where the recovery of ecological integrity, economic results, and the generation of local benefits are facilitated. Source: comunicacionrm.wixsite.com.
19. As summarized in FAO and RECOFTC (2016), p. 36, "[SE Asian] countries that have shown progress are the ones that have addressed the issues of tenure, use rights of forest resources and community participation, moving forestry from heavy state control to management increasingly undertaken by the private sector, individuals or corporations".
20. Murcia et al. (2017).
21. un-redd.org.
22. PROBOSQUE Law Decree 2-2015. ecologic.org.
23. Pham et al. (2013).
24. https://www.worldagroforestry.org/blog/2020/02/05/payments-ecosystem-services-philippines-win-win-approach-communities-businesses-and.
25. The analysis from Lewis et al. (2019) of current national restoration commitments made under the Bonn Challenge and other initiatives by 43 countries wholly or partly located in the tropics revealed that two-thirds of the area (292 Mha) was committed to global reforestation for carbon storage via tree plantations, mainly monocultures (45%) and agroforestry (21%). See also Péroches et al. (2022).
26. Although different policies in support of FLR exist, there are still gaps in the coordination with other sectors. Being agriculture the main driver of deforestation, more coherence in

policies is needed to avoid perverse or negative incentives that counteract the effect of FLR policies.

27. There is little articulation between the environmental sector (SEMARNAT, CONAFOR) and the other ministries (agriculture, welfare). In particular, the prominent role assumed by the Ministry of Welfare through its multi-billion dollar Sembrando Vida restoration program, requires stronger coordination with SEMARNAT and CONAFOR.
28. While forest restoration activities have been part of the core business of the National Directorate of Forests, this has been changing from the Ministry of Agriculture and Rural Development to the Ministry of Land and Environment. Environmental issues, including climate change, deforestation and forest restoration, have also been promoted by the National Directorate of Environment at the Ministry of Land and Environment. A technical unit at the National Fund for Sustainable Development (FNDS) has been established to implement the World Bank projects. This mixture of institutions makes institutional coordination a highly complicated task.
29. May et al. (2011).
30. Bekele et al. (2015).
31. This is the case of the Ecosystem Restoration Roundtable, an intersectoral, multi-stakeholder consultative platform functioning (2014–2019) as the operational body for consultation and actions for the restoration of ecosystems and landscapes. This coordination has allowed convening of specific sectoral groups for dialogue, mainly in the coffee, sugarcane, and cocoa production sectors, with unions and producers, the Biosphere Reserve, as well as the table of Indigenous Peoples.
32. FLR activities are coordinated by and discussed at this National Restoration Roundtable, a multi-stakeholder group established to design and support the National Forest Landscape Restoration Strategy.
33. FAO and UNECE (2019).
34. http://mocc.gov.pk/ProjectDetail/M2QzOWJmMjUtZTU3MC00NmFkLWE4YmMtZDFhMmRlOGU2NGRh?gdxufbjzynhojsia?vqnynhfbcvxlnydq.
35. http://portal.inab.gob.gt/index.php/component/content/article/112-servicios/182-probosque?Itemid = 437.
36. https://www.gob.mx/bienestar/acciones-y-programas/programa-sembrando-vida.
37. Ministerio de Medio Ambiente y Recursos Naturales.
38. Murcia et al. (2017); https://www.ambiente.gob.ec/programa-socio-bosque/.
39. Murcia et al. (2017); Banco2–BancO2.
40. Duchelle et al. (2014).
41. Pham et al. (2013, 2021).
42. Milne and Chervier (2014).
43. https://www.worldagroforestry.org/blog/2020/02/05/payments-ecosystem-services-philippines-win-win-approach-communities-businesses-and.
44. initiative20x20.org.
45. World Bank (2017).
46. FAO and RECOFTC (2016).
47. https://www.worldbank.org/en/programs/mozambiques-integrated-forest-and-landscape-management-portfolio.
48. https://unece.org/sites/default/files/2020-12/FR-Finacing%20WB-Nemova-Agostini.workshop%209%20December%202020.pdf.
49. Global Environmental Facility, Green Climate Fund, Inter-American Development Bank.
50. APFNet and AFoCO.
51. https://www.iucn.org/news/forests/201805/restoring-experience-myanmar#:~:text = As%20part%20of%20this%20process,environmental%20conditions%20of%20local%20communities.

52. FAO and RECOFTC (2016).
53. Pham et al. (2013, 2021).
54. FAO (2019).
55. FAO and UNECE (2019).
56. Mesa Nacional de Restauración del Paisaje Forestal de Guatemala (2018).
57. https://redcre.com/sobrelared/.
58. https://www.bonnchallenge.org/pledges/brazils-atlantic-forest-restoration-pact.
59. https://www.worldcocoafoundation.org/initiative/cocoa-forests-initiative/.
60. https://www.kefri.org/assets/publications/tech/Landscape%20Restoration%20in%20Kenya2.pdf.
61. https://www.fao.org/in-action/forest-landscape-restoration-mechanism/our-work/countries/pakistan/ru/.
62. https://unece.org/forests/events/ministerial-roundtable-forest-landscape-restoration-and-bonn-challenge-caucasus-and.
63. See, for instance, Chamshama & Nwonwu (2004), Szulecka et al. (2014), Péroches et al. (2022).
64. Rebugio et al. (2017).
65. Vargas Ríos et al. (2012), MADS (2015), Murcia et al. (2017).
66. This method basically consists of using mostly pioneer species to initiate successional development and more complex planting and direct seeding schemes (FAO and RECOFTC 2016).

References

Adeleke, W. 2018. "Regaining Ecological Functions and Enhancing Livelihood through Forest Landscape Restoration in West Africa: The West Africa Biodiversity and Climate Change (WABiCC) Approach." *Nature & Faune* (FAO) 32 (1): pp. 16–22.

Bai, Z. G., D .L. Dent, L. Olsson, and M. E. Schaepman. 2008. "Proxy Global Assessment of Land Degradation." *Soil Use and Management* 24: pp. 223–234.

Baker, S., and K. Eckerberg. 2016. "Ecological Restoration Success: A Policy Analysis Understanding." *Restoration Ecology* 24 (3): pp. 284–290.

Bampton, J. F. R., A. Ebregt, and M. R. Banjade. 2007. "Collaborative Forest Management in Nepal's Terai: Policy, Practice and Contestation." *Journal of Forest and Livelihood* 6 (2): pp. 30–43.

Bastin, J.- F., Y. Finegold, C. Garcia, et al. 2019. "The Global Tree Restoration Potential." *Science* 365: pp. 76–79.

Bekele, M., Y. Tesfaye, Z. Mohammed, et al. 2015. "The Context of REDD+ in Ethiopia: Drivers, Agents And Institutions." DOI: 10.17528/cifor/005654.

Besseau, P., S. Graham, and T. Christophersen (eds.). 2018. *Restoring Forests and Landscapes: The Key to a Sustainable Future.* Vienna: Global Partnership on Forest and Landscape Restoration.

Blaser, J., A. Sarre, D. Poore, and S. Johnson. 2011. "Status of Tropical Forest Management 2011." *ITTO Technical Series* no. 38: p. 420. Yokohama: International Tropical Timber Organization.

Boucher, D., P. Elias, K. Lininger, C. May-Tobin, S. Roquemore, and E. Saxon. 2011. *The Root of the Problem - What's Driving Tropical Deforestation Today?* Cambridge: Union of Concerned Scientists.

Bromhead, M.-A. 2012. *Forest, Trees, and Woodlands in Africa: An Action Plan for World Bank Engagement.* Washington: World Bank. https://openknowledge.worldbank.org/handle/10986/11927.

Burton, P. J., and S. E. MacDonald. 2011. "The Restorative Imperative: Challenges, Objectives and Approaches to Restoring Naturalness in Forests." *Silva Fennica* 45 (5): pp. 843–863. DOI:10.7939/R34746R3T.

Cabrera, E., G. Galindo, J. González, et al. 2020. "Colombian Forest Monitoring System. Assessing Deforestation in an Environmental Complex Country." In *Forest Degradation Around the World*, edited by M. N. Suratman, Z. A. Latif, G. de Oliveira, et al., pp. 1–18. IntechOpen. DOI:10.5772/intechopen.77433.

Carvalho, W. D., K. Mustin, R. R. Hilário, I. M. Vasconcelos, V. Eilers, and P. M. Fearnside. 2019. "Deforestation Control in the Brazilian Amazon: A Conservation Struggle Being Lost as Agreements and Regulations Are Subverted and Bypassed." *Perspectives in Ecology and Conservation* 17: pp. 122–130. DOI:10.1016/j.pecon.2019.06.002.

Caspari, T., S. Alexander, B. T. Brink, and L. Laestadius. 2013. Review of Global Assessments of Land and Ecosystem Degradation and Their Relevance in Achieving the Land-Based Aichi Biodiversity Targets. A Technical Report Prepared for the Secretariat of the Convention on Biological Diversity (Draft). Draft. https://www.isric.org/sites/default/files/cop-12-inf-18-en.pdf.

Chamshama, S. A. O., and F. O. C. Nwonwu. 2004. Forest Plantations in Sub-Saharan Africa. Report Prepared for the Project Lessons Learnt on Sustainable Forest Management in Africa. https://energypedia.info/images/f/fe/Forest_Plantation_in_Sub_Saharan_Africa.pdf.

Chazdon, R. L., P. H. S. Brancalion, D. Lamb, L. Laestadius, M. Calmon, and C. Kumar. 2017. "A Policy-Driven Knowledge Agenda for Global Forest and Landscape Restoration." *Conservation Letters* 10 (1): pp. 125–132.

Chazdon, R. L., T. Gutierrez, P. H. S. Brancalio, L. Laestadius, and M. R. Guariguata. 2020. "Co-Creating Conceptual and Working Frameworks for Implementing Forest and Landscape Restoration Based on Core Principles." *Forests* 11: p. 706. doi:10.3390/f11060706.

Chirwa, P. W., M. Larwanou, S. Syampungani, and F. D. Babalola. 2015. "Management and Restoration Practices in Degraded Landscapes of Eastern Africa and Requirements for Up-Scaling." *International Forestry Review* 17 (S3): pp. 20–30.

Curtis, P.G., C. M. Slay, N. L. Harris, A. Tyukavina, and M. C. Hansen. 2018. "Classifying Drivers of Global Forest Loss." *Science* 361: pp. 1108–1111.

DAR. 2015. "La transformación del bosque. Titulación de predios y cambio de uso de suelos en la Amazonía peruana." Lima: Derecho, Ambiente y Recursos Naturales.

Dave, R., C. Saint-Laurent, L. Murray, et al. 2019. *Second Bonn Challenge Progress Report. Application of the Barometer in 2018.* Gland: IUCN.

Duchelle, A. E., M. Greenleaf, D. Mello, M. F. Gebara, and T. Melo. 2014. "Acre's State System of Incentives for Environmental Services (SISA), Brazil." In *REDD+ on the Ground: A Case Book of Subnational Initiatives across the Globe*, edited by E. O. Sills et al., pp. 33–50. Bogor: CIFOR.

FAO. 2011. *Southeast Asia Regional Report. Asia-Pacific Forestry Sector Outlook Study II. Asia-Pacific Forestry Commission.* RAP PUBLICATION 2010/20. Bangkok: FAO Regional Office for Asia and the Pacific.

FAO. 2019. *Forest Futures – Sustainable Pathways for Forests, Landscapes and People in the Asia-Pacific Region*. Asia-Pacific Forest Sector Outlook Study III. Bangkok: FAO

FAO. 2020. *Global Forest Resources Assessment 2020: Main Report*. Rome: FAO. doi:10.4060/ca9825en.

FAO and APFNet. 2018. Regional Strategy and Action Plan for Forest and Landscape Restoration in Asia-Pacific: Promoting and Accelerating Forest and Landscape Restoration to Enhance Ecological Functioning and Human Well-Being in Degraded and Deforested Landscapes of the Asia-Pacific Region. Bangkok: FAO and APFNet.

FAO and RECOFTC. 2016. Forest Landscape Restoration in Asia-Pacific Forests. Bangkok: FAO and RECOFTC.

FAO and UNECE. 2019. Forest Landscape Restoration in the Caucasus and Central Asia. Background Study for the Ministerial Roundtable on Forest Landscape Restoration and the Bonn Challenge in the Caucasus and Central Asia (21–22 June 2018, Astana, Kazakhstan). ECE/TIM/DP/72 Forestry and Timber Section, Geneva, Switzerland. Geneva Timber and Forest Discussion Paper 72. New York: UN.

FAO and UNECE. 2021. Forest Landscape Restoration in Eastern and South-East Europe. Background study for the Ministerial Roundtable on Forest Landscape Restoration and the Bonn Challenge. United Nations and the Food and Agriculture Organization of the United Nations. Geneva Timber and Forest Discussion Paper 87. Geneva.

Fearnside, P. M. 2017. "Deforestation of the Brazilian Amazon." In *Oxford Research Encyclopedia of Environmental Science*, edited by H. Shugart, pp. 1–53. New York: Oxford University Press. DOI:10.1093/acrefore/9780199389414.013.102.

Fedele, G., C. I. Donatti, I. Bornacelly, and D. G. Hole. 2021. "Nature-Dependent People: Mapping Human Direct Use of Nature for Basic Needs across the Tropics." *Global Environmental Change* 71: p. 102368. DOI:10.1016/j.gloenvcha.2021.102368.

Gann, G. D., T. Mcdonald, B. Walder, et al. 2019. "International Principles and Standards for the Practice of Ecological Restoration." *Restoration Ecology* 27 (S1): pp. S1–S46. DOI:10.1111/rec.13035.

Gobierno de Mexico. 2019. Plan Nacional de Desarrollo 2019–2024. Ciudad de Mexico.

Gomes César, R., L. Belei, C. Giudice Badari, et al. 2020. "Forest and Landscape Restoration: A Review Emphasizing Principles, Concepts, and Practices." *Land* 10 (1): 28. DOI:10.3390/land10010028.

Hosonuma, N., M. Herold, V. De Sy, et al. 2012. "An Assessment of Deforestation and Forest Degradation Drivers in Developing Countries." *Environ. Res. Lett.* 7 (4): 044009. DOI:10.1088/1748-9326/7/4/044009.

IPBES. 2018. *Summary for Policymakers of the IPBES Assessment Report on Land Degradation and Restoration*, edited by L. Montanarella, R. Scholes, and A. Brainich. Bonn: Secretariat of the Intergovernmental Science-Policy Platform on Biodiversity and Ecosystem Services. DOI:10.5281/zenodo.3237392.

ITTO. 2002. *ITTO Guidelines for the Restoration, Management and Rehabilitation of Degraded and Secondary Tropical Forests*. ITTO Policy Development Series 13.

Kissinger, G., M. Herold, and V. De Sy. 2012. *Drivers of Deforestation and Forest Degradation. A Synthesis Report for REDD+ Policymakers*. Vancouver: Lexeme Consulting.

Laestadius, L., S. Minnemeyer, and A. Leach. 2012. *Assessment of Global Forest Degradation: Report to PBL*. Washington: World Resources Institute.

Lemenih, M., and H. Kassa. 2014. "Regreening Ethiopia: History, Challenges and Lessons." *Forests* 5: pp. 1896–1909.

Lewis, S. L., C. E. Wheeler, E. T. A. Mitchard, and A. Koch. 2019. "Regenerate Natural Forests to Store Carbon." *Nature* 568: pp. 25–28.

Lwanga, J.S. 2003. "Forest Succession in Kibale National Park, Uganda: Implications for Forest Restoration and Management." *African Journal of Ecology* 41: pp. 9–22.

MADS. 2015. Plan Nacional de Restauración: restauración ecológica, rehabilitación y recuperación de áreas disturbadas. Bogotá: Ministerio de Ambiente y Desarrollo Sostenible. minambiente.gov.co.

Maginnis, S., and W. Jackson. 2002. "Restoring Forest Landscapes." *ITTO Tropical Forest Update* 12 (4): pp. 9–11. https://www.itto.int/files/user/tfu/back_issues_pdf/TFU.2002.04.English.pdf?v=1235114658

Mansourian, S., N. Dudley, and D. Vallauri, 2017. "Forest Landscape Restoration: Progress in the Last Decade and Remaining Challenges." *Ecological Restoration* 35: pp. 281–288. DOI:10.3368/rec. 35.4.281.

Mansuy, N., P. J. Burton, J. Stanturf, et al. 2020. "Scaling Up Forest Landscape Restoration in Canada in an Era of Cumulative Effects and Climate Change." *Forest Policy and Economics* 116: p. 102177. DOI:10.1016/j.forpol.2020.102177.

May, P. H., B. Millikan, and M. F. Gebara. 2011. *The Context of REDD+ in Brazil: Drivers, Agents and Institutions. Occasional paper* 55. 2nd ed. Bogor: CIFOR.

Mesa Nacional de Restauración del Paisaje Forestal de Guatemala. 2018. "Oportunidades de restauración del paisaje forestal en Guatemala." https://www.inab.gob.gt/images/publicaciones/guatemala_oportunidades_de_restauracion_en_guatemala.pdf.

Milne, S., and C. Chervier. 2014. *A review of payments for environmental services (PES) experiences in Cambodia. Working Paper* 154. Bogor: CIFOR.

Misra, V. K., and S. N. Shabbeer. 2001. "Participatory Forest Management in India: An Overview and Some Emerging Issues." *Social Change* 31 (1&2): pp. 156–164.

Murcia, C. 2018. *Retos y Oportunidades de la Restauración del Paisaje Forestal en Colombia.* Quito: UICN-América del Sur. https://infoflr.org/sites/default/files/2020-04/challenges_and_opportunities_for_forest_landscape_restoration_in_colombia.pdf.

Murcia, C., M. R. Guariguata, M. Peralvo, and V. Gálmez. 2017. *La restauración de bosques andinos tropicales: Avances, desafíos y perspectivas del futuro. Documentos Ocasionales* 170. Bogor: CIFOR.

Pacheco, P., K. Mo, N. Dudley, et al. 2021. *Deforestation Fronts: Drivers and Responses in a Changing World.* Gland: WWF.

Payn, T., J.M. Carnus, P. Freer-Smith, et al. 2015. "Changes in Planted Forests and Future Global Implications." *Forest Ecology and Management* 352: pp. 57–67. DOI:10.1016/j.foreco.2015.06.021.

Péroches, A., H. Baral, M. Chesnes, A. Lopez-Sampson, and G. Lescuyer. 2022. "Suitability of Large-Scale Tree Plantation Models in Africa, Asia and Latin America for Forest Landscape Restoration Objectives." *Bois et Forêts des Tropiques* 351: pp. 29–44. DOI:10.19182/bft2022.351.a36870.

Pham, T. T., K. Bennet, T. P. Vu, J. Brunner, N. D. Le, and D. T. Nguyen. 2013. *Payments for Forest Environmental Services in Vietnam: From Policy to Practice. Occasional Paper* 93. Bogor: CIFOR.

Pham, T. T., N. T. B. Duong, T. Thürer, and E. O'Connell. 2021. *Payments for Forest Environmental Services in Viet Nam: Strengthening Effectiveness through Monitoring and Evaluation*. CIFOR Infobrief. https://www.researchgate.net/publication/351547762.

PRISMA. 2015. *Adaptation-Based Mitigation: Facing Climate Change in El Salvador and Central America*. San Salvador: PRISMA.

Ramankutty, N., Z. Mehrabi, K., Waha, et al. 2018. "Trends in Global Agricultural Land Use: Implications for Environmental Health and Food Security." *Annual Review of Plant Biology* 69: pp. 789–815.

Rebugio, L. L., J. M. Pulhin, A. P. Carandang, E. O. Peralta, L. D. Camacho, and N. C. Bantayan. 2017. "Forest Restoration and Rehabilitation in the Philippines." In Keep Asia Green Volume I Southeast Asia. *IUFRO World Series* Volume 20-I, pp. 125–170. https://www.researchgate.net/publication/285314978.

Redo, D. J., H. R. Grau, T. M. Aide, and M. L. Clark. 2012. "Asymmetric Forest Transition Driven by the Interaction of Socioeconomic Development and Environmental Heterogeneity in Central America." *PNAS* 109 (23): pp. 8839–8844.

République de Côte d'Ivoire. 2017. *Stratégie Nationale REDD+ Cote d'Ivoire adoptee par le Gouvernement*. https://reddplus.ci/la-strategie-nationale-redd-cote-divoire-adoptee-par-le-gouvernement/.

Rudel, T. K. 2013. "The national determinants of deforestation in sub-Saharan Africa." *Philosophical Transactions of the Royal Society B* 368: p. 20120405. DOI:10.1098/rstb.2012.0405.

Russell, M. 2020. Forests in South-East Asia. Can They Be Saved? European Parliament Briefing. EPRS (European Parliamentary Research Service). https://www.europarl.europa.eu/RegData/etudes/BRIE/2020/652068/EPRS_BRI(2020)652068_EN.pdf.

Sesnie, S. E., B. Tellman, D. Wrathall, et al. 2017. "A Spatio-Temporal Analysis of Forest Loss Related to Cocaine Trafficking in Central America." *Environmental Research Letters* 12 (5): p. 054015.

Sloan, S., and J. A. Sayer. 2015. "Forest Resources Assessment of 2015 Shows Positive Global Trends but Forest Loss and Degradation Persist in Poor Tropical Countries." *Forest Ecology and Management* 352: pp. 134–145. DOI:10.1016/j.foreco.2015.06.013.

Stanturf, J. A., and S. Mansourian. 2020. "Forest Landscape Restoration: State of Play." *Royal Society Open Science* 7: p. 201218. DOI:10.1098/rsos.201218.

Stavi, I., and E. Lal. 2014. "Achieving Zero Net Land Degradation: Challenges and Opportunities." *Journal of Arid Environments* 112: pp. 44–51. DOI:10.1016/j.jaridenv.2014.01.016.

Szulecka, J., J. Pretzsch, and L. Secco. 2014. "Paradigms in Tropical Forest Plantations: A Critical Reflection on Historical Shifts in Plantation Approaches." *International Forestry Review* 16 (2): pp. 128–143.

UNCCD. 2019. *The Global Land Outlook, West Africa Thematic Report*. Bonn: United Nations Convention to Combat Desertification.

UNEP and ACTO. 2009. *Geoamazonia: Environmental Outlook in Amazonia*. Brasilia: United Nations Environment Programme (UNEP), Amazon Cooperation Treaty Organization (ACTO).

Van Oosten, C., P. Gunarso, I. Koesoetjahjo, and F. Wiersum. 2014. "Governing Forest Landscape Restoration: Cases from Indonesia." *Forests* 5: pp. 1143–1162.

Vargas Ríos, O., J. E. Díaz Triana, S. P. Reyes Bejarano, and P. A. Gómez Ruiz. 2012. Guías Técnicas Para La Restauración Ecológica De Los Ecosistemas de Colombia. Convenio de Asociación No. 22 entre Ministerio de Ambiente, Vivienda y Desarrollo Territorial (MAVDT) y Academia de Ciencias Exactas, Físicas y Naturales (ACCEFYN). Bogotá.

World Bank. 2017. *Colombia: Programmatic Forest Country Note. Deep Dive into Forest & Landscape-Smart Investments.* Washington: World Bank. https://documents1.worldbank.org/curated/en/892731545423859022/pdf/133160-REPL-PUBLIC-Forest-Country-Note-Compressed.pdf.

Zeleke, A., and A. Vidal. 2020. Contributing to Scaling Up Forest Landscape Restoration in Ethiopia. Restoration Diagnostic Applied in Sodo Guragie (SNNPR) and Meket (Amhara region) woredas. Gland: IUCN.

6

Forest restoration for climate change mitigation and adaptation

Daniel Kübler and Sven Günter

6.1 Introduction

Climate change mitigation is urgently required to meet the Paris Agreement goal to limit global warming to well below 2°C above preindustrial levels and to pursue efforts to limit the increase to 1.5°C (UNFCCC 2016). Land-based options are key components of mitigation strategies. Griscom et al. (2017) identify 20 natural climate solutions (NCS) and estimate that this ensemble of conservation, restoration, and improved land management actions can contribute 37% of the cost-effective CO_2 mitigation needed through 2030 to keep global warming under 2°C. Two-thirds of this mitigation are offered by forest-based NCS, and an important strategy within forest-based NCS is forest restoration.

Reaching the Paris goal would significantly reduce climate risks globally, yet many regions would remain vulnerable. Moreover, the goal will not be reached under current national mitigation ambitions as submitted under the Paris Agreement (IPCC 2018). Hence the need for climate change adaptation.

There is increasing awareness in science, policy, and public opinion that forest restoration has the potential to contribute to climate change mitigation and adaptation, that it can have co-benefits for conservation, reversal of land degradation, and sustainable development (Anderegg et al. 2020; Arneth et al. 2019). Forest restoration is now part of the global agenda. The Bonn Challenge and the UN Decade on Ecosystem Restoration, along with regional initiatives worldwide, aim to restore millions of hectares of degraded ecosystems and deforested landscapes.

Land-based options face various implementation challenges. Alongside potential synergies with other approaches, forest restoration has potential trade-offs. Depending on the scale of deployment, there are risks of adverse side effects, especially if there is competition for available land (IPCC 2019b). While there is broad recognition that forest restoration can contribute to both limiting and adapting to global warming, there is ongoing debate about the potential magnitude of this contribution, what trade-offs are involved, and which specific measures should be prioritized.

This chapter assesses the relevance of forest restoration for climate change mitigation and adaptation. The various understandings of forest restoration (see Chapter 1) can be a challenge for evaluating its potential. This chapter applies the widely used framework of forest landscape restoration (FLR; Besseau et al. 2018). This includes reforestation and afforestation[1] through active or passive approaches, agroforestry practices, silvopastoral systems, plantation establishment, and the restoration of degraded forests (Sabogal et al. 2015). It explicitly does not include afforestation on

Daniel Kübler and Sven Günter, *Forest restoration for climate change mitigation and adaptation* In: *Restoring Forests and Trees for Sustainable Development.* Edited by: Pia Katila, Carol J. Pierce Colfer, Wil de Jong, Glenn Galloway, Pablo Pacheco, and Georg Winkel, Oxford University Press. © Oxford University Press 2024. DOI: 10.1093/9780197683958.003.0006 areas that have been occupied historically by a non-forested biome, such as grassland, savanna, non-forested wetland, or peatland (Di Sacco et al. 2021).

Section 6.2 discusses the mitigation potential of forest restoration. It presents a selection of current global estimates, focusing on four in particular to highlight variations; it then discusses how geographical differences and different forest restoration options lead to variations in estimates; and it looks at the role wood use plays in forest restoration for climate change mitigation. Section 6.3 presents an overview of risks that may limit the success of forest restoration mitigation, looking at natural, socioeconomic, and management-related factors. Section 6.4 explores the potential of forest restoration for climate change adaptation, looking at how this can improve the adaptive capacity of people and of forests. Section 6.5 links adaptation and mitigation and looks at the synergies and trade-offs this entails for people and for forests. Section 6.6 sums up our findings and discusses their implications.

6.2 Potential of forest restoration to contribute to climate change mitigation

The potential of forest restoration to contribute to climate change mitigation can be assessed from multiple perspectives. We discuss existing estimates of the global mitigation potential of forest-based NCS, some of which overlap with the concept of forest restoration. Illustrating four estimates in particular, we explore how differences in underlying assumptions and understandings of forest restoration impact the outcomes. We then highlight the variability of mitigation potentials between geographic regions and different forest restoration options. We further discuss the role of wood use and its interplay with forest restoration and climate change mitigation.

6.2.1 Global estimates

Figure 6.1 shows a compilation of global mitigation potential estimates for forest-based NCS from recent, prominently published literature. Estimates vary between studies, as they depend on assumptions about contextual conditions regarding a number of biophysical and socioeconomic factors. We follow Roe et al. (2019) and differentiate between "technical potentials" (i.e., the amount possible with current technologies), "economic potentials" (i.e., mitigation at different carbon prices) and "sustainable potentials" (i.e., technical or economic mitigation constrained by food security and environmental considerations).

Multiple strategies fit under the category forest-based NCS; Figure 6.1 shows estimates for the climate mitigation potentials of the five most prominent of them. The NCS "reduce deforestation" and "reduce degradation" are outside of forest restoration. The NCS "forest management" includes the restoration of existing degraded forests, that is, forest land remaining forest land (Smith et al. 2019), which is a part of FLR. The main focus, however, is to improve the climate change mitigation of existing natural production forests (Griscom et al. 2017; Roe et al. 2017); only a small share of its potential can be attributed to forest restoration. The NCS "agroforestry" is a component of FLR, but its potentials are possibly also included in some afforestation/

reforestation estimates (Roe et al. 2019). For example, Griscom et al. (2017) include agroforestry systems that exceed 25% tree cover and all silvopastoralism in the reforestation pathway. Attributing the mitigation potential of agroforestry to forest restoration entails the risk of double counting. The NCS "afforestation/reforestation" describes the conversion of non-forest land to forest land. Forest restoration is frequently equated with this category in public discussions about climate change mitigation. Estimates for this NCS are therefore highly relevant to assess the potential of forest restoration to limit global warming, and is hence where we turn our focus.

Figure 6.1 illustrates the considerable variation of estimates. For instance, the global mitigation potential for afforestation/reforestation ranges from 0.0 to 10.1 $GtCO_2\ yr^{-1}$ for the 2020–2050 period. To put this into perspective, IPCC (2018) estimates that the remaining carbon budget[2] to limit warming to 1.5°C at a 66% probability is 420 $GtCO_2$. Current emissions deplete this budget by 42 ± 3 $GtCO_2\ yr^{-1}$. Afforestation/reforestation could thus contribute significantly to limiting warming, especially if estimates at the upper limit can be realized.

Other forest-based strategies have large mitigation potentials (Figure 6.1). Griscom et al. (2020) report that avoided deforestation has the largest mitigation potential for NCS in the tropics, with more than twice as much of the cost-effective climate

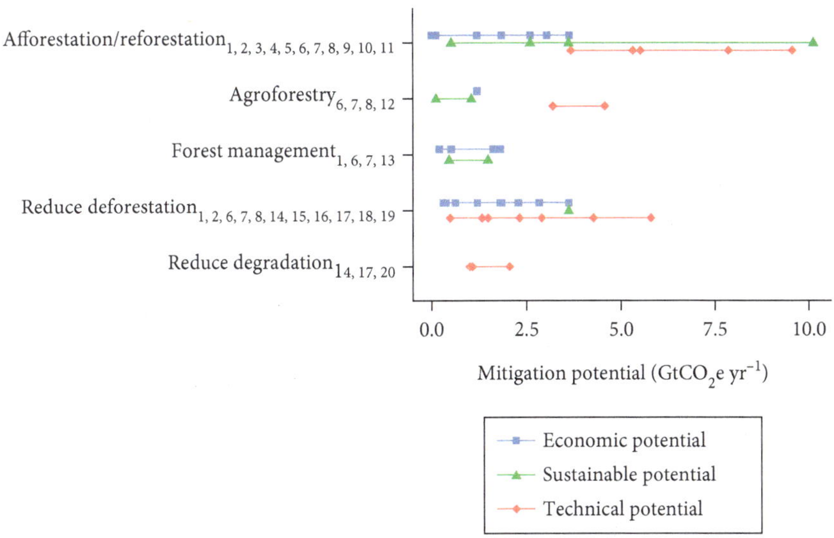

Figure 6.1 Mitigation potential of forest-based NCSs for 2020–2050. Data were compiled following methods from Roe et al. (2019). The wide range of estimates is due to studies differing in basic assumptions on a number of biophysical and socioeconomic factors. Technical, economic, and sustainable potentials are differentiated by color. References: 1: Austin et al. (2020), 2: Busch et al. (2019), 3: Cook-Patton et al. (2020), 4: Dooley and Kartha (2018), 5: Fuss et al. (2018), 6: Griscom et al. (2017), 7: Griscom et al. (2020), 8: Hawken (2017), 9: Houghton et al. (2015), 10: Kreidenweis et al. (2016), 11: Lenton (2010), 12: Dickie et al. (2014), 13: Sasaki et al. (2016), 14: Baccini et al. (2017), 15: Busch and Engelmann (2017), 16: Carter et al. (2015), 17: Federici et al. (2015), 18: Houghton and Nassikas (2018), 19: Smith et al. (2013), 20: Pearson et al. (2017).

mitigation potential as the NCS with the second largest potential, reforestation. It is thus important that afforestation/reforestation does not come at the expense of avoided deforestation. This could be the case if leakage occurs (see Section 6.3.2), for example, through the displacement of smallholders due to tree planting on marginal agricultural sites. In this context, agroforestry and forest management are attractive, despite their smaller mitigation potentials, because they have fewer trade-offs with existing land uses, creating less competition for land (Griscom et al. 2020).

Table 6.1 briefly summarizes four recent studies that estimate reforestation and afforestation mitigation potential. Methodological variations and differences in principle assumptions about land-use constraints give a better understanding of the wide range in estimates.

The studies have important limitations. Bastin et al. (2019) and Griscom et al. (2017) assume a global shift toward plant-based diets, making current grazing land in areas where forests naturally occur available for afforestation/reforestation. Bastin et al. (2019) have been criticized for including afforestation on historically non-forested biomes (Veldman et al. 2019). Most studies presented in Figure 6.1 are limited by not considering certain factors, such as social and political constraints or the impacts of future climate change (Jia et al. 2019).

Table 6.1 Overview of methods and principal results of recent studies that report estimates for the climate change mitigation potential of afforestation and/or reforestation

Busch et al. (2019)	**Austin et al. (2020)**	**Griscom et al. (2017)**	**Bastin et al. (2019)**
Modeling approach			
Land-use-change modeling assuming that 1) reforestation is partly a response of land-use decision makers to economic forces, and 2) responses to future changes in carbon prices will be the same as to past variations in agricultural prices.	Elaborate dynamic economic modeling considering aspects such as feedbacks between markets as well as future demand growth for agricultural products.	Bioclimatic modeling that maps areas where forest cover naturally occurs and that then removes certain areas (e.g., cropland and urban settlements) to estimate areas suitable for reforestation.	
Scenario approach			
Future forest area increase and associated mitigation were estimated for a business-as-usual (BAU) scenario and alternative scenarios. Additional mitigation above the BAU scenario was considered as the mitigation potential.			No scenario approach was used.
Geographic and temporal extent			
Pantropical; for 2020–2050	Pantropical; for 2025–2055	Temperate and tropical; for 2030	Global; no specific timeframe

Table 6.1 Continued

Busch et al. (2019)	Austin et al. (2020)	Griscom et al. (2017)	Bastin et al. (2019)
Resulting reforestation areas and mitigation potentials			
Under the BAU scenario, reforestation area was 387.8 Mha, resulting in the removal of 102.5 $GtCO_2$ by 2050. Additional mitigation compared to the BAU scenario over a period of 30 years under scenarios of payments for increased CO_2 removals at carbon prices of USD 20/tCO_2 and USD 50/tCO_2 was 5.7 and 15.1 $GtCO_2$, respectively.	Relative to the BAU scenario, forest area increased by 415–875 Mha by 2055 at prices of USD 35–USD 100/tCO_2. However, these values are for the combined effect of a reduction in deforestation and increase in afforestation/reforestation. Estimates for additional mitigation potentials compared to the BAU scenario were reported for afforestation/reforestation, resulting in 0.1–2.6 $GtCO_2\ yr^{-1}$ for assumed initial carbon prices of USD 5–USD 100/tCO_2 and price growth rates of 1%–3%.	A BAU baseline reforestation area was calculated by assuming a continuation of the historic mean forest "gain" rate for 2000–2012 (based on Hansen et al. 2013) up to 2030. This resulted in mitigation of 1.1 $GtCO_2\ yr^{-1}$. The maximum potential extent of reforestation was 678 Mha in 2030. This resulted in additional mitigation of 10.1 $GtCO_2\ yr^{-1}$ compared to the BAU baseline. For cost-constrained mitigation at marginal costs of USD 10/tCO_2 and USD 100/t CO_2, additional potentials were 0.0 and 3.0 $GtCO_2\ yr^{-1}$, respectively.	It was estimated that a global area of 1,700–1,800 Mha of potential forest land is currently not under forest cover. This corresponds to a global mitigation potential of 752.4 $GtCO_2$ if restored woodlands and forests in these areas were allowed to mature to a similar state of existing ecosystems in protected areas. This study was not included in Figure 6.1 because no specific timeframe was provided for the estimate of mitigation potential.

These examples illustrate the importance of critically assessing mitigation potential and to reflect on how estimates are produced. This is especially the case for estimates at the upper limit, which are largely theoretical and typically do not consider economic and political feasibility and constraints, nor issues of carbon permanence (see Section 6.3; Roe et al. 2019). Authors should be transparent about the assumptions behind these maximum potentials, which are frequently promoted in the literature. Moreover, they should be transparent in reporting potential minimum amounts and associated risks of failure. Despite these limitations, most studies agree that forest restoration can play an important role in climate change mitigation. What is lacking are clear, convincing implementation pathways.

6.2.2 Contextual conditions

Conditions for forest restoration vary significantly between and within geographic regions. Variations in biophysical and socioeconomic settings not only present different challenges for designing restoration programs (see Chapter 5) but they also result in different mitigation potentials. Likewise, variations in forest restoration options, ranging from passive approaches relying on natural regeneration to active approaches such as native or exotic plantations or agroforestry, present different challenges and result in different potentials. Better understanding of how these variations impact outcomes is important for successfully implementing restoration programs that target climate change mitigation.

Geographic regions

It is notable that forest restoration, as a mitigation strategy, focuses more on tropical and subtropical regions than temperate and boreal regions. This can be explained by a number of factors. Carbon sequestration rates, in association with forest restoration, exhibit large variations across the globe. For example, Cook-Patton et al. (2020) report an over 100-fold variation in potential aboveground carbon accumulation rates for the first 30 years of natural forest regrowth, with climatic factors explaining variation better than land-use history. Trees grow quicker in the tropics, compared to temperate and boreal regions; tropical forests therefore have a higher carbon sequestration rate per area (Lewis et al. 2019). This is the case for both natural regeneration (Cook-Patton et al. 2020) and intensively managed plantations (Silva et al. 2019). Additionally, land is relatively cheap and available in many places in the tropics (Lewis et al. 2019). Furthermore, increases in forest cover in tropical regions result in enhanced evapotranspiration, which causes additional cooling (IPCC 2019b). In contrast, increases in forest cover at high latitudes lead to the albedo effect, which can result in net warming despite the carbon sequestration of trees (Fuss et al. 2018; Veldman et al. 2019). Nevertheless, forest restoration has also been reported to be relevant for mitigation in temperate regions, with Russia, the United States, Canada, and China having large areas of tree restoration potential (Bastin et al. 2019; Roe et al. 2017).

For the tropical and subtropical regions, humid biomes generally show higher carbon sequestration rates in above-ground biomass (AGB) and below-ground biomass (BGB) than dry biomes (Bernal et al. 2018). For secondary forests (i.e., natural regeneration) younger than 20 years, region-specific average removal rates from the total plant biomass (i.e., AGB and BGB combined) ranged between 11.9–18.8 tCO_2 ha^{-1} yr^{-1} for tropical humid forests and 10.3–13.8 tCO_2 ha^{-1} yr^{-1} for tropical dry forests. Differences were more pronounced for older secondary forests (20–60 years); tropical humid forests of that age range had average rates between 5.2–17.3 tCO_2 ha^{-1} yr^{-1} and dry forests 0.2–3.5 tCO_2 ha^{-1} yr^{-1}. For planted forests and woodlots, differences between humid and dry regions were smaller. Excluding eucalyptus, average carbon sequestration rates from the total plant biomass across species were around 27 and 20 tCO_2 ha^{-1} yr^{-1}, respectively, for humid and dry tropical regions. Eucalyptus reached relatively consistent average rates across regions, with

40.7 tCO_2 ha^{-1} yr^{-1} in humid regions and 38.8 tCO_2 ha^{-1} yr^{-1} in dry regions (Bernal et al. 2018). Differences between tropical humid and dry forests in terms of mitigation potential are determined not only by carbon sequestration rates but also by the size of areas available for forest restoration (Cook-Patton et al. 2020). However, studies comparing the potential extents of forest restoration in tropical and subtropical humid and dry regions are scarce. For Latin America, Chazdon et al. (2016) estimate that existing young secondary forests (≤ 20 years of age) cover an area of 108.8 Mha and 42.5 Mha in humid and dry regions, respectively. Combining these estimates with aboveground carbon sequestration rates results in a net mitigation potential over four decades (2008–2048), respectively, of 21.85 and 3.23 $GtCO_2$ for humid and dry regions, if these secondary forests are permitted to regenerate naturally (Chazdon et al. 2016).

The described differences between regions should only be considered as approximative. Ultimately, forest cover change is linked with development and restoration options; strategies, feasibility, and mitigation potentials are largely influenced by socioeconomic contexts within geographical regions (see Chapter 10). This can range from subsistence-driven land use in one region to commercially driven land use in another, leading to different stakeholder configurations and opportunity costs for restoration. It is therefore important to tailor reforestation programs to specific contexts within each geographic region (IUCN and WRI 2014; see Chapter 5).

Forest restoration options

There is scientific debate not only on the total global mitigation potential of forest restoration but also on the relative contributions of different restoration options (e.g., Brancalion and Holl 2020; Lewis et al. 2019; Löf et al. 2019). Passive approaches relying on natural regeneration, active approaches such as native or exotic plantations, and agroforestry differ in terms of temporal dynamics of carbon stocks and maximum carbon storage per area of restored forest land. Assessing the mitigation potentials of these options depends on methods employed, the timeframes considered, which carbon pools are included, and scenarios of use/non-use of wood products. For example, Bernal et al. (2018) report, based on an analysis of empirical data from the literature, that across the globe and for the first 20 years of growth, planted forests have the highest CO_2 removal rates per unit land area from AGB and BGB combined (4.5–40.7 t CO_2 ha^{-1} yr^{-1}), followed by mangrove tree restoration (23.1 tCO_2 ha^{-1} yr^{-1}), natural regeneration (9.1–18.8 tCO_2 ha^{-1} yr^{-1}), and agroforestry (10.8–15.6 tCO_2 ha^{-1} yr^{-1}), with considerable overlap across these different restoration approaches. Similarly, Bonner et al. (2013) use a meta-analytical approach, reporting significantly higher AGB growth rates in plantations than secondary forests for young forests in the tropical region; these differences are less pronounced in older forests. The success of plantations, however, strongly depends on the selection of suitable species.

In stark contrast, in their commentary, Lewis et al. (2019) estimate for tropical and sub-tropical regions that naturally regenerated forests would be six times better than agroforestry and 40 times better than plantations at storing above and belowground carbon in the long term. These estimates are based on a number of assumptions. First, problems of permanence and leakage are ignored. It is assumed that natural regeneration is feasible in all areas with restoration opportunities and that carbon stocks of forests that naturally regenerate on disturbed land will be allowed to recover to their previous high-carbon state. This assumes that all direct and underlying drivers of deforestation, such as failures of governance, poverty, and economic development will disappear. Second, the substitution effect of wood harvested from plantations and agroforestry (see Section 6.2.3) is not considered. Third, to assess where restoration is likely to occur, an existing map of restoration opportunities is used (Laestadius et al. 2011). This map includes not only deforested areas but also degraded forests, which can store significant amounts of carbon. Based on this, the initial carbon stocks of restoration areas are then estimated. For agroforests and plantations, it is assumed that all initial carbon is lost during establishment, resulting in a negative carbon balance of these two options in some countries (Table S5 in Lewis et al. 2019).

Apart from this wide range of estimates, additional challenges remain in comparing restoration options. First, there is a positive site-selection bias in many meta-analyses comparing natural regeneration to active forest restoration (i.e., in contrast to studies on natural regeneration, studies on tree planting are conducted on sites that may or may not have regenerated without planting; Reid et al. 2018). Second, both plantations and natural regeneration are sensitive to site conditions; however, the relative contribution of different site-specific factors differs between studies (Bernal et al. 2018; Bonner et al. 2013; Cook-Patton et al. 2020). This constraint should be considered when estimating the global mitigation potential of forest restoration. Third, carbon stocks in plantations depend on the management regime. For example, Griscom et al. (2017) argue that increasing harvest rotation lengths in plantations from the economic optimum to the optimum for wood yield will not only increase timber production in the long run but also improve carbon stocks. They estimate that this has an unconstrained global mitigation potential of 0.44 $GtCO_2\,yr^{-1}$.

In addition to climate change mitigation, forest restoration measures usually consider additional goals, such as food security, poverty alleviation, and biodiversity conservation (Besseau et al. 2018). This is especially true for the restoration of multifunctional mosaic landscapes, which provide the greatest opportunity for FLR (Laestadius et al. 2011). Their restoration paths can include agroforestry systems, productive forests, and protected forests (Stanturf et al. 2019); this requires a broad palette of methods adapted to local contexts to overcome conflicting aims for land use (Löf et al. 2019; see Sections 6.5.1 and 6.5.2). Studies that focus only on mitigation potential thus have limited value to inform policy and practice. Balancing co-benefits and trade-offs in global large-scale forest restoration for climate change mitigation should play a greater role in the scientific debate. Instead of either–or discussions between active and passive restoration and between native and exotic species, it would be advantageous to focus research on assessing which restoration approach is best suited for which context and on limiting its respective potential negative impacts (see Chapter 10).

6.2.3 Wood use, forest restoration, and climate change mitigation

Wood use can contribute to climate change mitigation by reducing emissions in other sectors. Carbon storage in long-lived forest products (e.g., furniture or construction material) can substitute for emission-intensive materials (e.g., cement or steel), and woody biomass used for energy can substitute for fossil fuel (FAO 2013; IPCC 2019b). There is a trade-off when using wood products or woody biomass: unmanaged forests generally have higher carbon stocks, but their sequestration potential can eventually saturate (Arneth et al. 2019). In contrast, forests managed for wood production have lower carbon stocks, but due to the substitution effect they are carbon sinks that can be continuously used for mitigation if appropriate silvicultural practices are employed. This eventually leads to net savings, which can result in a positive net impact of wood production on atmospheric carbon (Jia et al. 2019). Accurately quantifying this impact is complex, as it depends on many factors: the materials and/or fuels substituted for, counterfactual assumptions on forest use, accounting timeframes, forest regrowth rates, and other site-specific contexts (IPCC 2019b; Olsson et al. 2019). Critical methodological differences in the literature result in a wide range of reported effects of wood substitution on greenhouse gas (GHG) emissions (Geng et al. 2017).

Global assessments of potential climate benefits of forest restoration typically do not include the substitution effect. This is likely due to the overall high uncertainty in the literature about the impact of wood use on GHG, but is also due to the fact that most substitution studies focus on nontropical countries, whereas forest restoration as a mitigation strategy generally focuses on the tropics. Ignoring the substitution effect can skew the outcomes of global assessments toward restoration options not primarily aimed at wood production or poverty alleviation, such as passive restoration using natural regeneration (e.g., Lewis et al. 2019).

Two other aspects are important for the wood use discussion. First, the long-term success of restoration depends on restored forests becoming an economic resource for local communities (Löf et al. 2019). Second, the demand for wood products has increased globally in line with growing populations and incomes; this trend is expected to continue in the coming decades (FAO 2020). Ignoring these two aspects risks exclusively focusing on ecological objectives. This could prevent the success of restoration efforts (Höhl et al. 2020), in turn negatively impacting climate change mitigation. Recognizing that wood production can have a positive impact on GHG emissions is important for forest restoration; it can increase the likelihood of successful implementation. Forest restoration approaches aiming to reconcile wood production, ecosystem enhancement, and socioeconomic development, such as FLR and "new generation plantations" (Silva et al. 2019), are thus promising.

6.3 Risks of forest restoration as a mitigation strategy

Forest restoration has the potential to remove large quantities of carbon from the atmosphere in the coming decades. However, carbon removals from forest restoration might not be realized as expected. This could lead to higher levels of global warming if relying on future carbon removals from forest restoration to reach global mitigation targets delays actions for near-term emission reductions (Dooley and Kartha 2018).

Factors that could prevent the successful deployment of forest restoration at scales relevant for climate change mitigation are discussed below, with a focus on natural and socioeconomic drivers of change and management-related factors. These factors could increase the risk that future carbon removals from forest restoration will not be feasible for technical or biophysical reasons, or because society deems ecological and societal impacts as unacceptable or carbon storage is not permanent (Dooley and Kartha 2018).

6.3.1 Natural factors

Restored forests are subject to natural disturbances such as drought, fires, and pests, which could trigger future loss of carbon stored in vegetation and soils (Fuss et al. 2018; IPCC 2019b). Climate change exacerbates this risk of sink reversal. Adaptation measures can decrease this risk (Section 6.4.2), but there are limitations (IPCC 2019b). Climate-driven tree mortality and disturbances change with time and are projected to increase with climate change (Figure 6.2; Anderegg et al. 2020). Additionally, temperature overshoot, that is, the temporary exceedance of a specified level of global warming (IPCC 2018), may lead to further disturbances by crossing important tipping points (Dooley and Kartha 2018; Reyer et al. 2015).

The potential future losses of forest carbon through natural disturbances and extreme events introduce considerable uncertainty in estimates of mitigation potentials of forest restoration (Arneth et al. 2019). For the tropics, Koch and Kaplan (2022)

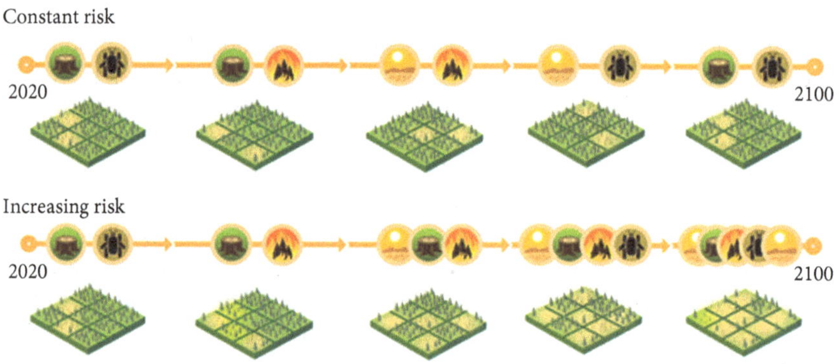

Figure 6.2 Conceptual diagram of impacts on forest carbon from climate-driven disturbance risk over time. The top of the diagram shows constant risk from disturbances at the landscape level; the bottom, increasing risk. Increasing risk impacts forest carbon much more severely. Disturbance events are illustrated in the circles and include fire, drought, biotic agents, and human disturbance. From Anderegg, W. R. L., A. T. Trugman, G. Badgley, et al. 2020. "Climate-Driven Risks to the Climate Mitigation Potential of Forests." Science 368 (6497). Reprinted with permission from AAAS.

report that carbon stored by forest restoration is largely preserved under the entire range of potential future climates.

A climate-driven increase in natural disturbances is not only a risk for restored forests but also for other land uses. This could reduce the future potential for agricultural intensification (Dooley and Kartha 2018), leading to a greater demand for agricultural land, thereby limiting land availability (Section 6.3.2).

These natural factors can impact land user preferences: risk-averse land users tend to prefer annual crops over trees, and short-rotation timber plantations over long-term multipurpose mixed-species forest restoration (Zhunusova et al. 2019). This constitutes an additional challenge for the implementation of large-scale forest restoration.

6.3.2 Socioeconomic factors

Implementing forest restoration on a scale relevant for climate change mitigation requires large amounts of land. Projected changes in global forest area range from −20 to + 720 Mha between 2010 and 2100 (Jia et al. 2019). Land is needed for various demands, especially food production. Particularly over the last few decades, population growth and changes of per capita consumption have led to increasing global demand in food, feed, fiber, and energy. Projected increases in population and income as well as changes in consumption patterns will further increase global demands (IPCC 2019b). Hence, forest restoration will likely have to compete with agriculture, and alternative land-based mitigation strategies (Delzeit et al. 2019; Fuss et al. 2018). Considering the various future land-use demands is important to assess the risk of forest restoration as a mitigation strategy.

For agricultural land, projections from Integrated Assessment Models (IAMs) result in a wide range of potential land-use futures for 2100 in the shared socioeconomic pathways (SSP), with global agricultural land decreasing by 743 Mha at the lower (SSP1) and increasing by 1,080 Mha (SSP3) at the upper end (Popp et al. 2017). The IAMs identify agricultural productivity, consumption preferences, and population development as the most influential drivers of change (Stehfest et al. 2019). If forest restoration is implemented on land used for food production, there is a risk that agricultural activities will be displaced to native forest areas, which would result in deforestation and additional CO_2 emissions, that is, leakage (Holl and Brancalion 2020; Olsson et al. 2019).

For alternative mitigation measures, estimates for land-use projections for 2100 vary, but overall show an increase in land demand. Recent studies estimating the "sustainable potentials" for bioenergy result in projected land requirements ranging from 10 to 1,300 Mha (Smith et al. 2019). Additionally, the effective implementation of Reducing Emissions from Deforestation and Forest Degradation (REDD +), which aims to reduce deforestation, would reduce future agricultural land expansion, further increasing land competition (Tabeau et al. 2017).

Another major factor on land-use demands relates to global diets and food waste. According to estimates, a global dietary change from meat and dairy to plant-based foods could free 400–2,500 Mha; a global reduction in food waste, as much as 700 Mha

(Smith et al. 2019). This change in consumption preferences is an influential driver in the IAM that leads to a reduction in projections of future global agricultural land (Stehfest et al. 2019). However, such demand-side measures have proceeded slowly because of limited awareness and political support, in addition to the difficulties of eliciting behavioral change (Roe et al. 2019). Nevertheless, many of the prominent studies estimating large mitigation potentials (e.g., Bastin et al. 2019; Cook-Patton et al. 2020; Griscom et al. 2017) assume a global shift toward plant-based diets, which would make current grazing land and cropland used to produce fodder available for reforestation.

The wide range of possible future land-use trajectories suggests that land competition is a significant risk for forest restoration as a mitigation strategy. If the optimistic assumptions are not met, future increases in land demand might lead to higher opportunity costs of land, which could be a significant barrier for the implementation and permanence of large-scale forest restoration (Brancalion et al. 2019). These concerns suggest that actual mitigation potential is at the lower end of reported estimates (Figure 6.1).

6.3.3 Management-related factors

Forest restoration frameworks aimed at creating multifunctional landscapes, such as FLR, require a broad palette of methods, including both active and passive approaches (Löf et al. 2019; Stanturf et al. 2019). This raises the important issue of availability of regeneration material for active restoration. In many parts of the tropics, it is currently very challenging to produce large amounts of reproductive material for native species due to knowledge and infrastructure limitations (Onyekwelu et al. 2011, Chapter 11). The lack of high-quality seeds and seedlings is an obstacle for the successful implementation of forest restoration with native species (Höhl et al. 2020; Löf et al. 2019). Consequently, reforestation with easily obtained non-native species might be the only option available in some tropical regions. This option, however, has potentially adverse ecological effects, such as the use of chemicals and fertilizers negatively impacting local biodiversity.

Additionally, many projects set short-term targets (number of planted trees) rather than considering whether restoration targets are achieved. This can result in a lack of long-term management and subsequent restoration failure (Holl and Brancalion 2020).

6.4 Climate change adaptation

As climate change unfolds it is crucial for forest restoration strategies to develop an adequate adaptation response (Pramova et al. 2019). Climate change adaptation for forests generally comprises two distinct, but related, components (Ciccarese et al. 2012; Locatelli 2011). The first, adaptation for forests, aims to strengthen forests' capacity to deal with climate change impacts. The second, forests for adaptation, aims to use forests to strengthen society's capacity to deal with climate change impacts.

In most forest restoration contexts the vulnerabilities of people and ecosystems are closely linked. This calls for an integrated socioecological adaptation perspective (Fedele et al. 2019; Pramova et al. 2019; Chapter 8).

6.4.1 For forests

Climate change significantly impacts environmental forest conditions and changes disturbance regimes, such as the frequency and intensity of storms, fires, and droughts as well as the spread of pests, disease, and invasive species (Section 6.3.1; Anderegg et al. 2020; Locatelli et al. 2015a). Changes in the frequency of wet or dry years may trigger increases or decreases in suitable land for long-term tree cover (Holmgren et al. 2013). Altered temperatures and precipitation rates, extreme events, and increased atmospheric CO_2 concentrations, as well as changes in plant-pollinator interactions, may change forest structure, species composition, seedling establishment, and tree growth and mortality rates (Anderson-Teixeira et al. 2013; Locatelli et al. 2015a). While scientists and practitioners widely agree that site properties will change due to climate change, future conditions predicted for a given region are uncertain. This uncertainty is a significant barrier for the effective implementation of adaptation measures (Jandl et al. 2019).

Forest restoration strategies need to provide desired ecosystem services not only under current climatic conditions but also under changing future conditions. There is wide consensus that a central adaptation strategy for forest restoration is to promote resilient ecosystems and forest landscapes (Stanturf et al. 2015), with resilience being "the capacity to cope with a hazardous event, trend or disturbance, responding or reorganising in ways that maintain their essential function, identity and structure" (IPCC 2019a, 822). Applying the concept of resilience in restoration practice is not straightforward, however, as there are numerous definitions in the literature. Three broad resilience conceptualizations are distinguished in forest science: engineering resilience, ecological resilience, and social-ecological resilience (Nikinmaa et al. 2020). The latter assumes that human and natural systems are coupled and focuses on the adaptive capacity of the social-ecological system as a whole (Folke et al. 2002). Social-ecological resilience is of particular importance in many forest restoration contexts where vulnerabilities of people and ecosystems are closely linked. For operationalization, the different resilience concepts can be considered as not mutually exclusive but rather complementary, with engineering resilience being nested within ecological resilience, which in turn is nested in social-ecological resilience (Nikinmaa et al. 2020).

Engineering resilience describes the ability of a given system to return to its equilibrium after a disturbance. This definition assumes that a stable state exists for that system (Nikinmaa et al. 2020). Engineering resilience can be fostered with buffering measures that focus on preventing perturbations (Locatelli et al. 2010). A prominent example is managing stand densities through the silvicultural intervention of thinning, which decreases the risk of disturbances such as storms and fire (Spathelf et al. 2018). Further intervention examples include reducing fuel loads in restored forests to reduce fire risk and phytosanitary treatments to prevent the spread of invasive species (Locatelli et al. 2015a; Stanturf et al. 2015). These interventions require intensive management; they become less effective and increasingly costly with accelerating climate change (Locatelli et al. 2010; Nikinmaa et al. 2020).

Ecological resilience does not assume that a single steady state exists for a given system. Instead it assumes that disturbances can move the system from one state to another (Nikinmaa et al. 2020). This resilience can be fostered by measures that do not aim to resist changes but facilitate a natural ecosystem shift toward a new state that should be socially acceptable (Locatelli et al. 2010). In forest restoration, these measures tend to focus on the landscape-level to reduce forest fragmentation and conserve genetic diversity; this can improve the resilience of species against climate change, thereby contributing to biodiversity conservation (Locatelli et al. 2015a). Increasing the diversity of tree species, planting material and age structures is one strategy to improve the adaptive capacity of planted forests (Locatelli et al. 2015a; Stanturf et al. 2015). Measures to achieve this include increasing forest cover in climate refugia to improve the long-term persistence of forest-dependent species, and removing ecological barriers to improve habitat connectivity with and seed dispersal from neighboring forests (Ciccarese et al. 2012; Stanturf et al. 2017). This latter strategy can facilitate species migration along climatic gradients and within human-modified landscapes (Brancalion et al. 2019).

Additionally, adaptation strategies for active forest restoration need to take projected climate change directions into account (Stanturf et al. 2017; Chapter 11). An important concept in this context is assisted migration, where species or reproductive material from provenances that are better adapted to future climate conditions are intentionally transferred to regions outside of their natural range (Spathelf et al. 2018).

6.4.2 For people

Increasing the resilience of local communities to climate change is an important target for forest reforestation. This is especially the case in the tropics: Erbaugh et al. (2020) estimate that over 1 billion people live on or near (within 8 km) land predicted to enable forest restoration from 2020 to 2050 if a moderate carbon-tax incentive of USD 20 tCO_2^{-1} is implemented.

Forest restoration can improve the adaptive capacity of local communities in a number of ways. It can contribute to livelihood diversification by providing alternative sources of food, material, and income, especially in times when agriculture is affected by drought (Locatelli et al. 2015a; Chapter 8). It can reduce the impacts of extreme weather events: for example, mangrove forests can stabilize coastal areas against storms and flood events (Pramova et al. 2019); other examples include forests' ability to regulate microclimatic conditions to reduce heat wave exposure for urban populations and protection of agricultural crops in dryer areas (Brancalion et al. 2019; Locatelli et al. 2015a). Additionally, forest restoration can mitigate water supply risks for agriculture and other human uses by stabilizing catchment hydrology, protecting water courses from siltation, enhancing rainfall infiltration into soils, regulating water distribution across the year, reducing flooding during rainfall events, and improving water quality (Brancalion et al. 2019; Locatelli et al. 2015a). Depending on the contextual conditions, however, forest restoration can negatively impact water yields (Feng et al. 2016; Filoso et al. 2017).

6.5 Linking mitigation and adaptation

Mitigation and adaptation are two separate discourses in international and national policy and practice. They are usually governed by different institutions, as they have different objectives on different spatial and temporal scales (Locatelli et al. 2015b). In the land sector, however, the link between mitigation and adaptation is particularly clear as land use has great potential to prevent GHG emissions and reduce CO_2 in the atmosphere (Section 6.2) and to increase the capacities of people and ecosystems to adapt to climate change (Section 6.4; Rizvi et al. 2015). Figure 6.3 illustrates this link in regard to forest restoration.

Forest-based mitigation projects could reduce the risk of non-permanence of carbon stocks by integrating complementary adaptation measures for people and ecosystems (Locatelli 2011). Moreover, both strategies share similar determinants, such as available technology and resources as well as credible information for policymakers (Arneth et al. 2019).

There have been recent calls in the land sector to integrate adaptation and mitigation strategies (Kongsager 2018; Locatelli et al. 2015b). This would reduce trade-offs and competition for limited resources and lead to policy coherence between adaptation and mitigation objectives as well as between climate change and development objectives. It could further contribute to vertical and horizontal policy structures that allow for effective mainstreaming of climate change into sectoral policies and cross-sectoral coordination, respectively (Arneth et al. 2019). In the following section,

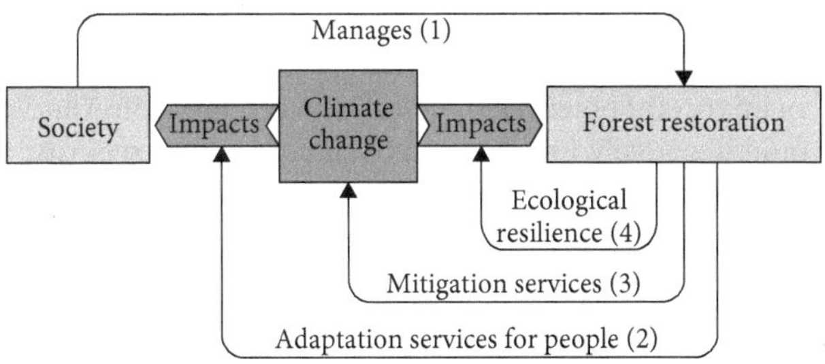

Figure 6.3 Conceptual framework of the integration of adaptation and mitigation into forest restoration (adapted from Locatelli et al. 2015a, 2015b). Forest restoration is managed by society (1) and contributes to the adaptation of people (2) and mitigation of climate change (3). Ensuring ecological resilience will result in forest restoration that is adapted to the impacts of climate change (4).

potential synergies and trade-offs between mitigation and adaptation as well as implications for forest restoration objectives are discussed.

6.5.1 Synergies and trade-offs for people

A significant number of people in the tropics and subtropics depend on ecosystem services (ESS) from forests such as fuelwood, charcoal, fodder and non-wood forest products for their livelihoods. Forest restoration has the potential to deliver these important provisioning ESS, which can contribute to strengthening and diversifying local livelihoods and thus improves adaptation for people (Gregorio et al. 2020). Additionally, restored forests provide regulating ESS that enhance nutrient and water fluxes, regulate pollinators and pests, and thus increase the adaptive capacity of adjacent agricultural land against disturbances (Lamb 2018). Forest restoration projects have the potential to improve infrastructure and social services, strengthen local institutions, generate carbon payments, and provide jobs in nurseries and in the establishment and management of the restored forest (Gregorio et al. 2020). Nevertheless, forest restoration potentially results in adverse outcomes for local communities. For example, a recent global meta-analysis reported that socioeconomic impacts of large-scale tree-plantations were characterized as predominantly negative across a number of categories that describe changes to human well-being (Malkamäki et al. 2018). Engaging local communities in decision-making and implementation is crucial for the success of forest restoration projects. Lack of such engagement risks resistance and conflict (Coleman et al. 2021; Erbaugh et al. 2020; Höhl et al. 2020; Holl and Brancalion 2020). If reforested areas are not protected and maintained, if they are used for livestock grazing or are cleared for other land uses then forest restoration projects will fail.

This illustrates that forest restoration for mitigation or adaptation has to consider the needs of local communities. Chapter 8 of this book provides an in-depth analysis of restoration for a resilient local development and discusses some of the potential trade-offs and synergies that can influence whether a specific forest restoration project will be a positive or negative resource for the local communities. For instance, forest restoration can compete with agriculture for available land, particularly if implemented at scales relevant for climate change mitigation (Section 6.3.2; Fuss et al. 2018). This is an important potential trade-off between mitigation and adaptation for people, as it could negatively impact local livelihoods, food security, and the long-term adaptive capacity of local communities. Agroforestry and trees on farms, strategies included in restoration concepts such as FLR, are gaining recognition because they can provide direct benefits to local people and contribute to biodiversity conservation (Somarriba et al. 2017).

6.5.2 Synergies and trade-offs for forests

Forest restoration projects primarily focusing either on mitigation or on bolstering the adaptation of people to climate change have potential benefits for the adaptation of forests to climate change. For example, establishing highly productive forests, plantations, or agroforestry systems around intact forests as a buffer can reduce anthropogenic pressures on native forests (Stanturf et al. 2017). Restored forests can also increase landscape connectivity and conserve biodiversity hotspots by aiding dispersal, migration, and gene flow among populations of plants and animals (Lamb 2018). This can considerably improve the adaptive capacity of old-growth forests and reduce their vulnerability to disturbances, thus contributing to carbon stock permanence.

However, tree planting that focuses on carbon storage and does not consider biodiversity can negatively impact ecosystem adaptation (Olsson et al. 2019). Native ecosystems can be harmed if afforestation takes places in grasslands and savannahs that historically have not been under forest cover (Holl and Brancalion 2020; Veldman et al. 2019). Additionally in such ecosystems, tree planting, especially with exotic species, can result in a reduction of soil organic carbon, which can partially negate additional carbon storage in biomass (Quiroz Dahik et al. 2021).

Restoration projects that primarily target mitigation can improve the permanence of carbon stocks by implementing adaptation measures to increase the resilience of restored forests against the impacts of a changing climate (Locatelli 2011; Stanturf et al. 2017). Restoration projects that primarily aim to increase the adaptive capacity of forests can result in increased or maintained carbon stocks; in that case, carbon payments could be an attractive additional funding opportunity. Nevertheless, forest restoration adaptation measures, such as introducing species that are more tolerant to future climate conditions or decreasing stand densities, can negatively affect the carbon sink capacity in the short-term. On the long-term, however, these measures lead to increased resilience and thus reduce the risk of future carbon loss (Jandl et al. 2015).

6.6 Conclusion

Forest restoration theoretically has the potential to globally sequester several $GtCO_2$ yr^{-1} over the coming decades, significantly mitigating climate change. To fully realize this potential, forest restoration will have to be deployed and maintained at a scale of several hundreds of millions of hectares. The feasibility of such large-scale restoration is challenged by critical issues across local, national, and global levels as described in this chapter and in Chapter 10.

Implementing large-scale forest restoration requires addressing the underlying economic and social drivers of deforestation and forest degradation (Stanturf et al. 2019). These drivers have presented important barriers to sustainable development for decades. Although frequently framed as a win-win solution, forest restoration has potential trade-offs and adverse outcomes (Holl and Brancalion 2020). Restored forests can be a positive or a negative resource for livelihoods (see Chapter 8) and, consequently, can contribute to addressing socioeconomic drivers of deforestation and forest degradation or can exacerbate them. This underlines that putting forest restoration on the political agenda is not a guarantee for successful climate change mitigation.

The design, implementation, and management of forest restoration should aim for climate-resilient land-use systems that combine production with conservation objectives, thereby contributing not only to climate change mitigation and adaptation for people and forests, but also to biodiversity conservation and sustainable development. This requires cross-sectoral policy approaches; for many countries these still have to be established on national and regional levels. On the landscape level, integrated approaches, especially FLR, are well-suited to achieve this by balancing co-benefits and trade-offs between different goals. Integrating the different human and ecological dimensions and upscaling landscape-level approaches to achieve global goals remains a challenge in FLR. This is partly due to the limited research on these highly complex aspects (Mansourian et al. 2020). Dewi et al. (2017) offer one of the first empirical approaches that systematically analyses portfolios of landscapes, but more research is required to address these challenges.

In view of the multiple interactions across sectors and scales and the need to overcome land-use conflicts and simultaneously address multiple interests and demands, it can be anticipated that effective mitigation contributions of forest-based and restoration measures on a global scale will take time, most likely more time than is available to meet the 1.5°C goal. Additionally, forest restoration, and the way it is implemented, cannot address all issues related to its large-scale deployment. This includes, first and foremost, the need for a substantial future reduction in land demand by agriculture, which depends strongly on agricultural intensification, lower population growth, and/or a global diet shift. The mitigation potential of forest restoration thus might not be as large as expected. Relying on future carbon removal from forest restoration, or other land-based negative emission technologies, instead of near-term emission reductions to reach global mitigation targets, could lead to higher levels of future warming and/or much higher costs to address climate change (Dooley and Kartha 2018; Sanderson and O'Neill 2020). Hence, forest restoration cannot be a substitute for rapid and drastic actions to reduce GHG emissions, especially from fossil fuel combustion (Anderegg et al. 2020; Cook-Patton et al. 2020; Griscom et al. 2017; Holl and Brancalion 2020), but it can be a worthwhile investment in sustainable land use aiming for multiple benefits including some, but not exclusively, climate change mitigation and adaptation effects.

Notes

1. We follow IPCC's 2014 definition of afforestation as the "direct human-induced conversion of land that has not been forested for a period of at least 50 years to forested land through planting, seeding and/or the human-induced promotion of natural seed sources," and reforestation as the "direct human-induced conversion of non-forested land to forested land through planting, seeding and/or the human-induced promotion of natural seed sources, on land that was forested but that has been converted to non-forested land." We distinguish this from the restoration of degraded forests, that is, forest land remaining forest land.
2. A remaining carbon budget expresses the "estimated cumulative net global anthropogenic CO_2 emissions from a given start date to the time that anthropogenic CO_2 emissions reach

net zero that would result, at some probability, in limiting global warming to a given level, accounting for the impact of other anthropogenic emissions" (IPCC 2018).

References

Anderegg, W.R.L., A.T. Trugman, G. Badgley, et al. 2020. "Climate-Driven Risks to the Climate Mitigation Potential of Forests." *Science* 368 (6497). DOI:10.1126/science.aaz7005.

Anderson-Teixeira, K.J., A.D. Miller, J.E. Mohan, et al. 2013. "Altered Dynamics of Forest Recovery under a Changing Climate." *Global Change Biology* 19 (7): pp. 2001–2021. DOI:10.1111/gcb.12194.

Arneth, A., F. Denton, F. Agus, et al. 2019. "Framing and Context." In *Climate Change and Land: An IPCC Special Report on Climate Change, Desertification, Land Degradation, Sustainable Land Management, Food Security, and Greenhouse Gas Fluxes in Terrestrial Ecosystems*, edited by P.R. Shukla, J. Skea, E. Calvo Buendia, et al., pp. 77–129. DOI:10.1017/9781009157988.003.

Austin, K.G., J.S. Baker, B.L. Sohngen, et al. 2020. "The Economic Costs of Planting, Preserving, and Managing the World's Forests to Mitigate Climate Change." *Nature Communications* 11 (5946). DOI:10.1038/s41467-020-19578-z.

Baccini, A., W. Walker, L. Carvalho, et al. 2017. "Tropical Forests Are a Net Carbon Source Based on Aboveground Measurements of Gain and Loss." *Science* 358 (6360): pp. 230–234. DOI:10.1126/science.aam5962.

Bastin, J.-F., Y. Finegold, C. Garcia, et al. 2019. "The Global Tree Restoration Potential." *Science* 365 (6448): pp. 76–79. DOI:10.1126/science.aax0848.

Bernal, B., L.T. Murray, and T.R.H. Pearson. 2018. "Global Carbon Dioxide Removal Rates from Forest Landscape Restoration Activities." *Carbon Balance and Management* 13 (1): p. 22. DOI:10.1186/s13021-018-0110-8.

Besseau, P., S. Graham, and T. Christophersen. 2018. *Restoring Forests and Landscapes: The Key to a Sustainable Future.* Vienna: IUFRO.

Bonner, M.T., S. Schmidt, and L.P. Shoo. 2013. "A Meta-Analytical Global Comparison of Aboveground Biomass Accumulation between Tropical Secondary Forests and Monoculture Plantations." *Forest Ecology and Management* 291: pp. 73–86. DOI:10.1016/j.foreco.2012.11.024.

Brancalion, P.H.S., and K.D. Holl. 2020. "Guidance for Successful Tree Planting Initiatives." *Journal of Applied Ecology* 57 (12): pp. 2349–2361. DOI:10.1111/1365-2664.13725.

Brancalion, P.H.S., A. Niamir, E. Broadbent, et al. 2019. "Global Restoration Opportunities in Tropical Rainforest Landscapes." *Science Advances* 5 (7): eaav3223. DOI:10.1126/sciadv.aav3223.

Busch, J., and J. Engelmann. 2017. "Cost-Effectiveness of Reducing Emissions from Tropical Deforestation, 2016–2050." *Environmental Research Letters* 13 (1): p. 15001. DOI:10.1088/1748-9326/aa907c.

Busch, J., J. Engelmann, S.C. Cook-Patton, et al. 2019. "Potential for Low-Cost Carbon Dioxide Removal through Tropical Reforestation." *Nature Climate Change* 9 (6): pp. 463–466. DOI:10.1038/s41558-019-0485-x.

Carter, S., M. Herold, M.C. Rufino, et al. 2015. "Mitigation of Agricultural Emissions in the Tropics: Comparing Forest Land-Sparing Options at the National Level." *Biogeosciences* 12 (15): pp. 4809–4825. DOI:10.5194/bg-12-4809-2015.

Chazdon, R.L., E.N. Broadbent, D.M.A. Rozendaal, et al. 2016. "Carbon Sequestration Potential of Second-Growth Forest Regeneration in the Latin American Tropics." *Science Advances* 2 (5): e1501639. DOI:10.1126/sciadv.1501639.

Ciccarese, L., A. Mattsson, and D. Pettenella. 2012. "Ecosystem Services from Forest Restoration: Thinking Ahead." *New Forests* 43 (5–6): pp. 543–560. DOI:10.1007/s11056-012-9350-8.

Coleman, E.A., B. Schultz, V. Ramprasad, et al. 2021. "Limited Effects of Tree Planting on Forest Canopy Cover and Rural Livelihoods in Northern India." *Nature Sustainability* 4 (11): pp. 997–1004. DOI:10.1038/s41893-021-00761-z.

Cook-Patton, S.C., S.M. Leavitt, D. Gibbs, et al. 2020. "Mapping Carbon Accumulation Potential from Global Natural Forest Regrowth." *Nature* 585 (7826): pp. 545–550. DOI:10.1038/s41586-020-2686-x.

Delzeit, R., J. Pongratz, J.M. Schneider, et al. 2019. "Forest Restoration: Expanding Agriculture." *Science* 366 (6463): pp. 316–317. DOI:10.1126/science.aaz0705.

Dewi, S., M. van Noordwijk, M.T. Zulkarnain, et al. 2017. "Tropical Forest-Transition Landscapes: A Portfolio for Studying People, Tree Crops and Agro-Ecological Change in Context." *International Journal of Biodiversity Science, Ecosystem Services & Management* 13 (1): pp. 312–329. DOI:10.1080/21513732.2017.1360394.

Dickie, A., C. Streck, S. Roe, et al. 2014. "Strategies for Mitigating Climate Change in Agriculture." Abridged Report. Climate Focus and California Environmental Associates. https://www.climateandlandusealliance.org/wp-content/uploads/2015/08/Abridged_Report_Mitigating_Climate_Change_in_Agriculture.pdf.

Di Sacco, A., K.A. Hardwick, D. Blakesley, et al. 2021. "Ten Golden Rules for Reforestation to Optimize Carbon Sequestration, Biodiversity Recovery and Livelihood Benefits." *Global Change Biology* 27 (7): pp. 1328–1348. DOI:10.1111/gcb.15498.

Dooley, K., and S. Kartha. 2018. "Land-Based Negative Emissions: Risks for Climate Mitigation and Impacts on Sustainable Development." *International Environmental Agreements: Politics, Law and Economics* 18 (1): pp. 79–98. DOI:10.1007/s10784-017-9382-9.

Erbaugh, J., N. Pradhan, J. Adams, et al. 2020. "Global Forest Restoration and the Importance of Prioritizing Local Communities." *Nature Ecology & Evolution* 4 (11): pp. 1472–1476. DOI:10.1038/s41559-020-01282-2.

FAO. 2013. *Climate Change Guidelines for Forest Managers*. FAO Forestry Paper 172. Rome: FAO.

FAO. 2020. *Global Forest Resources Assessment 2020*. Rome: FAO. DOI:10.4060/ca9825en.

Fedele, G., C.I. Donatti, C.A. Harvey, L. Hannah, and D.G. Hole. 2019. "Transformative Adaptation to Climate Change for Sustainable Social-Ecological Systems." *Environmental Science & Policy* 101: pp. 116–125. DOI:10.1016/j.envsci.2019.07.001.

Federici, S., F.N. Tubiello, M. Salvatore, H. Jacobs, and J. Schmidhuber. 2015. "New Estimates of CO2 Forest Emissions and Removals: 1990–2015." *Forest Ecology and Management* 352: pp. 89–98. DOI:10.1016/j.foreco.2015.04.022.

Feng, X., B. Fu, S. Piao, et al. 2016. "Revegetation in China's Loess Plateau is Approaching Sustainable Water Resource Limits." *Nature Climate Change* 6 (11): pp. 1019–1022. DOI:10.1038/nclimate3092.

Filoso, S., M.O. Bezerra, K.C.B. Weiss, and M.A. Palmer. 2017. "Impacts of Forest Restoration on Water Yield: A Systematic Review." *PloS One* 12 (8): e0183210. DOI:10.1371/journal.pone.0183210.

Folke, C., S. Carpenter, T. Elmqvist, et al. 2002. "Resilience and Sustainable Development: Building Adaptive Capacity in a World of Transformations." *Ambio* 31 (5): pp. 437–440. DOI:10.1579/0044-7447-31.5.437.

Fuss, S., W.F. Lamb, M.W. Callaghan, et al. 2018. "Negative Emissions—Part 2: Costs, Potentials and Side Effects." *Environmental Research Letters* 13 (6): p. 63002. DOI:10.1088/1748-9326/aabf9f.

Geng, A., H. Yang, J. Chen, and Y. Hong. 2017. "Review of Carbon Storage Function of Harvested Wood Products and the Potential of Wood Substitution in Greenhouse Gas Mitigation." *Forest Policy and Economics* 85: pp. 192–200. DOI:10.1016/j.forpol.2017.08.007.

Gregorio, N., J. Herbohn, R. Tripoli, and A. Pasa. 2020. "A Local Initiative to Achieve Global Forest and Landscape Restoration Challenge—Lessons Learned from a Community-Based Forest Restoration Project in Biliran Province, Philippines." *Forests* 11 (4): p. 475. DOI:10.3390/f11040475.

Griscom, B.W., J. Adams, P.W. Ellis, et al. 2017. "Natural Climate Solutions." *Proceedings of the National Academy of Sciences of the United States of America* 114 (44): pp. 11645–11650. DOI:10.1073/pnas.1710465114.

Griscom, B.W., J. Busch, S.C. Cook-Patton, et al. 2020. "National Mitigation Potential from Natural Climate Solutions in the Tropics." *Philosophical transactions of the Royal Society of London. Series B, Biological sciences* 375 (1794): p. 20190126. DOI:10.1098/rstb.2019.0126.

Hansen, M.C., P.V. Potapov, R. Moore, et al. 2013. "High-Resolution Global Maps of 21st-Century Forest Cover Change." *Science 342* (6160): pp. 850–853. DOI:10.1126/science.1244693.

Hawken, P. (ed.). 2017. *Drawdown: The Most Comprehensive Plan Ever Proposed to Reverse Global Warming*. New York: Penguin Books.

Höhl, M., V. Ahimbisibwe, J.A. Stanturf, et al. 2020. "Forest Landscape Restoration—What Generates Failure and Success?" *Forests* 11 (9): p. 938. DOI:10.3390/f11090938.

Holl, K.D., and P.H.S. Brancalion. 2020. "Tree Planting Is Not a Simple Solution." *Science* 368 (6491): pp. 580–581. DOI:10.1126/science.aba8232.

Holmgren, M., M. Hirota, E.H. van Nes, and M. Scheffer. 2013. "Effects of Interannual Climate Variability on Tropical Tree Cover." *Nature Climate Change* 3 (8): pp. 755–758. DOI:10.1038/nclimate1906.

Houghton, R.A., B. Byers, and A.A. Nassikas. 2015. "A Role for Tropical Forests in Stabilizing Atmospheric CO2." *Nature Climate Change* 5 (12): pp. 1022–1023. DOI:10.1038/nclimate2869.

Houghton, R.A., and A.A. Nassikas. 2018. "Negative Emissions from Stopping Deforestation and Forest Degradation, Globally." *Global Change Biology* 24 (1): pp. 350–359. DOI:10.1111/gcb.13876.

IPCC. 2014. *2013 Revised Supplementary. Methods and Good Practice Guidance Arising from the Kyoto Protocol*, edited by T. Hiraishi, T. Krug, K. Tanabe, N. Srivastava, J. Baasansuren, M. Fukuda, and T.G. Troxler. Geneva: IPCC.

IPCC. 2018. "Summary for Policymakers." In *Global Warming of 1.5°C. An IPCC Special Report on the Impacts of Global Warming of 1.5°C above Preindustrial Levels and*

Related Global Greenhouse Gas Emission Pathways, in the Context of Strengthening the Global Response to the Threat of Climate Change, Sustainable Development, and Efforts to Eradicate Poverty, edited by V. Masson-Delmotte, P. Zhai, H.-O. Pörtner, et al., pp. 3–24. Cambridge: Cambridge University Press. DOI:10.1017/9781009157940.001.

IPCC. 2019a. "Annex I: Glossary [van Diemen, R. (ed.)]." In *Climate Change and Land: An IPCC Special Report on Climate Change, Desertification, Land Degradation, Sustainable Land Management, Food Security, and Greenhouse Gas Fluxes in Terrestrial Ecosystems*, edited by P.R. Shukla, J. Skea, E. Calvo Buendia, et al., pp. 804–826. Cambridge: Cambridge University Press. DOI:10.1017/9781009157988.010

IPCC. 2019b. "Summary for policymakers." In *Climate Change and Land: An IPCC Special Report on Climate Change, Desertification, Land Degradation, Sustainable Land Management, Food Security, and Greenhouse Gas Fluxes in Terrestrial Ecosystems*, edited by P.R. Shukla, J. Skea, E. Calvo Buendia, et al., pp. 3–36. Cambridge: Cambridge University Press. DOI:10.1017/9781009157988.001.

IUCN and WRI. 2014. *A guide to the Restoration Opportunities Assessment Methodology (ROAM): Assessing Forest Landscape Restoration Opportunities at the National or Sub-National Level.* Working Paper (Road-test edition). Gland: IUCN.

Jandl, R., J. Bauhus, A. Bolte, A. Schindlbacher, and S. Schüler. 2015. "Effect of Climate-Adapted Forest Management on Carbon Pools and Greenhouse Gas Emissions." *Current Forestry Reports* 1 (1): pp. 1–7. DOI:10.1007/s40725-015-0006-8.

Jandl, R., P. Spathelf, A. Bolte, and C.E. Prescott. 2019. "Forest Adaptation to Climate Change—Is Non-Management an Option?" *Annals of Forest Science* 76 (48). DOI:10.1007/s13595-019-0827-x.

Jia, G., E. Shevliakova, P. Artaxo, et al. 2019. "Land–Climate Interactions." In *Climate Change and Land: An IPCC Special Report on Climate Change, Desertification, Land Degradation, Sustainable Land Management, Food Security, and Greenhouse Gas Fluxes in Terrestrial Ecosystems*, edited by P.R. Shukla, J. Skea, E. Calvo Buendia, et al., pp. 131–247. Cambridge: Cambridge University Press.

Koch, A., and J.O. Kaplan. 2022. "Tropical Forest Restoration under Future Climate Change." *Nature Climate Change* 12 (3): pp. 279–283. DOI:10.1038/s41558-022-01289-6.

Kongsager, R. 2018. "Linking Climate Change Adaptation and Mitigation: A Review with Evidence from the Land-Use Sectors." *Land* 7 (4): p. 158. DOI:10.3390/land7040158.

Kreidenweis, U., F. Humpenöder, M. Stevanović, et al. 2016. "Afforestation to Mitigate Climate Change: Impacts on Food Prices under Consideration of Albedo Effects." *Environmental Research Letters* 11 (8): p. 85001. DOI:10.1088/1748-9326/11/8/085001.

Laestadius, L., S. Maginnis, S. Minnemeyer, et al. 2011. "Mapping Opportunities for Forest Landscape Restoration." *Unasylva* 62 (238): pp. 47–48.

Lamb, D. 2018. "Undertaking Large-Scale Forest Restoration to Generate Ecosystem Services." *Restoration Ecology* 26 (4): pp. 657–666. DOI:10.1111/rec.12706.

Lenton, T.M. 2010. "The Potential for Land-Based Biological CO2 Removal to lower Future Atmospheric CO2 Concentration." *Carbon Management* 1 (1): pp. 145–160. DOI:10.4155/cmt.10.12.

Lewis, S.L., C.E. Wheeler, E.T.A. Mitchard, and A. Koch. 2019. "Restoring Natural Forests Is the Best Way to Remove Atmospheric Carbon." *Nature* 568 (7750): pp. 25–28. DOI:10.1038/d41586-019-01026-8.

Locatelli, B. 2011. "Synergies between Adaptation and Mitigation in a Nutshell." Cobam brief. CIFOR. DOI:10.17528/cifor/003619.

Locatelli, B., M. Brockhaus, A. Buck, et al. 2010. "Forests and Adaptation to Climate Change: Challenges and Opportunities." In *Forests and Society: Responding to Global Drivers of Change*, edited by G. Mery, P. Katila, G. Galloway, et al., pp. 21–42. Vienna: International Union of Forest Research Organizations.

Locatelli, B., C.P. Catterall, P. Imbach, et al. 2015a. "Tropical Reforestation and Climate Change: Beyond Carbon." *Restoration Ecology* 23 (4): pp. 337–343. DOI:10.1111/rec.12209.

Locatelli, B., C. Pavageau, E. Pramova, and M. Di Gregorio. 2015b. "Integrating Climate Change Mitigation and Adaptation in Agriculture and Forestry: Opportunities and Trade-offs." *Wiley Interdisciplinary Reviews: Climate Change* 6 (6): pp. 585–598. DOI:10.1002/wcc.357.

Löf, M., P. Madsen, M. Metslaid, J. Witzell, and D.F. Jacobs. 2019. "Restoring Forests: Regeneration and Ecosystem Function for the Future." *New Forests* 50 (2): pp. 139–151. DOI:10.1007/s11056-019-09713-0.

Malkamäki, A., D. D'Amato, N.J. Hogarth, et al. 2018. "A Systematic Review of the Socio-Economic Impacts of Large-Scale Tree Plantations, Worldwide." *Global Environmental Change* 53: pp. 90–103. DOI:10.1016/j.gloenvcha.2018.09.001.

Mansourian, S., J. Parrotta, P. Balaji, et al. 2020. "Putting the Pieces Together: Integration for Forest Landscape Restoration Implementation." *Land Degradation & Development* 31 (4): pp. 419–429. DOI:10.1002/ldr.3448.

Nikinmaa, L., M. Lindner, E. Cantarello, et al. 2020. "Reviewing the Use of Resilience Concepts in Forest Sciences." *Current Forestry Reports* 6 (2): pp. 61–80. DOI:10.1007/s40725-020-00110-x.

Olsson, L., H. Barbosa, S. Bhadwal, et al. 2019. "Land Degradation." In *Climate Change and Land: An IPCC Special Report on Climate Change, Desertification, Land Degradation, Sustainable Land Management, Food Security, and Greenhouse Gas Fluxes in Terrestrial Ecosystems*, edited by P.R. Shukla, J. Skea, E. Calvo Buendia, et al., pp. 345–436. Cambridge: Cambridge University Press. DOI:10.1017/9781009157988.006.

Onyekwelu, J.C., B. Stimm, and J. Evans. 2011. "Review Plantation Forestry." In *Silviculture in the Tropics*, edited by S. Günter, M. Weber, B. Stimm, and R. Mosandl, pp. 399–454. Berlin: Springer.

Pearson, T.R.H., S. Brown, L. Murray, and G. Sidman. 2017. "Greenhouse Gas Emissions from Tropical Forest Degradation: An Underestimated Source." *Carbon Balance and Management* 12 (3). DOI:10.1186/s13021-017-0072-2.

Popp, A., K. Calvin, S. Fujimori, et al. 2017. "Land-Use Futures in the Shared Socio-Economic Pathways." *Global Environmental Change* 42: pp. 331–345. DOI:10.1016/j.gloenvcha.2016.10.002.

Pramova, E., B. Locatelli, H. Djoudi, et al. 2019. *Adapting Land Restoration to a Changing Climate: Embracing the Knowns and Unknowns.* Info Brief. Center for International Forestry Research. DOI:10.17528/cifor/007261.

Quiroz Dahik, C., P. Crespo, B. Stimm, et al. 2021. "Impacts of Pine Plantations on Carbon Stocks of Páramo Sites in Southern Ecuador." *Carbon Balance and Management* 16 (1): p. 5. DOI:10.1186/s13021-021-00168-5.

Reid, J.L., M.E. Fagan, and R.A. Zahawi. 2018. "Positive Site Selection bias in Meta-Analyses Comparing Natural Regeneration to Active Forest Restoration." *Science advances* 4 (5): eaas9143. DOI:10.1126/sciadv.aas9143.

Reyer, C.P.O., N. Brouwers, A. Rammig, et al. 2015. "Forest Resilience and Tipping Points at Different Spatio-Temporal Scales: Approaches and Challenges." *Journal of Ecology* 103 (1): pp. 5–15. DOI:10.1111/1365-2745.12337.

Rizvi, A.R., S. Baig, E. Barrow, and C. Kumar. 2015. *Synergies between Climate Mitigation and Adaptation in Forest Landscape Restoration.* Gland: IUCN.

Roe, S., C. Streck, M. Obersteiner, et al. 2019. "Contribution of the Land Sector to a 1.5 °C World." *Nature Climate Change* 9 (11): pp. 817–828. DOI:10.1038/s41558-019-0591-9.

Roe, S., C. Streck, P.H. Weiner, and M. Obersteiner. 2017. "How Improved Land Use Can Contribute to the 1.5°C Goal of the Paris Agreement." Working Paper prepared by Climate Focus and the International Institute for Applied Systems Analysis.

Sabogal, C., C. Besacier, and D. McGuire. 2015. "Forest and Landscape Restoration: Concepts, Approaches and Challenges for Implementation." *Unasylva* 66 (245): pp. 3–10.

Sanderson, B.M., and B.C. O'Neill. 2020. "Assessing the Costs of Historical Inaction on Climate Change." *Scientific Reports* 10 (1): p. 9173. DOI:10.1038/s41598-020-66275-4.

Sasaki, N., G.P. Asner, Y. Pan, et al. 2016. "Sustainable Management of Tropical Forests Can Reduce Carbon Emissions and Stabilize Timber Production." *Frontiers in Environmental Science* 4 (50). DOI:10.3389/fenvs.2016.00050.

Silva, L.N., P. Freer-Smith, and P. Madsen. 2019. "Production, Restoration, Mitigation: A New Generation of Plantations." *New Forests* 50 (2): pp. 153–168. DOI:10.1007/s11056-018-9644-6.

Smith, P., H. Haberl, A. Popp, K.-H. et al. 2013. "How Much Land-Based Greenhouse Gas Mitigation Can Be Achieved without Compromising Food Security and Environmental Goals?" *Global Change Biology* 19 (8): pp. 2285–2302. DOI:10.1111/gcb.12160.

Smith, P., J. Nkem, K. Calvin, et al. 2019. "Interlinkages between Desertification, Land Degradation, Food Security and Greenhouse Gas Fluxes: Synergies, Trade-Offs and Integrated Response Options." In *Climate Change and Land: An IPCC Special Report on Climate Change, Desertification, Land Degradation, Sustainable Land Management, Food Security, and Greenhouse Gas Fluxes in Terrestrial Ecosystems*, edited by P.R. Shukla, J. Skea, E. Calvo Buendia, et al., pp. 551–672. Cambridge: Cambridge University Press.

Somarriba, E., G. Carreño-Rocabado, F. Amores, et al. 2017. "Trees on Farms for Livelihoods, Conservation of Biodiversity and Carbon Storage: Evidence from Nicaragua on this 'Invisible' Resource." In *Integrating Landscapes: Agroforestry for Biodiversity Conservation and Food Sovereignty*, edited by F. Montagnini, pp. 369–393. Cham: Springer International.

Spathelf, P., J. Stanturf, M. Kleine, et al. 2018. "Adaptive Measures: Integrating Adaptive Forest Management and Forest Landscape Restoration." *Annals of Forest Science* 75 (55). DOI:10.1007/s13595-018-0736-4.

Stanturf, J., P. Kant, S. Mansourian, and P. Madsen. 2017. "Climate Change Mitigation and Adaptation in Forest Landscape Restoration." In *Implementing Forest Landscape Restoration, A Practitioner's Guide*, edited by J. Stanturf, S. Mansourian, and M. Kleine, pp. 75–93. Vienna: IUFRO.

Stanturf, J.A., P. Kant, J.-P.B. Lillesø, et al. 2015. *Forest Landscape Restoration as a Key Component of Climate Change Mitigation and Adaptation*. Vienna: IUFRO.

Stanturf, J.A., M. Kleine, S. Mansourian, et al. 2019. "Implementing Forest Landscape Restoration under the Bonn Challenge: A Systematic Approach." *Annals of Forest Science* 76 (50). DOI:10.1007/s13595-019-0833-z.

Stehfest, E., W.-J. van Zeist, H. Valin, et al. 2019. "Key Determinants of Global Land-Use Projections." *Nature Communications* 10 (1): p. 2166. DOI:10.1038/s41467-019-09945-w.

Tabeau, A., H. van Meijl, K.P. Overmars, and E. Stehfest. 2017. "REDD Policy Impacts on the Agri-Food Sector and Food Security." *Food Policy* 66: pp. 73–87. DOI:10.1016/j.foodpol.2016.11.006.

UNFCCC. 2016. *The Paris Agreement. United Nations Framework Convention on Climate Change*. FCCC/CP/2015/10/Add.1. https://unfccc.int/sites/default/files/resource/parisagreement_publication.pdf

Veldman, J.W., J.C. Aleman, S.T. Alvarado, et al. 2019. "Comment on 'The Global Tree Restoration Potential.'" *Science* 366, eaay7976. DOI:10.1126/science.aay7976.

Zhunusova, E., T.H. Le Sen, J.-M. Schröder, et al. 2019. "Smallholder Decision-Making on Sawlog Production: The Case of Acacia Plantation Owners in Central Vietnam." *Forests* 10 (11): p. 969. DOI:10.3390/f10110969.

7

Forest restoration, biodiversity, and ecosystem services

Rens Brouwer, Frans Bongers, Marielos Peña-Claros, Pieter A. Zuidema, Pedro Brancalion, Madelon Lohbeck, Alejandra Hernández Guzmán, Alan Heinze, Joannès Guillemot, Koen Kramer, and Douglas Sheil

7.1 Introduction

7.1.1 Ecosystem services from forest restoration

Functioning ecosystems support life on earth. Between 2001 and 2005, over a thousand experts from various disciplines assessed the consequences of ecosystem change for human well-being, resulting in the Millennium Ecosystem Assessment (MEA 2005). These authors concluded that approximately 60% of ecosystem services (ES) are being degraded. Forests are among the most important ecosystems providing ES: they sustain millions of species, the air we breathe, and the fresh water we drink, among other things vital to livelihoods. Since 1990, global forest cover has been reduced by 420 million hectares; over half of the remaining forest is degraded (FAO and UNEP 2020). Forest restoration can play an important role in regaining and improving ES and thus contribute to human well-being. Many of the United Nations (UN) Sustainable Development Goals (SDGs) may be realized, in part, through forest restoration (Melo et al. 2021). Forest restoration goes beyond tree cover and can contribute to and sustain various ES (Figure 7.1). There is much research on the effects of forest restoration on ES, and this chapter aims to review the results of these studies.

7.1.2 Ecosystem services frameworks

ES are classified in various ways using different criteria. The Ecosystem Services MEA framework (MEA 2005) provided the basis from which the Common International Classification of Ecosystem Services (CICES) (Haines-Young and Potschin 2018). More recently, the Nature's Contributions to People (NCP) framework (e.g., Díaz et al. 2018) was derived by the Intergovernmental Science-Policy Platform on Biodiversity and Ecosystem Services (IPBES). We use the widely recognized CICES framework to review current findings of forest restoration effects on ES. CICES classifies ES as provisioning, regulating, maintaining, and cultural (Figure 7.1).

Rens Brouwer, Frans Bongers, Marielos Peña-Claros, Pieter A. Zuidema, Pedro Brancalion, Madelon Lohbeck, Alejandra Hernández Guzmán, Alan Heinze, Joannès Guillemot, Koen Kramer, and Douglas Sheil, *Forest restoration, biodiversity, and ecosystem services* In: *Restoring Forests and Trees for Sustainable Development.* Edited by: Pia Katila, Carol J. Pierce Colfer, Wil de Jong, Glenn Galloway, Pablo Pacheco, and Georg Winkel, Oxford University Press.
DOI: 10.1093/9780197683958.003.0007

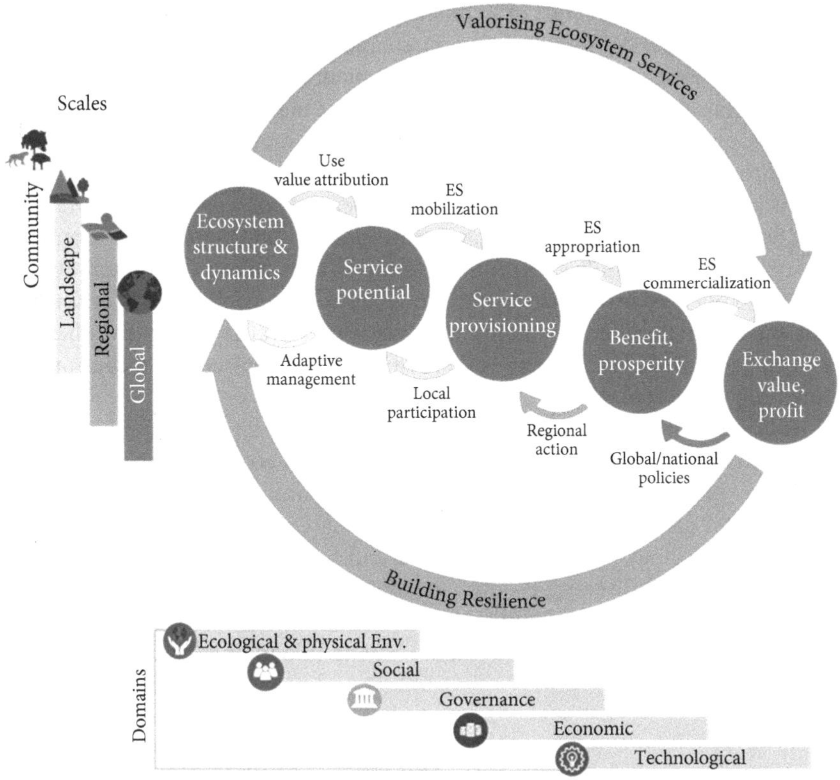

Figure 7.1 Illustration of the forest ecosystem service supply cascade. Figure is adapted from Haines-Young and Potschin (2011) and Kramer et al. (2021). (ES = Ecosystem Service.) The arrows going from left to right above the green circles represent the cascade or flow of ES from a forest ecosystem to societal domains; while the arrows from right to left indicate the transfer of stakeholder impacts and societal domains on the functioning and resilience of forest ecosystems. Horizontal and vertical bars illustrate the domains and spatial scales at which services and benefits are relevant.

These frameworks help identify specific services that can be targeted for restoration, which helps planning restoration for multiple benefits (Alexander et al. 2016). In this chapter we will address biodiversity, nutrient cycling, water cycling, provisioning and cultural services. Climate action and carbon dynamics is covered in Chapter 6.

7.1.3 Scope and methods

Reviewed journal articles on forest ES from 1994 to 2017 (Acharya et al. 2019). While findings show that ES is receiving increased attention, coverage in much of the tropics remains limited. An area of growing interest concerns augmenting and sustaining ES through forest restoration. As emphasized in the International Principles and Standards for the Practice of Ecological Restoration, forest restoration aims at recovering all or part of forest ecosystem characteristics (Gann et al. 2019). These characteristics range in scale from individual plant–soil interactions at the micro-level to community–environment interactions at the macro-level (Figure 7.1). Forest restoration provides an opportunity to alter these characteristics to address specific needs, through species selection and management (Figure 7.2). Forest restoration has a variety of approaches, and the approach depends on conditions, context, and goals (Figure 7.2). Some of these factors have been widely assessed; for example, a meta-analysis of 247 studies including 196 landscapes shows that restoration success is more likely when habitats are less fragmented (Crouzeilles and Curran 2016). While the distinction between active and passive restoration approaches has often been emphasized, this dichotomy is increasingly being replaced by a continuum of options (Chazdon et al. 2021).

This chapter summarizes current knowledge linking forest restoration with ecosystem characteristics. We examine a selection of ecosystem characteristics—properties, processes, functions, and services—linking forest restoration to environmental changes. We selected the following themes: biodiversity, water cycling, nutrient cycling, provisioning services, and cultural services. First, we describe the general role of biodiversity. Biodiversity is also considered throughout the chapter because of its interwoven relationship with ES recovery. Second, we review how forest restoration affects the four ES (Figure 7.1 and Infographic 7.1). Each of the four

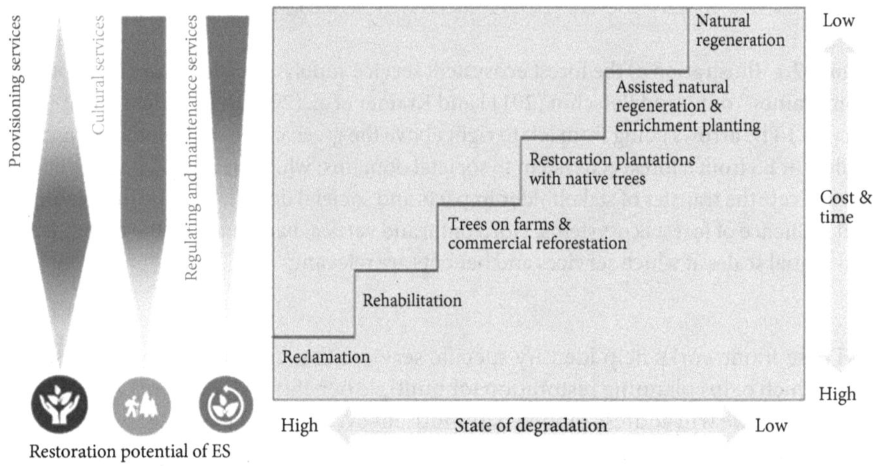

Figure 7.2 The restoration staircase. Depending on the state of degradation of an initially forested ecosystem, a range of management approaches can at least partially restore levels of biodiversity and ES given adequate time and financial investment (capital, infrastructure, and labor). Recovery potential of biodiversity and particular ES depends on the restoration method. See the infographic at the end of the chapter for a more detailed overview. Source: staircase adapted from Chazdon, R. L. 2008. "Beyond deforestation: Restoring forests and ecosystem services on degraded lands." Science 320 (5882): pp. 1458–1460. Reprinted with permission from AAAS.

ES topics is supported by a table which shows the general relationship between forest restoration and the ecosystem characteristics (in these tables, forest restoration is generalized as an increase in tree cover or tree diversity). Third, we discuss trade-offs between ES for different restoration methods and their applications in specific land uses. Fourth, we examine the disservices resulting from forest restoration and the various approaches to value ES. Fifth, we discuss restoration across time and space. Followed by a conclusion that looks at further research opportunities.

7.2 The role of biodiversity

7.2.1 Biodiversity as a regulator or a final service?

The relation between biodiversity and ES is complex and multifaceted. Biodiversity influences ES at all levels: it regulates ecosystem processes and contributes to ecosystem resilience. On the other hand, biodiversity is also valued as a final service in the ES frameworks, as it is important for the conservation of threatened species and of gene pools and contributes to the aesthetic values of forests (Mace et al. 2012). We equate biodiversity here to specific counts of specific taxa. We identify four established ecological theories of Biodiversity Ecosystem Functioning (BEF) relationships, which predict that biodiversity has a regulating and supporting relationship with ecosystem functioning and can affect ES supply. First, niche complementarity theory predicts that diversity in the number and characteristics of species (i.e., trait diversity or functional diversity) should increase resource use efficiency, thereby improving ES supply (Tilman et al. 1997). Second, mass-ratio theory predicts that the most dominant species and their characteristics, rather than the diversity of species, determine ecosystem processes (Grime 1998). Third, structural diversity (such as tree density or basal area) also affects ES supply (Finegan et al. 2015). Fourth, the insurance theory predicts that species respond differently to environmental changes: performance decline in some species may be offset by increases in others; thus, a species-rich community insures long-term ecosystem functioning under environmental change (Yachi and Loreau 1999). Only a few studies quantify biodiversity's contribution to forest carbon stocks and cycling (van der Sande et al. 2017); the role of biodiversity in supporting other ES remains unclear. Many studies indicate that restoring a portion of original biodiversity suffices to restore a diverse range of ES (Lamb 2018) and increase the chances of restoration success (Aerst and Honnay 2011), while a meta-analysis shows only a weak and variable role of biodiversity (Carrick and Forsythe 2020).

7.2.2 How restoration affects biodiversity

During restoration, local biodiversity and forest structure can be recovered by a variety of restoration methods (Figures 7.2 and 7.3). Natural regeneration is the process by which forests regrow from seeds and diaspores that arrive and germinate in situ, or through vegetative growth from preexisting cover or rootstock. It does not require active tree planting but does require seed sources (e.g., nearby forests) and dispersal possibilities (e.g., birds). If these conditions are in place, natural regeneration has an advantage over active planting because of its low costs and because it promotes local species, which are usually native and locally adapted (Chazdon and Guariguata 2016; Crouzeilles et al. 2019). Natural regeneration success depends on conditions (Figure 7.1). Meta-analyses show that the biodiversity in natural regeneration is more similar to that of old-growth forests when past disturbance was less severe (Jakovac et al. 2021). Tree species richness recovery can be surprisingly fast, as little as five decades (Poorter et al. 2021a; Rozendaal et al. 2019), while recovery (to 90% of old growth forest values) of tree species composition may take centuries (Poorter et al. 2021a). Non-plant functional groups essential to ecosystem functioning vary in recovery time: while soil biota can recover in less than two decades, fungi and lichens can take a century or two (Spake et al. 2015; Teixeira et al. 2020).

Natural regeneration is unlikely to be effective in heavily degraded areas or when local seed sources are unavailable (César et al. 2021; Lohbeck et al. 2020). In such cases, more active forms of natural regeneration might be more suitable, such as assisted natural regeneration (ANR), where some species are liberated of competing vegetation to kickstart regeneration. Uebel et al. (2017) found a fourfold-increase in native tree and shrub species richness compared to non-assisted regeneration sites. Similar results were found for applied nucleation (where only parts of an area are planted: "tree island planting"); experiments show that tree density and diversity is consistently higher in areas treated with applied nucleation compared to sites without (Corbin and Holl 2012). Farmer Managed Natural Regeneration (FMNR) is ANR that occurs on active agricultural land, a form of agroforestry (Lohbeck et al. 2020; Rinaudo 2007). FMNR traditionally occurs in the West African Sahel; although quantitative evidence on what ES are restored remains sparse, there is evidence FMNR contributes to on-farm regional diversity and services with direct benefits to farmer livelihoods (Chomba et al. 2020; Moore et al. 2020).

Enrichment planting is a means to overcome potential barriers to natural regeneration such as absence of seed sources or dispersal agents. Seeds or seedlings are planted in forest gaps, young secondary forests, and other sites to improve and recover biodiversity (Lamb et al. 2005). Enrichment planting often focuses on selected species that provide food, timber, or other raw materials or that improve soil properties. Such useful species could help maintain the system, as it improves how restoration is perceived; yet overall biodiversity could decline if few species are planted and/or if they are already present in the system and become more dominant.

Sometimes more drastic interventions are needed such as "restoration plantations" (Figure 7.2). A study of rehabilitation of former bauxite mining lands in the Brazilian Amazon compared natural forest regeneration, mixed commercial species plantings of mostly exotic timber trees, direct seeding with mostly native, early successional tree species, and mixed native species plantings of more than 70 tree species (Parrotta and Knowles 1999). Basal area developed best in the mixed commercial species plantations, but these were relatively poor in species richness compared to the other treatments. Interestingly, all treatments were structurally diverse, which is important for providing habitats for mammals and other animals (Deere et al. 2020). The mixed native species plantations had a lower risk of arrested succession compared to natural regeneration in this case, likely because natural regeneration was slowed by vines and grasses. Planting native species overcomes this while also providing a broad range of benefits, such as recreating linkages and showing natural succession dynamics (McNamara et al. 2006).

Agroforestry can restore biodiversity by planting useful tree species in annual cropping systems (Ordonez et al. 2014). However, agroforestry species planted often differ from the native forest species composition. A global meta-analysis shows that forest species richness (including trees, epiphytes, and forest animals) is significantly higher (46%) in more natural agroforestry systems compared to more intensively managed coffee/cacao plantations (De Beenhouwer et al. 2013). Similar diversity increases can be found in silvopastoral systems when native trees and shrubs are integrated in the system (Broom et al. 2013; Murgueitio et al. 2011).

In some cases active restoration efforts have a negative effect on the desired outcome (Coleman et al. 2021). For example, in the Andes *Alnus acuminata* (Andean alder) is often planted to restore forests; after 30 years, such areas have lower alpha and beta diversity compared to naturally regrown secondary forests of the same age (Murcia 1997). Furthermore, the planting technique may also negatively impact restoration success: the use of heavy machines to plant trees causes damage to naturally resprouting individuals (Sampaio et al. 2007). These examples show that there are different trade-offs to active and passive approaches (Holl and Aide 2011). Biodiversity can recover through forest restoration. A global review found that restoration has positive impacts on both biodiversity and ES provisioning, especially in the tropics (Rey Benayas et al. 2009). Plant and animal species abundance can recover quickly and completely, but diversity recovers more slowly. Recovery is slower in the tropics than

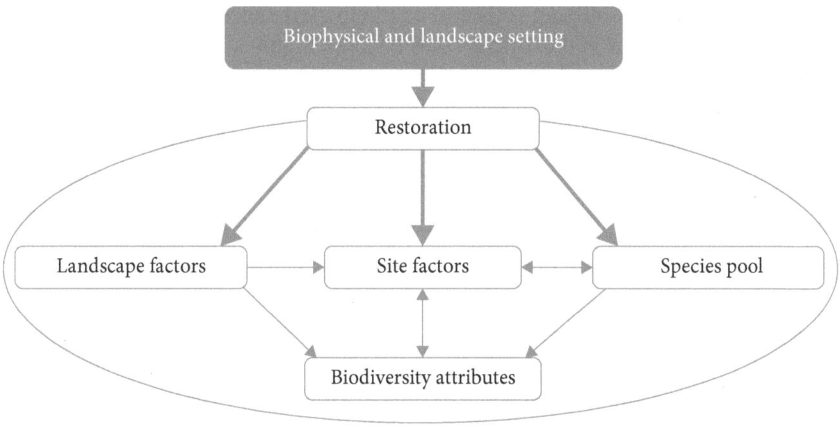

Figure 7.3 Conceptual model of biodiversity restoration. Restoration trajectory at a given site departs from the local biophysical and landscape setting (top box). Restoration might directly manipulate local site, landscape, or species pool (bold arrows). Biodiversity attributes, in turn, may be affected by local site, landscape, or historical factors that may or may not be directly influenced by restoration (light arrows). Biodiversity attributes can be defined at the taxonomic, functional, or structural level. Species pool and biodiversity attributes could also change site factors.

in temperate zones, possibly because a larger fraction of species was initially lost in the former (Meli et al. 2017). Moreover, the onset and speed of recovery of forests depends on climatic water availability (Poorter et al. 2021b). These generic results need to be interpreted cautiously because succession effectiveness depends on local context (Jakovac et al. 2021).

7.2.3 Native or exotic species?

In restoration, native species are generally preferred over exotic species. Justifications include their adaptation to the local environment and tendency to better support local biodiversity (Lamb et al. 2005). Wang et al. (2019) show that native forests support the highest overall diversity of arthropod species, followed by native/exotic mixed-species plantations, while exotic monocultures gave the lowest levels of support. A recent global synthesis of restoration approaches shows that carbon storage, water provisioning, erosion control, and biodiversity benefits are all supplied better by native forests, and that less complex (plantation) systems in dry regions supply these ES relatively poorly (Hua et al. 2022). Yet, conservationists argue that species should be assessed on their environmental impact rather than their origin (Davis et al. 2011). If environmental conditions have changed so significantly that an alternative state of the ecosystem is inevitable, then exotic species might facilitate the development of a novel ecosystem that better sustains ES provisioning (du Toit and Pettorelli 2019).

Local climate change might result in exotic species performing better than native ones (e.g., in response to drought). There are productive and fast-growing exotic species that are good at stabilizing soils and can help start succession (Vásquez-Castro et al. 2021). The danger is if exotic species become highly abundant and if they are functionally similar to co-occurring native species, then they risk becoming invasive (van der Sande et al. 2020). In summary, there are different risks to using native and exotic species in restoration, and careful species selection and studying their interaction with the environment is advised.

7.2.4 Diversity of biodiversity goals

People value different aspects of biodiversity and prioritize different services. Biodiversity goals for forest restoration take various forms; to evaluate success, it is important to define goals. These decisions have implications for what type of restoration interventions are used, and whether and when such efforts can be considered successful. Species richness, for example, is not always prioritized; aiming for full recovery of old-growth forest tree species richness can take over a century (Rozendaal et al. 2019). Some old-growth systems are naturally species poor and characterized by local endemics. In such cases, introducing new tree species will increase local diversity but threatens the uniqueness of the system. When restoration aims to support selected focal species, this can require keeping the system in a specific state that may not be the most diverse. For instance, restoring Dutch heathlands, a nutrient-limited ecosystem in a state of arrested succession, requires removing regenerating trees and intervening (e.g., controlled burning) to remove nutrients and unwanted grasses (Aerts and Heil 1993). Another example is the pied tamarin (*Saguinus bicolor*), one of the Amazon's most endangered primate species, which only occurs in highly degraded forest patches and city parks in central Manaus, Brazil's largest city in the Amazon forest. Maintaining a minimum habitat quality in these degraded forest patches and creating corridors between them is crucial for the conservation of this rare species, but aiming for the full recovery of these forest patches is unrealistic in this city forest (Barr 2016). As biodiversity and ES are closely linked, so are their goals.

7.3 Recovery of ecosystem services by means of forest restoration

7.3.1 Forest restoration, hydrological cycle, and water supply

Forests provide multiple functions and processes related to Earth's hydrological cycle, for example, by intercepting rainfall or mist (horizontal precipitation), and facilitating the infiltration and recharge of groundwater. Freshwater supply is a vital ES for all humans and other organisms; providing freshwater access is an important SDG. Maintaining freshwater supply is increasingly difficult; future estimates about global freshwater demand and supply are alarming. The UN World Water Development Report states that 3.6 billion people live in areas that suffer water scarcity at least one month per year, and that nearly 6 billion people will suffer from clean water scarcity by 2050 (WWAP/UN-Water 2018). Forest restoration changes the water dynamics both on the macro-level (i.e., between land surface and atmosphere) and micro-level (i.e., between soil and vegetation) (te Wierik et al. 2021) (Figure 7.4). The various mechanisms by which trees influence water availability remain only partly understood, but significant advances have been made in the last few decades. Until recently it was commonly accepted that increasing tree cover reduced local water availability we know now it is more complex (Ilstedt et al. 2016). A recent modeling study shows that sustainable reforestation in Europe will lead to increased summer precipitation (Meier et al. 2021).

In the past decade, significant advances have been made in understanding the effects of tree cover on the water cycle, especially the effect of increasing tree cover, an essential part of forest restoration. A systematic review on water yields after tree planting by Filoso et al. (2017) show that while most studies reported decreases in water yields following restoration, other studies observed hydrological benefits. They found that relatively few of the studies focus specifically on forest restoration, and fewer still assess the effect of native species; there is also a lack of large spatial or temporal-scale projects. Information is especially limited for the humid tropics and subtropics and is often context-specific. Many studies focus on the relationship between tree cover and groundwater recharge (Ellison et al. 2017; Ilstedt et al. 2016). Until recently, a commonly accepted theory was, simply put, more trees means less water. This theory is mainly based on case studies in plantations. The optimum cover theory has recently been proposed (Ilstedt et al. 2016), suggesting there

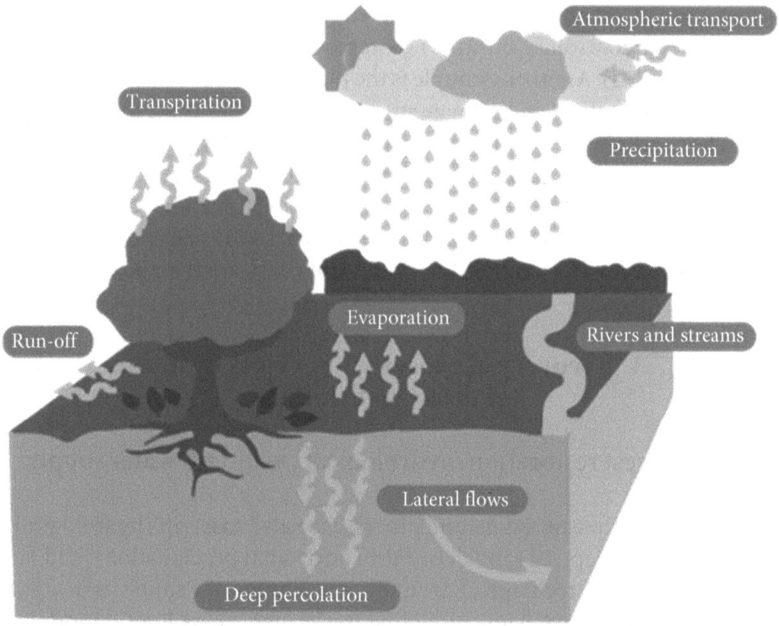

Figure 7.4 Representation between forests and water cycling. Atmospheric vapor is transported globally and regionally, falls as precipitation on forests, and is recycled by forests and other forms of vegetation. Locally within a forest, transpiration and evaporation occur. Infiltration, groundwater recharge through deep percolation and lateral flows, can be facilitated by trees. Runoff can be reduced by increased vegetation cover.

is an optimum for maximizing groundwater recharge, especially in the dry tropics (Table 7.1). A study in Burkina Faso shows that in treeless areas about 8 mm of precipitation per year recharged groundwater, but that for areas close to trees, groundwater recharge increased up to 39 mm due to improved soil infiltration capacity (Tobella et al. 2014). This capacity increased because of the presence of roots and soil fauna, which create macropores that act as water channels. This effect was observed up to 25 m away from tree stems, that even just a few trees per hectare can substantially improve groundwater recharge and reduce runoff. Other studies confirm there is an optimum in tree cover for groundwater recharge, but that it may result in less streamflow (van Meerveld et al. 2021). A study by Zimmermann et al. (2013) similarly found that where entire catchments undergo forest regrowth, there may be a substantial decrease in streamflow in the initial stages. Another comparison between a forested catchment and pasture catchment in Panama shows there was 35% less total runoff, smaller peak runoff rates and increased streamflow in dry periods in the forested catchment (Ogden et al. 2013). The effect of optimum tree cover on groundwater recharge in all these studies depends not only on improved soil properties but also on increased evapotranspiration through trees; combined, there still was a net positive effect. In another configuration (e.g., nonoptimal tree coverage) the result could be different.

Table 7.1 Effects of forest restoration on water cycling related ecosystem characteristics. Effects of an increase in tree cover and management practices are shown. The icons in the first column provide information on the scale of the effect of restoration on an ecosystem characteristic. Management effects refer to any restoration interventions, including selection of species with certain traits. The biodiversity effect is the product of the latter and to what extent certain species are promoted over others.

Ecosystem characteristics	Effects and influences of characteristic	Restoration and management effects
Infiltration 	Water entry in the soil, controlling surface runoff and recharging groundwater.	*Tree cover effect* = ↑ *Management*: tree spacing; soil management.
Groundwater recharge 	Water moves downward from surface water to groundwater through the soil.	*Tree cover effect* = ⌒ *Management*: species selection based on root system; soil management.
Transpiration 	Trees extract water from the soil or groundwater and transpire it to the atmosphere as vapor through their stomata in their leaves.	*Tree cover effect* = ↑ *Management*: species selection based on crown type and leaf traits; pruning.
Interception 	Interception of water by trees, leaves and other structures, preventing it from reaching the ground (and evaporating ultimately).	*Tree cover effect* = ↑ *Management*: species selection based on crown type and leaf traits; promoting complex forest structure to increase interception.
Soil evaporation 	Evaporation from soil surface, the barer the soil, the more evaporation. Vegetation cover can prevent this.	*Tree cover effect* = ↓ *Management*: understory management, soil *Management*; selection of fast-growing species to cover the soil.
Surface runoff 	Water leaving the system, running off the surface downstream. Bare soil increases runoff and stimulates soil erosion.	*Tree cover effect* = ↓ *Management*: understory management, soil treatments to decrease compaction; selection of fast-growing species to cover the soil.
Vapor and cloud capture 	Interception of fog and cloud provides significant amounts of moisture in certain locations/ seasons. Some plants extract water from humid air or water condensates on their leaves.	*Tree cover effect* = ↑ *Management*: species selection based on crown type and leaf traits; promoting more complex forest structure to increase the cloud intercepting surface.

(*continued*)

Table 7.1 Continued

Ecosystem characteristics	Effects and influences of characteristic	Restoration and management effects
Rainfall recycling 	An integrated property that results from many of the others, inside forest communities but also influences global atmospheric flow.	*Tree cover effect* = ↑ *Management*: not only forest type, but also density and structural complexity.
Biotic pump 	The theory that suggests that tree cover attracts atmospheric flows from elsewhere by favoring condensation to occur more frequently (a process that leads to lower air pressures).	*Tree cover effect* = ↑ *Management*: landscape management and planning, which impacts flow of air mass.
Flood moderation 	An integrated property that results from many of the others, tree cover allows for maintaining water longer in the system and releasing it more gradually which prevents floods.	*Tree cover effect* = ↑ *Management*: species selection based on root system; soil treatments such as gullies and barriers.
Water purification and fresh water supply 	Forests filter sediments and other pollutants from the water in the soil before they reach a water source, such as a stream, lake, or river.	*Tree cover effect* = ↑ *Management*: species selection based on root system; soil treatments that affect soil structure.

Legend:
Icons represent the **scale** of the effect.

tree level community level landscape level regional level global level
and the **direction** of the effect (**↑ increase, ↓ decrease, or ∩ here optimum).**

Other factors such as terrain, slope, amount of precipitation, previous and current land use, time, and species could affect groundwater recharge. For example, species with deep roots are able to connect the deep soil and groundwater with the atmosphere; the presence and number of deeply rooted species thus affects the water cycle (Fan et al. 2017). Much remains unknown about groundwater recharge, and optimal configurations are context- and location-specific.

It is increasingly recognized that much precipitation depends on moisture emitted by trees through evapotranspiration (te Wierik et al. 2021), yet there is still high uncertainty on where the tree-based moisture will fall as rain. Forest cover increases evapotranspiration, which can affect precipitation thousands of kilometers downwind on other continents (Sheil 2018; Van Der Ent et al. 2010). Though climate modelers still have difficulty simulating and predicting rainfall patterns in a consistent manner, most suggest that large-scale deforestation will reduce downwind precipitation, whereas maintaining or increasing forest cover can mitigate this reduction (Lawrence and Vandecar 2015). Recent empirical work confirms that deforestation negatively affects the hydrological cycle (Leite-Filho et al. 2021). Precipitation interception by the canopy might increase rapidly during forest succession and forest restoration; in as little as a decade throughfall volumes can approach typical mature-forest levels (Zimmermann et al. 2013). This could be explained by the theory that increased tree cover means intensified recycling of precipitation over land. More trees improve the recycling of this water, allowing it to evaporate and fall over land multiple times before it reaches the ocean again (Staal et al. 2018). The feedback between precipitation and tree cover has become an important focus for study. One recent theory, the "biotic pump," explains how tree cover influences pressure gradients that carry winds and moisture across continents; forest restoration could recover local moist climates and could bring back more reliable rainfall to drought-prone regions (Makarieva et al. 2013; Sheil and Murdiyarso 2009).

7.3.2 Forest restoration and nutrient cycling

Nutrient cycling has direct influence on local and global biogeochemical cycles and plays an important regulating and maintenance role in the provisioning of other ES. Many life forms depend on healthy soils. Fertile soils facilitate the production of food, timber, and other raw materials that people use daily. Humans obtain over 98% of their food from terrestrial lands; maintaining soil fertility is of great importance for human welfare (Pimentel and Burgess 2013). Forest cover typically plays an important role in maintaining the nutrient cycle (Figure 7.5), as forests can better retain soil and nutrients, and prevent runoff and leaching when compared to more open landscapes. Mineralization and leaf litter decomposition are the most commonly studied processes related to nutrient cycling in forests (de Bello et al. 2010); many other aspects, such as the interaction between soil fauna and vegetation, remain incompletely understood. Experiments show that the soil community is an important driver of forest restoration and can steer the direction of plant community development (Wubs et al. 2016). Forest restoration, for example, after intensive agricultural land use, can affect degraded soils by improving water infiltration, thereby preventing erosion and retaining nutrients.

A global meta-analysis of restoration of ES in tropical forests found that restoration activities contributed to a significant increase in soil nutrient attributes (Shimamoto et al. 2018). Recent evidence shows that on some abandoned agricultural fields and pastures, certain soil attributes (bulk density, carbon (C), and nitrogen (N)) recovered to 90% of old growth forest values in less than one year (Poorter et al. 2021a). In other cases, however, recovery took longer, for example, in Ghana after soil compaction from logging extraction machinery (Hawthorne et al. 2012). Recovery of other nutrient stocks may take longer as the soil gets extra nutrients from dying and

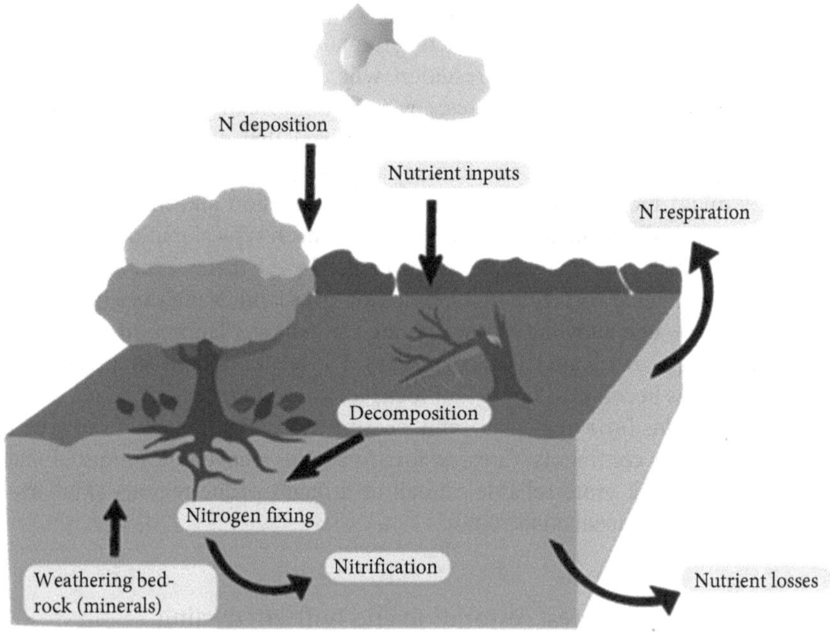

Figure 7.5 Representation between forests and nutrient cycling. Nutrients enter the forest ecosystem through deposition and by weathering bedrock. Within the system, nutrients are cycled through nutrient uptake by organisms and decomposition processes. Some trees form a mutually beneficial symbiotic relationship with soil bacteria that allow them to fix nitrogen. Microbes also facilitate the nitrification processes. Nutrients leave the system through leaching and respiration.

decomposing trees at later successional stages (van der Sande et al. 2022). Some people apply burning (e.g., in shifting cultivation) for a short-term increase in nutrient availability.

Soil recovery is strongly dependent on context. A recent analysis of 21 sites across the Neotropics shows that during secondary forest succession, bulk density decreased and C and N increased in forest regrowing on croplands and sites with high-activity clays; sites with annual lower precipitation generally decreased in pH, whereas sites with higher precipitation were more likely to show increased N and decreased C:N ratio (van der Sande et al. 2022).

Many studies show that live and dead biomass stocks can recover to mature forest levels within two or three decades. Soil functions and properties, however, do not recover as quickly. Soil recovery is positively linked to biomass increase (Gavito et al. 2021; Lohbeck et al. 2018); litter production and decomposition, are positively linked to standing vegetation biomass and are important aspects of restoring the nutrient cycle (Lohbeck et al. 2015). Organic matter can mediate key soil characteristics such as nutrient stocks and water-holding capacity (Larney and Angers 2012). Increasing soil organic matter through management and restoration measures could improve many other soil properties. Chen et al. (2020) show in a global meta-analysis that both soil organic C content and stock are on average 5% and 8% higher, respectively, in species mixtures than in monocultures. However, a study on riparian forest restoration in California, USA, found that soil carbon, N stocks and availability, and nutrient-use efficiency were not at mature forest levels after two decades (Matzek et al. 2016). Reforested sites might look similar to mature forests in terms of vegetation structure and biomass, but they do not function the same.

Soil function recovery is greatly influenced by prior land use (Jakovac et al. 2021). Gomes and Luizão (2012) show that previous land use, time since abandonment, and number of fire events explained 10–38% of the variation in the amount and configuration of soil macronutrients. The effect of prior land use varies with climate zones. In boreal forests it took soil 30 years to return to pre-disturbance soil C levels (Norris et al. 2009), while in secondary tropical forests it took 20–100 years (Martin et al. 2013).

Moreover, different restoration approaches return different results. A Costa Rican study of agricultural landscapes compared differences in litter production and macronutrients (C, N, Ca, Mg, P, Cu, Mn, and Fe) between natural regeneration, tree island plantations, patch plantations, and reference forests (Lanuza et al. 2018). They planted four species, of which two were N-fixing. After a decade, litter production and macronutrients did not differ between tree islands, patch plantation, or reference forests; however, all were greater than natural regeneration. In another study of agroforestry systems with multiple species, soil erosion rates were 50% lower than in monoculture systems (Muchane et al. 2020). This was probably due to through improved infiltration rates, reduced runoff, increased soil macro-aggregate proportion, and improved soil stability. What species are present, their characteristics and diversity, affects functions such as production, litter decay, and nutrient cycling (Hooper et al. 2005; Lohbeck et al. 2015). A meta-analysis of a large forestation program in Northern China illustrates that depending on the species, different soil attributes can be recovered (Gao and Huang 2020): one Pinus sp. increased soil organic C, while another increased total N content; a Robinia sp. was more effective at increasing available potassium content. In an experiment with 14 tree species, 30 years after establishing single-species plots, differences in litter calcium concentrations resulted in significant changes of soil chemistry and fertility (Reich et al. 2005). Other studies have similar results: certain pioneer species affected N availability and cycling more than others (Gomes et al. 2012); that native broadleaf species increased nutrient cycling more compared to exotic coniferous species (Ramírez et al. 2014), and that abundance of invasive species was a native broadleaf species increased nutrient cycling more than exotic coniferous species (Ramírez et al. 2014); and invasive species abundance is a good indicator for soil erosion restoration in degraded agricultural landscapes (Lohbeck et al. 2018). A study in China shows that plantations with only N-fixing species, after 25 years, resulted in 40–50% higher soil organic matter and 20–50% higher N concentration compared to non-N-fixing forests (Wang et al. 2010). In a 300-year chrono sequence in Panama, N-fixing tree species in young forests accumulated C up to nine times faster than non N-fixing species and showed species-specific differences in the amount and timing of fixation (Batterman et al. 2013).

These results mean that some species are more efficient than others at restoring the nutrient cycling processes in degraded lands, especially as species selection is an important aspect for forest restoration practitioners and since species selection is one of the easier restoration measures to influence. Interestingly, some studies report that prior to restoration often no soil indicators were analyzed, highlighting a gap in soil data and project preparation (Mendes et al. 2019). This can be problematic: while the manipulation of physical, chemical, or biological components of the soil system can benefit site restoration, it can also have unintended cascading effects on other (soil) ecosystem functions and processes (Heneghan et al. 2008). It is important to use a holistic approach when defining restoration goals, one that includes all aspects of soil, vegetation, and their interactions. Table 7.2 provides an overview of forest restoration's effects on nutrient cycling characteristics.

7.3.3 Forest restoration, tree-based resources and provisioning services

Forests and trees provide vital resources for 1.3 billion people, of which 300–350 million people (about half of whom are indigenous) living within or close to dense forests depend almost entirely on forests for subsistence (FAO and UNEP 2020). Forests and trees provide a wide range of materials, both timber and non-timber forest products (NTFPs). For instance, about 50% of arboreal species in lowland Amazon are classified as useful species (Coelho et al. 2021). The World Bank forecasts that global timber demand is set to quadruple by 2050, emphasizing the need for plantation forests and the inclusion of timber species in restoration efforts. Many NTFPs are used and traded locally and globally. Most forested regions produce resins, nuts, fruits, fibers, medicinal and ornamental plants, animal foods, and other products. Indigenous peoples especially have vast local ecological knowledge on useful species. A field study in East Kalimantan among traditional forest-dwelling communities shows that at least 1,457 plant species are being used combining 2,141 distinct uses or values (Sheil et al. 2009). Many people worldwide, particularly those living in extreme poverty, depend on forest products (Angelsen et al. 2014). Meeting these people's basic needs should receive sufficient emphasis in restoration (see Kaimowitz and Sheil 2007). While the demand and the trade of many products is increasing, the provisioning of many products is declining as a result of deforestation, forest degradation, and unsustainable exploitation practices. This unbalance leads to increased concerns about the long-term sustainability of these products and the species from which they are derived.

There is ample evidence of the declining supply of forest products, exacerbated by unsustainable extraction practices over recent decades (Arnold and Pérez 2001; Bongers et al. 2019). The global food system plays a large role in environmental degradation and forest deforestation. Restoration can assist in maintaining, even increasing, the supply of forest products, for example, through (enrichment) planting, assisted regeneration, and species management (Box 7.1). Many useful and often underutilized tree and plant species offer the potential for sustainably produced foods, providing livelihood benefits, including improved human nutrition and multiple other ES (Jansen et al. 2020).

Table 7.2 Effects of forest restoration on nutrient cycling-related ecosystem characteristics. Effects of an increase in tree cover and management practices are shown. The icons in the first column provide information on the scale of the effect of restoration on an ecosystem characteristic. Management effects refer to any restoration interventions, including selection of species with certain traits. The biodiversity effect is the product of the latter and determines to what extent certain species are promoted over others. See legend in table 7.1 for explanation of the icons.

Ecosystem Characteristics	Effects and Influences of Characteristics	Restoration and Management Effects
Nutrient uptake	Nutrient uptake influences plant growth and survival, affected by nutrient availability and tree traits.	*Tree cover effect* = ↓ *Management*: species selection based on relation with nutrient cycle (e.g., N fixers, mycorrhizal species).
Decomposition rate	Decomposition rate of plant biomass (e.g., leaves, trunks), is affected by tree traits, soil fauna, and microclimatic conditions.	*Tree cover effect* = ↑ *Management*: species selection based on wood and leaf traits; add decomposing bacterial/ fungi packages; promoting high diversity of species to increase decomposition rate.
Litter quantity and quality providing soil protection and nutrients	Leaf litter quantity and quality is affected by plant and community productivity (this indirectly by plant nutrient uptake), tree traits and leaf characteristics. Litter is the basis for decomposition: food for decomposers, and provisioning of nutrients.	*Tree cover effect* = ↑ Species diversity effect = ↑ *Management*: species selection based on litter production and decomposability; promoting high diversity of species to increase quality and quantity.
Soil organic matter: stock and decomposition	Holding water and nutrients, providing nutrients after decomposition, providing food for decomposers (fungi, animals, bacteria).	*Tree cover effect* = ↑ *Management*: species selection, decomposition management.
Soil nutrient stocks; fertility	Soil nutrient stocks are determined by past land use, atmospheric nutrient deposition, weathering of parent material, rate of leaching, and vegetation type and tree traits.	*Tree cover effect* = ↑ *Management*: species selection; promoting high diversity of species to increase nutrient stocks.
N-fixing	Some tree species live in symbiosis with bacteria, which allows them to fix nitrogen from the atmosphere into usable compounds in the soil through a chemical process. The proportion of such species present in a community can affect the amount of N-fixing at the community level.	*Tree cover effect* = depends on species *Management*: planting or promoting species that have this characteristic.

(*continued*)

Table 7.2 Continued

Ecosystem Characteristics	**Effects and Influences of Characteristics**	**Restoration and Management Effects**
Removal of material: • Timber • Fuel wood • Non-timber forest products (NTFPs) • Lopping and pruning 	All removal takes away nutrients out of the system. Removal in one system may decrease or prevent removal in another system, therefore removal of materials can have indirect effects on broad spatial ranges.	*Management*: control through regulations of what is and what is not allowed which affects the type of nutrients removed from the system (e.g., CITES – multilateral treaty to protect endangered plants and animals); sustainable management practices; proportion of useable species present in a community.

The supply of forest resources depends on forest production capacity (i.e., biomass production). This, in turn, depends on habitat and regulating functions. Maintenance of these ecosystem functions contributes to the (continuous) supply of forest and tree-based resources (Box 7.1).

The majority of tree plantations provide timber; a smaller portion produce fruit, resin, and other resources. Such plantations often consist of single species (monocultures), or a few commercial species (mixed plantations). Many of the (industrial) monocultures involve a limited number of species from a small number of genera, especially *Pinus*, *Eucalyptus*, *Tectona*, *Pseudotsuga*, and *Acacia*. These plantation forests are expanding to satisfy an increasing global demand for timber products.

There is an ongoing debate about whether monocultures can be considered forest restoration and what makes a forest a forest (see Chazdon et al. 2016). Monoculture plantations can supply large amounts of resources but hardly contribute to other forest restoration objectives (Hua et al. 2022). Moreover, they provide a very limited variety of resources compared to the variety found in original forests (e.g., bushmeat, medicine, and food). Studies show that plantations with two to four species can be more productive and have more advantages in biodiversity, economy, and forest health compared to monocultures (Liu et al. 2018). Careful design and planning are necessary in mixed plantations to combine trees with complementary traits that maximize positive and minimize negative interactions. Experience suggests that smallholders often prefer to include many tree species to diversify their production (Lamb and Gilmour 2005). Nonetheless, some smallholders participate in monoculture cash crop plantations such as oil palm. A household livelihood survey in multiple large-scale tree planting programs in India shows that tree planting supports little direct use by local people, indicating that large-scale tree planting does not take into account livelihood goals (Coleman et al. 2021).

Box 7.1 Brazil nut trees in smallholder Amazon forest restoration

Figure 7.6 Photo: Rens Brouwer

The Brazil nut tree (*Bertholletia excelsa*, locally also called Amazon nut, Castanha or Castaña) is an emergent tree species occurring solely in the Amazon region. In many parts of the Bolivian, Peruvian and Brazilian Amazon forest, Brazil nuts are one of the more important NTFPs for rural livelihoods. Since the beginning of the 20th century, commercial exploitation of Brazil nuts has risen rapidly due to the import by Europe, the United States and other countries. This has led to Brazil nut being one of the few globally traded nuts that is harvested from natural forests (Guariguata et al. 2017). Although evidence suggests that these 'natural' forest have been domesticated for a long time. In most cases this product is harvested from 'natural' Brazil nut tree populations, unfortunately recent studies show that the sustainable production of Brazil nuts is threatened by lack of natural regeneration, adult tree mortality, and land conversion. An option to sustain the nut production and to maintain the presence of this emergent tree species in the Amazon rainforest is to enhance regeneration through enrichment planting or using this species in other types of forest restoration. In Peru local farmers, or castañeros (photo), have been actively and successfully planting this species on their land, mainly with a socio-economic incentive, but with forest restoration as an additional benefit. Secondary forests seem to be especially good locations to be enriched and restored with this species, adding value and incentive to protect these forests from future deforestation and further degradation (Brouwer et al. 2021).

Market price differences means that an increase in low-value species lowers a plantation's economic value; yet, a greater diversity provides multiple benefits and services. There are intermediate restoration systems, such as the use of Eucalyptus as a transitional stage. Brancalion et al. (2020) show that biomass accumulation was nine times greater in mixed eucalypt-native species plantations compared to restoration plantations without the fast-growing eucalypt. Additionally, eucalypts did not affect the natural regeneration of native woody species before or after eucalypt logging in the stand. While both treatments had similar natural regeneration richness and planted non-pioneer growth, the income from eucalypt wood production offset 44–75% of implementation costs. The study shows how monocultures can aide forest restoration goals, and that intermediate transitional systems can provide multiple ES.

Natural regeneration has very diverse outcomes compared to monocultures, from species-rich forests to forests dominated by a few thriving species. Enrichment planting increases the populations of commercially important species in secondary forests (Lamb et al. 2005). Desirable species can be promoted through selective removal of competition and other silvicultural measures (Brouwer et al. 2021). FMNR allows farmers to enhance regeneration by managing certain species, mostly ones providing valuable resources (Lohbeck et al. 2020). Creating agroforestry systems with a mix of tree-crops, perennials, and annual plants is another intensive restoration method with high provisioning services output. Common perennial crops include mango, papaya, cocoa, coffee, rubber, and oil palm. Agroforestry systems have proven to be feasible tools for forest restoration and can achieve ecological and socioecological goals.

Species selection is crucial for restoring ES. Evidence shows that farmers prefer species that provide provisioning services (Cáceres et al. 2015; Heinze et al. 2020). Planting multi-purpose species can increase provisioning services while enhancing ecosystem function and biodiversity. The restoration of unproductive land with useful species is an important option to reduce the pressure on natural forests and to restore provisioning services. In Ecuador, for example, natural forests provide 120 timber species and many NTFPs for the market (Bendix et al. 2013); native tree restoration can help sustain the provision of these goods. More knowledge is needed on how restoration affects provisioning. An overview of the effects of restoration on provisioning ES is shown in Table 7.3.

7.3.4 Forest restoration and cultural services

Cultural services are different than the other ES. Cultural services provide mental, emotional, and physical benefits (MEA 2005). The most studied categories are recreation (e.g., ecotourism) and aesthetic values (e.g., artistic expression); less studied are cultural identity and heritage, as well as educational, spiritual, and religious values (Mengist et al. 2020; Schirpke et al. 2021). A recent meta-analysis found that 90% of 51 studies show at least one positive association between nature-based recreation and mental health (Lackey et al. 2021) (Box 7.2). In 2020, during the COVID lockdowns, there was a huge spike in daily forest visitors; many people turned to nature for recreation and relief (Derks et al. 2020). Psychological and social health, and other cultural services that can be linked to socioeconomic values, e.g., community economic

Table 7.3 Effects of forest restoration on provisioning ES characteristics. Effects of an increase in tree cover and management practices are shown. The icons in the first column provide information on the scale of the effect of restoration on an ecosystem characteristic. Management effects refer to any restoration interventions, including selection of species with certain traits. The biodiversity effect is the product of the latter and determines to what extent certain species are promoted over others. See legend in table 7.1 for explanation of the icons.

Ecosystem Characteristics	**Effects and influences of Characteristics**	**Restoration and Management Effects**
Timber production 	Presence or absence of species. Wood characteristics. Depends on the species selected, type of market (local vs international) and on market demand. Timber production demands some level of infrastructure, and through that also impacts on other issues (nutrient cycle, water cycle)	*Tree cover effect* = ↑⌒ bad for light-demanding species, good for slow-growing, old-growth species. *Management*: species selection, individual tree management towards high quality timber, reducing competition.
Non-timber forest products (NTFPs) production (food, medicinal, ornamental and raw material resources) 	Presence or absence of species providing NTFPs. Depends on the type of product and intensity of collection. The impact depends on the part of the plant being collected (leaves, fruits), and varies from low (leaves) to high (roots).	*Tree cover effect* = ↑⌒ depends on proportion of usable species in a community. *Management*: species selection, planting and management, depending on the target products.
Genetic resources 	Presence or absence of species with potentially useful genetic material "gene bank" (e.g., genes for resistance to pathogens).	*Tree cover effect* = ↑⌒ depends on species diversity. *Management*: selection for (genetic) diversity of species; control of dominating species.

development, are especially relevant in urban forests (Nesbitt et al. 2017). Forests are often connected to sacred natural sites, including particular trees and groves, that have a special spiritual significance for people and communities (Verschuuren et al. 2017). A study in Hawaii reveals how forests elicit spiritual, heritage, and identity-related values, and that these values vary with ethnicity and residence time (Gould et al. 2014). Sociocultural perspectives are not specific to cultural services, sociocultural

Box 7.2 Shinrin-yoku, or forest bathing

The wind flowing through leaves, birds chirping, the scent of trees and dirt, the sunlight scattering through the leaves, fresh air. These are all things that we experience in forests and for many of us they make us feel at ease and in comfort. Our stress reduces, we are more relaxed, feel refreshed and get back our energy. Going to nature and forests to enjoy these benefits is something many people from different cultures do, but the Japanese even have a specific term for this: Shinrin-yoku. Shinrin in Japanese means "forest," and yoku means "bath", so it translates as 'forest bathing'. The term was coined by the head of the Japanese Ministry of Agriculture, Forestry, and Fisheries, Tomohide Akiyama, in 1982 to encourage more visitors to forests, later it grew out to a therapeutic practice. Studies have shown that forest environments reduce stress, anger, anxiety, depression and improve immune functions among participants (Li, 2010; Park et al. 2010). It is not a sport activity; it is about taking in the forest atmosphere to gain therapeutic benefits. In practice, you already gain these benefits during a leisurely walk, but some people also prefer to go to the forest with a more focused mindset to actively take a 'forest bath' by focusing on their five senses. Finding calmness and relaxation of course differs from person to person, but the presence of forests and trees allows people to have access to the benefits of forest bathing. Forest restoration can aid in providing these places or maintaining them. Especially restoring forests near cities could help to restore the connection between people and nature.

Figure 7.7 Photo: Rens Brouwer

values show the importance that people and societies assign to a wide range of ES (Scholte et al. 2015). Understanding the cultural services gained by restoration projects could help capacity building and planning, and increase project support among multiple stakeholders.

The act of restoration in itself can be seen as a cultural service, in which people engage with nature for educational and inspirational purposes, giving some a sense of meaning (Bell 2001; Cramer 2008). Restoring urban and peri-urban forests enhances cultural services and well-being: people enjoy forests for the atmosphere and fresh air, relaxation and stress reduction, break from their hectic lifestyles, and emotional improvement (Hansen-Møller and Oustrup 2004; Schipperijn et al. 2010). Following a restoration project in the Atlantic forest of Brazil, people reported an appreciation for cultural ES such as aesthetic landscape improvement, tourism, recreation, as well as various religious, spiritual, and educational services (Brancalion et al. 2014). The support of local stakeholders in restoration projects increases the persistence of positive outcomes in the long term. FMNR, and likely other approaches that support local livelihoods, has the capacity to increase farmers' autonomy, which is an intangible cultural service (Moore et al. 2020).

Restoration does not necessarily mean all cultural services are improved. A survey in China shows that cultivated agricultural land and residential areas were evaluated as more important for providing cultural services than newly restored forests (Dou et al. 2019), probably because of local traditions of strong social neighborhood bonds in agricultural landscapes. Most interviewees believed that forest restoration had decreased religious and spiritual services, cultural heritage, and mental and physical health values. In contrast, aesthetic, recreational, inspirational, and education and science services were perceived as having increased in value after restoration. Forest restoration did not fit well in their current cultural values. This may be different for communities in landscapes where deforestation or forest degradation is more recent and where their cultural values are still somewhat centered around a forested landscape. Research on restoration effects on cultural services is scarce; most forest restoration research is focused on ecological aspects. An overview of the effects of restoration on cultural ES is shown in Table 7.4.

7.4 Trade-offs and disservices: One's gain can be another's loss

7.4.1 Trade-offs

Different restoration methods have different outcomes; even the same method can have different outcomes, depending on the former state of degradation and environmental conditions. Often there are trade-offs in the desired outcomes; there is no "silver bullet" method to restore all ES equally. Improving certain ES while supplying resources is difficult because trade-offs exist between forest product provisioning and ecological restoration (Lamb et al. 2005). Trying to achieve multiple goals can lead to hard choices and to compromises between conservation, biodiversity,

Table 7.4 Effects of forest restoration on cultural ES characteristics. Effects of an increase in tree cover and management practices are shown. The icons in the first column provide information on the scale of the effect of restoration on an ecosystem characteristic. Management effects refer to any restoration interventions, including selection of species with certain traits. The biodiversity effect is the product of the latter and determines to what extent certain species are promoted over others. See legend in table 7.1 for explanation of the icons.

Ecosystem Characteristics	Effects and Influences of Characteristics	Restoration and Management Effects
Recreation and maintaining mental and physical health	Positive impact on valuation of nature, also outside the restored areas, people taking more care of nature, potential trickling down effects of these in (local) political decisions and economic choices.	*Tree cover effect* = ↑ ⌒ high density will increase shelter but people also like parkland style with fewer but larger trees. *Management:* diversity of species can add to recreational values; structural and spatial diversification. Select species that do not commonly cause allergies.
Aesthetics	Aesthetic quality of the landscape, based on (e.g., structural diversity, plant diversity, 'greenness,' tranquillity, presence of flowering plants, streams, etc.	*Tree cover effect* = ↑ ⌒ create varying openness. *Management*: species selection; tree form; characteristics species; create structural diversity.
(Eco-)tourism	Happiness, health, and volunteer involvement in restoration increases acceptance and importance.	*Tree cover effect* = ↑ ⌒ depends on forest type and biome (e.g., savannas or forests are both valued touristic hotspots). *Management*: select species with a story (flagship species); select species that attract wildlife.
Natural heritage, sense of belonging, and traditional ecosystem knowledge (TEK)	Intergenerational connection, Increasing care of nature "future of our children," sacred sites are being protected.	*Tree cover effect* = ↑ ⌒ depends on historical forest cover, type, and biome. *Management*: highly diverse system increases connection and sense of belonging; opportunities of maintaining TEK through intergenerational connections; education, evaluation, and classification leads to valuation.

and production, which may result in suboptimal outcomes for each. Schwaiger et al. (2019) found a trade-off between increased productivity and decreased structural diversity in a production-oriented forest, and between improved groundwater recharge and loss in productivity in a multifunctional-oriented forest. Trade-offs in ES recovery can often be traced back to the species that were planted or regrew naturally.

Zeng et al. (2019) found significant effects of tree functional groups on ES (in this case, plant diversity, air quality regulation, soil fertility maintenance, global climate regulation, and timber provisioning). In naturally regenerated forests, deciduous broadleaves had significant positive effects on the measured ES, whereas conifers and broadleaf evergreens had negative effects.

7.4.2 Disservices

Forest restoration can have negative impacts or "disservices." Forests can have some harmful characteristics: they are sources of pests, diseases, and disease vectors. Well-known examples are malaria, ebola, river blindness, hendra, and possibly COVID-19 (Dunn 2010). Recent wildfires around the globe have shown that forests can become fuel for such devastating natural disasters. Forests can negatively impact freshwater supply because trees use water (Ilstedt et al. 2016). Also, within the ES frameworks disservices can be found. ES are defined in relation to human needs; as people's values, preferences, and expectations vary, so do their perceptions of ES. Therefore, how ES are perceived can differ between groups of people or individuals. An example of differing values between groups is that rural and indigenous people value the forest because of the spiritual importance of sacred forests and their dependence on its natural resources, while urban groups often value forests for employment and tourism. Also, at the individual level, perceptional differences occur: the hiker likely appreciates a healthy forest high in biodiversity with many animals to spot, while a farmer sees that same forest on his land as a limiting factor in the amount of farmable land. A study in southwestern France shows farmers' mixed views on the (material and nonmaterial) contributions of rural forests (Blanco et al. 2020). According to the farmers, trees provide some beneficial services to agriculture (e.g., erosion control, windbreak) and some disservices (e.g., reduced crop yield, pest source). Multiple values and perceptions coexist; the difficulty is in addressing them together. It is key for restoration decision-making and for understanding restoration effects that the perceptions of all relevant stakeholder groups are taken into account (Constant and Taylor 2020).

Land users seldom share the same priorities in terms of where, when, and how to address restoration, and who to involve. It is important to understand how livelihoods interact with different restoration methods (Crossland et al. 2018). Unsurprisingly, conflicts arise when single services for a few users are targeted in isolation, when other stakeholder interests are neglected (Bullock et al. 2011). This can undermine restoration efforts and influence long-term sustainability. In addition, the value of other land uses and vegetation types should not be neglected; forest 'restoration' can threaten unique vegetation such as open grassland and savannas, which have biodiversity value as well (Veldman et al. 2015). In fact, tree planting on savanna ecosystems would be afforestation, but there is a worrisome tendency to frame all tree planting as restoration nowadays. Scientific studies might help map stakeholder needs, which can advise policymakers and governing institutions on how to make restoration attractive and inclusive for all stakeholders (Lamb 2018). A study in the Brazilian Atlantic rainforest, for example, shows that investigating perceptions of restoration benefits could help promote consensus-building among stakeholders (Brancalion et al. 2014).

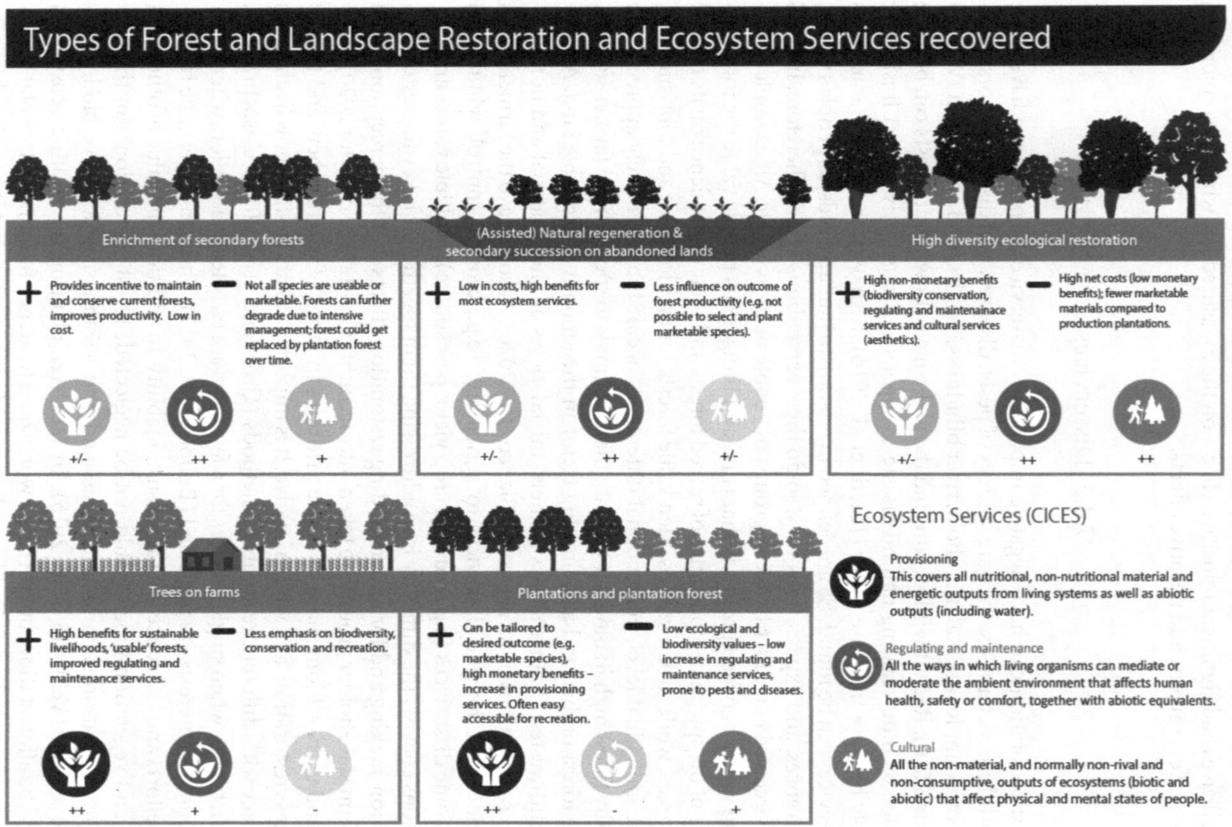

Infographic 7.1 ES provided by different forest restoration methods (also see Figure 7.2). The most common restoration methods are listed. (Infographic by R. Brouwer)

Understanding the scale and harm of disservices in forest restoration activities is key for success.

7.5 Restoration across scales of time and space

Global goals aim to restore millions of hectares of (formerly) forested land, yet more and more land is needed to produce our food, along with many other goods. Since 1970, the harvest of commercial timber has increased almost 50% while the value of agricultural crop production has increased about threefold (FAO and UNEP 2020). Forest restoration competes for land with monoculture timber plantations and agricultural land use among others. Therefore, restoring forest on former agricultural land may result in the displacement of agricultural activity. Increased efficiency of agricultural production, reduced food waste, and a more socially just distribution of food can help balance food demand and reduce agricultural land demand, providing opportunities for forest restoration (Bajželj et al. 2020). Some services (e.g., carbon sequestration and timber production) might be most easily provided by large industrial plantations rather than many small farmers restoring forests on their lands. Restoring forests with a production goal might then displace smallholders and generate significant social costs (Lamb et al. 2005). It could also lead to forest degradation or deforestation in the regions to which these smallholders are displaced.

High-diversity restoration forests are difficult to implement at large landscape scales, while they could thrive on smaller local scales (Lamb 2018). This discussion is tightly linked to the intensive land sparing versus land sharing debate (Sidemo-Holm et al. 2021; van Noordwijk 2021). Restoration by land sharing has the potential to enhance ES and biodiversity on agricultural sites at both the field and landscape scales. On the other hand, restoration by land sparing would provide these benefits only at the restored site and not at the agricultural site; but when zooming out to the landscape scale there, would be an overall net gain in restored ES (Rey Benayas and Bullock 2012). In a study on this topic in agro-ecosystems, land sparing led to similar biodiversity recovery but higher ecosystem recovery than land sharing (Barral et al. 2015). Other studies show that crop production and revenue can be enhanced alongside forest restoration and biodiversity conservation without the loss of other ES in the same landscape (Cavender-Bares et al. 2015; Marcilio-Silva et al. 2018).

Besides differences in ES supply on a spatial scale, there are also differences on a temporal scale. The outcome of forest restoration changes over time, and the supply of ES (and their interactions) is often lower at the start of restoration than after a few decades. Many studies show that ES recovery takes time: on rehabilitated mine lands, for example, environmental indicators such as soil chemical properties and species composition increased over time (Gastauer et al. 2020). Others also show that more mature forests have a greater capacity to supply higher and more evenly distributed ES benefits (Liu et al. 2019; Zeng et al. 2019). Many community characteristics, including diversity and biomass production, vary intrinsically with succession and community history (Sheil and Bongers 2020). Moreover, the temporal trajectories of

Table 7.5 Factors that drive variation in forest restoration outcomes. Source: adapted from Pickett et al. 1987

Factor	Processes or Aonditions Affecting Restoration Outcomes	Specific Characteristics of Factor
Local site and past land use	Disturbances, topography, climate, drainage	Type of past land use and disturbances, severity, frequency, resource availability
Time	Forest succession	Length of community development, length of past land use and disturbances
Species pool	Seed dispersal, resprouting ability, presence or absence of invasive species, type of species regenerating, types of species planted	Landscape configuration, presence of forest remnants and dispersal agents, remnant vegetation, prior vegetation, prior land use
Species traits	Functional and life history traits, ability to adapt to the current (changed) environment	Germination chance, establishment success, growth requirements, resource use
Species interactions	Intra- and interspecific species competition, diseases, herbivory, pollination, dispersal, defense mechanisms, mycorrhizae	Population size, structure and dynamics, recruitment, growth, mortality, trophic structure, facilitation
Human interaction with restored forest	Presence or absence of humans in restored landscape, use of restored forests	Policies, laws, enforcement, protection of restoration project, stakeholders' interest and needs, species usability

ecosystem structure and functions in restored ecosystems can, in some instances, be decoupled, which implies that the monitoring of restoration programs should include measurements of functions in addition to structural indices that can be remotely sensed (Ferraz et al. 2020).

Restoration happens across time and space; outcomes vary depending on multiple factors (Table 7.5). Before implementing a restoration intervention, it is important to map these factors' effects and the possible outcome trajectories.

7.6 Opportunities for further research and restoration projects

This chapter has summarized the known and expected effects of forest restoration on ecosystem characteristics. Despite the breadth of the literature, important geographic and thematic knowledge gaps remain. Our findings echo those of Howe et al. (2014), that is, the current literature focuses on particular geographic regions and individual ES and neglects trade-offs and synergies between ES. Current literature suggests that forest restoration bolsters biodiversity recovery, but that overall success is context-dependent. If restoration targets resilience, newly restored forests and their ES might be more resilient to future (climate) changes. Clearly, not everything is possible everywhere, but targeted measures can help achieve goals. Based on our evaluation, we suggest three research opportunities.

First, understanding the context of forest restoration activities is crucial to improving outcomes. Most studies focus on just a few regulating and maintenance services. A broader, multifunctional study approach is needed to identify trade-offs between ES in different forms of restoration.

Second, provisioning services and cultural services remain poorly represented in restoration projects. Balancing resource extraction such as timber production with other ES thus presents a challenge. Cultural services (e.g., recreation) are often neglected in restoration projects, despite being among the most tangible to local people. Provisioning and cultural services are both part of the socioeconomic benefits restoration provides. With good evaluation systems in place, these benefits offer a good opportunity for restoration research (Van Oudenhoven et al. 2012).

Third, very little research attention has been devoted to disservices. It is poorly understood how these (unintentional) restoration results balance with services, and how they differ between stakeholders. Clearly, the causes and impacts of disservices need more attention in restoration research (Grass et al. 2020).

Much can be gained through the effective communication of scientific results to stakeholders and the general public, in order to guide and inform restoration programs. One way to bolster interest and support from stakeholders is by targeting ES that are of sufficient interest to stakeholders.

Context matters in forest restoration projects. It defines appropriate methods, outcomes, and implications. Context is, by definition, created by different (potential) functions of restored forests. A broad, multifunctional focus in restoration is needed to identify and weigh services and disservices for them to be understood, and to evaluate total success. In addition, both short- and long-term goals need to be clear in order to design restoration projects that can succeed in changing circumstances. The generalized summary of studies in this chapter to evaluate restoration research can be helpful also in assessing the potential contribution of restoration projects to reaching SDGs and local societal goals and needs.

References

Acharya, R.P., T. Maraseni, and G. Cockfield. 2019. "Global Trend of Forest Ecosystem Services Valuation: An Analysis of Publications." *Ecosystem Services* 39 (July), p. 100979. DOI:10.1016/j.ecoser.2019.100979.

Aerts, R., and G.W. Heil (eds.). 1993. *Heathlands: Patterns and Processes in a Changing Environment*. Dordrecht: Kluwer Academic Publishers.

Aerts, R., and O. Honnay. 2011. "Forest Restoration, Biodiversity and Ecosystem Functioning." *BMC Ecology* 11 (29): pp. 1–21. DOI:10.1186/1472-6785-11-29.

Alexander, S., J. Aronson, O. Whaley, and D. Lamb. 2016. "The Relationship between Ecological Restoration and the Ecosystem Services Concept." *Ecology and Society* 21 (1): p. 34. DOI:10.5751/ES-08288-210134.

Angelsen, A., P. Jagger, R. Babigumira, et al. 2014. "Environmental Income and Rural Livelihoods: A Global-Comparative Analysis." *World Development* 64 (S1): pp. S12–S28. DOI:10.1016/j.worlddev.2014.03.006.

Arnold, J.E.M., and M.R. Pérez. 2001. "Can Non-Timber Forest Products Match Tropical Forest Conservation and Development Objectives?" *Ecological Economics* 39 (3): pp. 437–447. DOI:10.1016/S0921-8009(01)00236-1.

Bajželj, B., T.E. Quested, E. Röös, and R.P.J. Swannell. 2020. "The Role of Reducing Food Waste for Resilient Food Systems." *Ecosystem Services* 45 (May): p. 101140. DOI:10.1016/j.ecoser.2020.101140.

Barr, S. 2016. "Conservation efforts for Pied Tamarins (Saguinus bicolor) – Evaluating Ecological Corridors for Restoring the Forest Fragments of Urban Manaus, Brazil." Available at: https://lup.lub.lu.se/student-papers/search/publication/8518535.

Barral, M.P., J.M. Rey Benayas, P. Meli, and N.O. Maceira. 2015. "Quantifying the Impacts of Ecological Restoration on Biodiversity and Ecosystem Services in Agroecosystems: A Global Meta-Analysis." *Agriculture, Ecosystems and Environment* 202: pp. 223–231. DOI:10.1016/j.agee.2015.01.009.

Batterman, S.A., L.O. Hedin, M. Van Breugel, et al. 2013. "Key Role of Symbiotic Dinitrogen Fixation in Tropical Forest Secondary Succession." *Nature* 502 (7470): pp. 224–227. DOI:10.1038/nature12525.

Bell, A. 2001. "Engaging spaces: On school-based habitat restoration." *Canadian Journal of Environmental Education* 6 (1): pp. 139–154.

Bendix, J., E. Beck, A. Bräuning, et al. (eds.). 2013. *Ecological Studies. Ecosystem Services, Biodiversity and Environmental Change in a Tropical Mountain Ecosystem of South Ecuador*. Dordrecht: Springer.

Blanco, J., A. Sourdril, M. Deconchat, et al. 2020 "How Farmers Feel about Trees: Perceptions of Ecosystem Services and Disservices Associated with Rural Forests in Southwestern France." *Ecosystem Services* 42: p. 101066. DOI:10.1016/j.ecoser.2020.101066.

Bongers, F., P. Groenendijk, T. Bekele, et al. 2019. "Frankincense in Peril." *Nature Sustainability* 2 (7): pp. 602–610. DOI:10.1038/s41893-019-0322-2.

Brancalion, P.H.S., N.T. Amazonas, R.L. Chazdon, et al. 2020. "Exotic Eucalypts: From Demonized Trees to Allies of Tropical Forest Restoration?" *Journal of Applied Ecology* 57 (1): pp. 55–66. DOI:10.1111/1365-2664.13513.

Brancalion, P.H.S., I.V. Cardozo, A. Camatta, J. Aronson, and R.R. Rodrigues. 2014. "Cultural Ecosystem Services and Popular Perceptions of the Benefits of an Ecological Restoration Project in the Brazilian Atlantic Forest." *Restoration Ecology* 22 (1): pp. 65–71. DOI:10.1111/rec.12025.

Broom, D.M., F.A. Galindo, and E. Murgueitio. 2013. "Sustainable, Efficient Livestock Production with High Biodiversity and Good Welfare for Animals." *Proceedings of the Royal Society B: Biological Sciences* 280 (1771). DOI:10.1098/rspb.2013.2025.

Brouwer, R.G., P.A. Zuidema, F. Chiriboga-Arroyo, et al. 2021. "Establishment Success of Brazil Nut Trees in Smallholder Amazon Forest Restoration Depends on Site Conditions and Management." *Forest Ecology and Management* 498: p. 119575. DOI:10.1016/j.foreco.2021.119575.

Bullock, J.M., J. Aronson, A.C. Newton, R.F. Pywell, and J.M. Rey-Benayas. 2011. "Restoration of Ecosystem Services and Biodiversity: Conflicts and Opportunities" *Trends in Ecology and Evolution* 26 (10): pp. 541–549. DOI:10.1016/j.tree.2011.06.011.

Cáceres, D.M., E. Tapella, F. Quétier, and S. Díaz. 2015. "The Social Value of Biodiversity and Ecosystem Services from the Perspectives of Different Social Actors." *Ecology and Society* 20 (1): 62. DOI:10.5751/ES-07297-200162.

Carrick, P.J., and K.J. Forsythe. 2020. "The Species Composition-Ecosystem Function Relationship: A Global Meta-Analysis Using Data from Intact and Recovering Ecosystems." *PLoS One* 15 (7): pp. 1–23. DOI:10.1371/journal.pone.0236550.

Cavender-Bares, J., S. Polasky, E. King, and P. Balvanera. 2015. "A Sustainability Framework for Assessing Trade-Offs in Ecosystem Services." *Ecology and Society* 20 (1): 17. DOI:10.5751/ES-06917-200117.

César, R.G., V. de S. Moreno, G.D. Coletta, et al. 2021. "It Is Not Just About Time: Agricultural Practices and Surrounding Forest Cover Affect Secondary Forest Recovery in Agricultural Landscapes." *Biotropica* 53 (2): pp. 496–508. DOI:10.1111/btp.12893.

Chazdon, R.L. 2008. "Beyond Deforestation: Restoring Forests and Ecosystem Services on Degraded Lands." *Science* 320 (5882): pp. 1458–1460. DOI:10.1126/science.1155365.

Chazdon, R.L., P.H.S. Brancalion, L. Laestadius, et al. 2016. "When Is a Forest a Forest? Forest Concepts and Definitions in the Era of Forest and Landscape Restoration." *Ambio* 45 (5): pp. 538–550. DOI:10.1007/s13280-016-0772-y.

Chazdon, R.L., D.A. Falk, L.F. Banin, et al. 2021. "The Intervention Continuum in Restoration Ecology: Rethinking the Active–Passive Dichotomy." *Restoration Ecology* (August): p. e13535. DOI:10.1111/rec.13535.

Chazdon, R.L., and M.R. Guariguata. 2016. "Natural Regeneration as a Tool for Large-Scale Forest Restoration in the Tropics: Prospects and Challenges." *Biotropica* 48 (6): pp. 716–730. DOI:10.1111/btp.12381.

Chen, X., H.Y.H. Chen, C. Chen, et al. 2020. "Effects of Plant Diversity on Soil Carbon in Diverse Ecosystems: A Global Meta-Analysis." *Biological Reviews* 95 (1): pp. 167–183. DOI:10.1111/brv.12554.

Chomba, S., F. Sinclair, P. Savadogo, M. Bourne, and M. Lohbeck. 2020. "Opportunities and Constraints for Using Farmer Managed Natural Regeneration for Land Restoration in Sub-Saharan Africa." *Frontiers in Forests and Global Change* 3: p. 571679. DOI:10.3389/ffgc.2020.571679.

Coelho, S.D., C. Levis, F.B. Baccaro, et al. 2021. "Eighty-Four Per Cent of All Amazonian Arboreal Plant Individuals Are Useful to Humans." *PLoS One* 16 (10): p. e0257875. DOI:10.1371/journal.pone.0257875.

Coleman, E.A., B. Schultz, V. Ramprasad, et al. 2021. "Limited Effects of Tree Planting on Forest Canopy Cover and Rural Livelihoods in Northern India." *Nature Sustainability* 4 (11): pp. 997–1004. DOI:10.1038/s41893-021-00761-z.

Constant, N.L., and P.J. Taylor. 2020. "Restoring the Forest Revives Our Culture: Ecosystem Services and Values for Ecological Restoration Across the Rural-Urban Nexus in South Africa." *Forest Policy and Economics* 118: p. 102222. DOI:10.1016/j.forpol.2020.102222.

Corbin, J.D., and K.D. Holl. 2012. "Applied Nucleation as a Forest Restoration Strategy." *Forest Ecology and Management* 265: pp. 37–46. DOI:10.1016/j.foreco.2011.10.013.

Cramer, J.R. 2008. "Reviving the Connection between Children and Nature through Service-Learning Restoration Partnerships." *Native Plants Journal* 9 (3): pp. 278–286.

Crossland, M., L.A. Winowiecki, T. Pagella, K. Hadgu, and F. Sinclair. 2018. "Implications of Variation in Local Perception of Degradation and Restoration Processes for Implementing Land Degradation Neutrality". *Environmental Development* 28: pp. 42–54. DOI:10.1016/j.envdev.2018.09.005.

Crouzeilles, R., and M. Curran. 2016. "Which Landscape Size Best Predicts the Influence of Forest Cover on Restoration Success? A Global Meta-Analysis on the Scale of Effect." *Journal of Applied Ecology* 53 (2): pp. 440–448. DOI:10.1111/1365-2664.12590.

Crouzeilles, R., E. Santiami, M. Rosa, et al. 2019. "There is Hope for Achieving Ambitious Atlantic Forest Restoration Commitments." *Perspectives in Ecology and Conservation* 17 (2): pp. 80–83. DOI:10.1016/j.pecon.2019.04.003.

Davis, M.A., M.K. Chew, R.J. Hobbs, et al. 2011. "Don't Judge Species." *Nature* 474 (7350): pp. 153–154.

De Beenhouwer, M., R. Aerts, and O. Honnay. 2013. "A Global Meta-Analysis of the Biodiversity and Ecosystem Service Benefits of Coffee and Cacao Agroforestry." *Agriculture, Ecosystems and Environment* 175: pp. 1–7. DOI:10.1016/j.agee.2013.05.003.

De Bello, F., S. Lavorel, S. Díaz, et al. 2010. "Towards an Assessment of Multiple Ecosystem Processes and Services via Functional Traits." *Biodiversity and Conservation* 19 (10): pp. 2873–2893. DOI:10.1007/s10531-010-9850-9.

Deere, N.J., G. Guillera-Arroita, T. Swinfield, et al. 2020. "Maximizing the Value of Forest Restoration for Tropical Mammals by Detecting Three-Dimensional Habitat Associations." *Proceedings of the National Academy of Sciences of the United States of America* 117 (42): pp. 26254–26262. DOI:10.1073/pnas.2001823117.

Derks, J., L. Giessen, and G. Winkel. 2020. "COVID-19-Induced Visitor Boom Reveals the Importance of Forests as Critical Infrastructure." *Forest Policy and Economics* 118 (April): p. 102253. DOI:10.1016/j.forpol.2020.102253.

Díaz, S., U. Pascual, M. Stenseke, et al. 2018. "Assessing Nature's Contributions to People." *Science* 359 (6373): pp. 270–272. DOI:10.1126/science.aap8826.

Dou, Y., L. Zhen, X. Yu, et al. 2019. "Assessing the influences of Ecological Restoration on Perceptions of Cultural Ecosystem Services by Residents of Agricultural Landscapes of Western China." *Science of the Total Environment* 646: pp. 685–695. DOI:10.1016/j.scitotenv.2018.07.205.

Dunn, R.R. 2010. "Global Mapping of Ecosystem Disservices: The Unspoken Reality that Nature Sometimes Kills Us." *Biotropica* 42 (5): pp. 555–557. DOI:10.1111/j.1744-7429.2010.00698.x.

du Toit, J.T., and N. Pettorelli. 2019. "The Differences between Rewilding and Restoring an Ecologically Degraded Landscape." *Journal of Applied Ecology* 56 (11): pp. 2467–2471. DOI:10.1111/1365-2664.13487.

Ellison, D., C.E. Morris, B. Locatelli, et al. 2017. "Trees, Forests and Water: Cool Insights for a Hot World." *Global Environmental Change* 43: pp. 51–61. DOI:10.1016/j.gloenvcha.2017.01.002.

Fan, Y., G. Miguez-Macho, E.G. Jobbágy, R.B. Jackson, and C. Otero-Casal. 2017. "Hydrologic Regulation of Plant Rooting Depth." *Proceedings of the National Academy of Sciences of the United States of America* 114 (40): pp. 10572–10577. DOI:10.1073/pnas.1712381114.

FAO and UNEP. 2020. *The State of the World's Forests 2020, Forests, Biodiversity and People.* Rome: FAO. DOI:10.1515/9783035608632-002.

Ferraz, S., P.H.S. Brancalion, J. Guillemot, and P. Meli. 2020. "On the Need to Differentiate the Temporal Trajectories of Ecosystem Structure and Functions in Restoration Programs." *Tropical Conservation Science* 13: pp. 1–6. DOI:10.1177/1940082920910314.

Filoso, S., M.O. Bezerra, K.C.B. Weiss, and M.A. Palmer. 2017. "Impacts of Forest Restoration on Water Yield: A Systematic Review." *PLoS One* 12 (8): pp. 1–26. DOI:10.1371/journal.pone.0183210.

Finegan, B., M. Peña-Claros, A. de Oliveira, et al. 2015. "Does Functional Trait Diversity Predict Above-Ground Biomass and Productivity of Tropical Forests? Testing three alternative hypotheses." *Journal of Ecology* 103 (1): pp. 191–201. DOI:10.1111/1365-2745.12346.

Gann, G.D., T. McDonald, B. Walder, et al. 2019. "International Principles and Standards for the Practice of Ecological Restoration. Second edition." *Restoration Ecology* 27 (S1): pp. S1–S46. DOI:10.1111/rec.13035.

Gao, H., and Y. Huang. 2020. "Impacts of the Three-North Shelter Forest Program on the Main Soil Nutrients in Northern Shaanxi China: A Meta-Analysis." *Forest Ecology and Management* 458: p. 117808. DOI:10.1016/j.foreco.2019.117808.

Gastauer, M., C.F. Caldeira, S.J. Ramos, D.F. Silva, and J.O. Siqueira. 2020. "Active Rehabilitation of Amazonian Sand Mines Converges Soils, Plant Communities and Environmental Status to Their Predisturbance Levels." *Land Degradation and Development* 31 (5): pp. 607–618. DOI:10.1002/ldr.3475.

Gavito, M.E., H. Paz, F. Barragán, et al. 2021. "Indicators of Integrative Recovery of Vegetation, Soil and Microclimate In Successional Fields of a Tropical Dry Forest." *Forest Ecology and Management* 479: p. 118526. DOI:10.1016/j.foreco.2020.118526.

Gomes, A.C.S., and F.J. Luizão. 2012. "Leaf and Soil Nutrients in a Chronosequence of Second-Growth Forest in Central Amazonia: Implications for Restoration of Abandoned Lands." *Restoration Ecology* 20 (3): pp. 339–345. DOI:10.1111/j.1526-100X.2011.00773.x.

Gould, R.K., N.M. Ardoin, U. Woodside, et al. 2014. "The Forest Has a Story: Cultural Ecosystem Services in Kona, Hawai'i." *Ecology and Society* 19 (3). DOI:10.5751/ES-06893-190355.

Guariguata, M.R., P. Cronkleton, A.E. Duchelle, and P.A. Zuidema. 2017. "Revisiting the 'Cornerstone of Amazonian Conservation': A Socioecological Assessment of Brazil Nut Exploitation." *Biodiversity and Conservation* 26 (9).

Grass, I., C. Kubitza, V.V. Krishna, et al. 2020. "Trade-Offs between Multifunctionality and Profit in Tropical Smallholder Landscapes." *Nature Communications* 11 (1): pp. 1186. DOI:10.1038/s41467-020-15013-5.

Grime, J.P. 1998. "Benefits of Plant Diversity to Ecosystems: Immediate, Filter and Founder Effects." *Journal of Ecology* 86: pp. 902–910.

Haines-Young, R., and M. Potschin. 2018. *Common International Classification of Ecosystem Services (CICES) V5.1 and Guidance on the Application of the Revised Structure*. Nottingham: Common International Classification of Ecosystem Services (CICES). https://cices.eu/resources/.

Hansen-Møller, J., and L. Oustrup. 2004. "Emotional, Physical/Functional and Symbolic Aspects of an Urban Forest in Denmark to Nearby Residents." *Scandinavian Journal of Forest Research* Supplement 19 (4): pp. 56–64. DOI:10.1080/14004080410034065.

Hawthorne, W.D., D. Sheil, V.K. Agyeman, M. Abu Juam, and C.A.M. Marshall. 2012. "Logging Scars in Ghanaian High Forest: Towards Improved Models for Sustainable Production." *Forest Ecology and Management* 271: pp. 27–36. DOI:10.1016/j.foreco.2012.01.036.

Heinze, A., F. Bongers, N. Ramírez Marcial, L. García Barrios, and T.W. Kuyper. 2020. "The Montane Multifunctional Landscape: How stakeholders in a Biosphere Reserve Derive Benefits and Address Trade-Offs in Ecosystem Service Supply." *Ecosystem Services* 44: p. 101134. DOI:10.1016/j.ecoser.2020.101134.

Heneghan, L., S.P. Miller, S. Baer, et al. 2008. "Integrating Soil Ecological Knowledge into Restoration Management." *Restoration Ecology* 16 (4): pp. 608–617. DOI:10.1111/j.1526-100X.2008.00477.x.

Holl, K.D., and T.M. Aide. 2011. "When and Where to Actively Restore Ecosystems?" *Forest Ecology and Management* 261 (19): 1558–1563. DOI:10.1016/j.foreco.2010.07.004.

Hooper, D.U., F.S. Chapin, J.J. Ewel, et al. 2005. "Effects of Biodiversity on Ecosystem Functioning: A Consensus of Current Knowledge." *Ecological Monographs* 75 (1): pp. 3–35. DOI:10.1890/04-0922.

Howe, C., H. Suich, and B. Vira. 2014. "Creating Win-Wins from Trade-Offs? Ecosystem Services for Human Well-Being: A Meta-Analysis of Ecosystem Service Trade-Offs and Synergies in the Real World." *Global Environmental Change* 28: pp. 263–275. DOI:10.1016/j.gloenvcha.2014.07.005.

Hua, F., L.A. Bruijnzeel, P. Meli, et al. 2022. "The Biodiversity and Ecosystem Service Contributions and Trade-Offs of Forest Restoration Approaches." *Science* 376 (6595): pp. 839–844. DOI:10.1126/science.abl4649.

Ilstedt, U., A.B. Tobella, H.R. Bazié, et al. 2016. "Intermediate Tree Cover Can Maximize Groundwater Recharge in the Seasonally Dry Tropics." *Scientific Reports* 6 (1): pp. 1–12. DOI: 10.1038/srep21930.

Jakovac, C.C., A.B. Junqueira, R. Crouzeilles, et al. 2021. "The Role of Land-Use History in Driving Successional Pathways and Its Implications for the Restoration Of Tropical Forests." *Biological Reviews* 96 (4): pp. 1114–1134. DOI:10.1111/brv.12694.

Jansen, M., M.R. Guariguata, J.E. Raneri, et al. 2020. "Food for Thought: The Underutilized Potential of Tropical Tree-Sourced Foods for 21st Century Sustainable Food Systems." *People and Nature* 2 (4): pp. 1006–1020. DOI: 10.1002/pan3.10159.

Kaimowitz, D., and D. Sheil. 2007. "Conserving What and For Whom? Why Conservation Should Help Meet Basic Human Needs in The Tropics." *Biotropica* 39 (5): pp. 567–574. DOI:10.1111/j.1744-7429.2007.00332.x.

Kramer, K., L. Bouriaud, P. Feindt, et al. 2021. "Roadmap to Develop a Stress Test for Forest Ecosystem Services Supply." *One Earth* 5 (1): pp. 25–34. DOI: 10.1016/j.oneear.2021.12.009.

Lackey, N.Q., D.A. Tysor, G.D. McNay, et al. 2021. "Mental Health Benefits of Nature-Based Recreation: A Systematic Review." *Annals of Leisure Research* 24 (3): pp. 379–393. DOI:10.1080/11745398.2019.1655459.

Lamb, D. 2018. "Undertaking Large-Scale Forest Restoration to Generate Ecosystem Services." *Restoration Ecology* 26 (4): pp. 657–666. DOI:10.1111/rec.12706.

Lamb, D., P.D. Erskine, and J.A. Parrotta. 2005. "Restoration of Degraded Tropical Forest Landscapes." *Science* 310 (5754): pp. 1628–1632. DOI:10.1126/science.1111773.

Lamb, D., and D. Gilmour. 2005. "Rehabilitation and Restoration of Degraded Forests." *Restoration Ecology* 13 (3): pp. 578–579. DOI:10.1111/j.1526-100x.2005.00073.x.

Lanuza, O., F. Casanoves, R.A. Zahawi, et al. 2018. "Litterfall and Nutrient Dynamics Shift in Tropical Forest Restoration Sites after a Decade of Recovery." *Biotropica* 50 (3): pp. 491–498. DOI:10.1111/btp.12533.

Larney, F.J., and D.A. Angers. 2012. "The Role of Organic Amendments in Soil Reclamation: A Review." *Canadian Journal of Soil Science* 92 (1): pp. 19–38. DOI:10.4141/CJSS2010-064.

Lawrence, D., and K. Vandecar. 2015. "Effects of Tropical Deforestation on Climate and Agriculture." *Nature Climate Change* 5 (1): pp. 27–36. DOI:10.1038/nclimate2430.

Leite-Filho, A.T., B.S. Soares-Filho, J.L. Davis, G.M. Abrahão, and J. Börner. 2021. "Deforestation Reduces Rainfall and Agricultural Revenues in the Brazilian Amazon." *Nature Communications* 12 (1): pp. 1–7. DOI:10.1038/s41467-021-22840-7.

Li, Q. 2010. "Effect of Forest Bathing Trips on Human Immune Function." *Environmental Health and Preventive Medicine* 15 (1): pp. 9–17. DOI: 10.1007/s12199-008-0068-3

Liu, C.L.C., O. Kuchma, and K.V. Krutovsky. 2018. "Mixed-Species Versus Monocultures in Plantation Forestry: Development, Benefits, Ecosystem Services and Perspectives for the Future." *Global Ecology and Conservation* 15: e00419. DOI:10.1016/j.gecco.2018.e00419.

Liu, Y., Y. Lü, B. Fu, P. Harris, and L. Wu. 2019. "Quantifying the Spatio-Temporal Drivers of Planned Vegetation Restoration on Ecosystem Services at a Regional Scale." *Science of the Total Environment* 650: pp. 1029–1040. DOI:10.1016/j.scitotenv.2018.09.082.

Lohbeck, M., P. Albers, L.E. Boels, et al. 2020. "Drivers of Farmer-Managed Natural Regeneration in the Sahel. Lessons for Restoration." *Scientific Reports* 10 (1): pp. 1–11. DOI:10.1038/s41598-020-70746-z.

Lohbeck, M., L. Poorter, M. Martinez-Ramos, F. Bongers, and N.J.B. Craft. 2015. "Biomass is the Main Driver of Changes in Ecosystem Process Rates During Tropical Forest Succession." *Ecology* 96 (5): pp. 1242–1252. DOI:10.1890/14-0472.1.

Lohbeck, M., L. Winowiecki, E. Aynekulu, C. Okia, and T.G. Vågen. 2018. "Trait-Based Approaches for Guiding the Restoration of Degraded Agricultural Landscapes in East Africa." *Journal of Applied Ecology* 55 (1): pp. 59–68. DOI:10.1111/1365-2664.13017.

Mace, G.M., K. Norris, and A.H. Fitter. 2012. "Biodiversity and Ecosystem Services: A Multilayered Relationship." *Trends in Ecology and Evolution* 27 (1): pp. 19–25. DOI:10.1016/j.tree.2011.08.006.

Makarieva, A.M., V.G. Gorshkov, D. Sheil, A.D. Nobre, and B.L. Li. 2013. "Where Do Winds Come From? A New Theory on How Water Vapor Condensation Influences Atmospheric Pressure and Dynamics." *Atmospheric Chemistry and Physics* 13 (2): pp. 1039–1056. DOI:10.5194/acp-13-1039-2013.

Marcilio-Silva, V., M.C.M. Marques, and J. Cavender-Bares. 2018. "Land-Use Trade-Offs between Tree Biodiversity and Crop Production in the Atlantic Forest." *Conservation Biology* 32 (5): pp. 1074–1084. DOI:10.1111/cobi.13138.

Martin, P.A., A.C. Newton, and J.M. Bullock. 2013. "Carbon Pools Recover More Quickly Than Plant Biodiversity in Tropical Secondary Forests." *Proceedings of the Royal Society B: Biological Sciences* 280 (1773). DOI:10.1098/rspb.2013.2236.

Matzek, V., S. Warren, and C. Fisher. 2016. "Incomplete Recovery of Ecosystem Processes after Two Decades of Riparian Forest Restoration." *Restoration Ecology* 24 (5): pp. 637–645. DOI:10.1111/rec.12361.

McNamara, S., D.V. Tinh, P.D. Erskine, et al. 2006. "Rehabilitating Degraded Forest Land in Central Vietnam with Mixed Native Species Plantings." *Forest Ecology and Management* 233 (2–3): pp. 358–365. DOI:10.1016/j.foreco.2006.05.033.

MEA. 2005. *Ecosystems and Human Well-Being. A Framework for Assessment.* Millennium ecosystem assessment. Washington: Island Press.

Meier, R., J. Schwaab, S.I. Seneviratne, et al. 2021. "Empirical Estimate of Forestation-Induced Precipitation Changes in Europe." *Nature Geoscience* 14 (7): pp. 473–478. DOI:10.1038/s41561-021-00773-6.

Meli, P., K.D. Holl, J.M.R. Benayas, et al. 2017. "A Global Review of Past Land Use, Climate, and Active vs. Passive Restoration Effects on Forest Recovery." *PLoS One* 12 (2): pp. 1–17. DOI:10.1371/journal.pone.0171368.

Melo, F.P.L., L. Parry, P.H.S. Brancalion, et al. 2021. "Adding Forests to the Water–Energy–Food Nexus." *Nature Sustainability* 4 (2): pp. 85–92. DOI:10.1038/s41893-020-00608-z.

Mendes, M.S., A.E. Latawiec, J.B.B. Sansevero, et al. 2019. "Look Down—There Is a Gap—the Need to Include Soil Data in Atlantic Forest Restoration." *Restoration Ecology* 27 (2): pp. 361–370. DOI:10.1111/rec.12875.

Mengist, W., T. Soromessa, and G. Legese. 2020. "Ecosystem Services Research in Mountainous Regions: A Systematic Literature Review on Current Knowledge and Research Gaps." *Science of the Total Environment* 702: p. 134581. DOI:10.1016/j.scitotenv.2019.134581.

Moore, E., T. van Dijk, A. Asenga, et al. 2020. "Species Selection and Management under Farmer Managed Natural Regeneration in Dodoma, Tanzania." *Frontiers in Forests and Global Change* 3: pp. 1–9. DOI:10.3389/ffgc.2020.563364.

Muchane, M.N., G.W. Sileshi, S. Gripenberg, et al. 2020. "Agroforestry Boosts Soil Health in the Humid and Sub-Humid Tropics: A Meta-Analysis." *Agriculture, Ecosystems and Environment* 295: p. 106899. DOI:10.1016/j.agee.2020.106899.

Murcia, C. 1997. "Evaluation of Andean Alder as a Catalyst for the Recovery of Tropical Cloud Forests in Colombia." *Forest Ecology and Management* 99 (1–2): pp. 163–170. DOI:10.1016/S0378-1127(97)00202-8.

Murgueitio, E., Z. Calle, F. Uribe, A. Calle, and B. Solorio. 2011. "Native Trees and Shrubs for the Productive Rehabilitation of Tropical Cattle Ranching Lands." *Forest Ecology and Management* 261 (10): pp. 1654–1663. DOI:10.1016/j.foreco.2010.09.027.

Nesbitt, L., N. Hotte, S. Barron, J. Cowan, and S.R.J. Sheppard. 2017. "The Social and Economic Value of Cultural Ecosystem Services Provided by Urban Forests in North America: A Review and Suggestions for Future Research." *Urban Forestry and Urban Greening* 25: pp. 103–111. DOI:10.1016/j.ufug.2017.05.005.

Norris, C.E., S.A. Quideau, J.S. Bhatti, R.E. Wasylishen, and M.D. MacKenzie. 2009. "Influence of Fire and Harvest on Soil Organic Carbon in Jack Pine Sites." *Canadian Journal of Forest Research* 39 (3): pp. 642–654. DOI:10.1139/X08-207.

Ogden, F.L., T.D. Crouch, R.F. Stallard, and J.S. Hall. 2013. "Effect of Land Cover and Use on Dry Season River Runoff, Runoff Efficiency, and Peak Storm Runoff in the Seasonal Tropics of Central Panama." *Water Resources Research.* 49 (12): pp. 8443–8462. DOI:10.1002/2013WR013956.

Ordonez, J.C., E. Luedeling, R. Kindt, et al. 2014. "Constraints and Opportunities for Tree Diversity Management along the Forest Transition Curve to Achieve Multifunctional Agriculture." *Current Opinion in Environmental Sustainability* 6 (1): pp. 54–60. DOI:10.1016/j.cosust.2013.10.009.

Park, B.J., Y. Tsunetsugu, T. Kasetani, T. Kagawa, and Y. Miyazaki. 2010. "The Physiological Effects of Shinrin-yoku (Taking in the Forest Atmosphere or Forest Bathing): Evidence from Field Experiments in 24 Forests across Japan." *Environmental Health and Preventive Medicine* 15 (1): pp. 18–26. DOI: 10.1007/s12199-009-0086-9.

Parrotta, J.A., and O.H. Knowles. 1999. "Restoration of Tropical Moist Forests on Bauxite-Mined Lands in the Brazilian Amazon." *Restoration Ecology* 7 (2): pp. 103–116. DOI:10.1046/j.1526-100X.1999.72001.x.

Pickett, S.T.A., S.L. Collins, and J.J. Armesto. 1987. "Models, Mechanisms and Pathways of Succession." *The Botanical Review* 53 (3): pp. 335–371.

Pimentel, D., and M. Burgess. 2013. "Soil Erosion Threatens Food Production." *Agriculture* 3 (3): pp. 443–463. DOI:10.3390/agriculture3030443.

Poorter, L., D. Craven, C.C. Jakovac, et al. 2021a. "Multidimensional Tropical Forest Recovery." *Science* 374 (6573): pp. 1370–1376. DOI: 10.1126/science.abh3629.

Poorter, L., M.A. Rozendaal, F. Bongers, et al. 2021b. "Functional Recovery of Secondary Tropical Forests". *PNAS* 118 (49): e2003405118. DOI.org/10.1073/pnas.2003405118.

Potschin, M.B., and R.H. Haines-Young. 2011. "Ecosystem Services: Exploring a Geographical Perspective." *Progress in Physical Geography* 35 (5): pp. 575–594. DOI:10.1177/0309133311423172.

Ramírez, J.A., J.D. León-Peláez, D. Craven, et al. 2014. "Effects on Nutrient Cycling of Conifer Restoration in a Degraded Tropical Montane Forest." *Plant and Soil* 378 (1–2): pp. 215–226. DOI:10.1007/s11104-014-2024-x.

Reich, P.B., J. Oleksyn, J. Modrzynski, et al. 2005. "Linking Litter Calcium, Earthworms and Soil Properties: A Common Garden Test with 14 Tree Species." *Ecology Letters* 8 (8): pp. 811–818. DOI:10.1111/j.1461-0248.2005.00779.x.

Rey Benayas, J.M., and J.M. Bullock. 2012. "Restoration of Biodiversity and Ecosystem Services on Agricultural Land." *Ecosystems* 15 (6): pp. 883–899. DOI:10.1007/s10021-012-9552-0.

Rey Benayas, J.M., A.C. Newton, A. Diaz, and J.M. Bullock. 2009. "Enhancement of Biodiversity and Ecosystem Services by Ecological Restoration: A Meta-Analysis." *Science* 325 (5944): pp. 1121–1124. DOI:10.1126/science.1172460.

Rinaudo, T. 2007. "The Development of Farmer Managed Natural Regeneration." *Leisa Magazine* 23 (2): pp. 32–34. Available at: http://fmnrhub.com.au/wp-content/uploads/2013/09/Rinaudo-2007-Development-of-FMNR.pdf.

Rozendaal, D.M.A., F. Bongers, T.M. Aide, et al. 2019. "Biodiversity Recovery of Neotropical Secondary Forests." *Science Advances* 5 (3). DOI:10.1126/sciadv.aau3114.

Sampaio, A.B., K.D. Holl, and A. Scariot. 2007. "Regeneration of Seasonal Deciduous Forest Tree Species in Long-Used Pastures in Central Brazil." *Biotropica* 39 (5): pp. 655–659. DOI:10.1111/j.1744-7429.2007.00295.x.

Schipperijn, J., O. Ekholm, U.K. Stigsdotter, et al. 2010. "Factors Influencing the Use of Green Space: Results from a Danish National Representative Survey." *Landscape and Urban Planning* 95 (3): pp. 130–137. DOI:10.1016/j.landurbplan.2009.12.010.

Schirpke, U., G. Wang, and E. Padoa-Schioppa. 2021. "Editorial: Mountain Landscapes: Protected Areas, Ecosystem Services, and Future Challenges." *Ecosystem Services* 49: p. 101302. DOI:10.1016/j.ecoser.2021.101302.

Scholte, S.S.K., A.J.A. van Teeffelen, and P.H. Verburg. 2015. "Integrating Socio-Cultural Perspectives into Ecosystem Service Valuation: A Review of Concepts and Methods." *Ecological Economics* 114: pp. 67–78. DOI:10.1016/j.ecolecon.2015.03.007.

Schwaiger, F., W. Poschenrieder, P. Biber, and H. Pretzsch. 2019. "Ecosystem Service Trade-Offs for Adaptive Forest Management." *Ecosystem Services* 39: p. 100993. DOI:10.1016/j.ecoser.2019.100993.

Sheil, D. 2018. "Forests, Atmospheric Water and an Uncertain Future: The New Biology of the Global Water Cycle." *Forest Ecosystems* 5 (19). DOI.org/10.1186/s40663-018-0138-y.

Sheil, D., and F. Bongers. 2020. "Interpreting Forest Diversity-Productivity Relationships: Volume Values, Disturbance Histories and Alternative Inferences." *Forest Ecosystems* 7 (1). DOI:10.1186/s40663-020-0215-x.

Sheil, D., M. van Heist, N. Liswanti, I. Basuki, and M. Wan. 2009. "Biodiversity, Landscapes, and Livelihoods: A Local Perspective." In *The Decentralization of Forest Governance*, edited by M. Moeliono, E. Wollenberg, and G. Limberg, pp. 61–90. London: Earthscan.

Sheil, D., and D. Murdiyarso. 2009. "How Forests Attract Rain: An Examination of a New Hypothesis." *BioScience* 59 (4): pp. 341–347. DOI:10.1525/bio.2009.59.4.12.

Shimamoto, C.Y., A.A. Padial, C.M. Da Rosa, and M.C.M. Marques. 2018. "Restoration of Ecosystem Services in Tropical Forests: A Global Meta-Analysis." *PLoS One* 13 (12): pp. 1–16. DOI:10.1371/journal.pone.0208523.

Sidemo-Holm, W., J. Ekroos, and H.G. Smith. 2021. "Land Sharing versus Land Sparing—What Outcomes Are Compared between Which Land Uses?" *Conservation Science and Practice* 3 (11): pp. 1–11. DOI:10.1111/csp2.530.

Spake, R., T.H.G. Ezard, P.A. Martin, A.C. Newton, and C.P. Doncaster. 2015. "A Meta-Analysis of Functional Group Responses to Forest Recovery outside of the Tropics." *Conservation Biology* 29 (6): pp. 1695–1703. DOI:10.1111/cobi.12548.

Staal, A., O.A. Tuinenburg, J.H.C. Bosmans, et al. 2018. "Forest-Rainfall Cascades Buffer against Drought across the Amazon." *Nature Climate Change* 8 (6): pp. 539–543. DOI:10.1038/s41558-018-0177-y.

Teixeira, H.M., I.M. Cardoso, F.J.J.A. Bianchi, et al. 2020. "Linking Vegetation and Soil Functions during Secondary Forest Succession in the Atlantic Forest." *Forest Ecology and Management* 457: p. 117696. DOI:10.1016/j.foreco.2019.117696.

te Wierik, S.A., E.L.H. Cammeraat, J. Gupta, and Y.A. Artzy-Randrup. 2021. "Reviewing the Impact of Land Use and Land Use Change on Moisture Recycling and Precipitation Patterns." *Water Resources Research* 57 (7): e2020WR029234. DOI:10.1029/2020WR029234.

Tilman, D., J. Knops, D. Wedin, et al. 1997 "The Influence of Functional Diversity and Composition on Ecosystem Processes." *Science* 277 (5330): pp. 1300–1302. DOI:10.1126/science.277.5330.1300.

Tobella, A.B., H. Reese, A. Almaw, et al. 2014. "The Effect of Trees on Preferential Flow and Soil Infiltrability in an Agroforestry Parkland in Semiarid Burkina Faso." *Water Resources Research* 50: pp. 3342–3354. DOI:10.1002/2013WR015197.

Uebel, K., K.A. Wilson, and L.P. Shoo. 2017. "Assisted Natural Regeneration Accelerates Recovery of Highly Disturbed Rainforest." *Ecological Management and Restoration* 18 (3): pp. 231–238. DOI:10.1111/emr.12277.

Van Der Ent, R.J., H.H.G. Savenije, B. Schaefli, and S.C. Steele-Dunne. 2010. "Origin and Fate of Atmospheric Moisture over Continents." *Water Resources Research*, 46 (9): 1–12. DOI: 10.1029/2010WR009127.

van der Sande, M.T., H. Bruelheide, W. Dawson, et al. 2020. "Similar Factors Underlie Tree Abundance in Forests in Native and Alien Ranges." *Global Ecology and Biogeography* 29 (2): pp. 281–294. DOI:10.1111/geb.13027.

van der Sande, M.T., L. Poorter, L. Kooistra, et al. 2017. "Biodiversity in Species, Traits, and Structure Determines Carbon Stocks and Uptake in Tropical Forests." *Biotropica* 49 (5): pp. 593–603. DOI:10.1111/btp.12453.

van der Sande, M., J. Powers, T. Kuyper, et al. 2022. "Soil resistance and Recovery during Neotropical Forest Succession." *Philosophical Transactions B* 378 (1867): p. 20210074. DOI:10.1098/rstb.2021.0074.

van Meerveld, H.J., J.P.G. Jones, C.P. Ghimire, et al. 2021. "Forest Regeneration Can Positively Contribute to Local Hydrological Ecosystem Services: Implications for Forest Landscape Restoration." *Journal of Applied Ecology* 58 (4): pp. 755–765. DOI:10.1111/1365-2664.13836.

van Noordwijk, M. 2021. "Agroforestry-Based Ecosystem Services: Reconciling Values of Humans and Nature in Sustainable Development." *Land* 10 (7). DOI:10.3390/land10070699.

van Oudenhoven, A.P.E., K. Petz, R. Alkemade, L. Hein, and R.S. De Groot. 2012. "Framework for Systematic Indicator Selection to Assess Effects of Land Management on Ecosystem Services." *Ecological Indicators* 21: pp. 110–122. DOI:10.1016/j.ecolind.2012.01.012.

Vásquez-Castro, D.C., R.R. Rodrigues, P. Meli, et al. 2021. "Preliminary Results of Using Green Manure Species as a Cost-Effective Option for Forest Restoration." *Scientia Forestalis* 48 (127): pp. 1–7. DOI:10.18671/SCIFOR.V48N127.21.

Veldman, J.W., G.E. Overbeck, D. Negreiros, et al. 2015. "Where Tree Planting and Forest Expansion Are Bad for Biodiversity and Ecosystem Services." *BioScience* 65 (10): pp. 1011–1018. DOI:10.1093/biosci/biv118.

Verschuuren, B., R. Wild, and G. Verschoor. 2017. "Connecting Policy and Practice for the Conservation of Sacred Natural Sites." In *Indigeneity and the Sacred: Indigenous Revival and Sacred Sites Conservation in the Americas*, edited by F.O Sarmiento, and S. Hitchner, pp. 11–40. New York: Berghan. Available at: http://www.berghahnbooks.com/title/SarmientoIndigeneity.

Wang, F., Z. Li, H. Xia, et al. 2010. "Effects of Nitrogen-Fixing and Non-Nitrogen-Fixing Tree Species on Soil Properties and Nitrogen Transformation during Forest Restoration in Southern China." *Soil Science and Plant Nutrition* 56 (2): pp. 297–306. DOI:10.1111/j.1747-0765.2010.00454.x.

Wang, X., F. Hua, L. Wang, D.S. Wilcove, and D.W. Yu. 2019. "The Biodiversity Benefit of Native Forests and Mixed-Species Plantations over Monoculture Plantations." *Diversity and Distributions* 25 (11): pp. 1721–1735. DOI:10.1111/ddi.12972.

Wubs, E.R.J., W.H. Van Der Putten, M. Bosch, and T.M. Bezemer. 2016. "Soil Inoculation Steers Restoration of Terrestrial Ecosystems." *Nature Plants*, 2 (8): pp. 1–5. DOI:10.1038/NPLANTS.2016.107.

WWAP (United Nations World Water Assessment Programme)/UN-Water. 2018. *The United Nations World Water Development Report 2018: Nature-Based Solutions for Water.* Paris: UNESCO. Available at: https://unesdoc.unesco.org/ark:/48223/pf0000261424.

Yachi, S., and M. Loreau. 1999. "Biodiversity and Ecosystem Productivity in a Fluctuating Environment: The Insurance Hypothesis." *Proceedings of the National Academy of Sciences of the United States of America* 96 (4): pp. 1463–1468. DOI:10.1073/pnas.96.4.1463.

Zeng, Y., M. Gou, S. Ouyang, et al. 2019. "The Impact of Secondary Forest Restoration on Multiple Ecosystem Services and Their Trade-Offs." *Ecological Indicators* 104 (498): pp. 248–258. DOI:10.1016/j.ecolind.2019.05.008.

Zimmermann, B., A. Zimmermann, H.L. Scheckenbach, et al. 2013. "Changes in Rainfall Interception along a Secondary Forest Succession Gradient in Lowland Panama." *Hydrology and Earth System Sciences* 17 (11): pp. 4659–4670. DOI:10.5194/hess-17-4659-2013.

8

Making forest landscape restoration work for livelihoods and well-being of local communities

Houria Djoudi, Ida N.S. Djenontin, and Carol J. Pierce Colfer

8.1 Introduction

Forests, trees, and lands provide direct and indirect material, economic, social, and cultural benefits to people worldwide. The livelihoods of 90% of the world's 1.2 billion poorest people depend on forest, tree, and land resources in various ways (Fedele et al. 2021; World Bank 2016). The degradation and loss of these resources negatively impact agricultural productivity as well as other vital ecosystem services (ES), including water and energy resources (Fedele et al. 2021). This endangers the foundation of rural livelihoods and the economic, cultural, institutional, and financial systems that support them (Aronson et al. 2010; NGFS-INSPIRE 2022; UNEP 2021). Restoring degraded forest landscapes appears to be a natural way to regenerate and sustain poor people's livelihoods.

Forest landscape restoration (FLR) is seen as a prominent solution to the intertwined global socio-environmental challenges of deforestation, biodiversity loss, and climate change (Besseau et al. 2018; Chazdon and Brancalion 2019; Erbaugh and Oldekop 2018). Triggered by the Bonn Challenge, the New York Declaration on Forests, and amplified under the UN Decade on Ecosystem Restoration, FLR is fostered as a people-centered approach to restore millions of hectares of degraded forested landscapes by 2030 (Besseau et al. 2018; Djenontin et al. 2021; Elias et al. 2021a). We define FLR as a process that aims to regain ecological integrity and enhance human well-being in deforested or degraded forest landscapes. We adapt FLR's established definition (e.g., WWF and IUCN 2000; and Chapters 5 and 10) by removing the word "planned." We recognize that restoration can occur spontaneously (i.e., without explicit planning), especially when led by local communities (Djenontin et al. 2022). FLR aims to reverse degradation and improve multifunctional landscapes so as to support climate resilience and biodiversity recovery as well as cultural integrity and food and livelihood security. As a dynamic process, FLR unfolds within and across landscapes rather than being confined to isolated individual sites. Such landscapes encompass a diverse tapestry of interconnected land uses, which are often regulated by different tenure and governance systems. It is precisely at this expansive scale that the intricate web of ecological, sociocultural, and economic priorities can be effectively harmonized and balanced (Guariguata et al. 2021).

FLR conceptual framings go beyond eco-centric restoration. Emphasizing equal priority for ecological and sociocultural outcomes, FLR aims to balance natural

Houria Djoudi, Ida N.S. Djenontin, and Carol J. Pierce Colfer, *Making forest landscape restoration work for livelihoods and well-being of local communities* In: *Restoring Forests and Trees for Sustainable Development*. Edited by: Pia Katila, Carol J. Pierce Colfer, Wil de Jong, Glenn Galloway, Pablo Pacheco, and Georg Winkel, Oxford University Press. © Oxford University Press 2024. DOI: 10.1093/9780197683958.003.0008 resource-based livelihood and food security, adaptation needs, and poverty reduction with climate change mitigation and biodiversity conservation goals (Djenontin et al. 2018; Mansourian 2005). Nevertheless, FLR interventions continue to fail to adequately integrate and address local aspirations and expectations and to explicitly pursue livelihood outcomes. Cultural issues, such as values, symbolism, and existing lifestyles, are rarely highlighted (Elias et al. 2021a; 2021b).

There is a lack of empirical research on the impacts of FLR interventions on local livelihoods. Rigorous socioeconomic assessments are limited, and investigations into cultural elements are virtually nonexistent. The existing but limited evidence from past interventions that can be related to FLR actions shows that FLR implementation processes often face challenges due to insufficient consideration of complex multi-scale interactions (Mansourian 2017; Reed et al. 2017). Moreover, the predominant focus on production forests and carbon sequestration as primary FLR objectives narrows the potential of FLR to deliver a diverse array of ES and multiple benefits. In a top-down drive for expedited project execution, local realities, priorities, and issues of social equity are often marginalized during FLR design and implementation (Chazdon et al. 2021; Elias et al. 2021a; Löfqvist et al. 2022; Schultz et al. 2022). Notably, within the so-called Global South, areas targeted for national restoration pledges often coincide with rural regions characterized by high poverty rates, limited livelihood opportunities, inadequate infrastructure, and minimal access to markets and basic services (Erbaugh et al. 2020; Schultz et al. 2022). These communities typically possess limited political or economic influence to safeguard their resources.

Cultural and socioeconomic objectives in restoration initiatives are consistently underemphasized and under-recognized and frequently neglected (Adams et al. 2016; Aronson et al. 2010; Löfqvist et al. 2022).

Moreover, the perspectives and aspirations of local communities often diverge from those of other stakeholders involved in FLR. For instance, in dryland areas, local land users often hold distinct views on what constitutes land degradation, leading to varying priorities regarding specific areas, timing, and methods (Crossland et al. 2018). Nevertheless, many initiatives and projects to combat degradation still advocate for control-oriented approaches, such as establishing green belts and planting forests. This top-down perspective contrasts with an alternative approach that embraces uncertainty and leverages local variability, as exemplified by many traditional local systems such as pastoralism in drylands. Neglecting seasonal variability and relying solely on biophysical criteria to define degradation and potential land for restoration can jeopardize ancient ecosystems and disrupt livelihoods (Scoones 2018). Learning from past experiences can help to better align FLR with local socio-institutional and economic development goals (Elias et al. 2021a, 2021b; Stanturf 2021; Chapter 2).

This chapter uncovers intricate challenges and trade-offs that lie at the heart of addressing the diverse spectrum of understandings, goals, and varying visions concerning FLR—what it is and what it should strive to become. Our aim is to contribute to the knowledge and insights needed to make well-informed decisions regarding FLR, ones that not only acknowledge these trade-offs but also actively work towards mitigating them for the benefit of the most vulnerable communities. We also contribute to the collective endeavor of amplifying voices and redoubling efforts in the pursuit of rectifying the inequities stemming from the historical neglect of local social needs within FLR interventions. We achieve this by meticulously examining:

- What is the potential for FLR to provide benefits for local livelihoods and well-being?
- What are current challenges and trade-offs in linking restoration to local livelihoods and well-being?
- What are some ways forward to increase the local and global co-benefits of FLR interventions?

Moving forward, Section 8.2 examines the connections between FLR and local livelihoods and well-being and delves into specific local outcomes that FLR interventions can intentionally harness. Section 8.3 identifies major trade-offs and challenges, at various levels, in linking restoration to local livelihoods and well-being. We shed light on bottlenecks that must be recognized and navigated to address the critical needs of local communities, their cultures, and their livelihoods. Section 8.4 identifies "leveraging points" and recommendations to tackle these challenges and trade-offs through collaborative governance. These aim at enhancing attention and consideration to local concerns, cultural practices, and livelihoods in FLR processes and outcomes. Section 8.5 concludes the chapter.

8.2 FLR potential to provide local livelihoods and well-being benefits

FLR measures differ in what they deliver at the local level, but when properly designed and executed, they hold significant potential to provide a wide range of benefits to local communities in multiple ways (ITTO 2020; Kumar et al. 2015). Table 8.1 illustrates specific examples of various types of FLR schemes, their linkages, and their contributions to local livelihoods, based on existing evidence and anticipated benefits.

Agroforestry, for instance, supports food production, health and nutrition, wood-based energy, and income (Kuyah et al. 2020). It can enhance net primary productivity of crops—playing a crucial role in climate regulation—and deliver high-value products, thereby increasing income levels and diversifying both production and diets while securing energy needs. Additionally, it offers essential shade and seasonal fodder for livestock while providing vital ES, such as regulating water flow and mitigating soil erosion. In addition, it can enhance biodiversity, fostering pollinators and controlling natural pests and diseases. Furthermore, if legume trees are used, it aids in nitrogen fixing and enhances nutrient recycling, ultimately bolstering food security. Lastly, it enhances a vital safety net for local livelihoods, increasing overall resilience within these communities (IPCC 2019; Jose 2009; Muthee et al. 2022; Smith et al. 2019). A study assessing agroforestry systems in Sub-Saharan Africa (Muthee et al. 2022) shows that income generation emerged as the predominant provisioning service, accounting for 31% of the total, while greenhouse gas emission reduction took center stage as the primary regulatory service, also representing 31% of the total. Additionally, soil fertility management played a crucial role as a key support service.

Agroforestry systems are typically multifunctional, although the types of goods and services produced vary depending on the components of agroforestry and the way these are managed in the landscape.

Through multi-species reforestation, FLR holds the potential to yield local benefits that extend services provided by mono-species *planted forests*. Multi-species plantations can offer benefits like timber production, carbon sequestration, and wildlife habitat. However, realizing these benefits in the long term depends on two key factors: forest access and effective management. To ensure equitable access, it is advisable for the plantation industry to extend benefits—including employment, land lease income, and Corporate Social Responsibility (CSR) benefits—to all community members, irrespective of landownership, social status, or education level (Kull et al. 2019; World Bank 2016). For instance, in remote communities, road improvements and benefits from large-scale plantation forests were significant, particularly in the farthest community with limited livelihood options. Yet, there was uneven distribution, with elites gaining more benefits, such as employment and income, and playing a more significant role than the most vulnerable people (Kainyande et al. 2022). Also companies in remote areas are often seen as substitutes to states for local development (Pirard et al. 2017).

A global systematic review reveals predominantly negative impacts of large-scale tree plantations on the local level, with 41% of studies reporting negative effects on employment (81% predominantly negative), livelihoods (48% predominantly negative), and interconnected social impacts (69% predominantly negative) (Adams et al. 2016). These impact patterns mirror those observed in similar large-scale land-based investments. Importantly, the study observed that costs for local residents are often front-loaded, especially when plantations displace customary land uses, while potential benefits may accumulate over time, influenced by the development of local processing and complementary livelihood activities. A livelihoods-based intervention project that supports extensive agroecological practices (including shifting cultivation) and integrates forests into multifunctional landscapes, can recenter local interests in reforestation approaches.

Natural regeneration often leads to faster forest recovery, greater carbon storage, and more co-benefits for biodiversity and people (Fleischman et al. 2020). Naturally regenerating forests are an invaluable resource for both timber and non-timber products, playing a pivotal role in sustaining the livelihoods of local communities (Adams et al. 2016). Furthermore, these naturally regenerating forests yield a rich diversity of non-timber products, which hold a crucial place in preserving and enhancing the cultural traditions of rural communities. These products encompass bushmeat, a variety of food sources, palm thatch for shelter, cordage, palm leaves for the ornamental market, and a wide array of medicinal plants. This wealth of resources has been extensively documented in several studies (Chazdon and Coe 1999; Gavin 2004; Hernández-Barrios et al. 2015; Junqueira et al. 2011; Pulido and Caballero 2006; Voeks and Leony 2004).

Improved fallow systems have demonstrated success in revitalizing soil health, contributing positively to household welfare through agricultural and pastureland productivity. Various examples exist in western Kenya, southern Malawi, and eastern Zambia. Enhancing the prevalence of young secondary forests in tropical regions through extended fallow periods in shifting cultivation cycles has the potential to bolster essential hydrological ES, such as reducing overland flow, erosion, and flooding, all while ensuring consistent streamflow. Incorporating indigenous wisdom regarding optimal fallow lengths can improve crop productivity and nutritional outcomes as well as advance restoration objectives. Moreover, the restoration of plants used in ceremonies and craftsmanship holds the promise of revitalizing meaningful practical, spiritual, and artistic dimensions of local cultures (Colfer et al. 1997; 2008; Van Meerveld et al. 2021).

However, studies indicate that escalating land-use pressures are compelling farmers to abandon extended fallow periods (Cairns and Garrity 1999; Kouassi et al. 2021; Medrilzam et al. 2017). This shift has led to shorter fallow cycles, prolonged cropping seasons, and environmental degradation, particularly in Southeast Asia's mountainous regions. This trend results in declining crop yields, increased labor for weed control, and threats to household food security. Consequently, farmers are forced to seek new forested lands with more fertile soils for cultivation in some regions (Cairns and Garrity 1999).

Silvicultural management, as demonstrated in studies such as Sabastian et al. (2019) in Indonesia, offers substantial benefits for farmers. For instance, smallholder teak tree crop systems in Indonesia that embrace silvicultural practices such as pruning, thinning, maintaining germplasm quality, intercropping, and promoting biodiversity conservation have been shown to contribute significantly, constituting up to 40% of household income. Nevertheless, farmers encounter significant challenges stemming from limited resources, labor shortages, and a lack of technical information on silvicultural management. These hurdles have led to reduced productivity and decreased on-farm incomes among smallholder timber and non-timber forest product producers (Kallio et al. 2012; Sabastian et al. 2019). A study conducted in Nepal underscores forest user preferences when it comes to silviculture options (Cedamon et al. 2017). Nepalese forest users prioritize options that align with legal regulations, can be feasibly implemented with their current capabilities, and have the potential to yield multiple forest products. To achieve the goal of sustainable production of various forest products, it becomes imperative to harmonize traditional forest management practices with modern silviculture-based forest management systems. This integration ensures not only the preservation of local customs and knowledge but also the responsible and diversified utilization of forest resources, which is crucial for both environmental conservation and the livelihoods of the communities dependent on these forests (Cedamon et al. 2017).

Mangrove restoration is closely connected to the revival of valuable and edible species, which not only benefits local nutrition but also boosts income. Research reveals the global potential to restore coastal areas in 105 countries and territories. This effort could introduce over 60 trillion young fish and invertebrates of commercially valuable and edible species into coastal waters annually (Worthington and Spalding 2018). However, the method used for mangrove restoration is crucial, as different approaches can yield varying results. Studies suggest that planting should be considered only when natural recolonization fails, as it can potentially double the cost involved. As a notable illustration of the myriad advantages offered by mangrove forests, Barbier (2007) calculates that in Thailand, mangroves hold an estimated value of USD 8,966–10,821/ha for storm protection, USD 484–595/ha for products benefiting communities, and USD 708–987/ha in increased offshore fisheries production. Additionally, Menéndez et al. (2019) estimate that mangrove forests in the Philippines alone prevent approximately USD 1 billion in property damages annually through their floodwater protection capabilities.

Watershed protection encompasses various crucial benefits at the local and landscape levels, including the regulation of shade and water temperature, erosion control, safeguarding water quality and enhancing landscape aesthetics (Broadmeadow and Nisbet 2004). Existing literature underscores the significance of market-based incentives in providing substantial support for communities that promote sustainable income-generating activities such as fishing and irrigated agriculture. These incentives also create opportunities for local communities to explore alternative livelihoods (e.g., ecotourism). Moreover, communities can reap financial benefits from conservation practices that enhance freshwater ES, primarily through Payment for Watershed Services (PWS) programs facilitated by public or private sectors. Despite the growing awareness of markets for watershed protection services, they remain relatively rare and contribute only modestly to the income streams of local communities (Salzman et al. 2018; Zhang et al. 2023). A significant obstacle in implementing PWS programs is the necessity for clear resource rights, a requirement often lacking in many communities for watersheds.

FLR measures aimed at restoring biodiversity can also generate local benefits that go beyond the mere physical restoration of degraded ecosystems. For example, ecological restoration efforts can play a role in preserving cultural and spiritual values, as demonstrated by the restoration practices of Māori indigenous people guided by the principle of Kaitiakitanga (environmental guardianship). This approach goes beyond the restoration of ecosystem services, as it also focuses on healing the strained relationships between humans and the broader natural world (Parsons et al. 2021), as well as preserving cultural and spiritual values (Constant and Taylor 2020).

Involving Indigenous Peoples and local Communities (IPLCs) in the design and implementation of FLR schemes can create employment opportunities and enhance the skills and capacities of local people. In addition, FLR can have positive social impacts by strengthening local governance and empowering marginalized groups, such as women and indigenous peoples. Another potential of FLR can be realized by involving IPLCs in decision- making processes. Doing so can help ensure that their rights and interests are considered and respected. Such collaboration can reduce the likelihood of implementation bottlenecks due to inadequate understanding of complex local contexts. Furthermore, FLR can improve human health by providing clean water, reducing air pollution, and enhancing food security. This can help reduce the burden of disease and improve overall well- being.

Table 8.1 Extant and anticipated multiple benefits of FLR to local livelihoods

FLR Measure (Adapted from Sabogal et al. 2015)	Examples of Representative Local FLR Services
Planted forests and woodlands*	• North America: small Christmas tree plantations provided a nest egg for children's education and performed conservation functions. • Sumatra: Benzoin trees provided what locals symbolically considered 'mother's milk', along with longer term income (also often used for education) and biodiversity maintenance.
Agroforestry**	• African drylands: rainfall harvesting and re-greening helped replenish water resources and increase crop yields. • Burkina Faso: water-harvesting techniques (e.g., stone bunds) doubled grain yields; maize yields increased by 88–190% when grown in an agroforestry system under the canopies of *Faiderbia albida* trees. • Niger: new agroforestry systems on 5 Mha increased annual agricultural production by about 500,000 t. • Estimates suggest agroforestry has the potential to increase food security for 1.3 billion people; reduce soil erosion by 50% and increase soil carbon by 21%, inorganic nitrogen by 46%, and phosphorus by 11%.
Natural regeneration of formerly degraded land***	• African drylands: rainfall harvesting and small infrastructures to conserve soil and water replenished water resources and increased crop yields. • Benefits from increased climate change resilience and tree cover: sequesters carbon, mitigates climate change and provides rural communities with stocks of wood to boost incomes and offset poor harvests. • Sumatra: marginalized ex-hunter gatherer and immigrant households benefited from forest restoration (*Hutan Harapan;* lit. forests of hope), providing 36% of local annual income. • Burkina Faso: Women curbed food insecurity and increased resilience through small scale protection and natural regeneration of degraded land.
Improved fallow#	• Tropics: Increasing the abundance of secondary forest by increasing fallow lengths in shifting cultivation cycles, could improve locally important hydrological ecosystem services (specifically reducing overland flow, erosion and flooding, while maintaining streamflow). • Building on indigenous knowledge of fallow length requirements can enhance productivity of subsequent crops, good nutrition and restoration goals. •Restoring plants used in ceremonies and crafts can revitalize meaningful practical, spiritual and artistic aspects of local cultures.

(*continued*)

Table 8.1 Continued

FLR Measure (Adapted from Sabogal et al. 2015)	Examples of Representative Local FLR Services
Silviculture ##	• Borneo: governmental and industry reforestation efforts (1980s–1990s) replaced logged forests with *Acacia mangium* as a silvicultural restoration initiative with severe negative consequences. • Borneo: Kenyah managed their forest by protecting desired seedlings and managing fire, maintaining multiple stages of forest regrowth to maximize biological diversity for product availability. • New York: Residents maintain small plots of woodland, selecting trees to save (walnut, sugar maple, oak) – for color, flowers, firewood, nuts, apples, shade, conservation functions.
Mangrove restoration ###	• Restoring mangroves in 105 countries and territories could add over 60 trillion young fish and invertebrates of commercially valuable and edible species to coastal waters every year.
Watershed protection and erosion control ####	• Regulation of shade and water temperature, erosion control, protection of water quality, and improvement of landscape quality.

*Data from Shanley et al. 2002; https://agamerica.com/blog/tips-from-the-land-lenders-on-starting-a-christmas-tree-farm/; Colfer observations

** Data from IPCC 2019; Jose 2009; Muthee et al.2022; Smith et al. 2019

*** Data from Djenontin and Djoudi 2015; Widianingsih et al. 2016; Winterbottom et al. 2013

Data from Colfer et al. 1997, 2008; Van Meerveld et al. 2021

Data from Colfer et al. 1997; Colfer observations

Data from Worthington and Spalding 2018

Data from Broadmeadow and Nisbet 2004

8.3 Challenges and trade-offs in linking restoration to local livelihoods and well-being

This section highlights significant and persistent bottlenecks within FLR interventions at the local level. Specifically, it identifies recurring challenges and trade-offs that have had detrimental effects on local communities. These challenges encompass the potential for cultural, livelihood, income, and resource losses, as well as constraints on local decision-making and resource control, ultimately contributing to increased inequities. Table 8.2 summarizes the trade-offs and tensions of FLR design and implementation.

Table 8.2 Trade-offs in FLR framing, objectives, and needs

Trade-offs & Tensions	Examples
Varying priorities, beliefs, values	• Economic priorities vs. social welfare vs. conservation • Productivity-based vs. locally adapted agroecological food systems • Development vs. conservation • Indigenous value of nature as a living being vs. landscape as an exploitable resource
Time horizons	• Short term provisioning vs. long term regulation services • Food security today vs. food system resilience tomorrow • Fast-growing imported species for energy vs. slow-growing local species for livelihoods
Spatial scales & interactions	• Farmers' fields vs. vast areas for pastoralists • Farmer vs. global institutions' landscape perspectives • Local vs. national/global needs • Connectivity concerns at forest-farm interface
Biodiversity aspects	• Number of trees planted vs. ecosystem restoration • Varying motivation to protect by social group due to differing biodiversity needs • Indigenous views of species as living beings vs. resources to be optimized
Social groups	• Local elite capture can disadvantage community segments • Influential outsiders vs. marginalized groups • Varied interests across gender, ethnicity, age, religion
Ecosystem services	• Provisioning vs. regulating services • Carbon vs. adaptation services • Ecosystem needs vary by subsistence system

8.3.1 Contested definitions, understandings, and interests among actors and across scales

One of the first challenges in considering the livelihoods dimension of restoration starts with the contested understandings of what forest landscapes mean—a situation that is often based on disciplinary perspectives. One particularity of FLR is its explicit landscape approach. Contemporary forest restoration interventions share the defining feature of implementation at the landscape scale. Such a landscape scale enables integration of disciplinary-based social and ecological dimensions of ecosystem-based management as well as different, but linked, sectoral resource management goals (Freeman et al. 2015; Scherr et al. 2013). To address contestations surrounding views on forest landscapes that may continue to render FLR practice sterile and obscure its potentials as enumerated above (Section 8.2), we draw attention to the following critical analytical lenses on landscape and forest landscapes. Box 8.1 offers a socio-biology perspective to understanding landscapes, especially as relevant for FLR.

Box 8.1 Defining landscape

A landscape is a holistic geographical concept that includes natural and human-altered ecosystems, along with cultural and institutional attributes. It represents the socio-ecological history of a specific area and emphasizes the interconnections between various land-use types. In the context of restoration, landscapes are dynamic and encompass political, economic, sociocultural, and ecological systems. Forest landscapes, for instance, consist of interconnected patches with spatial and temporal dimensions along with cultural and institutional attributes. (Adapted from Djenontin et al. 2021)

Second, differing perceptions on forest roles can be detrimental. The roles of forests as components of FLR within mosaic landscapes varies between stakeholders. This variability is shaped by stakeholders' unique perceptions, requirements, and objectives for forest use and management within a landscape. These diverse perspectives view forests as rich ecosystems teeming with diverse biological life, as sources of various products and commodities for companies, as vital habitats for indigenous communities, as crucial carbon storage sinks, and/or as providers of numerous regulating ES. The distinction between the actual characteristics of forested lands and their legal designations further complicates the issue. For instance, a forest that has undergone degradation may still maintain its legal classification as a forest, even when it no longer possesses sufficient tree cover.

Similarly, differing perceptions on forest functional status are sources of challenges. For instance, while a fundamental principle of FLR is its emphasis on restoring degraded ecosystems, there exists a substantial diversity in the definitions, applications, and assessments of the term "degraded" (Crossland et al. 2018; Hobbs 2016; Scoones 2018). Box 8.2 illustrates that perceptions of degradation are in the eye of the beholder, underscoring here different and unique perspectives on what degradation entails for local rangeland users.

Third, misreading of landscapes can sustain nonconsensual designation of areas for restoration potential. Recent quantitative estimations of "available" areas for restoration have overlooked locally adapted systems. These estimations often label large treeless areas as "empty" and ready for restoration, disregarding the economic, social, and cultural significance of these areas for local communities, including pastoralists. Consequently, many sparsely populated regions have been categorized as available for restoration without conducting a thorough assessment of their current uses or considering claims of local inhabitants (Bond 2016). Examples exist in tropical rainforests where practitioners of swidden agriculture have been marginalized in favor of carbon markets, with the aim of preserving forests in an idealized pristine state. This approach often neglects the potential for sustainable, cyclical land use practices that local communities may already employ (Cairns 2015; Colfer et al. 1997). Such disregard for local knowledge and practices can lead to adverse outcomes and the failure of FLR initiatives. To truly integrate the intricacies of local contexts and draw from the rich experiences of IPLCs, a paradigm shift within FLR is necessary.

Box 8.2 Degradation narratives in the drylands

Failure to consider seasonal variability and relying solely on biophysical criteria to determine degradation and the potential for Forest Landscape Restoration (FLR) can put at risk specific ancient ecosystems like rangelands and disrupt the livelihoods of those who depend on these traditional grazing areas for their livestock (Scoones 2018).

Drylands literature vividly highlights the unique perspectives of local land users when it comes to defining land degradation in their respective landscapes. This divergence in viewpoints leads to differing priorities regarding where, when, and how to address land degradation. Top-down and remote sensing studies often label these areas as degraded land in need of restoration, promoting a control-oriented approach that includes measures like establishing green belts and forest planting. This perspective contrasts with an alternative approach known as adaptive management, which embraces uncertainty and makes use of local vegetation and seasonal and interannual variations, much like the traditional local systems found in drylands, such as pastoralism.

8.3.2 Nonaligned objectives among stakeholders

Trade-offs and tensions often arise from varying objectives between different actors operating at different scales. These conflicts can be a result of disparate goals set across different levels. For instance, governments may prioritize ecological and environmental outcomes or national income, while local communities strive for improved livelihoods, and businesses focus on economic gains. Such conflicts can also manifest within the same level, spanning different sectors, national strategies, or between various social groups, all of whom expect a restored area to deliver a varying array of ES.

In cases of top-down and poorly designed FLR approaches, these trade-offs can escalate into conflicts and tensions. This occurs when distinct objectives, such as livelihood improvement, climate mitigation and adaptation, or biodiversity preservation are pursued independently without adequately recognizing and integrating the goals of different stakeholders at different scales. The recognition of the potential negative outcomes arising from these conflicts and negative externalities has spurred calls for integrated FLR approaches (Djenontin and Zulu 2021; Mansourian and Parrotta 2018). Achieving a balance among these competing priorities requires trade-offs. Consequently, fostering negotiations through engagement with multiple stakeholder platforms is imperative.

Socio-ecological systems present a complex array of challenges and opportunities for FLR (Carmenta and Vira 2018; Yang et al. 2018). Understanding the intricate interactions between the multiple ES that various FLR interventions can provide for livelihoods is paramount. However, those interactions remain difficult to grasp due to several factors, including the diverse spatial and temporal scales involved, the intricate web of land tenure and access regulations that differ across regions, and the varying perspectives and priorities of different stakeholders (de Jong et al. 2018;

Larson et al. 2010). This complexity is further compounded by the differentiated value attributed to ES at different scales and by different actors (Alexander et al. 2016; Fedele et al. 2021). For instance, regulating and supporting services may be highly regarded at larger scales and by a broader global audience, while cultural or provisioning services with direct livelihood benefits hold paramount importance at the local level. These dynamics often lead to trade-offs, where conflicting interests arise, such as those between commercial wood production and local demands for various forest goods. These dynamics are exacerbated by the role of the private sector, capitalism, market-driven approaches, and the development of professional bureaucracies and modern states. Their detailed exploration extends beyond the scope of this discussion.

8.3.3 Power imbalances between different actors

While restoration is implemented at the local level, it does not happen in a power vacuum. It is shaped by national and global governance structures. Forests and forest transitions hold distinct meanings for different individuals. Kull (2017) suggests that forest transitions should not be viewed solely as an increase in tree cover, but rather, as a process intertwined with power dynamics, shedding light on the intricate web of interests entangled in negotiations surrounding forest change. This perspective underscores the importance of comprehending why various facets of quantitative and qualitative change carry significance and for whom. Similarly, when evaluating FLR it should be examined within the same framework, considering the diverse perspectives, interests, and powers at play.

Power dynamics significantly shape the outcomes of restoration efforts, stemming from varying interests and priorities within forest landscapes. These power imbalances, both between different actors (communities, state, NGOs, private sector, etc.) and different scales (global, national, local), often wield significant influence over decision-making processes related to FLR interventions. The roles of various actors at each stage of the process, from planning to execution and monitoring, dictate whose interests take precedence and can exacerbate conflicts (Brockhaus et al. 2021). Despite the growing recognition of the need to consider trade-offs and power relations in implementing FLR (Elias et al. 2021b), the necessary expertise and skills to address such issues and support negotiations among different actors to seek out synergies are rarely available among FLR organizations and personnel.

One evident manifestation of power imbalances and recurrent tensions in FLR pertains to the disparities in interests between developed nations, which possess the financial resources to support FLR, and developing tropical countries where these interventions are carried out (Brockhaus et al. 2021). Additionally, disparities in power across different scales and between decision-making actors can further intensify conflicts. For example, local farmers may prioritize interventions that provide essential services such as food, firewood, fodder, and water, while international funders and private companies may emphasize carbon sequestration. In such scenarios, local communities have the least influence but stand to lose the most.

Moreover, in some instances, local populations may be displaced to create protected areas, large-scale tree plantations, or natural regeneration projects. Despite the growing acknowledgment of the necessity to consider trade-offs and power dynamics in FLR implementation (Elias et al. 2021b), organizations and personnel involved in FLR often lack the expertise and skills needed to address these issues and facilitate negotiations among diverse stakeholders to discover synergies.

8.3.4 Lack of effective local participation from design to monitoring

FLR efforts cannot succeed without the participation of local stakeholders (Erbaugh et al. 2020; Fleischman et al. 2022). However, studies show that most restoration interventions have lacked an effective bottom-up consultative process for engaging communities (Duguma et al. 2020). Decisions related to FLR that exclude local actors are likely to lead to an inequitable distribution of costs and benefits for local communities. Local communities must deal with the consequences of FLR actions, such as changes in the availability and flow of crucial ES generated by decisions and activities in more distant parts of the landscape and at national or global levels. Local stakeholders may have different livelihood goals for their traditional customarily owned land and land under overlapping regimes (customary/state), and they may not benefit from large-scale and top-down FLR actions. Recent studies show, for instance, that large-scale tree plantations rarely provided any benefits to local communities and their livelihoods (e.g., Li and Semedi 2021; Malkamäki et al. 2018). Furthermore, the lack of effective participation can lead to an inequitable distribution of costs and benefits for local communities, exacerbating tensions and engendering conflicts at the local level.

Efforts to engage with indigenous management systems for FLR implementation have been attempted in some countries. These endeavors have encountered numerous obstacles and challenges rooted in centralized governance structures (Winter et al. 2023). IPLC knowledge continues to be undervalued and excluded from restoration programming (Uprety et al. 2012; Wehi and Lord 2017). One of the obstacles to effectively involve IPLCs is the historical suppression of their invaluable knowledge, a legacy of colonialism that continues to persist due to factors such as postcolonial state policies and a disproportionate reliance on one-sided academic knowledge. We argue that Western scientists engaged in restoration may not fully acknowledge the influence of colonial and neocolonial structures on contemporary FLR design and implementation. However, a significant number of scientists from across the globe including indigenous researchers, emphasize the enduring impact of colonial legacies on various aspects directly connected to FLR and challenge the Eurocentrism of dominant restoration projects and narratives (de Leeuw and Hunt 2018; Parsons et al. 2021; Todd 2014; Uprety et al. 2012; Watts 2013; Wehi and Lord 2017; Winter 2018; Winter et al. 2023). These legacies have disrupted the social fabric in many societies in the Global South, and a substantial portion of conflicts related to land tenure can be traced back to profound changes imposed by colonial powers on local institutions, land tenure systems, and the imposition of colonial conservation paradigms (Anderson 2006).

Settler-colonialism has prevented Indigenous Peoples from carrying out their traditional restoration practices for generations. The injustices endured by IPLCs extend beyond acts of violence and dispossession; they include policies and strategies designed to marginalize and exclude local knowledge and values. Moreover, postcolonial governments have continued to perpetuate such historical patterns by asserting ownership over forests and other lands, even when situated on indigenous territories. For instance, recent analysis on Malaysia, Sierra Leone, and Zambia, underscores the importance of precolonial institutions in establishing the frameworks for land redistribution, transferring ownership from native populations to colonial settlers, and shaping land ownership patterns. Consequently, there is an ongoing struggle to reclaim, rediscover, and reaffirm related identities in the face of current restoration approaches (Parsons 2021). Engaging in decolonization practices of land tenure is imperative to comprehensively address the challenges of secure rights in FLR.

8.3.5 Lack of meaningful, locally relevant incentives

Local stakeholders often have their own priorities, which may include addressing immediate subsistence needs or dealing with other pressing social, cultural, or economic issues. These concerns should take precedence over the top-down designed objectives of FLR (Crossland et al. 2022). Absence of congruent incentives that align with the interests of different stakeholders and encourage active participation within local communities can become a challenge (Djenontin et al. 2020). Such incentives play a critical role in compensating for livelihood losses, such as the loss of land for agriculture during ecological restoration or restricted access and land use limitations, particularly concerning activities like livestock grazing, which may be necessary for tree growth and ecosystem restoration, or when lands previously used for farming are designated for restoration. These can enhance the likelihood of long-term success for FLR initiatives, as they play a pivotal role in building trust and fostering cooperation among various stakeholders. In the absence of incentives, local communities may fail to recognize the benefits of FLR and could prioritize activities that offer more immediate gains. For instance, a notable case in Madagascar involved a multimillion-dollar native tree planting project that failed to deliver the anticipated carbon revenue, resulting in significant challenges and setbacks (Peña Valderrama 2023).

However, it can be challenging to design appropriate incentives with critical attention to potential trade-offs. Incentives must be customized to align with practical mechanisms and the genuine needs of local communities. It is vital to stress that when governments and other organizations are driven by substantial financial rewards linked to ambitious targets for land area or the quantity of saplings planted, their attention frequently becomes fixated solely on the act of planting, thus neglecting the crucial necessity of continuous maintenance to ensure the survival and growth of these trees (Duguma et al. 2020). Furthermore, insights gained from incentive-based approaches suggest that nurturing nature-inclusive rural livelihoods over the long term offers more significant potential than solely depending on short-term incentives, like compensating landowners for a specific project duration (Höhl et al. 2020). For instance, a study conducted in Ghana reveals that incentives aimed at promoting tree plantations successfully motivated farmers to participate in tree-planting activities in the short term. However, the absence of sufficient mechanisms for long-term benefits deterred the majority of farmers from sustained involvement in restoration efforts (Kumeh et al. 2023).

Restoration initiatives carried out through PES programs, operating within a market-oriented framework that involves compensating for environmental contributions, have encountered significant challenges. Focusing payments exclusively on a single ES, such as carbon sequestration, is unlikely to alleviate poverty (Jindal et al. 2013). While carbon payments can augment income, they have not yielded the expected substantial impact on household livelihoods. Furthermore, PES programs exhibit varying outcomes from one location to another, contingent on their design and the benefits they offer, which can significantly differ (Wunder et al. 2008). State-led PES initiatives may achieve poverty reduction to some extent, but only in areas where government capacity is robust, funding for implementation is adequate, and local contextual factors are thoughtfully considered (Erbaugh and Oldekop 2018). Conversely, user-financed PES programs are more likely to attain efficiency, as they are better targeted and tailored to local contexts and people's needs, with enhanced monitoring and fewer conflicting secondary objectives (Wunder et al. 2008). An illustrative case can be found in Costa Rica's longstanding PES program, where local communities contested the approach. In this example, local residents advocated for a fundamental shift in the PES mechanisms, proposing a perspective that views it as "support for stewards." This alternative framing aims to bolster collective management, uphold social values, and foster deeper connections to nature rather than undermining them (Chapman et al. 2020).

8.3.6 Downplaying local priorities: When community priorities are not taken seriously

The FLR literature is replete with examples vividly demonstrating the adverse consequences of neglecting the needs and priorities of local communities, which can jeopardize the anticipated social and environmental outcomes of FLR efforts (Bakkegaard and Vongvisouk 2019; Feintrenie et al. 2019; Nurfatriani et al. 2021).

For instance, in Nigeria, a restoration project planted non-fruit trees, which were subsequently felled for fuelwood. In contrast, fruit trees thrived. Similar patterns emerged in central Java, Indonesia (Liswanti et al. 2020), where women were excluded from the consultation process regarding agroforestry activities. Consequently, they resorted to cutting down trees. The program increased their workload without yielding additional economic benefits. In Bolivia, a forest conservation program intended to safeguard and restore forests by offering economic incentives to local communities faced resistance from certain communities. They felt their land rights were being compromised and that they were not adequately consulted or involved in the decision-making process. In Madagascar, a community-based forest management project struggled to secure support from local communities due to a lack of trust in the project implementers and concerns about potential restrictions on access to forest resources. In Kenya, a reforestation project encountered resistance from local communities who believed that the project did not prioritize their immediate needs, such as access to water and livestock grazing. In India, a forest conservation project faced opposition from local communities who perceived it as a threat to their traditional livelihoods, including the collection of non-timber forest products (Höhl et al. 2020).

8.3.7 Mismatched timeframes: Short- and long-term objectives

Tree-based systems require time to be established or restored, while local communities often face urgent livelihood needs. It is crucial to emphasize the time gap between planting and the availability of tangible benefits or ecological functioning, which means immediate livelihood benefits are not available. If the immediate livelihood needs of the most vulnerable are not integrated into the design (e.g., through diversification of livelihoods or incentives), marginalized communities may hesitate and lose interest in participating. Additionally, project-based FLR interventions operate within relatively short timeframes, while many tree species, especially in drylands, require considerable time to grow and yield expected livelihood outcomes (Duguma et al. 2020).

Numerous instances in the literature underscore FLR initiatives that faced opposition from local communities. These communities believed that their immediate essentials, including food, fuelwood, access to water, livestock grazing, and other vital services, were not prioritized (Höhl et al. 2020). For example, in a restored Atlantic forest of Brazil, the prospect of future long-term timber production failed to persuade the most economically disadvantaged members of the community to engage, as their immediate challenges were more pressing.

8.3.8 Rigid FLR design and lack of adaptive management

Resilience and adaptive learning are critical aspects of FLR (Spathelf et al. 2018), and inadequate consideration of climate change adaptation negatively impacts landscape resilience (DeRose and Long 2014). This, in turn, affects local communities' ability to cope with climate events, particularly when their rights or livelihoods are undermined, increasing vulnerabilities (Djoudi et al. 2022). A successful FLR demands adaptability not only to climate but also due to its implementation within intricate socio-ecological systems influenced by various factors, such as cultural differences, land-use shifts, and economic development. Therefore, a flexible, learning-oriented approach is essential to accommodate changing circumstances and ensure the sustainability of restoration efforts. Rigid FLR initiatives often fail to account for local socioeconomic and ecological conditions, potentially rendering them unsuitable or ineffective in achieving restoration goals. They may be fixated on predetermined objectives that do not align with the priorities and needs of local communities, potentially resulting in conflicts, objective failures, or livelihood losses. Furthermore, inflexible FLR approaches often cannot integrate multiple objectives or new knowledge, hindering their ability to adapt to evolving circumstances and address emerging challenges. An example from Ecuador's Chocó Andino region illustrates the consequences of rigid FLR funding allocation. Instead of considering farmers' nature-inclusive suggestions for cattle raising, agriculture, and vital economic and cultural activities in the region, funds were exclusively earmarked for predetermined approaches, such as tree planting or natural regeneration. This failure to invest in restorative farmland practices led to the inability of the National Forest Restoration plan to stimulate economic activities and reduced farmers' willingness to participate (Wiegant et al. 2020).

8.4 Leveraging points and recommendations for FLR to deliver livelihoods and well-being for local communities

To ensure the successful realization of local-level outcomes, restoration efforts must effectively address trade-offs (Tedesco et al. 2023). They need to prioritize the continuous and sustainable provision of the myriad ES offered by well-functioning ecosystems, directly and indirectly enhancing human well-being through subsistence, livelihoods, and food security (Alexander et al. 2016; Fedele et al. 2021). In this context, we delve into perspectives and suggestions spotlighting seven domains, where reconciliation and integration are imperative to catalyzing adaptive learning in FLR implementation (Reinecke and Blum 2018). It is essential to recognize that these leverage points differ from principles; they should be regarded as cautionary signals. While principles are designed for universal application and often represent a top-down perspective, the challenges discussed here must be addressed within the diverse contexts encountered in the field, encompassing both human and biophysical factors. This necessitates a more bottom-up approach. Our aim is not to introduce additional principles to the existing array of FLR principles. Rather, our goal is to pinpoint leverage points and domains of action to enhance the outcomes of FLR at the local level.

8.4.1 Building on IPLC knowledge systems

Efforts to strengthen IPLC's involvement in FLR governance, spanning design, implementation, monitoring, and evaluation, should take center stage. Additionally, it is essential to implement strategies that enhance connectivity across various scales and sectors, and among diverse social actors, to design and implement participatory and user-friendly monitoring systems together with communities. Ensuring transparency and accountability of key stakeholders is of paramount importance (McLain et al. 2021).

Indigenous ontologies often establish enduring, reciprocal connections between human and more-than-human communities, sometimes rooted in kinship-based relationships (De Leeuw and Hunt 2018; Parsons et al. 2021; Todd 2014; Watts 2013; Winter 2018). Land as a sacred entity deeply intertwined with culture and spirituality challenges the anthropocentric perspective that predominates in many settler worldviews, which often see land as a tradable commodity (Hird et al. 2023; Price et al. 2022). It is imperative to not only recognize but also include ontologies, social structures, and cultural contexts in the design, implementation, and monitoring of FLR. In this regard, it is vital to incorporate indigenous ontologies, particularly principles emphasizing the interconnectedness of ecosystems and people. The narrow focus on isolated ES restoration should be broadened to focus on the protection and restoration of life.

To truly integrate the intricacies of local contexts and draw from the rich experiences of IPLCs, a paradigm shift within FLR is necessary. Engaging with, embracing, and learning from ancestral management systems and decentralized, holistic approaches of indigenous stewardship in natural resource management are crucial for a paradigm shift towards equitable and sustainable FLR (Winter et al. 2023). The inclusion of IPLC is indispensable for the decolonization of restoration initiatives, particularly those initiated or overseen by external actors, such as researchers from the Global North (Pashkevich et al. 2022).

It is crucial to engage with numerous examples around the world that illustrate how communities have revitalized indigenous and other local small-scale stewardship approaches. Many have developed place-based and holistic landscape management practices that yield multiple benefits for local livelihoods (Garnett et al. 2018).

To bolster FLR benefits for IPLCs, it is imperative to construct context-specific FLR design and implementation that enhances land governance processes models by embracing a multitude of values, interpretations, and "ways of knowing" (Tengö et al. 2017). This approach can help alleviate the challenges typically associated with top-down policy agendas that primarily adhere to a single dominant value system, a factor often leading to conflicts and opposition (Temper et al. 2015).

Box 8.3 presents examples of successful FLR initiatives firmly grounded in traditional knowledge systems and networks. These initiatives have yielded significant advantages for local livelihoods and well-being. They also shed light on why we eliminated the term "planned" from the definition.

8.4.2 Fostering multiple benefits

Though difficult, constructing restoration activities that optimize gains in livelihood benefits, biodiversity, and ecosystem health will be fundamental to the long-term success of any FLR action. To achieve multiple objectives effectively, it may be beneficial to employ various options at different locations within the landscape mosaic and at different times (Lamb et al. 2005). Proactive participation in agroforestry projects has been shown to generate substantial additional and consistent income, especially among the poorest households (Jindal et al. 2013) and women (Crossland et al. 2022; Mukasa et al. 2022), which can significantly impact environmental landscape outcomes.

In Tanzania FLR activities have garnered acceptance for forest conservation goals due to tangible improvements in local livelihoods supported by FLR projects (Höhl et al. 2020). Furthermore, recent evaluations of FLR interventions in India confirm improved ecological indicators in areas where villagers have been provided with financial resources and technologies (such as fuel-efficient stoves and

Box 8.3 Successful FLR initiatives using traditional knowledge systems and networks

Examples of successful FLR initiatives rooted in traditional knowledge systems and networks that have brought tangible benefits to local livelihoods and well-being are listed below.

Forest seed local supply networks in Brazil (Urzedo et al. 2020). Diverse community networks in Brazil have effectively boosted local seed production to support a thriving restoration economy across Amazonia, the Cerrado, and the Atlantic Forest. Over a decade, 1,016 collectors exchanged 416.91 t of native tree seeds, averaging 31.41 kg annually and generating an average of USD 256.5 in household income annually. Depending on the adoption of various restoration methods, this could create work opportunities for 13–57 thousand collectors annually, generating a total income of USD 34–146 million, leading to significant socioeconomic benefits.

Indonesia, Cameroon, New-Zealand: Indigenous eco-cultural beliefs and practices used to restore degraded forest landscapes (Asteria et al. 2022; Parsons et al. 2021; Wanjiru Ng'ang'a 2022). Indigenous communities worldwide employ a range of cultural practices to restore their forest landscapes. For example, the Baduy community in Indonesia practices sustainable forest management, engaging local communities to improve their well-being while upholding Baduy traditions, including customary prohibitions known as *pikukuh*. In New Zealand, specific Māori communities incorporate many of their restoration initiatives within the framework of *kaitiakitanga* (environmental guardianship). This concept revolves around interconnectedness and healing relationships between humans and more-than-human entities. The Māori communities also blend local wisdom with Western scientific approaches to achieve their objectives. Similarly, in Cameroon, the Akamba community employs sacred forests and spiritual significance to restore crucial ecosystem services in degraded areas like the Kivaa ecosystem, which suffered biodiversity loss due to human activities such as logging and habitat destruction. Over the past decade, community-led initiatives, guided by indigenous resource governance and sacred forest practices (*Ithembo*), including taboos and indigenous tree planting, have revitalized the ecosystem, greatly benefiting local communities with minimal government intervention.

Philippines: Restoration of native forest (Tolentino 2008). In the Philippines, successful native forest restoration and exotic species reduction have centered on indigenous tree species with local communities playing a vital role. Those communities carefully select from a diverse range of indigenous trees, providing crucial insights into their uses, the locations of seed trees, ecological characteristics, and growth cycles. This collaborative effort leverages traditional knowledge and extends to such activities as nursery establishment, propagation, domestication, and tree management. Notably, using seeds and wildlings as germplasm sources highlights indigenous communities' capacity to overcome the challenge of limited native planting materials, gradually replacing exotic species with the rich biodiversity of their native forests. This holistic, community-driven approach not only restores valuable services but also preserves the cultural heritage and wisdom of the local people.

Tanzania: Restoration of large areas of acacia and miombo woodlands that were transformed into semidesert (Monela et al. 2004). In Tanzania, efforts have been dedicated to revitalizing acacia and miombo woodlands that had deteriorated into semidesert environments. This endeavor encompasses several essential elements needed to enhance local livelihoods, including careful species selection, the identification of livelihood requirements and preferences, and the active engagement of local communities through traditional institutions. Moreover, traditional techniques, such as pasture management, fodder production, and grazing control, are integrated with non-timber forest products, contributing significantly to the enhancement of local livelihoods. The success of this restoration initiative can be attributed to its bottom-up approach and the alignment of incentives for local inhabitants and the establishment of supportive legal frameworks, encompassing both traditional and institutional aspects.

cooking cylinders) to reduce their reliance on fuelwood for cooking. This reduction in pressure on tree resources has been particularly effective in targeted areas (Osuri et al. 2019).

Studies highlight that reforestation with monoculture and exotic species can generate financial benefits relatively quickly, which may be particularly appealing to impoverished communities (Le 2012). However, this approach often comes at the cost of biodiversity and sociocultural losses. It raises the question of whether these immediate financial gains are the most beneficial for impoverished communities. In landscapes where poverty is prevalent, achieving both immediate and long-term objectives simultaneously is imperative. As Alexander et al. (2016) argues, it is possible to make incremental changes over time to achieve a degree of success in all these outcomes. In line with recent studies on forest conversion, industrial tree plantations, and multifunctional landscapes, we assert that there is a compelling argument for FLR to give more prominence to agroecological practices, such as shifting cultivation, integrating forests into multifunctional landscapes while prioritizing the recentering of local interests within reforestation approaches (Pichler and Ingalls 2021).

8.4.3 Reconciling biodiversity and food security

A significant challenge that FLR faces is mitigating conflicts between food production and the conservation of native ecosystems (Edwards et al. 2021). One promising approach to address this challenge is agroforestry, which has the potential to support food production, enhance food security, and improve nutrition within the context of FLR. Agroforestry achieves this by integrating food-providing tree species and diversifying crop cultivation, ultimately contributing to both local livelihoods and environmental sustainability (Chazdon and Uriarte 2016; Vieira et al. 2009). Properly managed fallows within agroforestry systems can further generate additional income for food provision and enhance ES and biodiversity (Djenontin and Djoudi 2015; Koffi et al. 2017).

Ideally, restoration efforts should aim to improve and diversify agricultural production while respecting local sociocultural systems and avoiding adverse impacts (Brancalion et al. 2012; Latawiec et al. 2015). Unfortunately, food security considerations have often been overlooked in FLR interventions, underscoring the need for further research to provide guidance to decision-makers on integrating food security into restoration programs. Feasibility studies and impact assessments should consistently incorporate food security considerations throughout the implementation phase, including site selection, tree selection, choice of FLR methods, and collaborative monitoring.

Moreover, the selection of FLR measures and tree species should involve a thoughtfully designed participatory approach that encompasses a diverse range of locally relevant livelihoods and social groups, including youth, women, and men (Black 2019). Indigenous agroforestry systems, such as swidden systems and food-tree parklands in West Africa, already incorporate valuable food trees into their rotational practices (Djoudi et al. 2022; Ickowitz et al. 2022,). These systems offer an opportunity to merge traditional wisdom with modern science, enhancing the sustainability of FLR efforts in humid tropical forests (Colfer et al. 1997).

8.4.4 Addressing local bioenergy needs

Fuelwood serves as a primary energy source for cooking and heating in many regions of the Global South, meeting the energy needs of approximately 3–4 billion people who heavily rely on solid fuels, especially traditional options like firewood and charcoal (Jagger et al. 2019). This heavy reliance on fuelwood, often harvested from natural forests and woodlots, significantly contributes to forest degradation. However, biomass-based energy plays a pivotal role in local livelihoods and economies, with approximately 20% of Africa's population engaged part-time in commercial fuelwood production (Zulu and Richardson 2013). Impressively, fuelwood contributes to 35% of forest income, both subsistence and cash, in 25 Global South countries (Angelsen et al. 2014).

Despite the recognized importance of fuelwood, its potential contribution to FLR initiatives has been largely overlooked. Sustainable fuelwood production can offer various livelihood and health advantages to local communities. For example, it can reduce the time, physical effort, or expenses typically spent by rural women and children on gathering wood (Arnold et al. 2006). Prioritizing fuelwood can also facilitate income-generating activities that require energy access, such as baking or brickmaking. Strategies for integrating sustainable fuelwood production into FLR initiatives have been available for some time (Stanturf 2015) but often receive insufficient attention as a critical FLR outcome (Besseau et al. 2018).

Beyond promoting sustainable fuelwood production, FLR initiatives can also advocate for the adoption of clean fuels and efficient stoves. These technologies emit fewer pollutants and reduce smoke exposure, thereby enhancing the health of rural households (Grieshop et al. 2011; Iiyama et al. 2014; Mendum and Njenga 2018). It is important to acknowledge the health risks associated with biomass energy for women and children due to smoke inhalation from traditional stoves (Bede-Ojimadu and Orisakwe 2020; Smith 2008). Additionally, improving the sustainability of fuelwood production can bolster food security (Mendum and Njenga 2018) and provide essential employment and income opportunities for rural households, particularly those located near urban markets (Zulu and Richardson 2013). Aligning FLR initiatives with the bioenergy needs of local communities is crucial for their success. By endorsing sustainable fuelwood production, clean fuels, and efficient stoves, FLR initiatives can deliver livelihood and health benefits to local communities while simultaneously contributing to food security and rural development.

8.4.5 Generating income and local (youth) employment

In many regions of the Global South where FLR is being implemented, there is a stark shortage of employment opportunities for the increasing number of young people. These areas often grapple with high rates of unemployment and underemployment, leading to significant outmigration to urban centers (Aker 2019; Baah-Boateng 2016; Metelerkamp et al. 2019; Zulu et al. 2021). This challenging scenario presents an opportunity for FLR to enhance local motivations to address ecological issues by expanding the landscape of employment opportunities (Lindsjö et al. 2020; Sumberg and Okali 2013; Sumberg et al. 2019), particularly for youth, thereby reducing their out-migration from rural areas. Nevertheless, addressing the employment challenge requires a comprehensive and integrated approach with a strong emphasis on capacity building for stakeholders. This should encompass broader national and local programs aimed at enhancing the capabilities and knowledge of stakeholders in areas such as local and subnational economic development, as well as equipping them with practical skills to develop business models and secure financing for more sustainable landscape restoration efforts (van Oosten et al. 2021).

Identifying economic opportunities for restoration while supporting them through appropriate land-use planning (van Oosten et al. 2021) could significantly contribute to community well-being. Recent studies on FLR identify a few promising business opportunities but note a lack of investors. Box 8.4 provides four ways in which FLR interventions can absorb, harness, and positively transform the demographic "youth bulge" while simultaneously advancing environmental goals.

In recent decades, a compelling debate has arisen, exploring the interconnectedness of migration, remittances, and their impact on the local landscapes. Notably, remittances have played a substantial role in fostering local economic growth and alleviating poverty (De Haas 2007; Hecht et al. 2015). Additionally, out-migration can open up opportunities and aid in forest restoration by reshaping household economies and reducing reliance on forest resources. In a migration-prone watershed in Nepal, changing demographics and household economies led to natural afforestation due to reduced labor needs linked to migration, albeit potentially reducing biological diversity (Robson and Berkes 2011). In Guatemala, despite an initial forest decline caused by cash remittances invested in cattle and pasture, some level of forest recovery ensued. Cash remittances reduced dependence on land-based livelihoods

Box 8.4 Promising FLR approaches to support employment

1. *Youth-led or youth-oriented projects.* The Malawi Youth Forest Restoration Program, launched in 2018, aims to promote youth environmental stewardship by providing training and employment opportunities. The objective is to engage young people in restoring at least 50,000 ha of degraded agrarian and forested areas, offering a daily wage of US$1.25 for various restoration tasks (Reytar et al. 2018).
2. *Strongly integrate employment within design.* In India employment opportunities were made available for collecting seeds and wildings, planting, weeding, fire line cutting, and monitoring (Stanturf and Mansourian 2020). Under the Land Accelerator program, the World Resource Institute (WRI) matches aspiring FLR small and medium-sized enterprises (SMEs) with would-be investors. Also as implemented in Mozambique, a carbon forestry program aims to extend beyond short-term opportunities (Jindal et al. 2012).
3. *Enhance current working conditions for farmers.* In Tanzania an indigenous in-situ natural resource management system known as Ngitili has significantly improved woodlands and pasture regeneration, leading to enhanced human well-being. This FLR initiative has resulted in households saving up to 6 hours per day in the collection of fuelwood, thatch, or fodder and added a monthly value of up to USD 14 per person. Additionally, local households can collect medicinal herbs from the restored lands (Malunguja et al. 2020; Safari et al. 2019; Wilson et al. 2012).
4. *Livelihood diversification.* The Sloping Land Conversion Program in China had a positive impact on reducing income inequality among households participating in the employment schemes within the project; however, it did not yield the anticipated shift of employment and labor away from farming activities (Li et al. 2011).

for certain households but enabled intensive cattle raising and ongoing land clearing for others (Taylor et al. 2016).

Additionally, for FLR schemes centered on plantation and timber activities to realize the employment opportunities they offer, critical enabling factors and conditions identified by Clarke (2018) must be considered. These encompass effective land reform, improved governance in communal lands, financial support for forestry development, stronger public-private collaborations, political commitment, enhanced transport infrastructure, advocacy for legality and certification, and the expansion of biomass energy markets and technology networks.

8.4.6 Enhancing local resilience

Enhancing natural capital, protecting ecosystems, and ensuring the availability of ES for adaptation constitute an important discussion among ES and climate change scholars (Koffi et al. 2017; Munera-Roldan et al. 2022; Pramova et al. 2012). Recent studies suggest that conserving or restoring ecosystems may not enhance adaptive capacity for everyone when multiple potential beneficiaries with different capacities and livelihood needs are competing for services and benefits from natural resources. This may result in winners and losers. Furthermore, as trees are long-lived, the benefits may be claimed by the current land user as well as future generations, making any inequities that emerge or are reinforced during restoration very long-lasting (Goffner et al. 2019). Therefore, FLR projects must carefully consider the distribution of benefits and involve a diversity of stakeholders in decision-making processes (see e.g., Crossland et al. 2022; Sarmiento Barletti 2022).

From the perspective of local livelihood adaptation, it is crucial that FLR initiatives recognize and incorporate the twofold nature of local climate change impacts on both people and ecosystems. Research on how valuable ES delivery will be impacted by climate change shows that impact is predominantly negative. Relatively few studies identified ecosystem solutions that were robust to future uncertainty (Runting et al. 2017). Furthermore, studies of loss and damage from climate change tend to overlook the mediating role of ES to people, livelihoods, and society (van der Geest et al. 2019).

For instance, in food insecure regions, the safety net role of ecosystems during food insecurity periods after climatic and non-climatic shocks is vital. In those harsh times of crisis, vulnerable people benefit from both provisioning and regulating services provided by forests and trees, as the restoration farmers are likely to manage usually leads to the emergence of a variety of tree species that can provide alternative subsistence, foods, and income opportunities (Koffi et al. 2017). Examples include shrubs that produce feed during the dry season, fruit trees that produce harvests at different times of the year, and wood products and nuts that are sold for income during hardship or any other time of the year. Also, medicines, fibers, and wild game (bushmeat) are important (Colfer et al. 2006; Ickowitz et al. 2022, Koffi et al 2017; Place et al. 2016; Reij and Garrity 2016). Droughts are shown to have fewer negative impacts on the livelihoods of communities in regions where communities restored their forest resources through farmer managed natural regeneration (FMNR) approaches (Chomba et al. 2020; Sendzimir et al. 2011). One of these prominent examples is shown in Box 8.5.

8.4.7 Leveraging adaptive collaborative learning

Adaptive Collaborative Management (ACM) has demonstrated the ability to foster equity, empower local communities, and encourage collective action at the grassroots level. ACM emphasizes the principles of equity, learning, and collaboration and their great promise for guiding restoration efforts, especially in the evolving landscape of climate change. Given the pervasive impact of global changes, adaptive learning, as implemented with ACM, becomes imperative to avoid repeating past maladaptive FLR patterns and to ensure that future FLR interventions align with and are responsive to local contexts and effectively achieve their intended objectives. A significant limitation lies in their limited capacity to scale up and engage a wider array of actors and institutions (Colfer et al. 2022; Colfer and Prabu 2023a).

Box 8.5 Farmer Managed Natural Regeneration (FMNR) in the Sahel

Sahel farmers were heavily affected by droughts in the early 1970s and 1980s. By creatively adapting traditional agroforestry practices as well as water and soil management techniques to restore soil fertility and health, farmers initiated a positive feedback loop that catalyzed a broad ecosystem recovery cycle over three decades. Practices such as *zaï* planting pits in Burkina Faso and FMNR of valuable trees in Niger have spread from farmer to farmer through knowledge-sharing venues, schools, and networks supported by NGOs and research organizations.

Formerly a desert fringe characterized by erosion and heavy degradation, landscapes in Niger, Burkina Faso, and Mali have undergone a remarkable transformation into productive and biodiverse land, reversing degradation and desertification across 6 Mha. More than 200 million new trees have grown, with a production value exceeding USD 260 million. This transformation provides food security, increases resilience against droughts, and potentially increases income 18–24% for approximately 3 million people.

Research indicates that regions where communities have restored their forest resources through FMNR approaches are less negatively affected by droughts (Chomba et al. 2020; Sendzimir et al. 2011).

At the policy level, changes in forestry laws and government decentralization play a crucial role in enabling this transformation, encouraging greater local ownership and control of natural resources. Farmers, who were once passive victims of droughts, resource loss, and conflict, have transformed into agents of change and true stewards of the landscape.

Local stakeholder platforms are indispensable for preventing the repetition of maladaptive FLR actions. They serve several crucial purposes in addressing the challenges associated with top-down approaches (Section 8.3). Firstly, they foster trust and collaboration among diverse stakeholders, potentially including governments, businesses, communities, and civil society organizations. Secondly, these platforms play a pivotal role in identifying and prioritizing FLR needs relevant to local communities facilitating the development and implementation of FLR plans tailored to local conditions and priorities. Thirdly, they contribute significantly to enhancing monitoring and evaluating FLR actions to ensure desired outcomes at the local level and sustained FLR efforts over the long term. Nevertheless, merely establishing an FLR platform is insufficient. For effective planning, moderating, and monitoring of FLR activities, multi-stakeholders need to be supported by capacity building and financial and human resources, including skilled facilitators (Hagmann et al. 2023; Stanturf 2021).

8.5 Conclusion

This chapter highlights the substantial potential of FLR to foster local development. Nevertheless, socioeconomic and cultural objectives are often overlooked in FLR initiatives, leading to adverse local impacts. To effectively promote sustainable local livelihoods and well-being, FLR must deepen its engagement with social complexities and address the underlying conditions that perpetuate marginalization in development practices. Research underscores the importance of operational flexibility, holistic approaches, long-term sustainability assessments, and cross-sectoral collaboration as prerequisites for FLR success. This requires explicit consideration of trade-offs and co-benefits from the design phase, prioritizing actions that enhance local livelihoods. A deep understanding of the complexities and context-specific linkages between productivity, ecosystem services, cultural preservation, livelihood improvement, and adaptive capacities is key. Building a participatory grassroots movement for FLR can help navigate local complexities, integrate diverse local actors, and recognize existing restoration efforts and potential. The involvement of local stakeholders, especially those disproportionately affected by degradation, biodiversity loss, and climate change, is essential in FLR planning and management. This engagement should ensure tangible economic returns through localized restoration projects, innovative financial methods, and supportive legislation. Moreover, lasting incentives for equitable cost and benefit distribution in FLR efforts are key to supporting long-term success.

Tailoring FLR efforts to unique local contexts, encompassing sociocultural, ecological, and economic factors, is essential. Failing to do so can lead to unequal distribution of benefits and costs, potentially exacerbating existing environmental and social issues. Local community involvement in FLR project design and execution is crucial to ensure that local needs, aspirations, and knowledge are considered. Furthermore, interdisciplinary collaboration and the integration of traditional and scientific wisdom are vital to address the complexities of socio-ecological systems and develop effective FLR strategies. Recognizing and incorporating the knowledge and practices of IPLCs throughout FLR implementation and monitoring is paramount. Integrating rights-based approaches into FLR becomes a fundamental and nonnegotiable starting point, as this promotes solutions that also address the needs of women, youth, and indigenous groups, resulting in documented positive outcomes for both human well-being and the environment. Therefore, FLR efforts should prioritize rights-based approaches clarifying rights and responsibilities to land and resources in all FLR phases. These approaches should empower IPLCs by granting them the rights to effectively manage and oversee FLR design, implementation, and monitoring, including transparency and control of funds, while establishing robust governance mechanisms.

While trade-offs are inherent in the complex system of FLR use and management, the necessary skills to assess and evaluate them are often lacking among FLR organizations and stakeholders. This chapter proposes several components to mitigate trade-offs and create conditions for multiple wins. Existing success stories from community led FLR initiatives offer valuable insights into how to mitigate trade-off risks. Therefore, prioritizing the development of tools and negotiation skills to facilitate trade-off discussions and bolstering community capacity for FLR engagement at all stages is essential. Establishing structures that prevent elite capture and promote connectivity across scales, sectors, and social actors can address accountability and transparency issues, fostering trust among local communities in FLR.

Overall, promoting a place-based and people-centered restoration paradigm and practice is essential for the success of FLR. This requires significant efforts in building human, natural, physical, and social livelihood assets while flexibly protecting cultural integrity, to move beyond business-as-usual and poverty-trap local trajectories.

References

Adams, C., S.T. Rodrigues, M. Calmon, and C. Kumar. 2016. "Impacts of Large-Scale Forest Restoration on Socioeconomic Status and Local Livelihoods: What We Know and Do Not Know." *Biotropica* 48 (6): pp. 731–744.

Aker, J.C. 2019. "Information and Communication Technologies and Rural Youth." *IFAD Research Series* 52. DOI:10.2139/ssrn.3523207.

Alexander, S., J. Aronson, O. Whaley, and D. Lamb. 2016. "The Relationship between Ecological Restoration and the Ecosystem Services Concept." *Ecology and Society* 21 (1): pp. 34. DOI:10.5751/ES-08288-210134.

Anderson, M.K. 2006. *Tending The Wild: Native American Knowledge and the Management of California's Natural Resources*. University of California Press.

Angelsen, A., P. Jagger, R. Babigumira, et al. 2014. "Environmental Income and Rural Livelihoods: A Global Comparative Analysis." *World Development* 64 (Supplement 1): pp. S12–28.

Arnold, J.E.M., G. Köhlin, and R. Persson 2006. "Woodfuels, Livelihoods, and Policy Interventions: Changing Perspectives." *World Development* 34 (3): pp. 596–611.

Aronson, J., J.N. Blignaut, S.J. Milton, et al. 2010. "Are Socioeconomic Benefits of Restoration Adequately Quantified? A Meta-analysis of Recent Papers (2000–2008) in Restoration Ecology and 12 Other Scientific Journals." *Restoration Ecology* 18: pp. 143–154.

Asteria, D., P. Alvernia, B.N. Kholila, S.I. Husein, and F.W. Asrofani. 2022. "Forest Conservation by the Indigenous Baduy Community in the Form of Customary Law." *Journal of Cultural Heritage Management and Sustainable Development* 12 (1): pp. 175–189.

Baah-Boateng, W. 2016. "The Youth Unemployment Challenge in Africa: What Are the Drivers?" *Economic and Labour Relations Review* 27 (4): pp. 413–431. DOI:10.1177/1035304616645030.

Bakkegaard, R.K., and T. Vongvisouk. 2019. "An Assessment of Attitudes towards Forest Landscape Restoration in Laos." *Forests* 10 (11): p. 960.

Barbier, E.B. 2007. "Natural Capital and Labor Allocation: Mangrove-Dependent Households in Thailand." *The Journal of Environment & Development* 16 (4): pp. 398–431.

Bede-Ojimadu, O., and O.E. Orisakwe. 2020. "Exposure to Wood Smoke and Associated Health Effects in Sub-Saharan Africa: A Systematic Review." *Annals of global health* 86 (1): 32.

Besseau, P., S. Graham, and T. Christophersen (eds.). 2018. *Restoring Forests and Landscapes: the Key to a Sustainable Future*. Vienna: Global Partnership on Forest and Landscape Restoration. https://www.forestlandscaperestoration.org/images/gpflr_final%2027aug.pdf.

Black, V.C. 2019. "Enhancing Food Security through Landscape Restoration: Case Studies from Cameroon and Peru." Master's thesis, Roma Tre University, 2019. https://www.foreststreesagroforestry.org/wp-content/uploads/2020/03/Black-Violet-Chanza-Master-thesis.pdf.

Bond, W. 2016. "Seeing the Grasslands through the Trees—Response." *Science* 351 (6277): pp. 1036–1037.

Brancalion, P.H., R.A. Viani, B.B. Strassburg, and R.R. Rodrigues. 2012. "Finding the Money for Tropical Forest Restoration." *Unasylva* 63 (239): pp. 41–50.

Broadmeadow, S, and T.R. Nisbet. 2004. "The Effects of Riparian Forest Management on the Freshwater Environment: A Literature Review of Best Management Practice." *Hydrology and Earth System Sciences* 8 (3): pp. 286–305.

Brockhaus, M., M. Di Gregorio, H. Djoudi, M. Moeliono, T.T. Pham, and G.Y. Wong. 2021. "The Forest Frontier in the Global South: Climate Change Policies and the Promise of Development and Equity." *Ambio* 50: pp. 2238–2255.

Cairns, M., and D.P. Garrity. 1999. "Improving Shifting Cultivation in Southeast Asia by Building on Indigenous Fallow Management Strategies." *Agroforestry Systems* 47 (1–3): pp. 37–48.

Cairns, M.F. (ed.). 2015. *Shifting Cultivation and Environmental Change: Indigenous People, Agriculture and Forest Conservation.* New York: Earthscan.

Carmenta, R., and B. Vira. 2018. "Integration for Restoration: Reflecting on Lessons Learned from the Silos of the Past." In *Forest Landscape Restoration: Integrated Approaches to Support Effective Implementation*, edited by S. Mansourian, and J. Parrotta, pp. 16–36). Abingdon-on-Thames: Routledge. DOI:10.4324/9781315111872.

Cedamon, E., I. Nuberg, G. Paudel, M. Basyal, K. Shrestha, and N. Paudel. 2017. "Rapid Silviculture Appraisal to Characterise Stand and Determine Silviculture Priorities of Community Forests in Nepal." *Small-scale Forestry* 16: pp. 195–218.

Chapman, M., T. Satterfield, H. Wittman, and K.M. Chan. 2020. "A payment by Any Other Name: Is Costa Rica's PES a Payment for Services or a Support for Stewards?" *World Development* 129: p. 104900.

Chazdon, R., and P. Brancalion. 2019. "Restoring Forests as a Means to Many Ends." *Science* 365 (6448): pp. 2425.

Chazdon, R.L., and F.G. Coe. 1999. "Ethnobotany of Woody Species in Second-Growth, Old-Growth, and Selectively Logged Forests of Northeastern Costa Rica." *Conservation biology* 13 (6): pp. 1312–1322.

Chazdon, R.L., and M. Uriarte. 2016. "Natural Regeneration in the Context of Large-Scale Forest and Landscape Restoration in the Tropics." *Biotropica* 48 (6): pp. 709–715.

Chazdon, R.L., S.J. Wilson, E. Brondizio, M.R. Guariguata, and J. Herbohn. 2021. "Key Challenges for Governing Forest and Landscape Restoration across Different Contexts." *Land Use Policy* 104: p. 104854.

Chomba, S., F. Sinclair, P. Savadogo, M. Bourne, and M. Lohbeck. 2020. "Opportunities and Constraints for Using Farmer Managed Natural Regeneration for Land Restoration in Sub-Saharan Africa." *Frontiers in Forests and Global Change* 3: p. 571679.

Clarke, J.M. 2018. Job Creation in Agriculture, Forestry and Fisheries in South Africa: An Analysis of Employment Trends, Opportunities and Constraints in Forestry and Wood Products Industries. Working Paper 52. Bellville: Institute for Poverty, Land and Agrarian Studies (PLAAS), University of the Western Cape.

Colfer, C.J.P. 2005. *The Complex Forest: Communities, Uncertainty, and Adaptive Collaborative Management.* Washington, D.C.: Resources for the Future and Center for International Forestry Research (CIFOR).

Colfer, C.J.P. 2008. *The Longhouse of the Tarsier: Changing Landscapes, Gender and Well Being in Borneo.* Phillips: Borneo Research Council.

Colfer, C.J.P., R.G. Dudley, and R. Gardner. 2008. "Forest Women, Health and Childbearing" in Human Health and Forests: A Global Overview of Issues, Practice and Policy" In *Human Health and Forests. A Global Overview of Issues, Practice and Policy,* edited by C.J.P. Colfer, pp. 113–133. London: Earthscan.

Colfer, C.J.P., N. Peluso, and C.S. Chung. 1997. *Beyond Slash and Burn: Building on Indigenous Management of Borneo's Tropical Rain Forests.* New York: New York Botanical Gardens.

Colfer, C.J.P., and R. Prabhu. 2023a. *Responding to Environmental Issues through Adaptive Collaborative Management: From Forest Communities to Global Actors.* London: Earthscan.

Colfer, C.J.P., and R. Prabhu. 2023b. *Forest Landscape Restoration: A Comparison of Two Approaches.* Occasional Paper 239. Bogor: CIFOR.

Colfer, C.J.P., R. Prabhu, and A. Larson. (eds.). 2022. *Adaptive Collaborative Management in Forest Landscapes: Villagers, Bureaucrats and Civil Society.* London: Earthscan.

Colfer, C.J.P., D. Sheil, M., and Kishi. 2006. Forests and Human Health: Assessing the Evidence. Occasional Paper 45. Bogor: CIFOR.

Constant, N.L. and P.J. Taylor. 2020. "Restoring the Forest Revives Our Culture: Ecosystem Services and Values for Ecological Restoration across the Rural-Urban Nexus in South Africa." *Forest Policy and Economics* 118: p. 102222.

Crossland, M., A.M. Paez Valencia, G. Adeyiga, et al. 2022. *Gender Transformative Approaches for Land Restoration: Lessons Learnt from a Multi-stakeholder Co-design Process in Makueni County, Kenya.* Nairobi: World Agroforestry.

Crossland, M., L.A. Winowiecki, T. Pagella, K. Hadgu, and F. Sinclair. 2018. "Implications of Variation in Local Perception of Degradation and Restoration Processes for Implementing Land Degradation Neutrality." *Environmental development* 28: pp. 42–54.

De Haas, H. 2007. "Remittances, Migration and Social Development. A Conceptual Review of the Literature." United Nations Research Institute for Social Development. *Social Policy and Development Programme Paper* 34: pp. 1–46.

De Jong, W., M. van der Zon, A.F. Urushima, J.C. Youn, J. Liu, and N. Li. 2018."Tenure, Property Rights and Forest Landscape Restoration." In *Forest Landscape Restoration: Integrated Approaches to Support Effective Implementation.* edited by S. Mansourian, and J. Parrotta, pp. 158–175. Abingdon-on-Thames: Earthscan. DOI:10.4324/9781315111872.

De Leeuw, S., and S. Hunt. 2018. "Unsettling Decolonizing Geographies." *Geography Compass* 12 (7): e12376.

DeRose, R.J., and J.N. Long. 2014. "Resistance and Resilience: A Conceptual Framework for Silviculture." *Forest Science* 60 (6): pp. 1205–1212.

Djenontin, I., and H. Djoudi. 2015. "From Degraded to Functional Restored Forest Land: Smallholder Farmers Curbing Food Insecurity in Central Burkina Faso." In *Enhancing food security through forest landscape restoration: Lessons from Burkina Faso, Brazil, Guatemala, Viet Nam, Ghana, Ethiopia and Philippines* edited by C. Kumar, C. Saint-Laurent, S. Begeladze, and M. Calmon, pp. 18–41. Gland: IUCN. DOI:10.2305/IUCN.CH.2015.FR.2.en.

Djenontin, I.N., and L.C. Zulu. 2021. "The Quest for Context-Relevant Governance of Agro-Forest Landscape Restoration in Central Malawi: Insights from Local Processes." *Forest Policy and Economics* 131: p. 102555.

Djenontin, I.N.S., S. Foli, and L.C. Zulu. 2018. "Revisiting the Factors Shaping Outcomes for Forest and Landscape Restoration in Sub-Saharan Africa: A Way Forward for Policy, Practice and Research." *Sustainability* 10 (4): p. 906.

Djenontin, I.N.S., L.C. Zulu, and D. Etongo. 2021. "Ultimately, What is Forest Landscape Restoration in Practice? Embodiments in Sub-Saharan Africa and Implications for Future Design." *Environmental Management* 68: pp. 619–641. DOI:10.1007/s00267-020-01360-y.

Djenontin, I.N.S., L.C. Zulu, and A. Ligmann-Zielinska. 2020." Improving Representation of Decision Rules in LUCC-ABM: An Example With an Elicitation of Farmers' Decision Making for Landscape Restoration in Central Malawi." *Sustainability* 12 (13): p. 5380.

Djenontin, I.N.S., L.C. Zulu, and R.B. Richardson. 2022. "Smallholder Farmers and Forest Landscape Restoration in Sub-Saharan Africa: Evidence from Central Malawi." *Land Use Policy* 122: 106345. DOI:10.1016/j.landusepol.2022.106345.

Djoudi, H., B. Locatelli, C. Pehou, et al. 2022. "Trees as Brokers in Social Networks: Cascades of Rights and Benefits from a Cultural Keystone Species." *Ambio* 51 (10): pp. 2137–2154.

Djoudi, H., B. Locatelli, C. Vaast, et al. 2016. "Beyond dichotomies: Gender and intersecting inequalities in climate change studies." *Ambio* 45: pp. 248–262.

Duguma, L., P.A. Minang, B.E. Aynekulu, et al. 2020. From Tree Planting to Tree Growing: Rethinking Ecosystem Restoration Through Trees. ICRAF Working Paper. No. 304. https://apps.worldagroforestry.org/downloads/Publications/PDFS/WP20001.pdf.

Edwards, D.P., G.R. Cerullo, S. Chomba, et al. 2021. "Upscaling Tropical Restoration to Deliver Environmental Benefits and Socially Equitable Outcomes." *Current Biology* 31 (19): pp. R1326–R1341.

Elias, M., D. Joshi, and R. Meinzen-Dick. 2021b. "Restoration for Whom, by Whom? A Feminist Political Ecology of Restoration." *Ecological Restoration* 39 (1–2): pp. 3–15.

Elias, M., M. Kandel, S. Mansourian, et al. 2021a. "Ten People-Centered Rules for Socially Sustainable Ecosystem Restoration." *Restoration Ecology* 30 (4): e13574.

Erbaugh, J.T., and Oldekop, J. 2018. "Forest Landscape Restoration for Livelihoods and Well-Being." *Current Opinion in Environmental Sustainability* 32: pp. 76–83.

Erbaugh, J.T., N. Pradhan, J. Adams, et al. 2020. "Global Forest Restoration and the Importance of Prioritizing Local Communities." *Nature Ecology & Evolution* 4: pp. 1472–1476. DOI:10.1038/s41559-020-01282-2.

Fedele, G., C.I. Donatti, I. Bornacelly, and D.G. Hole. 2021. "Nature-Dependent People: Mapping Human Direct Use of Nature for Basic Needs across the Tropics." *Global Environmental Change* 71: 102368. DOI:10.1016/j.gloenvcha.2021.102368.

Feintrenie, L., F. Claeys, and P. Levang. 2019. "Smallholders' Resistance to Tropical Forest Restoration: The Role of Livelihood Strategies and Social Networks in Central Kalimantan, Indonesia." *Land Use Policy* 86: pp. 120–131.

Fleischman, F., S. Basant, A. Chhatre, E.A. Coleman. 2020. "Pitfalls of Tree Planting Show Why We Need People-Centered Natural Climate Solutions. *BioScience* 70 (11): pp. 947–950.

Fleischman, F., E. Coleman, H. Fischer, et al. 2022. "Restoration Prioritization Must Be Informed by Marginalized People." *Nature* 607 (7918): pp. E5–E6.

Freeman O, L. Duguma, and P. Minang. 2015. "Operationalizing the Integrated Landscape Approach in Practice." *Ecology and Society* 20 (1): 24.

Garnett, S.T., N.D. Burgess, J.E. Fa, et al. 2018. "A Spatial Overview of the Global Importance of Indigenous Lands for Conservation." *Nature Sustain*ability 1 (7): pp. 369–374.

Gavin, M.C. 2004. "Changes in Forest Use Value through Ecological Succession and Their Implications for Land Management in the Peruvian Amazon." *Conservation Biology* 18 (6): pp. 1562–1570.

Giacomini, G. 2022. "Participatory Rights, Conservation and Indigenous Customary Law." In *Indigenous Peoples and Climate Justice: A Critical Analysis of International Human Rights Law and Governance*, edited by G. Giacomini, pp. 227–314. Cham: Springer International Publishing.

Goffner, D., H. Sinare, and L.J. Gordon. 2019. "The Great Green Wall for the Sahara and the Sahel Initiative as an Opportunity to Enhance Resilience in Sahelian Landscapes and Livelihoods." *Regional Environmental Change* 19: pp.1417–1428.

Grieshop, A.P., J.D. Marshall, and M. Kandlikar. 2011. "Health and Climate Benefits of Cookstove Replacement Options." *Energy Policy* 39: pp. 7530–7542.

Guariguata, M., S. Atmadja, H. Baral, et al. 2021. *Forest and Landscape Restoration.* Bogor: CGIAR Research Program on Forests, Trees and Agroforestry. DOI:10.17528/cifor/008214.

Hagmann, J., E. Chuma, J. Ramaru, et al. 2023. "Herding Cats: Facilitation in Social Learning Processes." In *Responding to Environmental Issues through Adaptive Collaborative Management: From Forest Communities to Global Actors*, edited by C.P.J. Colfer and R. Prabhu, pp. 135–61. London: Earthscan.

Hecht, S., A.L. Yang, B.S. Basnett, C. Padoch, and N.L. Peluso. 2015. People in motion, forests in transition: trends in migration, urbanization, and remittances and their effects on tropical forests. Occasional Paper 142. Bogor: CIFOR.

Hernández-Barrios, J.C., N.P. Anten, and M. Martínez-Ramos. 2015. "Sustainable Harvesting of Non-Timber Forest Products Based on Ecological and Economic Criteria." *Journal of Applied Ecology* 52 (2): pp. 389–401.

Hird, C., D.M. David-Chavez, S.S. Gion, and V. van Uitregt. 2023. "Moving beyond Ontological (Worldview) Supremacy: Indigenous Insights and a Recovery Guide for Settler-Colonial Scientists." *Journal of Experimental Biology* 226 (12): jeb245302. DOI:10.1242/jeb.245302.

Hobbs, R.J. 2016 "Degraded or Just Different? Perceptions and Value Judgements in Restoration Decisions." *Restoration Ecology* 24 (2): pp. 153–158.

Höhl, M., V. Ahimbisibwe, J.A. Stanturf, P. Elsasser, M. Kleine, and A. Bolte. 2020. "Forest Landscape Restoration—What Generates Failure and Success?" *Forests* 11 (9): p. 938.

Ickowitz, A., S. McMullin, T. Rosenstock, et al. 2022. "Transforming Food Systems with Trees and Forests." *The Lancet Planetary Health* 6 (7): pp. e632–e639.

Iiyama, M., H. Neufeldt, P. Dobie, M. Njenga, G. Ndegwa, and R. Jamnadass. 2014. "The Potential of Agroforestry in the Provision of Sustainable Woodfuel in Sub-Saharan Africa." *Current Opinion in Environmental Sustainability* 6: pp. 138–147.

IPCC. 2019. Climate Change and Land: An IPCC Special Report on Climate Change, Desertification, Land Degradation, Sustainable Land Management, Food Security, and Greenhouse Gas Fluxes in Terrestrial Ecosystems. https://www.ipcc.ch/srccl/download.

ITTO. 2020. "Guidelines for Forest Landscape Restoration in the Tropics." ITTO Policy Development Series No. 24. Yokohama: International Tropical Timber Organization.

Jagger, P., R. Bailis, A. Dermawan, N. Kittner, and R. McCord. 2019. "SDG 7: Affordable and Clean Energy–How Access to Affordable and Clean Energy Affects Forests and Forest-Based Livelihoods." In *Sustainable Development Goals: Their Impacts on Forests and People*, edited by P. Katila, C.J.P. Colfer, W. de Jong, G. Galloway, P. Pacheco, and G. Winkel, pp: 206–236. Cambridge: Cambridge University Press.

Jindal, R., J.M. Kerr, and S. Carter. 2012. "Reducing Poverty through Carbon Forestry? Impacts of the N'hambita Community Carbon Project in Mozambique." *World development* 40 (10): pp. 2123–2135.

Jindal, R., J.M. Kerr, P.J. Ferraro, and B.M. Swallow. 2013. "Social Dimensions of Procurement Auctions for Environmental Service Contracts: Evaluating Trade-Offs between Cost-Effectiveness and Participation by the Poor in Rural Tanzania." *Land Use Policy* 31: pp. 71–80.

Jose, S. 2009. "Agroforestry for Ecosystem Services and Environmental Benefits: An Overview." *Agroforestry Systems* 76: p. 110. DOI:10.1007/ s10457-009-9229-7.

Junqueira, A.B., G.H. Shepard, and C.R Clement. 2011. "Secondary Forests on Anthropogenic Soils of the Middle Madeira River: Valuation, Local Knowledge, and Landscape Domestication in Brazilian Amazonia." *Economic Botany* 65: pp. 85–99.

Kainyande, A., E.F. Auch, and A.D. Okoni-Williams. 2022. "The Socio-Economic Contributions of Large-Scale Plantation Forests: Perceptions of Adjacent Rural Communities in the Northern Province of Sierra Leone." *Trees, Forests and People* 10: p.100329.

Kallio, M.H., M. Kanninen, and H. Krisnawati. 2012. "Smallholder Teak Plantations in Two Villages in Central Java: Silvicultural Activity and Stand Performance." *Forests, Trees and Livelihoods* 21 (3): pp. 158–175.

Koffi, C.K., H. Djoudi, and D. Gautier. 2017. "Landscape Diversity and Associated Coping Strategies during Food Shortage Periods: Evidence from the Sudano-Sahelian Region of Burkina Faso." *Regional Environmental Change* 17: pp. 1369–1380.

Kouassi, J.L., A. Gyau, L. Diby, Y. Bene, and C. Kouamé. 2021. "Assessing Land Use and Land Cover Change and Farmers' Perceptions of Deforestation and Land Degradation in South-West Côte d'Ivoire, West Africa." *Land* 10 (4): p. 429.

Kull, C.A. 2017. "Forest Transitions: A New Conceptual Scheme." *Geographica Helvetica* 72 (4): pp. 465–474.

Kull, C.A., S.L. Harimanana, A.R. Andrianoro, and L.G. Rajoelison. 2019. "Divergent Perceptions of the 'Neo-Australian' forests of Lowland Eastern Madagascar: Invasions, Transitions, and Livelihoods." *Journal of Environmental Management* 229: pp. 48–56.

Kumar, C., S. Begeladze, M. Calmon, and C. Saint-Laurent. 2015. *Enhancing Food Security through Forest Landscape Restoration: Lessons from Burkina Faso, Brazil, Guatemala, Viet Nam, Ghana, Ethiopia and Philippines.* Gland: IUCN. DOI:10.2305/IUCN.CH.2015.FR.2.en.

Kumeh, E.M., B. Kyereh, J. Asante, et al. 2023. “Retooling Incentive Mechanisms for Effective Smallholder Tree Growers’ Contributions to Global Landscape Restoration.” *Development in Practice* 33 (1): pp. 94–109.

Kusumanto, T., G.A.K. Surtiari, C. Zevenbergen, et al. 2023. “ACM as a Pathway to Mitigate Jakarta’s Flood Impacts in a Changing Climate”. In *Responding to Environmental Issues through Adaptive Collaborative Management: From Forest Communities to Global Actors*, edited by C.J.P. Colfer, and R. Prabhu, pp. 81–112. London: Earthscan.

Kuyah, S., G.W. Sileshi, E. Luedeling, et al. 2020. “Potential of Agroforestry to Enhance Livelihood Security in Africa.” *Agroforestry for Degraded Landscapes: Recent Advances and Emerging Challenges* 1: pp. 135–167.

Lamb, D., P.D. Erskine, and J.A. Parrotta. 2005. “Restoration of Degraded Tropical Forest Landscapes.” *Science* 310 (5754): pp. 1628–1632.

Larson, A., D. Barry, G.R. Dahal, and C.J.P. Colfer. (eds.). 2010. *Forest for People: Community Rights and Forest Tenure Reform.* London: Earthscan.

Latawiec, A.E., B.B. Strassburg, P.H. Brancalion, R.R. Rodrigues, and T. Gardner. 2015. “Creating Space for Large-Scale Restoration in Tropical Agricultural Landscapes.” *Frontiers in Ecology and the Environment* 13 (4): pp. 211–218.

Le, H.D., C. Smith, J. Herbohn, and S. Harrison. 2012. “More than Just Trees: Assessing Reforestation Success in Tropical Developing Countries.” *Journal of Rural Studies* 28 (1): pp. 5–19.

Li, J., M.W. Feldman, S. Li, and G.C. Daily. 2011. “Rural Household Income and Inequality under the Sloping Land Conversion Program in Western China.” *Proceedings of the National Academy of Sciences* 108 (19): pp. 7721–7726.

Li, T.M., and P. Semedi. 2021. *Plantation Life: Corporate Occupation in Indonesia’s Oil Palm Zone.* Durham: Duke University Press.

Lindsjö, K., W. Mulwafu, A. Andersson Djurfeldt, and M.K. Joshua. 2020. “Generational Dynamics of Agricultural Intensification in Malawi: Challenges for the Youth and Elderly Smallholder Farmers.” *International Journal of Agricultural Sustainability* 19 (5–6): pp. 423–436. DOI:10.1080/14735903.2020.1721237.

Liswanti, N., Tamara, A. and Djoudi, H., 2020. *Climate Finance and Gender in the Ground: Insights from Mitigation and Adaptation Interventions in Indonesia.* Occasional Paper 306. Bogor: CIFOR.

Löfqvist, S., F. Kleinschroth, A. Bey, et al. 2022. “How Social Considerations Improve the Equity and Effectiveness of Ecosystem Restoration.” *BioScience* 73 (2): pp. 134–148.

Malkamäki, A., D. D’Amato, N.J. Hogarth, et al. 2018. ”A Systematic Review of the Socio-Economic Impacts of Large-Scale Tree Plantations, Worldwide.” *Global Environmental Change* 53: pp. 90–103.

Malunguja, G.K., C.K. Rubanza, and A. Devi. 2020. “An Assessment of the Current Status and Regeneration Potential of the Traditional Conserved Forests (Ngitili) in Kishapu District, Tanzania.” *The Journal of the Society for Tropical Plant Research* 7 (2): pp. 336–356.

Mansourian, S. 2005.” Overview of Forest Restoration Strategies and Terms.” In *Forest Restoration in Landscapes: Beyond Planting Trees*, edited by S. Mansourian, and D. Vallauri, pp. 8–13. New York: Springer.

Mansourian, S. 2017. “Governance and Forest Landscape Restoration: A framework to Support Decision-Making.” *Journal for Nature Conservation* 37: pp. 21–30.

Mansourian, S., and J. Parrotta. 2018. "The Need for Integrated Approaches to Forest Landscape Restoration." In *Forest Landscape Restoration*, edited by S. Mansourian, and J. Parrotta, pp. 3–15. London: Routledge.

McLain, R., M.R. Guariguata, and S. Lawry. 2017. Implementing Forest Landscape Restoration Initiatives. Tenure, Governance, and Equity Considerations. A Paper Prepared for a Workshop on "Accelerating Restoration of Degraded Forest Landscapes: The Role of Tenure Security and Local Forest Governance in Catalyzing Global Restoration Initiatives" held in Bonn, Germany on 3 November 2017. https://www.cifor.org/wp-content/uploads/2017/11/Implementing%20FLR.pdf.

McLain, R., S. Lawry, M.R. Guariguata, and J. Reed. 2021. "Toward a Tenure-Responsive Approach to Forest Landscape Restoration: A Proposed Tenure Diagnostic for Assessing Restoration Opportunities." *Land Use Policy* 104: 103748.

Medrilzam, M., C. Smith, A.A. Aziz, J. Herbohn, and P. Dargusch. 2017. "Smallholder Farmers and the Dynamics of Degradation of Peatland Ecosystems in Central Kalimantan, Indonesia." *Ecological Economics* 136: pp. 101–113.

Mendum, R.M., and M. Njenga. 2018. "Integrating Woodfuels into Agriculture and Food Security Agendas and Research in sub-Saharan Africa (SSA)." *Facets* 3 (1): 11.

Menéndez, P., I.J. Losada, S. Torres-Ortega, A. Toimil, and M.W. Beck. 2019. "Assessing the Effects of Using High-Quality Data and High-Resolution Models in Valuing Flood Protection Services of Mangroves." *PloS One* 14 (8): p. e0220941.

Metelerkamp, L., S. Drimie, and R. Biggs. 2019. "We're Ready, the System's Not–Youth Perspectives on Agricultural Careers in South Africa." *Agrekon* 58 (2): pp. 154–179. DOI:10.1080/03031853.2018.1564680.

Monela, G.C., S.A.O. Chamshama, R. Mwaipopo, and D.M. Gamassa. 2004. A Study on the Social, Economic and Environmental Impacts of Forest Landscape Restoration in Shinyanga Region, Tanzania. Ministry of Natural Resources and Tourism, Forestry and Beekeeping Division, Dodoma and IUCN – The World Conservation Union Eastern Africa Regional Office, Nairobi.

Mukasa, C., A. Tibazalika, A. Banana, E. Mwangi, and T. Mutimukuru-Maravanyika. 2022. "Strengthening Women's Tenure Rights and Participation in Community Forestry in Central Uganda" In *Adaptive Collaborative Management of Forest Landscapes: Villagers, Bureaucrats and Civil Society*, edited by C.J.P. Colfer, R. Prabhu, and A. Larson, pp. 106–132. London: Earthscan.

Munera-Roldan, C., M.J. Colloff, B. Locatelli, and C. Wyborn. 2022. "Engaging with the Future: Framings of Adaptation to Climate Change in Conservation." *Ecosystems and People* 18 (1): pp. 174–188.

Muthee, K., Duguma, L., C. Majale, M. Mucheru-Muna, P. Wainaina, and P. Minang. 2022. "A Quantitative Appraisal of Selected Agroforestry Studies in the Sub-Saharan Africa." *Heliyon* 8 (9): e10670.

NGFS-INSPIRE. 2022. Central Banking and Supervision in the Biosphere: An Agenda for Action on Biodiversity Loss, Financial Risk and System Stability. Final Report of the NGFS-INSPIRE Study Group on Biodiversity and Financial Stability. NGFS Occasional Paper. https://www.ngfs.net/sites/default/files/medias/documents/central_banking_and_supervision_in_the_biosphere.pdf.

Nurfatriani, F., N. Khasanah, and A. Wibowo. 2021. "The Role of Rural Livelihoods in Forest Landscape Restoration: An Analysis of Community-Based Initiatives in Central Java, Indonesia." *Forest Policy and Economics* 128: 102499.

Osuri, A.M., S. Kasinathan, M.K. Siddhartha, D. Mudappa, and T.S. Raman. 2019. "Effects of Restoration on Tree Communities and Carbon Storage in Rainforest Fragments of the Western Ghats, India." *Ecosphere* 10 (9): p.e02860.

Parsons, M., K. Fisher, and R.P. Crease. 2021. "Decolonising River Restoration: Restoration as Acts of Healing and Expression of Rangatiratanga." In *Decolonising Blue Spaces in the Anthropocene: Freshwater management in Aotearoa New Zealand*, edited by M. Parsons, K. Fisher, and R.P. Crease, pp: 359–417. Melbourne: Palgrave Macmillan Cham.

Pashkevich, M.D., F. d'Albertas, A.A.K. Aryawan, et al. 2022. "Nine Actions to Successfully Restore Tropical Agroecosystems." *Trends in Ecology & Evolution* 37 (11): pp. 963–975. DOI10.1016/j.tree.2022.07.007. PMID: 35961912.

Pehou, C., H. Djoudi, B. Vinceti, and M. Elias. 2020." Intersecting and dynamic gender rights to néré, a food tree species in Burkina Faso". *Journal of Rural Studies* 76: pp. 230–239.

Peña Valderrama, S. 2023. "Disappearing Waste and Wasting Time: From Productive Fallows to Carbon Offset Production in Madagascar's Forests." *Ethnos* 88 (3): pp. 491–511, DOI:10.1080/00141844.2020.1796737.

Pichler, M., and M. Ingalls. 2021. "Negotiating between Forest Conversion, Industrial Tree Plantations and Multifunctional Landscapes. Power and Politics in Forest Transitions." *Geoforum* 124: pp. 185–194.

Pirard, R., H. Petit, and H. Baral. 2017. "Local Impacts of Industrial Tree Plantations: An Empirical Analysis in Indonesia across Plantation Types." *Land Use Policy* 60: pp. 242–253.

Place, F., D. Garrity, and P. Agostini 2016. "Tree-Based Systems: Multiple Pathways to Boosting Resilience." In *Confronting Drought in Africa's Drylands: Opportunities for Enhancing Resilience*, edited by R. Cervigni, and M. Morris, pp. 101–114. Washington, DC: World Bank.

Pramova, E., B. Locatelli, H. Djoudi, and O.A. Somorin. 2012. "Forests and Trees for Social Adaptation to Climate Variability and Change." *Wiley Interdisciplinary Reviews: Climate Change* 3 (6): pp. 581–96.

Price, M.J., A. Latta, A. Spring, et al. 2022. "Agroecology in the North: Centering Indigenous Food Sovereignty and Land Stewardship in Agriculture 'Frontiers.'" *Agriculture and Human Values* 39 (4): pp. 1191–1206.

Pulido, M.T. and J. Caballero. 2006. "The Impact of Shifting Agriculture on the Availability of Non-Timber Forest Products: The Example of Sabal Yapa in the Maya Lowlands of Mexico." *Forest Ecology and Management* 222 (1–3): pp. 399–409.

Reed, J., J. van Vianen, J. Barlow, and T. Sunderland. 2017. "Have Integrated Landscape Approaches Reconciled Societal and Environmental Issues in the Tropics?" *Land Use Policy* 63: pp. 481–492.

Reij, C., and D. Garrity. 2016. "Scaling Up Farmer-managed natural regeneration in Africa to Restore Degraded Landscapes." *Biotropica* 48 (6): pp. 834–843.

Reinecke, S., and M. Blum. 2018. "Discourses across Scales on Forest Landscape Restoration." *Sustainability* 10 (3): p. 613.

Reytar, K., S. Ray, and L. Toh. 2018. Malawi is Putting its Money Where its Forests Are. World Resources Institute. https://www.wri.org/insights/malawi-putting-its-money-where-its-forests-are?trk=public_post_comment-text.

Robson, J.P., and F. Berkes. 2011. "Exploring Some of the Myths of Land Use Change: Can Rural to Urban Migration Drive Declines in Biodiversity?" *Global Environmental Change* 21 (3): pp. 844–854.

Runting, R.K., B.A. Bryan, L.E. Dee, et al. 2017. "Incorporating Climate Change into Ecosystem Service Assessments and Decisions: A Review." *Global Change Biology* 2: pp. 28–41.

Sabastian, G.E., A. Yumn, J.M. Roshetko, P. Manalu, E. Martini, and A. Perdana. 2019. "Adoption of Silvicultural Practices in Smallholder Timber and NTFPs Production Systems in Indonesia." *Agroforestry Systems* 93: pp. 607–620.

Sabogal, C., C. Besacier, and D. McGuire. 2015. "Forest and Landscape Restoration: Concepts, Approaches and Challenges for Implementation." *Unasylva* 245 (66): pp. 3–10.

Safari, J., I. Singu, Z. Masanyiwa, and C. Hyandye. 2019. "Social Perception and Determinants of Ngitili System Adoption for Forage and Land Conservation in Maswa district, Tanzania." *Journal of Environmental Management* 250: p. 109498.

Salzman, J., G. Bennett, N. Carroll, A. Goldstein, and M. Jenkins. 2018. "The Global Status and Trends of Payments for Ecosystem Services." *Nature Sustainability* 1: pp. 136–44. DOI:10.1038/s41893018-0033-0.

Sarmiento Barletti, J.P. 2022. "Learning from Adaptive Collaborative Management: A Participatory Tool to Support Adaptive and Reflective Learning in Multi-Stakeholder Forums." In *Adaptive Collaborative Management of Forest Landscapes: Villagers, Bureaucrats and Civil Society*, edited by C.J.P. Colfer, R. Prabh, and A. Larson, pp. 159–84. London: Earthscan.

Scherr, S.J., S. Shames, and R. Friedman. 2013. "Defining Integrated Landscape Management for Policy Makers." Ecoagriculture Policy Focus 10. http://ecoagriculture.org/publication_details.php?publicationID=547.

Schultz, B., D. Brockington, E.A. Coleman, et al. 2022. "Recognizing the Equity Implications of Restoration Priority Maps." *Environmental Research Letters* 17 (11): p. 114019. DOI:10.1088/1748-9326/ac9918.

Scoones, I., 2018. "Review of The End of Desertification? Disputing Environmental Change in the Drylands by Roy H. Behnke and Michael Mortimore." *Pastoralism* 8: 26.

Shanley, P. (ed.). 2002. *Tapping the Green Market: Certification and Management of Non-Timber Forest Products*. London: Earthscan.

Sendzimir, J., C.P. Reij, and P. Magnuszewski. 2011. "Rebuilding Resilience in the Sahel: Regreening in the Maradi and Zinder Regions of Niger." *Ecology and Society* 16 (3). DOI:10.5751/ES-04198-160301.

Smith, K.R. 2008. "The Fuel that Warms You Twice." In *Human Health and Forests: A Global Overview of Issues, Practice and Policy*, edited by C.J.P. Colfer, pp. 97–111. London: Earthscan.

Smith, P., J. Nkem, K. Calvin, et al. 2019. "Interlinkages between Desertification, Land Degradation, Food Security and Greenhouse Gas Fluxes: Synergies, Trade-Offs and Integrated Response Options." In *Climate Change And Land: An IPCC Special Report on Climate Change, Desertification, Land Degradation, Sustainable Land Management, Food Security, and Greenhouse Gas Fluxes in Terrestrial Ecosystems*, edited by P.R. Shukla, J. Skea, E. Calvo Buendia, et al., pp. 551–672. Geneva: IPCC.

Spathelf, P., J. Stanturf, M. Kleine, R. Jandl, D. Chiatante, and A. Bolte. 2018. "Adaptive Measures: Integrating Adaptive Forest Management and Forest Landscape Restoration." *Annals of Forest Science* 75: pp. 1–6.

Stanturf, J.A. (ed.). 2015. *Restoration of boreal and temperate forests* (Vol. 13). Boca Raton: CRC Press.

Stanturf, J.A. 2021. "Forest Landscape Restoration: Building on the Past for Future Success." *Restoration Ecology* 29 (4): e13349.
Stanturf, J.A., and S. Mansourian. 2020. "Forest Landscape Restoration: State of Play." *Royal Society Open Science* 7 (12): 201218.
Sumberg, J., J. Chamberlin, J. Flynn, D. Glover, and V. Johnson. 2019. "Landscapes of Rural Youth Opportunity." *IFAD Research Series* 47. DOI:10.2139/ssrn.3521380.
Sumberg, J., and C. Okali. 2013. "Young People, Agriculture, and Transformation in Rural Africa: An "Opportunity Space" Approach." *Innovations: Technology, Governance, Globalization* 8 (1–2): pp. 259–269. DOI:10.1162/inov_a_00178.
Taylor, M.J., M. Aguilar-Støen, E. Castellanos, M.J. Moran-Taylor, and K. Gerkin. 2016. "International Migration, Land Use Change and the Environment in Ixcán, Guatemala." *Land Use Policy* 54: pp. 290–301.
Tedesco, A.M., S. López-Cubillos, R. Chazdon, et al. 2023. "Beyond Ecology: Ecosystem Restoration as a Process for Social-Ecological Transformation." *Trends in Ecology & Evolution* 38 (7): pp. 643–653.
Temper, L., D. Del Bene, J. and Martinez-Alier. 2015. "Mapping the Frontiers and Front Lines of Global Environmental Justice: The EJAtlas." *Journal of Political Ecology* 22 (1): pp. 255–278.
Tengö, M., R. Hill, P. Malmer, et al. 2017. "Weaving Knowledge Systems in IPBES, CBD and beyond—Lessons Learned for Sustainability." *Current Opinion in Environmental Sustainability* 26: pp. 17–25.
Todd, Z. 2014. "Fish Pluralities: Human-Animal Relations and Sites of Engagement in Paulatuuq, Arctic Canada." *Études/Inuit/Studies* 38 (1–2): pp. 217–238.
Tolentino, E.L. 2008. "Restoration of Philippine Native Forest by Smallholder Tree Farmers." In: Advances in Agroforestry 5. Smallholder Tree Growing for Rural Development and Environmental Services, edited by D.J. Snelder, and R.D. Lasco, pp. 319-346. Dordrecht: Springer. https://doi.org/10.1007/978-1-4020-8261-0_15.
UNEP. 2021. Becoming #GenerationRestoration: Ecosystem Restoration for People, Nature and Climate. Nairobi: UNEP. https://wedocs.unep.org/bitstream/handle/20.500.11822/36251/ERPNC.pdf.
Uprety, Y., H. Asselin, Y. Bergeron, F. Doyon, and J.F. Boucher. 2012. "Contribution of Traditional Knowledge to Ecological Restoration: Practices and Applications." *Ecoscience* 19 (3): pp. 225–237.
Urzedo, D.I.D., F.C. Piña-Rodrigues, R. Feltran-Barbieri, R.G. Junqueira, and R. Fisher. 2020. "Seed Networks for Upscaling Forest Landscape Restoration: Is It Possible to Expand Native Plant Sources in Brazil?" *Forests* 11 (3): p. 259.
van der Geest, K., A. De Sherbinin, S. Kienberger, et al. 2019. The Impacts of Climate Change on Ecosystem Services and Resulting Losses and Damages to People and Society. In *Loss and Damage from Climate Change: Concepts, Methods and Policy Options*, edited by R. Mechler, L. M. Bouwer, T. Schinko, S. Surminski, and J.-A. Linnerooth-Bayerpp, pp: 221–236. Cham: Springer.
Van Meerveld, H.J., J.P. Jones, C.P. Ghimire, B.W. Zwartendijk, et al. 2021. "Forest Regeneration Can Positively Contribute to Local Hydrological Ecosystem Services: Implications for Forest Landscape Restoration." *Journal of Applied Ecology* 58 (4): pp. 755–765.

van Oosten, C., H. Runhaar, and B. Arts. 2021. "Capable to Govern Landscape Restoration? Exploring Landscape Governance Capabilities, Based on Literature and Stakeholder Perceptions." *Land Use Policy* 104: p. 104020.

Veit, P. 2021. 9 Facts About Community Land and Climate Mitigation. WRI. https://files.wri.org/d8/s3fs-public/2021-10/9-facts-about-community-land-and-climate-mitigation.pdf.

Vieira, D.L., K.D. Holl, and F.M. Peneireiro. 2009. "Agro-Successional Restoration as a Strategy to Facilitate Tropical Forest Recovery." *Restoration ecology* 17 (4): pp. 451–459.

Voeks, R.A. and A. Leony. 2004. "Forgetting the Forest: Assessing Medicinal Plant Erosion in Eastern Brazil." *Economic Botany* 58 (1): pp. S294–S306.

Wanjiru Ng'ang'a, J., E.M. Kioko, S.C. Wairuri, and I. Ngare. 2022. "The 'Bald' Ecosystem: Indigenous Resource Governance and Restoration of Kivaa Forest, Kenya." *East African Journal of Environment and Natural Resources* 5 (2): pp. 111–123.

Watts, V. 2013. "Indigenous Place-Thought and Agency Amongst Humans and Non Humans (First Woman and Sky Woman Go On a European World Tour!)." *Decolonization: Indigeneity, Education & Society* 2 (1): pp. 20–34. https://jps.library.utoronto.ca/index.php/des/article/view/19145.

Wehi, P.M., and J.M. Lord. 2017. "Importance of Including Cultural Practices in Ecological Restoration." *Conservation Biology* 31 (5): pp. 1109–1118.

Widianingsih, N.N., I. Theilade, and M. Pouliot. 2016. "Contribution of Forest Restoration to Rural Livelihoods and Household Income in Indonesia." *Sustainability* 8 (9): p. 835.

Wiegant, D., M. Peralvo, P. van Oel, and A. Dewulf. 2020. "Five Scale Challenges in Ecuadorian Forest and Landscape Restoration Governance." *Land Use Policy* 96: p.104686.

Wilson, K.A., M. Lulow, J. Burger, and M.F. McBride. 2012. "The Economics of Restoration." In *Forest Landscape Restoration: Integrated Approaches to Support Effective Implementation*, edited by S. Mansourian, and J. Parrotta, pp. 215–231. Dordrecht: Springer.

Winter, C.J. 2018. The Paralysis of Intergenerational Justice: Decolonising Entangled Futures. Master's thesis, University of Sydney. https://ses.library.usyd.edu.au/handle/2123/18009.

Winter, K., M.B. Vaughan, N. Kurashima, et al. 2023. "Indigenous Stewardship through Novel Approaches to Collaborative Management in Hawaiʻi." *Ecology and Society* 28 (1): 26.

Winterbottom. R., C. Reij, D. Garrity, et al. 2013. *Improving Land and Water Management.* Washington: World Resources Institute.

World Bank. 2016. Why Forests are Key to Climate, Water, Health, and Livelihoods. World Bank. https://www.worldbank.org/en/news/feature/2016/03/18/why-forests-are-key-to-climate-water-health-and-livelihoods.

Worthington, T., and M. Spalding. 2018. *Mangrove Restoration Potential: A Global Map Highlighting A Critical Opportunity.* Apollo – University of Cambridge Repository. https://doi.org/10.17863/CAM.39153.

Wunder, S., S. Engel, and S. Pagiola. 2008. "Taking Stock: A Comparative Analysis of Payments for Environmental Services Programs in Developed and Developing Countries." *Ecological Economics* 65 (4): pp. 834–852.

WWF (Worldwide Fund for Nature), and IUCN (International Union for Conservation of Nature). 2000. Forests Reborn: A Workshop on Forest Restoration. In Proceedings of the WWF/IUCN International Workshop on Forest Restoration, Segovia, Spain, 3–5 July 2000; p. 3–5.

Yang, A., I. Bellwood-Howard, and M. Lippe. 2018. "Social-Ecological Systems and Forest Landscape Restoration." In *Forest Landscape Restoration: Integrated Approaches to Support Effective Implementation*, edited by S. Mansourian, and J. Parrotta, pp. 65–82. London: Earthscan. DOI:10.4324/9781315111872.

Zhang, W., H. ElDidi, Y.J. Masuda, R.S. Meinzen-Dick, et al. 2023. "Community-Based Conservation of Freshwater Resources: Learning from a Critical Review of the Literature and Case Studies." *Society & Natural Resources* 36 (6): pp. 733–754. DOI: 10.1080/08941920.2023.2191228 pp.1–22.

Zulu, L.C., I.N.S. Djenontin, and P. Grabowski. 2021. "From Diagnosis to Action: Understanding Youth Strengths and Hurdles and Using Decision-Making Tools to Foster Youth-Inclusive Sustainable Agriculture Intensification." *Journal of Rural Studies* 82: pp. 196–209.

Zulu, L.C., and R.B. Richardson. 2013. "Charcoal, Livelihoods, and Poverty Reduction: Evidence from Sub-Saharan Africa." *Energy for Sustainable Development* 17 (2): pp.127–137.

9

An economic view on the costs and benefits of forest restoration

Pablo Pacheco, Craig Beatty, and Jwalin Patel

9.1 Introduction

We are facing an unprecedented scale of nature loss due in part to the continuous process of land and forest degradation (IPBES 2018). The loss of natural ecosystems leads to increasing carbon emissions that contribute to climate change and translate in diminishing ecosystem services (IPCC 2019). This has raised stark concerns given the expected growing global demand for food, animal feed, and materials, which tend to rely on these services. The impacts from land and forest degradation are escalating, leading to growing calls not only to improve the efficient and sustainable use of existing land but to invest in restoring the integrity of degraded natural ecosystems, which would lead to broad financial, social, and environmental benefits (UNEP 2021).

Actions taken to restore ecological productivity thus far, especially in working lands and the natural ecosystems that support forest landscapes, form the basis of the initial Bonn Challenge. This global initiative aims to place 350 Mha of degraded and deforested land under restoration by 2030. Flowing from this work and the recognition of the role that restored ecosystems can play in addressing declines in nature and human well-being, the United Nations declared 2021–2030 the Decade on Ecosystem Restoration. In addition, the Kunming-Montreal Global Biodiversity Framework embraces the target to "ensure that by 2030 at least 30% of areas of degraded terrestrial, inland water, and coastal and marine ecosystems are under effective restoration, in order to enhance biodiversity and ecosystem functions and services, ecological integrity and connectivity" (CBD 2022, 2).

Growing support for restoration is founded on the understanding that restoring natural ecosystems, including forests, is not only critical to address biodiversity loss (IPBES 2018) but can also contribute to carbon removals from the atmosphere (IPCC 2019). Reforestation and restoration of degraded forests are considered among the most effective land-based strategies for sequestering carbon emissions to combat climate change (Griscom et al. 2017), and tree restoration might have positive impacts on food security (Ickowitz et al. 2022). Such perspectives implicitly assume that restoration efforts resulting in ecosystem services recovery—including climate regulation, clean air, fresh water, and soil fertility—translate into social, economic, and financial gains (Sutton et al. 2016).

These potential gains, however, might materialize under different time horizons and might differ across places since ecological gains from restoration depend on local climate conditions (Zhou et al. 2022) and the landscape's ecological context (Gilby

Pablo Pacheco, Craig Beatty, and Jwalin Patel, *An economic view on the costs and benefits of forest restoration*
In: *Restoring Forests and Trees for Sustainable Development*. Edited by: Pia Katila, Carol J. Pierce Colfer, Wil de Jong, Glenn Galloway, Pablo Pacheco, and Georg Winkel, Oxford University Press. DOI: 10.1093/9780197683958.003.0009

et al. 2018). In addition, depending on the social context in which restoration occurs, and the types of restoration interventions undertaken, externalities may arise. For example, forest restoration may displace agricultural activities in working lands that produce food, or may affect the livelihoods of poor landholders with insecure tenure rights. Forest restoration may also affect local economies by disrupting land values and prices, leading to conflict over resources. When forest restoration leads to financial gains, it may result in inequitable distribution of the associated costs and benefits among involved stakeholders. While there is a relatively good understanding of the environmental benefits of ecosystem restoration (IPBES 2018), the economic and financial gains of tree and forest restoration, as well as the potential net gains across contexts and disparate restoration options, are not well determined (Bodin et al. 2021).

This chapter focuses on forest restoration using trees, which we will refer to simply as forest restoration. It is noteworthy that forest restoration is only one of many types of restorative actions that increase ecological integrity in the farms and landscapes beyond forests (IUCN and WRI 2014). Our objective is to contribute to the discussion on economic and financial costs and benefits of trees and forest restoration across options. We do not provide an exhaustive review of available studies assessing these costs and benefits but highlight some key findings and knowledge gaps. The assessment of costs and benefits from an economic and financial perspective is undertaken through cost–benefit analysis (CBA). CBA consists, first, of the financial analysis of the costs incurred and benefits accrued by private actors or enterprises (e.g., farmers, firms, investors) undertaking restoration and, second, of the wider economic analysis of the returns of forest restoration to society at large. Financial analyses measure the financial performance of a particular restoration activity but do not consider externalities. Economic analyses involve comprehensive cost and benefit assessments for the whole society and tend to include the monetary value of ecosystem services, which is estimated based on their market value if they have one; otherwise it is estimated using different valuation methods. These assessments increasingly rely on spatially explicit analysis to better capture the spatial dimension of environmental services' provision, which tends to differ across places depending on the ecosystems currently under analysis.

The costs of forest restoration include not only short- to mid-term direct costs to establish and maintain trees and forests but also transaction costs, opportunity costs, and external costs. In turn, the benefits of forest restoration are associated with direct benefits to employment and income generated by restoration actions, as well as the incremental benefits from the recovery of forest ecosystem services to people, which are typically classified as provisioning, regulating, supporting, and cultural services. The most common forest restoration options (also known as interventions, actions, techniques, modalities, or systems) comprise natural regeneration under passive and assisted modalities, agroforestry, agrosilvopastoral systems, and tree plantations and woodlots (IUCN and WRI 2014). These restoration options may occur in lands under different management regimes including public lands, community lands and territories, and individual landholdings; many combinations may occur depending on the context, who initiates and performs the restoration actions, and the scale.

Forest restoration takes place within diverse ecological, social, economic, and institutional contexts and across landscapes in different states of degradation. As such, the costs and benefits of forest restoration vary widely. Given the lack of consistent methodologies and comparable data, it is difficult to determine the factors across all contexts that explain this variation (Cook et al. 2017). This limits our ability to make evidence-informed selections of restoration actions that are more likely to perform financially, and that yield economic and environmental returns that better serve the context needs. Previous work has highlighted the lack of meaningful data on restoration costs (TEEB 2010). Calls have been made to improve the ways in which economic data is gathered and compared across contexts; this requires adopting common frameworks for data collection and analysis (Bodin et al. 2021).

Our analysis here relies on a review of several studies focusing on the economic CBA of forest restoration, complemented by six country studies including detailed financial CBA with comparable data of 41 cases of forest restoration. Our analysis is not exempt from the limitations of representation described above. We focus on general trends and knowledge gaps arising from the challenges of CBA with a particular emphasis on forest restoration options, making necessary caveats linked to the data as explicitly as possible in our analysis and discussion.

Understanding the costs and benefits of different forest restoration options is critical for guiding investments and designing incentives that meet social and ecological objectives under diverse social, institutional, and biophysical conditions informed by the best available knowledge. The main issues to consider when weighing different forest restoration options are their financial, socioeconomic, and environmental benefits; how to balance short- and long-term costs and benefits; and the distribution of costs and benefits among stakeholders to ensure equity in access to benefits. These issues must be assessed within the local context, embracing the perspectives of all relevant stakeholders, particularly indigenous peoples and local communities, smallholders, farmers, and other diverse forest users who make a living on, or whose businesses depend on, the sustainable management of land and forest resources. Erbaugh et al. (2020) estimate that 294.5 million people live on tropical forest land identified as having restoration opportunities; thus a look at the costs and benefits of restoration should take wider social and economic considerations seriously.

This chapter is organized in five sections including this introduction. Section 2 provides elements for understanding the full range of costs and benefits associated with trees and forest restoration. Section 3 examines both financial and economic CBA of forest restoration. The fourth section discusses factors that influence the variability of costs and benefits across and within different forest restoration options, the resulting trade-offs, and ways to deal with these trade-offs. And Section 4 pulls together this chapter's main conclusions.

9.2 Costs and benefits of forest restoration

Despite costs being a major concern about restoration, it is inappropriate to look at them without considering potential benefits. The costs and benefits of forest restoration are multiple and include two dimensions, private and social. Private costs are those costs incurred by farms, firms, and investors undertaking restoration actions, while social costs are those corresponding to the society at large. In a similar way, private benefits are those accrued by private agents undertaking forest restoration, whereas social benefits are those that benefit society at large (IPCC 2007).

Figure 9.1 provides a simplified description of the costs and benefits of forest restoration. The upper side consists of the full range of (four) categories of costs, and the lower side includes the full range of (five) categories of benefits. The central portion of the diagram depicts the scales at which forest restoration can occur, either at the farm and/or land management unit (or groups of them) or at the whole landscape scale, the latter delineated by either ecological or administrative boundaries at the subnational, national, or even regional level.

We identify six main forest restoration options: 1) natural forest regeneration, 2) assisted natural forest regeneration, 3) agroforestry with annual and/or perennial crops, 4) agrosilvopastoral systems, 5) tree plantations with annual crops, and 6) tree plantations in monocrops. These six options are included in Figure 9.1 and outlined in detail in Appendix 9A. Furthermore, context-specific factors, such as those related to ecological conditions, political economies, and cultures, not only influence which options are undertaken but also the associated costs and benefits of the different restoration options at either the farm and land management units—depending on the land ownership and management type—or landscape levels.

Restoration costs include the direct costs that correspond not only to the initial investments for undertaking physical restoration activities but also the operational expenses incurred for long-term tree, tree-crop, and forest management. Direct costs are a function of the biophysical conditions where forest restoration takes place, in that poorer biophysical conditions will tend to either increase the scope of initial restoration investment or extend the timeline over which restoration takes place (e.g., soil quality and water availability). As a result, the greater the degradation, the higher the cost.

Forest restoration is nested within institutional processes aimed at planning, providing access to information, generating social and group incentives for action, enabling participation, and building capacity for long-term sustainable management (Verdone 2015). These constitute enabling conditions shaping the outcomes of forest restoration. These conditions entail transaction costs, which are seldom documented systematically (Bodin et al. 2021). These costs are strongly influenced by context-specific conditions associated with not only biophysical factors of the landscapes where restoration takes place but also socio-institutional characteristics of the groups undertaking forest restoration. These conditions vary depending on the scale at which restoration actions are undertaken.

In addition, opportunity costs and external costs are two types of costs that should be considered in planning and evaluating forest restoration. Opportunity costs of restoration consist of the foregone economic benefits of alternative land-use choices where restoration takes place (Verdone 2015). Most typical alternative land uses to forest restoration are those linked to non-forest land uses, such as commercial agriculture and livestock production. Land-use choices depend not only on land suitability but also on availability of labor and access to capital and inputs (Barbier 2007).

External costs, when taking a wider social perspective, are those costs generated by the externalities of restoration. These costs are borne by society at large, beyond the

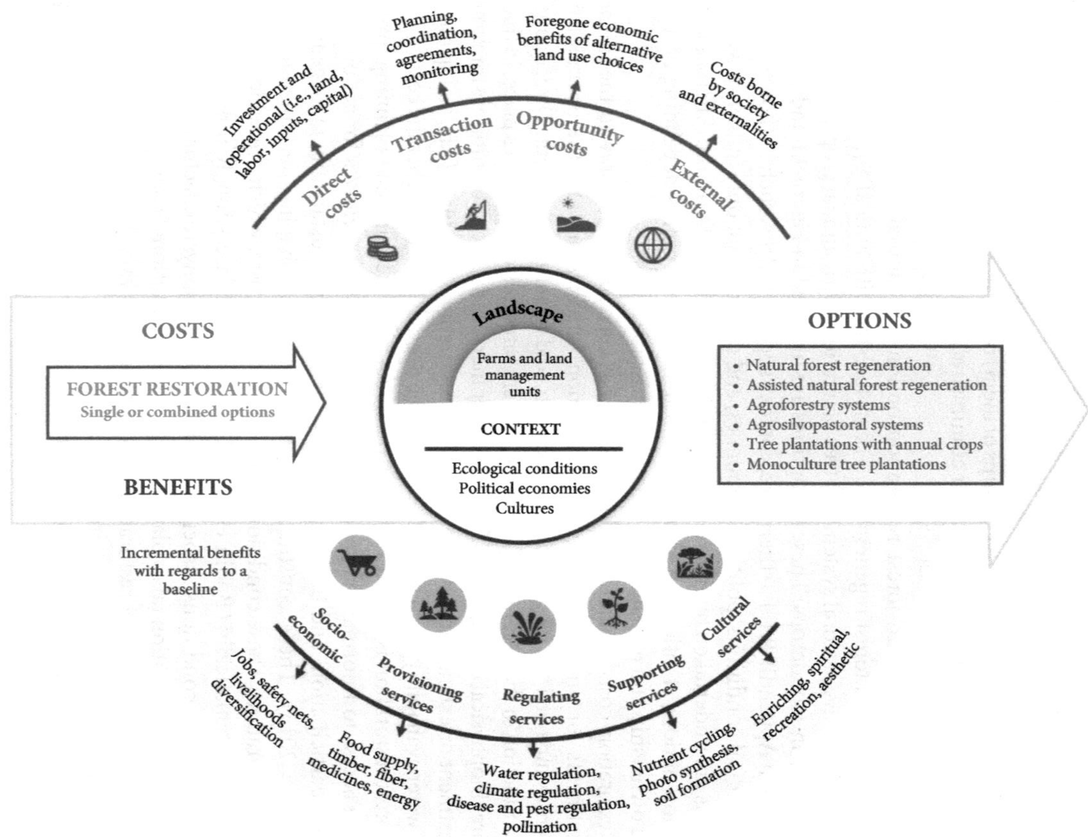

Figure 9.1 Private and social costs and benefits of forest restoration (Source: Authors' own depiction)

private costs of farmers, landowners, or firms conducting restoration. The external costs differ depending on the restoration option and scale (Hebling 2010), and social externalities may occur. For example, tree plantations could demand a large supply of water, or tree planting may compete with productive agricultural land, thereby affecting the supply of food. Restoration through monoculture plantations may exclude local populations from the benefits of restoration or restrict them access to land that is critical for their livelihoods.

The benefits of forest restoration consist of the incremental socioeconomic and environmental gains when compared with a degradation baseline. The socioeconomic benefits from restoration often translate into jobs and employment, safety nets, and livelihood diversification for landholders undertaking restoration. In addition, the commonly used typology of forest ecosystem services to people has four additional categories of benefits: provisioning services (e.g., food supply, timber, fiber, medicines, energy), regulating services (e.g., water regulation, climate regulation, disease and pest regulation, pollination), supporting services (e.g., nutrient cycling, photosynthesis, soil formation), and cultural services (e.g., enriching, spiritual and religious, recreation, aesthetics) (MEA 2005). In this classification there is no explicit mention of biodiversity, yet biodiversity can be considered a fundamental supporting service that underpins the delivery of all other services (OECD 2018).

Many of the benefits from forest restoration take time to accrue, except for some financial gains from annual crop production and income for labor invested in restoration activities. For example, annual crops intercropped with perennial crops generate returns in shorter timeframes compared with timber for logging, depending on the tree species planted. In addition, social benefits from restoration, such as improved safety nets or social resilience accrue over longer timeframes. This is the case of environmental benefits as well, including regulating services (e.g., climate and water regulation, erosion control) and supporting services (e.g., nutrient cycling). While some benefits of forest restoration accrue directly to farmers and firms, particularly provisioning services, several benefits are not confined to restoration lands and involve more distant climatic and ecological interactions.

The six forest restoration options vary in the socioeconomic and environmental benefits they generate. Options oriented to improve provisioning services, including the production of monetizable goods such as crops, timber, fuelwood, and nontimber forest products, will deliver greater private economic benefits. Proximity to markets and the mix of goods, which is also linked to the tree species selected for restoration, can determine on what scale and over what timespan forest restoration benefits materialize. Options that are more labor-intensive are expected to generate additional jobs and income, though the quality of those jobs may vary (Raes et al. 2021). Options such as agroforestry, agrosilvopastoral systems, and tree plantations with crops can contribute to income generation and livelihood diversification for local farmers, with the potential to satisfy social needs (e.g., health, education) (Raes et al. 2017; Sunderland and Rowland 2019).

Furthermore, natural regeneration leads to greater environmental benefits through carbon capture, recovery of biodiversity, and water regulation—benefits that accumulate when lower-cost forest restoration allows implementation at larger spatial scales (Chazdon and Uriarte 2016; Crouzeilles et al. 2020). The potential environmental benefits of agroforestry are extensive, including soil fertility recovery, soil salinity control, erosion reduction, water quality improvement, and biodiversity and pollinator habitat protection and recovery. Restoration interventions designed to satisfy fuelwood supply may reduce pressures on natural forests (Montagnini 2017; Raes et al. 2017). There is a diversity of agroforestry systems; the precise array of benefits depends on the system under consideration, its site characteristics, local economies, and management approaches. Monoculture plantations, instead, may contribute to higher impacts on carbon sequestration (Bernal et al. 2018) yet have limited or negative impacts on nature (Albert et al. 2021) and use excessive amounts of water (Vanclay 2009).

Context-specific factors, including the local biophysical settings and the state of degradation of the land and forest where restoration takes place, influence the costs incurred in forest restoration activities. Restoration outcomes, and the benefits they deliver, are influenced by the speed of ecological recovery, which can depend on climate stress and biotic agents (Jones et al. 2018), percentage of forest cover (Arroyo-Rodríguez et al. 2023), temperature and precipitation conditions (Zhou et al. 2022), and others factors including variability in soil type, density and depth, and drainage conditions. Soil recovery is a critical ingredient of successful ecosystem restoration (van der Sande et al. 2023). Political economy factors—including infrastructure, market development, price distortions, and resource competition—impact the costs of land, labor, and inputs, thereby influencing restoration costs. Furthermore, cultural systems, including values toward and conceptualizations of nature, may influence locally desired forest restoration options, acknowledging that diverse cultural values may coexist in landscapes (Schultz et al. 2022).

9.3 Economic and financial costs and benefits

The most common approach for assessing the private and social gains from forest restoration actions is CBA. Typical CBA can be conducted from a wider social perspective by considering the costs and benefits for society at large or from a purely private perspective to evaluate the financial profitability of the investments. While the latter may also have wider socio-environmental implications, these are often omitted from financial analysis (Campbell and Brown 2012). When looking at CBA for restoration, financial CBA helps to understand the net financial gains across restoration options from a private perspective for farms, firms, and investors, with limited or no consideration of externalities or social costs of these activities. In turn, CBAs that include whole landscapes or wider administrative jurisdictions (e.g., countries, states, provinces) require adopting a social perspective, accounting for economic costs and benefits to society, such as net changes in ecosystem services. This expands the CBA scope to include public goods not only with a monetary value but also with a nonmonetary value. Economic analyses tend to group these services in *use values* and *nonuse values* or those that are more intangible. Several methods are applied to conduct the economic valuation used in ecosystem services valuation but not without challenges (OECD 2018). Table 9.1 shows the differences between economic and financial cost–benefit approaches. See also Wainaina et al. (2020) for a more detailed explanation of these differences.

Table 9.1 Comparison of economic and financial CBA applied to forest restoration

	Economic CBA of Forest Restoration	**Financial CBA of Forest Restoration**
Scale	Landscapes or administrative jurisdictions	Individual or groups of farms and/or land management units
Purpose	Efficiency of public investment in restoration actions	Efficiency of private investments in restoration activities
Focus	Net return to society at large	Net returns to the enterprise (e.g., farmers, firms and/or investors)
Costs	Social costs to society at large, includes externalities	Costs to the private enterprise (e.g., farmers, firms and/or investors)
Benefits	Market and nonmarket goods and services	Market good and services
Discount rate	Social discount rate	Private opportunity cost of capital
Balance on investment	Uses economic prices	Uses market prices
Distribution of benefits	Not a principal concern	Not considered an issue

Source: Elaborated by authors based on Campbell and Brown (2012).

9.3.1 Economic cost–benefit analyses

Economic CBA of restoration is used to determine the net returns from reversing land degradation, which is typically categorized as leading to significant social and economic benefits. Land degradation is an incremental process that is estimated to have affected around 75% of the global land surface, including forests (IPBES 2019). Some estimate the global impact of deforestation and deteriorating agricultural productivity at about USD 6.3 trillion per year, arising from the loss of ecosystem services including climate regulation, clean air, fresh water, and soil fertility (Sutton et al. 2016). Adverse socioeconomic outcomes linked to land degradation include the loss of livelihoods and land productivity (Barbier and Hochard 2018). Land degradation also affects a range of cultural elements arising from community interactions with nature that give meaning to life and contribute to well-being (Reyes-García et al. 2013). Indeed, climate change and the looming biodiversity crisis reflect the consequences of the long and well-examined history of human-driven nature and habitat loss.

Several global comparative reviews have been undertaken to understand the economic costs and benefits of ecosystem restoration. For example, De Groot et al. (2013), reviews 94 studies of costs and 225 studies of benefits associated with ecosystem restoration projects across nine different biomes. While acknowledging data limitations and static estimates, he argues that the majority of restoration projects analyzed have provided net benefits and should be considered profitable. In this analysis, costs include capital investment and maintenance of restoration interventions, and benefits correspond to the total monetary value of the ecosystem services provided by the restored ecosystem. When looking at tropical forests (96 estimates), the total economic value (TEV) of all services equals USD 5,264/ha (SD = 6,526), with USD 1,588/ha (SD = 317) for woodlands. The TEV of forest-related restoration is lower compared with those in coral reefs and coastal wetlands. The analysis on monetary values builds upon De Groot et al.'s (2012) study that estimated values for 22 ecosystem services. This analysis, however, is not exempt from issues, among others, such as data heterogeneity, monetary estimate generalization for context-specific environmental services, value transference across socioeconomic contexts, as well as value estimate selection bias.

A systematic review of CBA conducted by Wainaina et. al. (2020) based on 31 articles across Latin America, Southeast Asia, and sub-Saharan Africa finds that most papers conducting economic CBA (84%) rely on estimates of use values (direct and indirect) (e.g., erosion control, air regulation, temperature regulation, recreational value, avoided nutrient loss, nitrogen fixation, and soil fertility improvement), and only a few include nonuse values (e.g., aesthetic value, bequest (inheritance) values, existence value). This analysis shows that most of the cases analyzed report positive net present values (NPVs) for the different restoration options considered (i.e., agroforestry, soil and water conservation, mangrove restoration). However, the number of studies that report positive NPVs for reforestation and afforestation is equal to the ones that report negative NPVs. For other options (i.e., assisted natural regeneration (ANR), woodlot establishment), the number of studies that report positive NPVs is higher than the number that report negative NPVs. Wainaina et al. (2020) highlight that the different studies assessed use different methods to valuate the respective benefits. There is an increasing use of spatial analysis for determining some environmental values, including net carbon sequestration. A few studies also apply contingent valuation methods for benefits, for example, of air pollution and recreational values.

Birch et al. (2010) assess the cost-effectiveness of dryland forest restoration using spatial analysis of ecosystem services in four study areas in Latin America. This study estimates the NPVs of carbon sequestration, non-timber forest products, timber, tourism, and livestock production. The cost-effectiveness of restoration is analyzed by estimating the net social benefit of restoration. The study finds that passive natural restoration is a cost-effective option, while the benefits from active restoration are outweighed by the relatively higher costs involved. This is likely due to the higher opportunity costs associated with the loss of income from livestock production—the main alternative land use in the study areas—and that only a few environmental benefits are considered. Similarly, Newton et al. (2012), assessing the cost–benefits of ecological networks based on a case in Dorset, England, argue that establishing these networks in landscapes dominated by agricultural land uses may not deliver net economic benefits given relatively higher land-use opportunity costs.

9.3.2 Financial cost–benefit analyses

There is a growing number of studies analyzing the financial performance for private farms, firms, and investors by conducting CBA of specific forest restoration options.

Table 9.2 presents data on financial costs and benefit-cost ratios (CBR) across tree and forest restoration options for 41 different cases originating from six country studies. These studies were conducted by the International Union for Conservation of Nature (IUCN) and the World Resources Institute (WRI) using the Restoration Opportunities Assessment Methodology (ROAM) at national and subnational jurisdictional scales (IUCN and WRI 2014). In the ROAM studies, costs and benefits are estimated through financial CBA at the country or landscape level, and some conduct multi-criteria analysis (MCA) including the spatially explicit optimization of economic outcomes across multiple objectives, including the provision of ecosystem services, under different social and biophysical constraints (Beatty et al. 2018a; Robbins and Daniels 2012). Some of these studies consist of CBA estimates for a time horizon of 20 years, while others comprise 30 years. Most of these estimates calculate NPV using a 10% discount rate. To facilitate comparison, estimates are adjusted for inflation in their respective currencies to 2020, and then converted to 2020 USD. The extent of comparability of these figures is subject to the comparability of the contexts and objectives of restoration in each place.

Overall, these estimates suggest that forest restoration options with the lowest financial costs are those linked to the natural regeneration of forests; costs are somewhat higher when assisted practices are used compared to passive natural forest regeneration. Restoration using monoculture plantations or woodlots have relatively higher short-term direct costs than natural regeneration, but lower than other options. Natural regeneration and tree plantations are likely implemented at larger spatial scales than agroforestry or agrosilvopastoral approaches and, thus, are likely to benefit from economies of scale, while the latter two have higher short- and long-term direct costs. The options with the highest short-term direct costs involve agroforestry, particularly when including high-value tree crops such as cacao. These estimates, however, would be lower when restoration actions aim at enhancing existing systems, including agro-successional systems (Vieira et al. 2009), rather than establishing new ones.

In line with our findings, a study focusing on agroforestry systems in Latin America suggests that agroforestry systems are profitable in most cases (Current et al. 1995). The study finds that agroforestry, which includes intercropping with perennials, perform better, as it diversifies income, adds value per unit of land, and improves cash streams. In addition, an analysis of restoration costs in Brazil, comprising the Amazon, Cerrado, and Atlantic Forest biomes, suggests that passive and ANR are comparatively cheaper (USD 48/ha and USD 344/ha, respectively) but are used less since most restoration interventions use seedling planting and direct seeding that are more costly (USD 2,328/ha and USD 1,754/ha, respectively), with a larger portion of the tree planting costs incurred during the implementation phase (Brancalion et al. 2019). The authors suggest that these costs depend on local market conditions, among other factors.

The financial CBA estimates encompass only the direct monetary costs and benefits to farms, firms, and investors and opportunity costs of the foregone financial benefits from likely alternative land uses. For comparison, we present BCR to provide an indication of the performance of expected financial benefits over costs. BCR are greater than one in most of the cases, with a few exceptions, which suggests that the financial benefits exceed their cost counterparts across most of the restoration options. It is hard to generalize in terms of BCR within specific response options or across them since they can be influenced by the specificity of countries and contexts, and there are no observable patterns across options based on the information shown in Table 9.2. Moreover, as each case constitutes an average estimate for that specific option in each country, it is not possible to observe the variation in costs within each country, which limits a meaningful comparison of how these different BCRs perform with each other or the factors explaining the variations. It is noteworthy that BCRs do not provide complete information on the feasibility of any restoration option in particular; this is because the "success" of a restoration activity or project should be interpreted contextually in line with restoration objectives, motivation, the time horizon over which benefits accrue, how are they distributed and to whom, and implementation experience that go beyond financial metrics.

9.3.3 Social and environmental benefits

The six tree and forest restoration options, beyond their disparate financial profitability, each tend to deliver differing benefits depending on specific socio-ecological conditions, as already mentioned, which merit consideration. As part of wider economic assessments of restoration, some ROAM studies have also conducted analysis—although in a less systematic way—of the likely social and environmental benefits of tree and forest restoration interventions, particularly those that can be more easily assessed due to available estimates and existing prediction models. Some of these rely on average national estimates, yet others draw on spatially explicit modeling outcomes such as erosion and nutrient exports. Table 9.2 compares the expected socioeconomic and environmental benefits associated with six tree and forest restoration options from our review of six ROAM country studies.

The previous comparison suggests that all forest restoration options have the potential to generate positive socioeconomic and environmental impacts in the right circumstances, though these impacts differ depending on the restoration option, as well as context-specific conditions (e.g., establishment location, management system, site productivity). For example, natural forest regeneration options tend to have more long-term environmental benefits than direct short-term economic gains, since their most common purpose is to restore missing or degraded environmental functions that result from forest loss or degradation. In addition, while plantations may deliver short-term economic benefits and contribute to carbon removal and erosion reduction, their impacts on biodiversity are questionable (Wang et al. 2022). Yet, benefits from plantations also depend on the choice of crops when combined with annual crops. Furthermore, agroforestry systems—including intercropping of trees, high-value trees, and annual crops—deliver important short-term benefits derived from jobs as well as long-term economic benefits from fuelwood and food supply, and environmental benefits through reversing soil erosion and enhancing sediment retention, if managed appropriately (see FAO 2017).

Different restoration options often deliver multiple benefits simultaneously (monetary and nonmonetary) at the farm or land management units and landscape scales. Using MCA has become a common approach to analyzing financial, socioeconomic,

Table 9.2 Compiled costs and benefits for six tree and forest restoration options (41 cases) in six countries

Restoration Options	Countries	COSTS			BENEFITS					
		No. of cases	Cost range per hectare in USD 2020 (1)	Benefit-Cost Ratio (1)	Socioeconomic		Provisioning Services	Regulating Services	Supporting Services	
					Income Diversification	Jobs (2)	Fuelwood (3)	CO_2 Sequestration (4)	Erosion Reduction	Sediment Retention
Natural forest regeneration	Peru, Kenya, Brazil	3	579–2,384	1.0–5.3	No	NA	NA	NA	High	Moderate to high
Assisted natural forest regeneration	El Salvador, Honduras, Peru	8	515–5,279	1.0–6.1	Yes	NA	NA	Moderate	High	Moderate to high
Agroforestry with annual and/or perennial crops	El Salvador, Honduras, Peru, Madagascar, Kenya	14	2,754–28,058	1.2–4.3	Yes	Low to high	Moderate	Low to high	Moderate to high	Moderate to high
Agrosilvopastoral systems	El Salvador, Honduras, Kenya	4	3,192–20,234	1.3–3.6	Yes	Low to moderate	Low	Low	Moderate to high	Moderate
Tree plantations with annual crops	Peru	4	1,621–3,562	1.7–4.0	Yes	Low	High	Moderate to high	High	Moderate to high
Monoculture tree plantations	Honduras, Madagascar, Kenya, Brazil	8	781–3,829	0.8–6.9	No	Low	High	Low to moderate	High	Moderate

Notes: (1) Mixes estimates using 20- and 30-year time horizons and 10% discount rate. These estimates are based on projected financial estimates based on available data gathered from secondary data sources and validated in workshops with relevant stakeholders in each country/landscape; (2) Low < 0.5 jobs/ha/yr, Moderate 0.5–7.5 jobs/ha/yr, High > 1 jobs/ha/yr; (3) Low <1 m^3/ha/yr/, Moderate 1–5 m^3/ha/yr, High > 5 m^3/ha/yr; (4) Low < 25 TCO_2/ha/yr, Moderate 25–75 TCO_2/ha/yr, High > 50 TCO_2/ha/yr. Data extracted from Cheboiwo et al. (2018) for Kenya, Lacroix et al. (2016) for Madagascar, Nello et al. (2019) for Honduras, Nunes et al. (2017) for Brazil, Quiñones et al. (2018) for Peru, and Raes et al. (2017) for El Salvador. NA = Not available because they were not assessed in the studies reviewed, which does not mean that those options do not deliver benefits across those categories as well. The benefits included in the table are all those that were included in the studies reviewed.

and environmental indicators altogether, particularly to support restoration planning while considering different social and biophysical constraints, under the understanding that it may improve the cost-effectiveness of restoration investments (Beatty et al. 2018a; Robbins and Daniels 2012). For example, Strassburg et al. (2019) argue that strategic approaches to ecosystem restoration, while considering spatial socio-environmental constraints and opportunities, can triple conservation gains and halve the costs; the increase in benefits results from compromise solutions that simultaneously deliver the maximum outcome for both biodiversity conservation and carbon sequestration, and the reduction in costs is associated with efficiencies of scale and the prioritization of areas with lower opportunity costs.

9.4 Discussion

Forest restoration and increased tree cover are acknowledged as promising options to deliver social, economic, and environmental benefits (IUCN and WRI 2014). As our analysis shows, in line with what others have determined (see Strassburg and Latawiec 2014), in many circumstances and locations, forest restoration benefits outweigh the costs involved in the establishment and management of trees; yet those benefits depend on what forest restoration options are adopted and by whom. The establishment and management of trees—sometimes in combination with crops under different forest restoration options differ across places and their diverse social, biological, and geographic conditions. There are many contextual factors that influence the private and social costs and benefits of forest restoration efforts and whether these efforts lead to successful outcomes in the long run (Mansourian et al. 2021). In addition, forest restoration is always associated with socio-environmental trade-offs, depending on the objectives and the timeframes in which benefits materialize. Thus, the costs and benefits of forest restoration activities can be better understood by taking into consideration the disparity of costs and benefits across options, their likely trade-offs under different time horizons, and the contextual factors shaping outcomes from the farm to the landscape.

9.4.1 Disparity of costs and benefits

Significant disparities across forest restoration options are found in the financial costs and benefit analysis, from a private perspective, for the actors (e.g., farms, firms, investors) who undertake forest restoration. Overall, costs seem to rise with increasing complexity of the option adopted in terms of active tree planting, combination of trees, requirements of labor and inputs, and the associated management; these costs are also distributed over time. For example, restoration options favoring natural regeneration are the less costly option, followed by ANR. Moreover, the options that require lower establishment and management costs (e.g., monoculture timber plantations or woodlots) demand lower costs than more complex systems (e.g., agroforestry including different combinations of trees, and perennial and annual crops). Some options like agroforestry require higher upfront investments for establishment; yet, as demonstrated by those cases where there is a large use of perennial crops (e.g., cacao, coffee, banana), the expectation to obtain higher benefits can be a strong incentive for adopting these systems. However, it is likely that the objectives of restoration will limit the available restoration options, leaving decision-making as less of a process of selecting from a large number of types and more toward the specific considerations within a small subset of restoration options that deliver on the selected restoration objectives.

While the total costs vary across options depending on the labor, inputs, and materials that are employed, they are also influenced by the previous land use under which restoration takes place. In this regard, evidence from a meta-analysis suggests ecological recovery from landscape restoration is more successful on land regimes with less intensive previous disturbance (Crouzeilles et al. 2016). In general, restoration activities aimed at enhancing the environmental and economic performance of preexisting systems may demand lower financial investment and establishment costs compared to those entailing more drastic shifts in land use. Nonetheless, additional restorative actions may be needed in intervened working lands where farmers depend on their land for their livelihoods. Thus, restoration options that deliver subsistence and commercial gains may be preferred despite their relatively higher costs. Furthermore, economies of scale (i.e., reductions in the average cost of restoration as area is increased) are unlikely to arise in most forest restoration options. Possible exceptions include production forestry, where the dominant practice of monoculture tree plantations may allow easier planting and harvesting cycles that meet the expected financial returns, subject to site productivity.

Forest restoration also results in wider costs and benefits for the society at large, as evidenced by economic CBA. Here too, significant disparities are found across restoration options. However, these disparities arise due to both methodological and restoration-based, context-driven reasons. Available studies differ in the range of benefits from environmental services that are considered in analysis. Additionally, results vary based on the valuation methods employed. In general, when a larger number of environmental services are monetized, analyses suggest that most restoration actions deliver net benefits to society. However, these analyses should be considered carefully, given the uncertainty in estimates. Valuation methods informed by data that are not spatially relevant and reflective of local ecological conditions may generate misleading estimates of the impact of restoration on ecosystem services. Furthermore, it is not entirely clear to what extent the varying positive or negative net gains from forest restoration are associated with the type of environmental services valued or with the differences across the biophysical and ecological contexts of restoration landscapes.

Economic externalities from forest restoration options tend to receive less attention. For example, demand for water is expected from new tree plantations, which may impose externalities on irrigated agriculture and wetlands, depending on the context (Nordblom et al. 2012). Forest restoration, mainly large-scale tree planting, could result in the displacement of agricultural lands, thus affecting food security (IPCC 2019), which may eventually lead to growing pressures on indigenous peoples and local community lands. These impacts may be further pronounced if investments in tree plantations neglect local people's rights or do not meaningfully engage stakeholders. Negative impacts must be avoided when planning restoration actions; social safeguards that support rather than erode the rights and tenure of indigenous people and local communities should not be negotiable. Socio-ecological gains from restoration will be inextricably linked to the strengthening of local organizations, capacities, and improvements in local livelihoods, among others (Erbaugh et al. 2020).

9.4.2 Trade-offs, time horizons, and uptake

A common trade-off in forest restoration is between the economic benefits and the environmental benefits, which also have a temporal dimension associated with the timeframes in which these benefits accrue. How economically attractive an option is should depend on the overall benefits for society, including the incremental monetary and nonmonetary benefits on ecosystem provisioning, regulating, supporting, and cultural services. Overall, there is a net gain from restoration for society when a larger number of environmental benefits are valued during this comparison, yet some of these benefits are not immediate and may only accrue in the long run. For example, benefits associated with carbon removals, water regulation, and soil recovery take time to accumulate; in the long term these benefits can help regulate local climate and sustain agricultural yields. Most immediate gains from restoration are associated with income generation from direct restoration activities; they can also include food supply when restoration options involve combinations with annual crops through agroforestry, silvopastoral, or intercropping systems.

Managing trade-offs may be possible at the farm and land management unit level but could also be desirable at wider geographical scales, such as whole landscapes, jurisdictions, or watersheds. It is only at this scale that more complex portfolios of forest restoration options could develop to minimize ecological and economic trade-offs, while considering implications on cost sharing and benefit distribution. Some argue that more strategic approaches to restoring ecosystems at a wider landscape level can triple conservation gains and halve the costs (Strassburg et al. 2019). Indeed, habitat fragmentation is a landscape-level process (Fahrig 2003; McGarigal and Cushman 2002) and is a key driver of biodiversity loss (Fischer and Lindenmayer 2007; Haddad et al. 2015). Forest landscapes around the world face rising competition for land use; addressing the connectivity of their habitats is understood as a critical ingredient in conserving biodiversity (Rudnick et al. 2012). Given this context, forest restoration options that hope to accrue benefits in biodiversity conservation may be better designed at a wider landscape level (Beatty et al. 2018b; Tambosi et al. 2014).

While it is hard to generalize, it is likely that the most cost-effective options for ecological restoration (including forests as well as non-forest ecosystems) are natural and ANR. This may not be the case in productive lands where there is high competition for agricultural production, where more financially profitable options are privileged. For example, tree plantations combined with annual crops may constitute an attractive option given their lower establishment and management cost; however, the benefits from timber harvesting take longer to materialize. In contrast, natural forest regeneration options hold a larger potential to restore nature and its benefits—for soils, water, carbon, biodiversity—but may not yield short-term incremental economic returns. It is expected in some future contexts that there will be some compensations for securing water flows and for forest carbon removals, thus enhancing the financial attractiveness of (passive or assisted) natural forest restoration. Agroforestry systems have a demonstrated potential to minimize the economic and environmental trade-offs given the potential of these systems to generate additional jobs and income streams for smallholder farmers, while simultaneously generating positive impacts on soils, water and nutrient retention, biodiversity, and carbon. Yet, whether or not agroforestry yields benefits—and whether they be ecological, social, or economic—depends on the specific agroforestry system and how it is managed and designed (Pashkevich et al. 2022).

A major challenge to the buy-in and uptake of restoration is associated with the time horizon over which benefits accrue to farmers, firms, and investors undertaking restoration at their discount rates. The uptake of restoration options to achieve restoration outcomes at larger scales is influenced by several factors. The cost and supply of labor naturally pose a significant constraint to the uptake of several restoration options, particularly when restoration involves both labor-intensive planting and labor-intensive management over time, which includes not only harvesting but also cultural practices, thus limiting restoration at larger scales. In addition, capital becomes a significant constraint for options that either require relatively large upfront establishment costs (e.g., agroforestry) or that are aimed at the management of larger areas (e.g., monoculture plantations). Furthermore, large-scale plantations, for example, may also require substantial investments in equipment for management practices in established plantations that are beyond the reach of communities and smallholders involved in restoration.

9.4.3 Factors shaping costs and benefits

There are significant differences in the costs and benefits across forest restoration options. These differences may result from how these costs and benefits are estimated but, more importantly, may be related to the biophysical and socio-institutional contextual factors in the places where forest restoration occurs. Some of those factors are connected to biophysical attributes, including the initial stages of degradation of soil and/or forests in a specific place, as well as other biophysical conditions, including topography, soil, and climate variability. A different set of factors have to do with market conditions that influence the price of seedlings, inputs, labor, commodities, and capital. To some extent these prices are related to access, transport, logistics, and labor availability. In addition, institutional conditions may also explain the variability in costs and benefits, as well as inflation trends, which to some degree are related to the economic and financial stability across countries. Examining these factors is therefore critical for understanding the variability of costs and benefits for restoration. These contextual factors will ultimately explain the degree of permanence that can be expected of these options in the long run.

Furthermore, the preferences of local stakeholders engaging in forest restoration, as influenced by cultural values, are often left out of CBA. Nonetheless, these are critically important. What option is chosen by the actors implementing restoration on the ground is influenced by more than profit maximization; motivations include social preferences with regard to what type of system is desirable for achieving local needs, providing income streams, supporting human health, and managing labor–leisure trade-offs. These different factors tend to influence the farmer-managed natural restoration practices adopted (Chomba et al. 2020). Moreover, when restoration options build on social preferences and cultural values at a local level, this can lead to the reinforcement of those values and the associated institutions. However, restoration may have the opposite effect when forest restoration options are determined by top-down perspectives and interests, or goals beyond the interest of local actors.

The complexity of dynamics between demographic aspects of local stakeholders and the changing forest landscape is also worthy of attention (Vincent et al. 2021). Household size and the movement of forest communities within the forest landscape introduce heterogeneity in their impact on forest landscapes (Carr 2009). Restoration can alter livelihoods via changes in the use of forest resources and agricultural practices. Livelihoods can change over time and have novel impacts on the forest landscape. Literature on migration and forest landscapes shows that networks bridging households near forests to their out-migrated members can introduce new pressures on forest resources; thus there is motivation to integrate these interactions in studying the impacts of restoration (Vincent et al. 2021). Overall, preferences of local stakeholders cannot be viewed as static as the other many factors impacting on restoration are also changing. Social preferences should be incorporated and involved in the codesign and implementation of restoration programs throughout their projected time horizon. The analysis of costs and benefits of restoration needs to consider more seriously the dynamic cultural values and practices (especially those associated with forest use) of communities, and their wider implications on (re)shaping the interactions between people and nature.

9.5 Conclusion

This chapter reviews select studies undertaking economic and financial CBA for tree and forest restoration. The economic CBA findings rely on meta-analysis at the global or regional levels, and the financial CBA findings rely on a review of studies at the national level of forest restoration options. Overall, economic CBAs agree on the net gains for society from restoration, given the large number of benefits that are considered, while costs for society—particularly transaction and external costs—may be underestimated. When looking at financial benefits, almost all forest restoration options are financially attractive, but there are differences across options. Natural regeneration options (passive and assisted) are the least costly options, followed by tree plantations in monocultures and woodlots. In addition, agroforestry systems—including combinations of trees with tree-crops and annual crops—are most costly and require higher upfront investments, yet may yield higher economic and ecological benefits, depending on the agroforestry system.

The variation in costs within similar forest restoration options suggests that several contextual factors, including biophysical and political economic conditions, may explain some of this variability. While not analyzed here, factors other than financial attractiveness may explain the adoption of specific forest restoration options, and those decisions are likely influenced by cultural, aesthetic, and spiritual values as well. There are several trade-offs associated with forest restoration, mainly regarding restoration's financial, socioeconomic, and environmental benefits, which also have a temporal dimension since benefits accrue over different timeframes, and most benefits may only accrue in the long run. Likely, the options, or combination of options, that better deliver benefits for the society at large are going to be those that will endure. Private actors may tend to prioritize short-term benefits over long-term or societal benefits, leading to different perspectives on the value of forest restoration for economic versus conservation goals. Reconciling these goals will be central for a successful restoration at scale.

Appendix 9A: Description of six main types of forest restoration options identified

Restoration Options	Description	Restoration Activities	
		Establishment	Management
Natural forest regeneration	Recovery of native tree species that were disturbed due to natural or human related causes	Fencing to prevent livestock grazing	Fire management
Assisted natural forest regeneration	Forest regeneration assisted through human interventions including enrichment planting to increase the canopy cover or species richness by planting seedlings or sparing seeds	Planting enrichment trees, fencing to control livestock grazing	Weed control, fire management
Agroforestry with annual and/or perennial crops	Agroforestry combines trees with perennial crops (e.g., coffee, cacao) and annual crops often through inter-cropping in different vegetation successions	Land cleaning, distribution of the vegetal material, alignment of trees, and planting of trees and perennials (tree crops), sowing of annual crops	Fertilizing, weeding, phytosanitary control, pruning, and thinning
Agrosilvopastoral systems	Combination of trees and pasture for raising livestock, often by planting trees in pasture lands and improving the quality of pastures	Ploughing, purchase tree seedlings, purchase grass seeds, transport seeds/seedlings, planting of grass, trees fencing	Weeding, harvesting grass seed, livestock management, pruning, thinning
Tree plantations with annual crops	Tree planting (monoculture plantations of exotic and native species) combined with the establishment of annual crops (e.g., maize, rice)	Land cleaning (e.g., slash and burn practices), simultaneous planting of annual crops	Fertilizing, weeding, pruning, thinning
Monoculture tree plantations	Planted forests through medium- and large-scale monocrop systems often to produce commercial wood, pulp, and paper and biomass for energy	Planting of seedlings or through direct seeding	Thinning and burning

Notes: Elaborated by authors based on analysis from Andradi-Brown et al. (2013), Cheboiwo et al. (2018), FAO (2020), Quinones et al. (2018), SEAMA-ES (2017), Wojtkowski (1998).

References

Albert, G., S.C. Gallegos, K.A. Greig, et al. 2021. "The Conservation Value of Forests and Tree Plantations for Beetle (Coleoptera) Communities: A Global Meta-Analysis." *Forest Ecology and Management* 491: p. 119201. DOI:10.1016/j.foreco.2021.119201.

Andradi-Brown, D.A., C. Howe, G.M. Mace, and A.T. Knight. 2013. "Do Mangrove Forest Restoration or Rehabilitation Activities Return Biodiversity to Pre-Impact Levels?" *Environmental Evidence* 2 (1): p. 20. DOI:10.1186/2047-2382-2-20.

Arroyo-Rodríguez, V., K.F. Rito, M. Farfán, et al. 2023. "Landscape-Scale Forest Cover Drives the Predictability of Forest Regeneration across the Neotropics." *Proceedings of the Royal Society B: Biological Sciences* 290 (1990): p. 20222203. DOI:10.1098/rspb.2022.2203.

Barbier, E.B. 2007. "Valuation of Ecosystem Services Supporting Aquatic and other Landbased Food Systems." In *Comparative Assessment of the Environmental Costs of Aquaculture and Other Food Production Sectors: Methods for Meaningful Comparisons*, edited by D.M. Bartley, C. Brugère, D. Soto, P. Gerber, and B. Harvey, pp. 71–86. Vancouver: FAO/WFT.

Barbier, E.B., and J.P. Hochard. 2018. "Land Degradation and Poverty." *Nature Sustainability* 1 (11): pp. 623–631. DOI:10.1038/s41893-018-0155-4.

Beatty, C., L. Raes, A.L. Vogl, et al. 2018a. Landscapes, At Your Service: Applications of the Restoration Opportunities Optimization Tool (ROOT). Gland: IUCN. DOI:10.2305/IUCN.CH.2018.17.en.

Beatty, C., N.A. Cox, and M. Kuzee. 2018b. Biodiversity Guidelines for Forest Landscape Restoration Opportunities Assessments. Gland: IUCN. DOI:10.2305/ IUCN.CH.2018.10.en.

Bernal, B., L.T. Murray, and T.R.H. Pearson. 2018. "Global Carbon Dioxide Removal Rates from Forest Landscape Restoration Activities." *Carbon Balance and Management* 13 (1): p. 22. DOI:10.1186/s13021-018-0110-8.

Birch, J.C., A.C. Newton, C.A. Aquino, et al. 2010. "Cost-Effectiveness of Dryland Forest Restoration Evaluated by Spatial Analysis of Ecosystem Services." *Proceedings of the National Academy of Sciences* 107 (50): pp. 21925–21930. DOI:10.1073/pnas.1003369107.

Bodin, B., V. Garavaglia, N. Pingault, et al. 2021. "A Standard Framework for Assessing the Costs and Benefits of Restoration: Introducing The Economics of Ecosystem Restoration." *Restoration Ecology* 30 (3): e13515. DOI:10.1111/rec.13515.

Brancalion, P.H.S., P. Meli, J.R.C. Tymus, et al. 2019. "What Makes Ecosystem Restoration Expensive? A Systematic Cost Assessment of Projects in Brazil." *Biological Conservation* 240: p. 108274. DOI:10.1016/j.biocon.2019.108274.

Campbell, H.F., and R. Brown. 2012. *Benefit-Cost Analysis: Financial and Economic Appraisal using Spreadsheets.* Cambridge: Cambridge University Press.

Carr, D. 2009. "Population and Deforestation: Why Rural Migration Matters." *Progress in Human Geography* 33 (3): pp. 355–378. DOI:10.1177/0309132508096031.

CBD. 2022. *Final text of Kunming-Montreal Global Biodiversity* (Framework available in all languages). Convention on Biological Diversity. Montreal, Canada, 7–19 December 2022. https://www.cbd.int/conferences/2021-2022/cop-15/documents.

Chazdon, R.L., and M. Uriarte. 2016. "Natural Regeneration in the Context of Large-Scale Forest and Landscape Restoration in the Tropics." *Biotropica* 48 (6): pp. 709–715. DOI:10.1111/btp.12409.

Cheboiwo, J., D. Langat, M. Muga, and J. Kiprop. 2018. *Economic Analysis of Forest Landscape Restoration Options in Kenya.* Nairobi, Kenya: Ministry of Environment and Forestry.

Chomba, S., F. Sinclair, P. Savadogo, M. Bourne, and M. Lohbeck. 2020. "Opportunities and Constraints for Using Farmer Managed Natural Regeneration for Land Restoration in Sub-Saharan Africa." *Frontiers in Forests and Global Change* 3 (122): p. 15 DOI:10.3389/ffgc.2020.571679.

Cook, C.N., A.S. Pullin, W.J. Sutherland, G.B. Stewart, and L.R. Carrasco. 2017. "Considering Cost Alongside the Effectiveness of Management in Evidence-Based Conservation: A Systematic Reporting Protocol." *Biological Conservation* 209: pp. 508–516. DOI:10.1016/j.biocon.2017.03.022.

Crouzeilles, R., H.L. Beyer, L.M., Monteiro, et al. 2020. "Achieving Cost-Effective Landscape-Scale Forest Restoration through Targeted Natural Regeneration." *Conservation Letters* 13 (3): p. e12709. DOI:10.1111/conl.12709.

Crouzeilles, R., M. Curran, M.S. Ferreira, D.B. Lindenmayer, C.E.V. Grelle, and J.M. Rey Benayas. 2016. "A Global Meta-Analysis on the Ecological Drivers of Forest Restoration Success." *Nature Communications* 7 (1): p. 11666. DOI:10.1038/ncomms11666.

Current, D., E. Lutz, and S. Scherr. 1995. *Costs, Benefits, and Farmer Adoption of Agroforestry: Project Experience in Central America and the Caribbean.* Washington, DC: The World Bank.

De Groot, R., J. Blignaut, S. van eer Ploeg, J. Aronson, T. Elmqvist, and J. Farley. 2013. "Benefits of Investing in Ecosystem Restoration." *Conservation Biology* 27 (6): pp. 1286–1293. DOI:10.1111/cobi.12158.

De Groot, R., L. Brander, L., S. van der Ploeg, et al. 2012. "Global Estimates of the Value of Ecosystems and Their Services in Monetary Units." *Ecosystem Services* 1 (1): pp. 50–61. DOI:10.1016/j.ecoser.2012.07.005.

Erbaugh, J.T., N. Pradhan, J. Adams, et al. 2020. "Global Forest Restoration and the Importance of Prioritizing Local Communities." *Nature Ecology & Evolution* 4 (11): pp. 1472–1476. DOI:10.1038/s41559-020-01282-2.

Fahrig, L. 2003. "Effects of Habitat Fragmentation on Biodiversity." *Annual Review of Ecology, Evolution, and Systematics* 34 (1): pp. 487–515. DOI:10.1146/annurev.ecolsys.34.011802.132419.

FAO. 2017. *Agroforestry for Landscape Restoration: Exploring the Potential of Agroforestry to Enhance the Sustainability and Resilience of Degraded Landscapes.* Rome: FAO. https://www.fao.org/3/i7374e/i7374e.pdf.

FAO. 2020. *Sustainable Forest Management (SFM) Toolbox: Silviculture in Natural Forests.* Rome: FAO. http://www.fao.org/sustainable-forest-management/toolbox/modules/silviculture-in-natural-forests/in-more-depth/en/.

Fischer, J., and D.B. Lindenmayer. 2007. "Landscape Modification and Habitat Fragmentation: A Synthesis." *Global Ecology and Biogeography* 16 (3): pp. 265–280. DOI:10.1111/j.1466-8238.2007.00287.x.

Gilby, B.L., A.D. Olds, R.M. Connolly, C.J. Henderson, and T.A. Schlacher. 2018. "Spatial Restoration Ecology: Placing Restoration in a Landscape Context." *Bioscience* 68 (12): pp. 1007–1019. DOI:10.1093/biosci/biy126.

Griscom, B.W., J. Adams, P.W. Ellis, et al. 2017. "Natural Climate Solutions." *Proceedings of the National Academy of Sciences* 114 (44): pp. 11645–11650. DOI:10.1073/pnas.1710465114.

Haddad, N.M., L.A. Brudvig, J. Clobert, et al. 2015. "Habitat Fragmentation and Its Lasting Impact on Earth's Ecosystems." *Science Advances* 1 (2): p. e1500052. DOI:10.1126/sciadv.1500052.

Hebling, T. 2010. "What Are Externalities? What Happens When Prices Do Not Fully Capture Costs." *Finance and Development* (December 2010) 47 (4): pp. 48–49.

Ickowitz, A., S. McMullin, T. Rosenstock, et al. 2022. "Transforming Food Systems with Trees and Forests." *The Lancet Planetary Health* 6 (7): p. e632–639. DOI:10.1016/S2542-5196(22)00091-2.

IPBES. 2018. *The IPBES Assessment Report on Land Degradation and Restoration*, edited by L. Montanarella, R. Scholes, and A. Brainich. Bonn: Secretariat of the Intergovernmental Science-Policy Platform on Biodiversity and Ecosystem Services. DOI:10.5281/zenodo.3237392.

IPBES. 2019. *Global Assessment Report on Biodiversity and Ecosystem Services of the Intergovernmental Science-Policy Platform on Biodiversity and Ecosystem Services*, edited by E.S. Brondizio, J. Settele, S. Díaz, and H.T. Ngo. Bonn: IPBES secretariat. DOI:10.5281/zenodo.3831673.

IPCC. 2007. *Climate Change 2007. Fourth Assessment Report.* Geneva, Switzerland: Intergovernmental Panel on Climate Change (IPCC). https://www.ipcc.ch/assessment-report/ar4/.

IPCC. 2019. *Climate Change and Land: IPCC Special Report on Climate Change, Desertification, Land Degradation, Sustainable Land Management, Food Security, and Greenhouse gas fluxes in Terrestrial Ecosystems.* Summary for Policymakers. Geneva, Switzerland: Intergovernmental Pable on Climat Change (IPCC). https://www.ipcc.ch/site/assets/uploads/2019/08/Edited-SPM_Approved_Microsite_FINAL.pdf.

IUCN and WRI. 2014. *A guide to the Restoration Opportunities Assessment Methodology (ROAM): Assessing Forest Landscape Restoration Opportunities at the National or Sub-National Level.* Working Paper. Gland: IUCN. https://portals.iucn.org/library/sites/library/files/documents/2014-030.pdf.

Jones, H.P., P.C. Jones, E.B. Barbier, et al. 2018. "Restoration and Repair of Earth's Damaged Ecosystems." *Proceedings of the Royal Society B: Biological Sciences* 285 (1873): 20172577. DOI:10.1098/rspb.2017.2577.

Lacroix, E., S. Carodenuto, F. Richter, et al. 2016. *Restauration des paysages forestiers Evaluation des potentialités dans le contexte des engagements de Bonn 2.0 et de la Déclara- tion de New York sur les forêts. Méthodologie et résultats pour Madagascar.* Freiburg, Germany: UNIQUE Forestry and Land Use. https://afr100.org/sites/default/files/06-MDG_RPF_MEOR_Madagascar_Final.pdf.

Mansourian, S., N. Berrahmouni, J. Blaser, et al. 2021. "Reflecting on Twenty Years of Forest Landscape Restoration." *Restoration Ecology* 29 (7): p. e13441. DOI:10.1111/rec.13441.

McGarigal, K., and S.A. Cushman. 2002. "Comparative Evaluation of Experimental Approaches to the Study of Habitat Fragmentation Effects." *Ecological Applications* 12 (2): pp. 335–345. DOI:10.1890/1051-0761(2002)012[0335:CEOEAT]2.0.CO.

Millennium Ecosystem Assessment (MEA). 2005. *Ecosystems and Human Well-Being: Synthesis.* Washington, DC: Island Press.

Montagnini, F. 2017. *Integrating Landscapes: Agroforestry for Biodiversity Conservation and Food Sovereignty.* New York: Springer International Publishing AG.

Nello, T., L. Raes, A. Wong, O. Chacon, and A. Sachun. 2019. *Análisis económico de acciones para la restauración de paisajes productivos en Honduras.* San José, Costa Rica: IUCN-ORMACC. https://portals.iucn.org/library/sites/library/files/documents/2019-008-Es.pdf.

Newton, A.C., K. Hodder, E. Cantarello, et al. 2012. "Cost–Benefit Analysis of Ecological Networks Assessed through Spatial Analysis of Ecosystem Services." *Journal of Applied Ecology* 49 (3): pp. 571–580. DOI:10.1111/j.1365-2664.2012.02140.x.

Nordblom, T.L., J.D. Finlayson, and I.H. Hume. 2012. "Upstream Demand for Water Use by New Tree Plantations Imposes Externalities on Downstream Irrigated Agriculture and Wetlands." *Australian Journal of Agricultural and Resource Economics* 56 (4): pp. 455–474. DOI:10.1111/j.1467-8489.2012.00593.x.

Nunes, S., G. Gatti, A. Diederichsen, and D.S.A. Pinto. 2017. *Oportunidades para Restauração Florestal no Estado do Pará.* Belem, Para: IMAZON. https://imazon.org.br/oportunidades-para-restauracao-florestal-no-estado-do-para/.

OECD. 2018. *Cost-Benefit Analysis and the Environment: Further Developments and Policy Use* Paris: OECD Publishing. DOI:10.1787/9789264085169-en.

Pashkevich, M.D., F. d'Albertas, A.A.K. Aryawan, et al. 2022. "Nine Actions to Successfully Restore Tropical Agroecosystems." *Trends in Ecology & Evolution* 37 (11): pp. 963–975. DOI:10.1016/j.tree.2022.07.007.

Quiñones, L., M. Reyes, J. Chero, et al. 2018. *Opciones de Restauración Del Paisaje Forestal En Padre Abad: Implementación de ROAM a Nivel Sub-Nacional Como Insumo a La Formulación de Una Estrategia de Restauración.* Lima: ICRAF Oficina Regional para América Latina.

Raes, L., D. Mittempergher, M. Piaggio, and J. Siikamäki. 2021. *Nature-Based Recovery Can Create Jobs, Deliver Growth and Provide Value for Nature.* Nature-based Recovery Initiative.Technical Paper No. 3. Gland: IUCN Nature-based Recovery Initiative. https://www.iucn.org/sites/default/files/2022-06/iucn-nbr-tp-3-compressed.pdf

Raes, L., T. Nello, M. Nájera, O. Chacón, K.M. Prado, and A. Sanchún. 2017. *Análisis económico de acciones para la restauración de paisajes productivos en El Salvador.* Gland: IUCN. https://portals.iucn.org/library/node/47177.

Reyes-García, V., M. Guèze, A.C. Luz, et al. 2013. "Evidence of Traditional Knowledge Loss among a Contemporary Indigenous Society." *Evolution and Human Behavior* 34 (4): pp. 249–257. DOI:10.1016/j.evolhumbehav.2013.03.002.

Robbins, A.S.T., and J.M. Daniels. 2012. "Restoration and Economics: A Union Waiting to Happen?" *Restoration Ecology* 20 (1): pp. 10–17. DOI:10.1111/j.1526-100X.2011.00838.x.

Rudnick, D., S.J. Ryan, P. Beier, et al. 2012. "The Role of Landscape Connectivity in Planning and Implementing Conservation and Restoration Priorities." *Issues in Ecology* 16: pp. 1–20.

Schultz, M., H. Jamieson, E. Piontek, J.J. Coon, and S.J. Halsey. 2022. "The Importance of Cultural Values in Ecological Restorations: A Systematic Review." *Society & Natural Resources* 35 (9): pp. 1021–1039. DOI:10.1080/08941920.2022.2073625.

SEAMA-ES. 2017. *Avaliação das oportunidades da restauração de paisagens e florestas para o Estado do Espírito Santo.* Espirito Santo, Brasil: WRI, TNC, IIS. https://seama.es.gov.br/Media/Seama/Documentos/ROAM-ES%202018.pdf.

Strassburg, B., and A.E. Latawiec. 2014. The Economics of Restoration: Costs, Benefits, Scale and Spatial Aspects. Rio de Janeiro, Brazil: International Institute for

Sustainability (IIS). https://www.cbd.int/doc/meetings/ecr/cbwecr-sa-01/other/cbwecr-sa-01-iis-en.pdf.

Strassburg, B.B.N., H.L. Beyer, R. Crouzeilles, et al. 2019. "Strategic Approaches to Restoring Ecosystems Can Triple Conservation Gains and Halve Costs." *Nature Ecology & Evolution* 3 (1): pp. 62–70. DOI:10.1038/s41559-018-0743-8.

Sunderland, T.C.H., and D. Rowland. 2019. "Forests, Land Use, and Challenges to Climate Stability and Food Security." In *Sustainable Food and Agriculture*, edited by C. Campanhola, and S. Pandey, pp. 95–116. Rome: FAO and Academic Press.

Sutton, P.C., S.J. Anderson, R. Costanza, and I. Kubiszewski. 2016. "The Ecological Economics of Land Degradation: Impacts on Ecosystem Service Values." *Ecological Economics* 129: pp. 182–192. DOI:10.1016/j.ecolecon.2016.06.016.

Tambosi, L.R., A.C. Martensen, M.C., Ribeiro, and J.P. Metzger. 2014. "A Framework to Optimize Biodiversity Restoration Efforts Based on Habitat Amount and Landscape Connectivity." *Restoration Ecology* 22 (2): pp. 169–177. DOI:10.1111/rec.12049.

TEEB. 2010. *The Economics of Ecosystems and Biodiversity: Ecological and Economic Foundations.* London: Earthscan.

UNEP. 2021. *Ecosystem restoration for people, nature and climate.* Nairobi: UNEP. https://wedocs.unep.org/bitstream/handle/20.500.11822/36251/ERPNC.pdf.

Vanclay, J.K. 2009. "Managing Water Use from Forest Plantations." *Forest Ecology and Management* 257 (2): pp. 385–389. DOI:10.1016/j.foreco.2008.09.00.

Van der Sande, M.T., J.S. Powers, T.W. Kuyper, et al. 2023. "Soil Resistance and Recovery during Neotropical Forest Succession." *Philosophical Transactions of the Royal Society B: Biological Sciences* 378 (1867): p. 20210074. DOI:10.1098/rstb.2021.0074.

Verdone, M. 2015. *A Cost-Benefit Framework for Analyzing Forest Landscape Restoration Decisions.* Gland: IUCN. https://portals.iucn.org/library/sites/library/files/documents/2015-018.pdf.

Vieira, D.L.M., Holl, K.D., and Peneireiro, F.M. 2009. "Agro-Successional Restoration as a Strategy to Facilitate Tropical Forest Recovery." *Restoration Ecology* 17 (4): pp. 451–459. DOI:10.1111/j.1526-100X.2009.00570.x.

Vincent, J.R., S.R. Curran, and M.S. Ashton. 2021. "Forest Restoration in Low- and Middle-Income Countries." *Annual Review of Environment and Resources* 46 (1): pp. 289–317. DOI:10.1146/annurev-environ-012220-020159.

Wainaina, P., P.A. Minang, E. Gituku, and L. Duguma. 2020. "Cost-Benefit Analysis of Landscape Restoration: A Stocktake." *Land* 9 (11): p. 465. DOI:10.3390/land9110465.

Wang, C., W. Zhang, X. Li, and J. Wu. 2022. "A Global Meta-Analysis of the Impacts of Tree Plantations on Biodiversity." *Global Ecology and Biogeography* 31 (3): pp. 576–587. DOI:10.1111/geb.13440.

Wojtkowski, P.A. 1998. *The Theory and Practice of Agroforestry Design. A Comprehensive Study of the Theories, Concepts and Conventions that Underlie the Successful Use of Agroforestry.* Enfield: Science Publishers.

Zhou, G., X. Zhou, D.J. Eldridge, et al. 2022. "Temperature and Rainfall Patterns Constrain the Multidimensional Rewilding of Global Forests." *Advanced Science* 9 (18): p. 2201144. DOI:10.1002/advs.202201144.

10

Financial, ecological, political, and social feasibility of forest restoration targets

Rodney J. Keenan, Bas Louman, David Brand, Hemant Ojha, and Luo Xi

10.1 Introduction

Forests are under increasing pressure from a growing global population, increasing consumption and rapid climate change. The risk of global pandemics and their associated economic impacts, population shifts, increased poverty and reduced tourism are driving further change in forest-people interactions. Despite these challenges, the opportunities to restore forests in the coming decades are increasing (Seymour and Busch 2016) as more people recognize the many benefits of healthy, resilient forests, including for meeting climate policy objectives (Griscom et al. 2017).

Forest restoration aims to increase forest area and improve the ecological conditions and functions of forests. Large-scale international commitments have been made to restore degraded forest land and to increase tree cover in agricultural lands (see Chapter 1; Stanturf and Mansourian 2020). For example, the Bonn Challenge and the New York Declaration on Forests (NYDF) have set goals to restore 350 Mha by 2030; in January 2020, the World Economic Forum (WEF) launched 1 Trillion Trees, a global initiative to grow, restore, and conserve forests around the world to reduce biodiversity loss and help fight climate change. The United Nations (UN) recently declared the 2020s the "Decade on Ecosystem Restoration." At COP26 of the UN Framework Convention on Climate Change (UNFCCC) in Glasgow in 2021, over 145 countries committed to working collectively to halt and reverse forest loss and land degradation by 2030 while delivering sustainable development and promoting an inclusive rural transformation (UN Climate Summit 2021). While these commitments to forest restoration indicate growing political will, their feasibility is uncertain. Feasibility is an assessment of whether, and how, a proposed project can be successfully implemented (McLeod 2021). We consider feasibility in terms of extent of suitable land, technical knowledge and tools, finance, and social factors. The importance of each of these factors differs according to the local context. Feasibility assessment is therefore complicated by these local differences in goals, costs, and potential benefits of restoration.

10.2 Context and related studies

Forests are dynamic at the global and local scales. For millennia, humans have been converting forest to other land uses to meet our needs, with forest loss and recovery that parallels expansion and contraction of populations, changes in settlements and

Rodney J. Keenan, Bas Louman, David Brand, Hemant Ojha, and Luo Xi, *Financial, ecological, political, and social feasibility of forest restoration targets* In: *Restoring Forests and Trees for Sustainable Development.* Edited by: Pia Katila, Carol J. Pierce Colfer, Wil de Jong, Glenn Galloway, Pablo Pacheco, and Georg Winkel, Oxford University Press. © Oxford University Press 2024.
DOI: 10.1093/9780197683958.003.0010

food production systems, and movement in response to disease or climate change (Williams 2008). At smaller scales, for centuries people have cleared forests and allowed recovery in cycles of shifting (or swidden) cultivation (Dressler et al. 2016). Gains made by restoring forests in some places will be diminished unless we slow or stop permanent forest loss and degradation in others (Gitz et al. 2020). However, loss of humid primary tropical forests has accelerated since 2014 (NYDF Assessment Partners 2019).

Restoring forest means different things to different people. Many past restoration practices have involved establishing large areas of single, or a few, tree species. Many now feel restoration should focus on reestablishing more diverse "natural" forests. Concerns have therefore been expressed about the type of restoration proposed in meeting global targets. One analysis indicated nearly half of the Bonn Challenge commitments involve tree plantations, with only one-third (34%) aiming to restore forests through natural regeneration. Trees integrated in agricultural and pastoral areas account for the remainder (NYDF Assessment Partners 2019). In assessing feasibility of international targets, it is important to acknowledge different forms of forest restoration. This includes forest landscape restoration—restoring trees in a landscape mosaic with agricultural and other land uses with the aim of improving landscape function and providing for human livelihoods.

Others have explored the feasibility of large-scale forest restoration. For example, Brancalion et al. (2019) investigated restoration opportunities in 66 tropical countries, 27 of which committed to restoring 117 Mha of forest under the Bonn Challenge. They focused on restoration to improve habitat for vulnerable species, climate change mitigation potential, and contributions to adaptation, including reduced water security risks, and using three indicators of restoration feasibility: (i) opportunity and implementation costs, (ii) potential restoration success, and (iii) probability of persistence of restored forests (based on the relative rate of recent tree cover loss in the region). Opportunity costs of alternative land uses focused on returns from mechanized agriculture (but not for subsistence users). Benefits and feasibility were combined to generate a "restoration opportunity score." These two elements tend to be inversely correlated, that is, landscapes with higher potential benefits were found to have lower feasibility. Countries with large areas of "high restoration opportunity" include Brazil (61 Mha), Indonesia (23 Mha), India (15 Mha), Madagascar (14.7 Mha), and Colombia (12 Mha) (Brancalion et al. 2019). This analysis excluded higher altitude areas (above 1000 msl) and wetlands, due to lack of data, and human and social factors to achieve successful restoration (Elias et al. 2022) were also not addressed. Our approach is not as detailed in spatial analysis but involves a broader geographic view and considers these human and social factors as well as available finance, not just costs.

We begin the chapter with an overview of forest restoration and describe current forest extent, trends in gains and losses, and estimates of suitable land. In Section 10.5 we describe the technical requirements, in Section 10.6 finance and in Section 10.7 the social and policy aspects of achieving large-scale restoration. We conclude with a synthesis and discuss the utility of setting "top-down" area based global targets for forest restoration.

10.3 What is forest restoration?

As indicated above, forest restoration means different things to different people. To some, it is *ecological restoration*, that is, reestablishing a self-organizing ecosystem on a trajectory to "full recovery" for conservation and habitat values. This is not necessarily recovery to a historic state or condition since climate change or other factors may alter the ecosystem trajectory. Others have focused restoration of trees on deforested landscapes, establishing single tree species for commercial use to generate local economic benefits, such as employment or regional industry development, but also to protect soils and water resources. While these often intensively-managed, monoculture forests improve landscape ecological functions compared with say, intensive agriculture, they generally do not deliver the full range of benefits or functions of more diverse natural forests (Hua et al. 2022). Other restoration approaches encompass a broader range of goals and involve planting a broader suite of species or use different techniques, including facilitating natural regeneration (Stanturf and Mansourian 2020). Goals include improved livelihoods, food and fuel production, water security, and climate change adaptation and mitigation benefits.

Forest restoration practices also include restoring Indigenous forest management to improve forest health and condition, such as small patch, cool burning used by Indigenous people in North America and Australia (Stanturf and Mansourian 2020). In Australia, climate finance to use these practices has reconnected Indigenous communities with their Country, improved human health, and provided many other social and economic benefits (Long and Lake 2018; Russell-Smith et al. 2017). In other models, restoration involves removing unwanted (usually invasive) vegetation to restore vegetation condition and ecological functions (Kimiti et al. 2017).

The term *forest landscape restoration* (FLR) has also become widely used (Maginnis et al. 2012). It describes "a planned process that aims to regain ecological integrity and enhance human wellbeing in deforested or degraded landscapes" (Stanturf and Mansourian 2020). FLR involves multiple land uses with forests restored by combinations of tree planting, managed or natural regeneration, agroforestry and trees on farms, improved management of riparian zones and wildlife corridors. In setting global and national forest restoration targets, policymakers need to be clear about the types of restoration approaches and land uses incorporated in these targets and the scale of restoration involved ("forest" versus "landscape").

10.4 Global forests and land use

Assessing whether sufficient land area is suitable and potentially available for achieving global forest restoration targets requires understanding the current state of global forests. Estimates of global forest area, and change in area, are confounded by varying definitions of forest, including thresholds and land use designations, and different technologies used to assess forest cover (Keenan et al. 2015). The Food and Agriculture Organization of the United Nations (FAO) uses a relatively low threshold of 10% tree canopy cover to assess forest extent. It also includes a requirement that the land is under "forest use" (thus excluding agriculture or urban land use), although this definition is not uniformly applied across countries in their reporting (Keenan et al. 2015).

Using this definition, the current extent of forest is 4.06 billion ha, or 31% of the total global land area (FAO 2020). This forest estate is categorized in the Global Forest Resources Assessment by biome, regeneration type, degree of disturbance, ownership, coverage by a management plan, and designation. Over half of global forests (56%) are in tropical or subtropical biomes, while 27% are in the boreal biome. In terms of regeneration type, 7% of global forests are planted (with 45% of this area being intensively managed plantations), 93% are naturally regenerated, and 27% are considered primary (undisturbed) forests. Nearly three-quarters of global forests are publicly owned, and the remainder are owned by corporate, small private, or community entities. Almost half of the global forest area (including public, private, and community forests) are reported to be under a management plan, and 18% (0.73 billion ha) are in formally designated protected areas (FAO 2020).

Forest cover is also dynamic. Humans have cleared forests throughout their history, and forest cover is reduced through fire, geological activity, and drought. Forests naturally regenerate and recover from these disturbances or are restored through human intervention. Historically much clearing occurred in temperate and boreal regions. More recently, clearing has mostly been in the tropics, with rates of human-induced forest loss about 16 Mha per year in the 1990s but have progressively declined to 10 Mha per year over the last decade (FAO 2020). Recent rates of forest loss are highest in countries with a high proportion of forest cover but lower per capita national income, particularly in Africa (Keenan et al. 2015) due to growing human populations needing land for their subsistence. Regional initiatives in Africa are resulting in increased tree cover in some places. Government policies also drive forest loss. For example, in the Brazilian Amazon, deforestation declined in the 2000s due to increased government regulation, but recently increased due to relaxing of these regulations (Carvalho et al. 2019). Drivers also change in character, with more forest loss due to activities of small-scale actors (Kalamandeen et al. 2018). Significant areas of forest in the Amazon are being degraded by unregulated harvesting and understory impacts from fires (Matricardi et al. 2020).

While forest loss is occurring in some countries, forest cover is increasing in others, mostly in temperate regions. Drivers of increased forest cover are agricultural land abandonment followed by natural regeneration (mostly in Europe and North America) and active tree planting in Asian countries, such as China and Vietnam. Afforestation in the latter two countries was accelerated by government policies to address environmental problems (see Box 10.1). The global rate of forest expansion increased to 10 Mha per year in the 2000s but has slowed to 5 Mha per year in the 2010s (FAO 2020), due to a halving of rates of tree planting in East and South East Asia and declines in natural forest expansion in Europe to about one-third of the rate in earlier periods.

The shift from a period of forest loss to a period of increasing forest cover is described as a "forest transition" (Mather 1992). Some countries (e.g., South Korea, China, Costa Rica) experienced forest transitions in the 1970s and 1980s. These were connected to socioeconomic and demographic transitions, with rural depopulation occurring as younger people left for better economic opportunities in the cities. Other countries, such as Vietnam, started such transitions in the 1990s (Keenan et al. 2015; Meyfroidt and Lambin 2011).

10.5 Global forest restoration targets and suitable land areas

Global initiatives and targets for forest restoration have been documented by Stanturf and Mansourian (2020). The most prominent initiative, the Bonn Challenge, launched by the Government of Germany and the International Union for the Conservation of Nature (IUCN) in 2011, set a commitment of 350 Mha of deforested and degraded land under restoration by 2030. This was an aspirational commitment, with signatories subsequently confirming target areas in "pledges" over the last decade. As of January 2023, 74 pledges had been registered in 61 countries to restore 210.1 Mha by 2030. These are mostly in Africa (31 pledges) and the Americas (29 pledges). Interestingly, given Germany was the originating government partner, only one pledge has been registered from Europe (for 30 Mha under restoration by 2030). In another example, a recent report by the Intergovernmental Panel on Climate Change proposed increasing global forest area by 1 billion ha by 2050 (IPCC 2018). Other targets have taken different forms. For example, in January 2020, the WEF launched a campaign to support the UN Decade on Ecosystem Restoration 2021–2030 through a global movement to conserve, restore, and grow *one trillion trees* by 2030. A multi-stakeholder group (1t.org) formed to facilitate leadership of companies, nonprofits, governments, ecopreneurs, academia, communities, grassroots organizations, and individuals to bring about concrete change on local, national, and global levels.

Forest restoration can happen by restoring forests and tree cover on formerly forested agricultural land, by integrating trees with agricultural land use (FLR) or by improving the condition of degraded forests. Globally, about 5.1 billion ha of land are currently used for agriculture (Table 10.1); this represents 50% of the habitable land area (i.e., excluding glaciers, desert, exposed rock, beaches, and dunes). Of the agricultural land, 1.71 billion ha may have been previously forested and hence suitable for reforestation (WRI 2014).

Land suitability for meeting these targets depends on existing land cover and use, extent of land degradation and the opportunity costs of replacing current land uses with trees and forests. Agricultural land is considered degraded if soil fertility and productivity has been lowered due to erosion, repeated cultivation or use of agricultural chemicals or fertilizers, or hydrological regimes have been altered by drainage, or fire regimes altered through fire suppression or grazing, impacting on vegetation composition or habitat values. High levels of soil degradation, for example through mining, or repeated fire disturbance may mean land is no longer suitable for the original tree species or other biota.

The WEF goal was supported by a study by Bastin et al. (2019), who suggested a potential to restore tree cover on 0.9 billion ha of "degraded" land; but this study has been heavily criticized for including environments with obvious environmental or socioeconomic constraints, for example towns and cities (Luedeling et al. 2019), and for assuming natural savannas, grasslands, and shrublands might be suitable for forest restoration (45% of identified areas have ecological conditions that favor grasslands) (Bond et al. 2019; Veldman et al. 2019). Carbon sequestration potential was a significant driver of the goal but some argued the study significantly overestimated this potential (Friedlingstein et al. 2019; Veldman et al. 2019).

While agricultural land is a target for restoration, the extent to which trees can replace productive agriculture will be restricted for national food security and local, social, and political reasons. Large-scale replacement of agriculture with forests in one country or region may also mean more forests are cleared elsewhere with no overall gain in forest cover (Holl 2017). On the other hand, integrating more trees or forest areas into agricultural landscapes can potentially increase agricultural production and improve landscape functions (Guariguata et al. 2021; Latawiec et al. 2015; Sunderland et al. 2019). More trees in agricultural landscapes can also improve biodiversity outcomes. For example, Borda-Niño et al. (2017) found that in a human modified landscape in Mexico, adding tree components to the agricultural matrix improved connectivity between isolated forest fragments more than expanding the existing closed forest fragments individually.

While integrating trees with agriculture is a potentially significant opportunity to increase global tree cover, the way forest cover statistics are collected make it difficult to confirm this potential. Zomer et al. (2014) estimated that 43% of agricultural land (more than 1 Bha) already contain greater than 10% tree cover but much of this area is not included in estimates of forest because land is not considered under "forest use."

Forest restoration may also occur in "degraded" forests, with certain forms of forest restoration aiming to increase tree density and forest quality through improved management and reduced degrading activities. However, defining "forest degradation" has proved challenging (Ghazoul et al. 2019), and the extent of degraded forests is uncertain. The International Tropical Timber Organisation recently estimated 900 Mha of tropical forests may be degraded (ITTO 2020). Findings from the latest Global Forest Resources Assessment (FAO 2020; Table 10.1) indicate that 0.95 billion ha of forest may be "unmanaged" and potentially degraded. This may make them suitable for bringing under management and "restoring," although care is needed in considering the role of Indigenous people in these "unmanaged" areas and their views on whether areas are degraded (Fletcher et al. 2021).

Assuming that a part of the previously forested agricultural land is degraded and suitable for restoration, the extent of suitable land is not a constraint to achieving global forest restoration targets of up to 1 billion ha. This does not mean this area is "available" for restoration. The condition, value of current uses to communities and extent of tree cover on this land may mean that trees are not the "highest and best use" for these areas. Availability will therefore depend on the local context, people, and land uses. For example, in their recent assessment, Brancalion et al. (2019) found that while 834 Mha across the humid tropics is "potentially restorable," because of socioeconomic constraints only 27% was considered to have a high feasibility score, according to their criteria.

As discussed above, the type of forest restoration to include in these global targets for tree or forest restoration is a matter of debate, with concerns raised, for example,

Table 10.1 Indicative global areas in different land use/land cover classes (derived from FAO 2020). The extent of degraded forests was calculated as the forested area outside intact natural forest and not covered by a management plan.

Land Cover Class	Area (billion hectares)
Agriculture and cropland	5.1
Agricultural land previously forested	1.7
Total forest area	4.06
Intact natural forest (primary)	1.11
Managed forests (inc. planted)	2.0 (0.3)
Unmanaged (potentially degraded) forest	0.95

about the extent of forest plantations in the targets and researchers arguing prioritizing restoration of natural forests (Lewis et al. 2019).

The desired location of forest restoration is also debated. Much recent focus has been on restoring forests in the humid tropics, due to high biodiversity in these regions (Lewis et al. 2019). Dryland regions present an alternative opportunity. They are home to over 2 billion people, including many of the world's poorest, and deforestation in these regions, particularly in Africa, is increasing (Malhi et al. 2014). While restoring dense forest cover in these regions is not ecologically feasible given the climate conditions, nor socially desirable due to potential impacts on water availability (Fagan 2020; Jackson et al. 2005), restoring low-to-medium density tree cover in these landscapes can improve many ecological and social values. Ecosystem services benefits, such as carbon sequestration, or biodiversity generated by restoring forests in these regions may be lower per hectare than in the humid tropics, but overall a greater scale of environmental and social benefits may be realized because of the land area involved (Sheil and Bargués-Tobella 2020). However, restoration in these regions is understudied, with only 5% of publications in the decade before 2017 addressing restoration in dryland areas (Svejcar and Kildisheva 2017).

In summary, international targets for forest restoration have been specified of 350 Mha by 2030, or 1 billion ha by the middle of the century. Suitable land may exist, but to meet these targets, about 35 Mha of new area needs to come under some form of forest restoration each year by 2050, seven times higher than the current rate of forest expansion. Whether much of the currently "degraded" should be reforested will be contested. Most agricultural land may be needed for food production and local livelihoods or may already have substantial tree cover. Local people and industries are likely to be using degraded forests. Restoration targets will therefore need to be met by either integrating more trees on agricultural land, while maintaining food and fiber production or by restoring ecological functions of degraded forests, while maintaining production of timber and non-timber forest products. This is referred to as "productive restoration."

10.6 Technical factors: Where and how to restore?

Achieving global forest restoration targets will clearly be a complex task, given it will involve a mix of restoration approaches and land uses (Holl and Brancalion 2020). Activities will need to engage many different stakeholders, each with different goals and perspectives on land use and condition. Multiple forms of ecological and technical knowledge will be needed to inform these restoration decisions. Expanding forest cover by tree planting can build on the large body of knowledge from centuries of experience and scientific investigation of tree propagation and planting, knowledge that has evolved over time with changing objectives of tree growers (Bae et al. 2012). For other practices, such as facilitating natural regeneration or manipulating forest succession, many of the technical requirements are well understood (Stanturf and Mansourian 2020) but more research is needed (Chazdon et al. 2017). Important technical and social issues facing restoration practitioners include (adapted from Mansourian et al. 2017):

i. Agreeing on goals of restoration;
ii. Deciding where to restore and choosing the restoration approach;
iii. Quantifying and monitoring provision of ecosystem services or impacts of restoration (e.g., on water or biodiversity); and
iv. Incorporating climate change implications.

We consider each of these issues in the following sections.

10.6.1 Agreeing on goals

Agreeing on goals for restoring forests is a social as well as technical process. Goal setting depends, in part, on understanding the distribution of restoration costs and benefits among key actors arising from a mix of land uses in a landscape. Managing trade-offs between ecological goals and broader needs for human well-being is challenging but not impossible (Chazdon and Brancalion 2019). For ecologically focused forms of forest restoration, the goal may be to restore land to an historical or reference ecological condition. This is difficult to achieve when knowledge of the "original" condition has been lost, when opinions differ on the right historical period used to determine the original state, or where altered environmental conditions (such as a changing climate), mean it is impossible to achieve this past state (Stanturf et al. 2019). In such cases, an alternative future state needs to be decided. Inclusive goal setting, involving multiple perspectives about this future state, is important for the long-term success of restoration projects.

Goals for many forest restoration efforts in the past have largely been determined by top-down government decisions. While these efforts have led to the "restoration" of about 300 Mha of planted forests, they have often been composed of a limited number of species and have not provided for a broad suite of values. Consequently, they have often not been widely supported. Allowing local people to decide goals, approaches, and locations for restoration and considering their needs is more likely to lead to more widely supported restoration programs (Koech et al. 2020).

10.6.2 Deciding on where to restore and choosing the restoration approach

Deciding where to restore forests is often considered a trade-off between tree cover and other land uses. In particular, replacing agricultural land with trees is often socially contested. This replacement may be appropriate where economic and social factors are driving agricultural land abandonment, if agricultural land is highly degraded and no longer economically productive, or where the overall benefits of increased tree cover exceed those of continuing agricultural land uses (Pannell 2008). Taking a landscape perspective can avoid such trade-offs by integrating different forms of tree cover with other land uses, or by matching optimal sites for different uses. Doing so can meet the needs of a wider group of stakeholders and bring a larger area of land into play for restoration (Stanturf et al. 2014). This "mosaic" landscape restoration maintains areas of sustainable agriculture and incorporates trees in integrated agroforestry, natural regeneration, riparian plantings, and/or wildlife corridors. Using the FLR model, these different land uses can be integrated at a landscape scale with managed plantations or areas of protected natural forests (Maginnis et al. 2012).

Restoration planning decisions can be supported by tools for identifying and prioritizing areas for restoration, such as the Restoration Opportunities Assessment Methodology (ROAM) (Maginnis et al. 2014). This methodology combines spatial analysis and multi-stakeholder assessments of ecological and socioeconomic values to inform restoration decisions (African Forestry and Wildlife Commission 2020; Li et al. 2018). Suggested improvements to the ROAM include incorporating understanding of local socioecological systems, assessing the impacts of forest restoration on different values and including use rights and gender in decisions (Koech et al. 2020; McLain et al. 2018; Sacande et al. 2020; Sheil and Bargués-Tobella 2020; Siles and Prebble 2018).

Deciding on the restoration approach depends on the goals, the financial resources available, and the potential future returns from products or services (Stanturf et al. 2019). Tree planting and direct seeding are commonly used but face many challenges in scaling up to meet ambitious global targets. Large-scale plantings have often been of exotic, fast-growing species because their seed is readily available and easy to store, and knowledge of their propagation is extensive and readily available. These single species plantations improve certain ecosystem functions (such as wood supply, carbon sequestration, reduced soil erosion, or improved water quality). However, they often do not contribute significantly to conservation or social objectives (Lamb et al. 2012; Malkamäki et al. 2018). But in the right configuration with other vegetation, monoculture plantations can support conservation objectives (Brockerhoff et al. 2013; Nájera and Simonetti 2010); for example, they can be used to "facilitate" the regeneration of native tree and understory species on harsh or degraded sites (Keenan et al. 1997; Lamb 2014). Many rural landowners also prefer to plant commercial tree species because they grow rapidly, and they can sell the products while also getting shade, shelter, and other on-farm benefits.

Direct seeding requires large amounts of seed and investment in site preparation. Seedling survivorship in this method is a function of seed size and conditions meeting seed germination requirements (Palma and Laurance 2015). To increase the scale and success of this method needs more research on seed production and germination, including on phenotypic and genotypic plasticity in plant selection, and site preparation techniques, such as ripping or incorporating biochar to improve soil water-holding capacity and plant productivity (Palma and Laurance 2015).

These "active" restoration methods, planting or direct seeding, are relatively expensive and labor intensive. Natural regeneration can be cheaper, but success depends first on removing activities suppressing regeneration (such as grazing or fire) and the presence of regeneration capacity in soil-stored seeds or vegetative propagules and/or proximity to seed sources. Further, suitable site conditions for regeneration and growth are needed (Borda-Niño et al. 2020; Rinaudo et al. 2020). In highly degraded lands with few residual native trees, or no seeds or vegetative material in the soil, potential for this form of restoration is limited, in which case there may be little option but to plant or sow seeds.

10.6.3 Quantifying and monitoring restoration benefits

Quantifying potential benefits and services that potentially arise from different forms of restoration is important to inform restoration decisions. Expected benefits from current restoration projects are wider than in the past, when the focus was mainly on producing timber or reducing soil erosion (de Jong et al. 2021). The capacity to quantify ecosystem services from planted forests and other forms of forest restoration has improved in recent years (Baral et al. 2016). New methods can assess habitat provision, flood protection, recreation, social or cultural services, and benefits (Paudyal et al. 2018) and link them to relevant indicators (Olander et al. 2018). Assessment is also needed of potential negative impacts of restoring forests on water resources, conservation values, and social and cultural factors.

Monitoring and reporting of these benefits and impacts are important to demonstrate success and show progress toward desired outcomes. Trialing different approaches and monitoring outcomes is also used to determine practice changes needed as part of adaptive management, social learning, and continuous improvement (Sanders et al. 2020). Ideally, monitoring starts with baseline information of current land and vegetation condition (Fagan 2020). Monitoring and reporting need to go beyond simple indicators, such as hectares planted, and include measures of forest growth, carbon sequestration, vegetation condition, target ecosystem services, and social benefits (Viani et al. 2017). For example, Zamora Cristales et al. (2020) developed a sustainability index for landscape restoration to measure ecological (water quality, water quantity, vulnerability reduction, soil quality, carbon equivalent, and landscape biodiversity) and socioeconomic aspects (landscape governance, labor provision, and adverse community impacts).

The financial value of ecosystem services is increasingly being recognized in science and policy (Costanza et al. 2017). With the right data, increases in ecosystem services values can be integrated into public policy decisions and guide private capital investment (Kareiva et al. 2015). However, better tools are needed to identify trade-offs and synergies between different services (Dade et al. 2019), including those generated from different restoration approaches (Hua et al. 2022).

10.6.4 Climate change implications

Climate change has implications for assessing the suitability of sites for restoration and their capacity to deliver ecosystem services in the longer term (Anderegg et al. 2020). Relatively small future shifts in disturbances, like fire and storms, can impact on establishment success, forest growth, and survival (Pugh et al. 2019). These implications may not all be negative: for example, in a study of restoration potential in China, Zhang et al. (2021) found that, while climate change will cause shifts in the distribution of suitable climate for different types of forests, overall area suitable for reforestation will increase. Other research indicates that carbon in restored tropical forests can be maintained over the long term under a range of potential future climates (Koch and Kaplan 2022).

Climate change also impacts on tree species or provenances chosen for restoration. Given shifts in climate zones, those with the best prospects of survival and growth may not be in nearby forests, or those species used in the past. Those planning restoration therefore need to consider including a mix of species or provenances, including those from climatic zones with conditions more like those of projected future climates (Broadhurst et al. 2017a; McDonald et al. 2021). This will require public engagement to build support for using nonlocal species (Keenan 2015).

Young, recently restored forests with lower species diversity and a higher proportion of trees in a more vulnerable growth stage are at greater risk from climate-related disturbances than older, mature forests (Frelich et al. 2020). These new forests may require more protection from fire or pests until they are established and have capacity to regenerate after disturbances. Other strategies to support the adaptation of restored forests to climate change include more corridors between remnant forests to allow plant and animal movement (Imbach et al. 2013; Krosby et al. 2010) and planting more diverse forests (Keenan 2015). Generating knowledge needed to adapt to a changing climate requires new types of collaboration among evolutionary biologists, conservation scientists, forest managers, restoration practitioners and community stakeholders (Broadhurst et al. 2017b; Cook et al. 2021).

In summary, while progress has been made addressing each of these challenges, much is experimental and/or at project scale. More research effort and finance are needed to apply and test them at larger landscape or global scales.

10.7 Financial feasibility: What is the cost, who will pay, and how?

Financing the scale of activity required is a key challenge in meeting global restoration targets. Costs that need to be covered include planning, community engagement, providing information and tools, acquiring and preparing land, propagating and planting seedlings, and ongoing management and protection. Mechanisms are also needed to offset the opportunity costs of alternative land uses. Types of costs vary with the restoration approach: passive removal of pressures on natural forests have low direct costs but opportunity costs may be significant, while planting to restore high diversity of natural tropical forests in remote locations or coastal areas is expensive

(Chazdon et al. 2020). Estimates of restoration costs range from less than USD 10/ha in some tropical forest regions to hundreds of thousands of USD/ha in coastal wetlands (De Groot et al. 2013). This high variation means estimates of the total funding needed to meet global targets vary widely. The Convention on Biodiversity (CBD) estimated that achieving the Bonn Challenge would cost USD 6 billion and 65 billion per year (CBD 2012), while FAO and the Global Mechanism of the United Nations Convention to Combat Desertification (UNCCD 2015) estimated up to USD 36 billion would be needed to meet the Bonn Challenge of 150 Mha by 2020 and USD 49 billion to meet the target of 350 Mha by 2030.

These amounts are beyond the capacity of public funding and international development assistance. During the UNFCCC COP 26 in Glasgow in 2021, the private sector and governments made large pledges for financing sequestration in land use sectors, but not all these pledged funds are for restoration, with much targeted at avoiding deforestation and forest degradation. Public financing for forest restoration, including climate finance, therefore currently falls well short of that required to achieve stated global restoration targets. For example, in 2019–2020, annually only USD 16.3 billion (about 2.5% of total climate finance) was invested in land-use projects (CPIC 2022), with only a fraction of this funding for ecological or forest restoration projects.

Mobilizing sufficient finance is therefore a major challenge to the feasibility of meeting the international targets. Investor financial decisions generally depend on returns expected by investors, expressed as risk adjusted rates of return on investment (ROI). The acceptable ROI depends on the type of investor. These fall into the following broad groups:

1. Public sector finance and development assistance—these may have little expectation of a financial return and invest for other benefits such as employment, improved local incomes, poverty alleviation, gender benefits, and enhancement of ecosystem services.
2. Public finance institutions (such as rural or international development banks) that require a low threshold ROI and social and environmental benefits.
3. Private investors seeking social or environmental "impact" will want a higher ROI than public institutions but also demonstrated environmental or social benefits.
4. Corporate social responsibility programs or foundations supported by commercial investors invest in projects with social or environmental outcomes but may not expect a financial return.
5. Conventional commercial investors who make decisions primarily on the risk adjusted ROI, although many have other motivations for their investment choices.

Beyond rates of returns, these investors have different requirements in terms of scale of investment (often starting at hundreds of millions of USD), tolerable risks, and social and/or ecological impacts. These requirements are a major challenge for investors in the agriculture and forest sectors in general (e.g., Louman et al. 2020), but are particularly challenging for forest restoration due to the complex settings in which it takes place, with a diversity of practices, land uses, and landowners. Project developers in these settings need to understand these specific investor needs and make their business case accordingly (Pistorius and Gromko 2019).

To better understand these challenges, this section briefly describes the economics of land use, the role of public and private finance in forest restoration, emerging trends in financial markets, and the challenges of making the business case for restoration. We conclude the section by describing innovative approaches for integrating funding from different types of investors.

10.7.1 Economics of land use and payments for ecosystem services

In a market economy, financial factors generally determine private land use. Whether it is a local landowner clearing a small area of forest for cropping or grazing, or a large-scale agribusiness enterprise clearing hundreds of hectares of forest for commodity production, the decision to clear or reforest land is driven by the need to generate income from the land. In these land use decisions, certain goods are seen to have market value (which fluctuate over time with demand and supply), while other environmental, social or cultural benefits generally do not. The decision to restore a forest, or incorporate trees into the current land use, will therefore depend on the "value" generated for the landowner in doing so.

Some environmental benefits from restoring forests, such as carbon or water quality or yield, are now being valued in environmental services markets, and landowners are being rewarded for their provision. However, these markets can be difficult to access, particularly by poor rural people, who are often the current users of land targeted for restoration (Börner et al. 2017). These poor rural people may not have secure ownership of land or trees, or the right to transact in these services. Many payment programs are short duration, with future payments uncertain, and payments often do not exceed the opportunity costs of foregoing current land uses (Louman et al. 2016). Other cultural or social factors may ultimately have a stronger influence on their land use decisions.

10.7.2 Public sector finance

Most historical forest restoration in temperate regions has been funded by the public sector (de Jong et al. 2021). Examples include the US Conservation Reserves Program (USDA 2021), similar programs in Europe, and large-scale plantations programs in China (see Box 10.1) and in Vietnam (Dang et al. 2019). Governments also support new forests indirectly through tax concessions and other forms of support (Enters et al. 2006; Ferguson 2014; Hyde 2012).

This direct or indirect investment has often been driven by concerns among policy makers about falling prices of agricultural commodities due to overproduction or lower availability and rising costs of timber and fuelwood, as natural sources are exhausted. Rapid expansion of single-species plantations typically supported by past public programs have not been universally popular. They have caused social conflict due to altered rural communities or takeover of land from marginal rural producers

(Leys and Vanclay 2010; Malkamäki et al. 2018). Focus has shifted in recent years to investment of public funds to providing public goods, such as environmental or habitat benefits. For example, a recent study of large restoration efforts (>10,000 ha) reported that governments most commonly support forest restoration for water conservation and to meet international goals, while erosion control, soil conservation, and securing energy sources were less common objectives (Mansourian et al. 2022). Most of these cases were at least partially paid for from national budgets. Some national programs, such as in Costa Rica and Vietnam, used schemes for payment for environmental services to incentivize the participation of private land holders, channeling money from private sources (such as a tax on transport fuels in Costa Rica) and toward restoration for public good objectives.

New, people-centered programs can address these concerns and build greater public support for forest expansion. For example, the Trees for Global Benefit program in Uganda uses public funds to strengthen farmer associations, create financial literacy, and support small-scale tree planting for carbon sequestration and biodiversity conservation. These small plantations are owned by landowners and complement their income from agricultural production with carbon credits bought by private investors. This carbon income allows farmers to access additional finance from local banks to diversify their production and set up small businesses based on non-timber tree products (Byakagaba et al. 2021; Gitz et al. 2020). Carbon fund managers are increasingly looking for partnerships with community development organizations to help support or create such initiatives.

10.7.3 Private sector investment in forests

Corporate investment in tree and shrub crops, such as rubber, tea, coffee, and copra was part of commercial colonial strategies in tropical regions from the 1800s onward, often involving clearing of native forests. Many of these estates were later taken over by smallholders after decolonization and land reform (Byerlee 2014). Corporate investment in timber forests began more recently, driven by changes to tax rules in the United States. In the late 1990s and 2000s, these investors expanded to forestry assets in Latin America, Australia, and New Zealand and later, to "riskier" settings in Asia and Africa (Brand 2019). From 1995, global investments in tree plantations (90% for wood production) increased from USD 4.7 billion per year to USD 5.7 billion per year (Tomaselli 2012), with 21% in Latin American countries (LAC) mainly in Brazil, Chile, Argentina, and Uruguay. A relatively small proportion of institutional investments is oriented toward sustainable management of natural forests: in the LAC this was approximately 0.5% in the period between 2006 and 2011 (Tomaselli 2012).

This increased investment in timberlands has been based on relatively stable, risk-adjusted returns from timber sales, combined with the long-term return profile of forests. This is attractive to investors seeking to diversify portfolios because it has low correlation with other asset classes and future returns match long-dated liabilities of pension funds and similar institutions. Overall, the current value of investable global timberland (depending on how it is defined) is estimated between USD 200–400

Box 10.1 The China experience

When the Chinese Civil War ended in 1949 and the People's Republic of China was established, the country had only 8.6% forest cover. The effects of floods, landslides, and dust storms in the 1970s to the 1990s were attributed to widespread deforestation, and these concerns drove large-scale afforestation programs (Zhang et al. 2017). The first—the Three-North Shelterbelt Program (1978–2050)—was launched in 1978 to address desertification, drought, and soil erosion in northeast, northern and northwest China. At that time, China was one of the world's poorest countries. After a flood in the Yangtze River Basin in 1998 killed 4,150 people and caused large-scale economic damage, China launched other programs, including the Grain for Green Program (1999–2020) and the Natural Forest Conservation Program (1998–2020). These aimed to protect existing vegetation, retire croplands on slopes to prevent soil erosion, and improve farmer livelihoods. Sandstorms in Beijing and surrounding regions in the early 2000s triggered the Sand Source Control Program in Beijing and Tianjin. As a result, forest cover in China has grown significantly since the 1980s, with the annual increase peaking at 4.1 Mha in the early 2000s. By the end of 2020, forest cover reached 221 Mha, 23.0% of the total land area. These figures are based on China's forest definition which includes shrubs and bamboo forests.

Current national commitment and targets

In December 2020 President Xi Jinping committed to increase forest cover to 24.1% (231 Mha) of the land area by 2025 and 26% (250 Mha) by 2035, and to increase forest growing stock to 19 billion m^3 by 2025 and 21 billion m^3 by 2035. In addition to their role in preventing disasters such as floods, sandstorms, and landslides, forests are now part of China's solution to climate change.

Cost of restoration

In total, it has been estimated the national Chinese government spent USD 378.5 billion on sustainability programs from 1978 to 2015, of which USD 136.1 billion was related to forest restoration (Bryan et al. 2018). Increasing government investment in forestry programs was matched by unpaid labor contribution, with most of the investment (74.9%) in this between 1978 and 2009 (Feng et al. 2010). With rapid economic growth, government investment greatly increased, and the unpaid labor policy was cancelled in the early 2000s. Central government payments for afforestation to private actors increased rapidly, from the equivalent of USD 27 per ha 1978–1985 to USD 187 per ha 2001–2008 to USD 428 per ha in 2009–2015 and to USD 1,250 per ha from 2016 onward (Chinese Academy of Science 2018). This payment to farmers only aimed to cover direct planting costs and excluded ongoing costs of silviculture and management. Actual cost per hectare was estimated to be more than USD 2,800 per ha (Chinese Academy of Science 2018; Feng et al. 2010; Gui 2015; Liu et al. 2014; State Forestry Administration 2008). Additional funds were spent on research and restoration planning (Zhu and Zheng 2019). The survival rate in early programs was only 53%, further increasing the cost per hectare that was successfully reforested. This indicated initial systemic problems with species selection, seedling production, transport, planting, and management. Survival improved, but the average rate under the Three-North Shelterbelt Program was only 65% overall between 1978 and 2018. A stronger process to learn from experience might have enhanced program success (Cassells et al. 2020).

Quality after quantity

China faces technical and financial challenges to further expand its forest cover. Early programs focused on the easiest areas. According to the 2013 Eighth National Forest Inventory, two-thirds of remaining land suitable for restoration is in semi-arid and arid regions (State Forestry Administration 2013), regions with the lowest survival rates after planting (Mei 2019). Consequently, China's forest expansion targets for 2025 and 2035 are less ambitious than those in early 2000s: 2 Mha of new forest per year until 2025 and 1.8 Mha per year between 2025 and 2035.

Of the planted forests, 85% were monocultures, and 65% were young or middle-aged when assessed in 2013 (State Forestry Administration 2013). Fast-growing species were favored in early programs because they could quickly meet national targets and were better accepted by local people because they could provide short-term income. Five genera make up 60% of planted forests: Abies, Populus, Eucalyptus, Larix, and Pinus (Liu et al. 2014). In northeast China, 98% of the shelterbelts are Populus sp. (Zhu and Zheng 2019), that are now highly susceptible to pests and diseases. Many are at the end of their natural life spans.

China's efforts to restore its deforested and degraded lands has been a process of learning by doing. With an inadequate science base, limited finance, and the urgent need to prevent disasters, early efforts focused on forest expansion with little attention paid to management and silviculture. A study in 2013 assessed forest quality, with 18 indicators used, including tree stocking, growing stock per unit area, annual increment growing stock per unit area, crown density, mean DBH (diameter at breast hight), and stand species diversity. Only 19% of the forests were rated "good" quality (State Forestry Administration 2013), with most of the estate delivering poorly on timber production and ecosystem services. This assessment found that, while forest cover increased, the forest growing stock and quality were low—89.8 m^3 per ha, 69% of the global average (Xu 2014).

Realizing the need to improve forest quality to maintain or increase ecosystem services, the central government now provides subsidies to encourage silviculture and to restore degraded forests. Since 2014, young and middle-aged protection forests have been eligible for grants of USD 224 per ha for stand improvement (The State Council of the People's Republic of China 2014). In 2015, the earliest forestry program, the Three-North Shelterbelt Program, which had more degraded or unsuccessful areas received a higher figure of USD 448 per ha for piloting the restoration of degraded forests and shrublands across 90 counties (National Forestry and Grassland Administration 2019).

billion; mostly in the United States, Europe, Oceania, and Latin America, in countries with clear property rights that define land and tree ownership (Brand 2019).

While the focus has been on timber returns, investment drivers are changing due to two important trends: concern about climate change and loss of biodiversity. These drivers are being incorporated into requirements of "responsible" investors, a rapidly growing investment class that involves scrutiny of social and environmental risks from investment and the desire to invest in assets with verified social and environmental benefits. Demand for these types of investment has accelerated since the 2015 Paris Agreement and associated net zero targets for 2050, with governments and businesses driving demand to reduce emissions (Vivid Economics 2020). Past institutional investment in land has often driven emission increases and species loss through deforestation and conversion to agriculture, but now portfolio allocations are shifting from emissions intensive sectors toward sectors with low or even negative emissions. This is a major structural change. For example, a group of 33 institutional investors[1] representing over USD 5 trillion in assets under management have recently committed to achieve net zero emissions from their portfolios by 2050 (UNEP 2022). In 2021, more than 100 countries signed the Glasgow declaration, committing to halt deforestation and forest degradation by 2030 (UN Climate Summit 2021).

Consequently, "nature-based solutions" or "natural climate solutions" for climate change are central in the global agenda toward sustainability (Griscom et al. 2017), with organizations like the Coalition for Private Investment in Conservation mainstreaming investment in conservation and restoration opportunities (CIPC 2023). This demand may result in a global carbon market of USD 167 billion per annum by 2030 and USD 347 billion per annum by 2050, with land use-based mitigation being a substantial part of that market (IETA University of Maryland and CPLC 2019). The investment market is evolving to capture opportunities that integrate these different values, for example: carbon offsets, conservation, wetland and endangered species, and recreation benefits. Maturing market-based mechanisms are shifting the underlying economics of land management toward conservation and restoration (Brand and Singh 2014). For restoration initiatives to be attractive to investors, they will need to meet requirements for scale and rates of return AND demonstrate climate and nature benefits through transparent monitoring and reporting systems. They will also need to demonstrate sound management related to the risk of losing standing carbon. This will require assessment and monitoring to demonstrate claims for environmental and social outcomes to investors and regulators. For example, a Task Force on Nature-Related Financial Disclosures (TFND 2022) aims to provide transparency on the impacts of business on nature. This is a form of market self-regulation to minimize negative impacts on biodiversity, water quality, and aesthetic values.

With rising value of carbon and biodiversity benefits, returns from restoration initiatives could be very attractive to investors. The Economics of Ecosystems and Biodiversity initiative estimated returns of 20–50% on long-term investment, depending on the forest ecosystems to be restored (Kumar 2012). Many investors see opportunity in assets that will increase in value from rising carbon or biodiversity prices, which are likely to become a primary focus of investment in land management, rivaling traditional agricultural and forestry markets. More land may be converted from agriculture to production forests, because these have higher carbon stocks, or to permanent carbon plantings. Timber plantation management may also be modified to provide more habitat and increase biodiversity and ecosystem services. This can generate returns from a mix of products and services (Silva et al. 2018). Investment managers need to assess the regulatory uncertainty, price volatility, and market risks associated with these different income streams.

Another challenge is to set appropriate rules and standards to mobilize this investment without causing community displacement or impoverishment. Rising carbon or other values for reforestation and forest restoration will be capitalized into land value. The value of suitable degraded or marginal land will increase because it has an "option value" to capture these returns in the future. Without constraints, investment finance may largely go to corporate entities that buy land use rights, resulting in loss of sovereignty, with negative impacts on smallholder farmers with insecure tenure or communities with informal land rights.

Rising land values, changes in land use and increased corporate ownership are likely to generate negative reactions from rural communities and political backlash. In Laos, for example, upon declaring independence in 1975, the government declared that forests used by shifting cultivators were owned by the state. The government then promoted large private tree plantations in "degraded" parts of these state lands, reducing traditional access for the local people (Pichler et al. 2022). While not primarily driven by carbon prices, the community backlash about timber plantations led to a moratorium on plantation expansion and significant change in plantation policy (Smith et al. 2021). To be successful at scale and over time, large-scale investors in forest restoration need to invest in ways that align with local community interests and values and generate local benefits.

10.7.4 Making the business case for restoration

To grow investment in restoration initiatives, they need to be packaged in a way that's attractive to investors. Investors often lack knowledge of the agricultural and forest sector in general, and knowledge of local conditions. Projects need to be framed in language investors unfamiliar with forests can understand, with clear goals and rigorous measurements of progress (Burrows 2014). Nonprofit organizations and small and medium-scale farmers and landowners have different priorities that need to be aligned to those of international investors. Restoration project business cases must clearly document all costs and benefits for investors and participating landowners or communities. Tools are now available to support these analyses (e.g., Hanson et al. 2015).

Restoration investment is not without risk. As we saw in the Chinese example (Box 10.1), creating new forests takes time and long-term investment. Even with effective monitoring, desired outcomes take time to realize, not all activities are successful, and changes to projects will be required in response to experience. Developing a business case is particularly difficult in mosaic landscapes with fragmented or unclear ownership and user rights and with projects involving a variety of land uses. In new initiatives, or with innovative approaches, for example, managing agents working across groups of small landowners who may not have credit records and business structures, it is difficult for investors to understand and estimate risks.

In many cases, support is required to develop good governance, build local organizations, write business cases, and coordinate between agencies and private landowners to align land uses and manage potential conflicts, particularly when not all landowners may want to participate in a forest restoration initiative. These enabling activities do not generate direct financial returns. Commercial investors generally are often unwilling to cover these costs, unless returns from commercial activities are particularly high. Business cases can be enhanced by showing that these issues are being resolved through co-investment from government development assistance.

10.7.5 Financial innovations and blended finance models

Meeting the need to support the enabling investment environment, public and philanthropic funds can act as a "catalyst," to facilitate private investment and help manage risks associated with restoration projects. These **blended finance** structures (Figure 10.1) are widely used in other sectors. They allow investors (usually private) with financial goals and those with community development or conservation goals (usually public or philanthropic organizations) to combine their resources to jointly finance initiatives. Public sector funds can invest in enabling conditions indicated above and strengthen financial literacy of civil society organizations. This can reduce risks for private sector investment and overcome barriers to finance for smallholders or communities (Louman et al. 2020). When a mix of benefits are generated from a project, public and philanthropic investors can invest in social and ecological benefits from restoration while private sector investors can receive direct financial returns from timber, carbon, or other commodities while also demonstrating their investment has broader environmental and social benefits. This collaboration can build knowledge in financial institutions about the land sector, helping prepare a pipeline of investible projects and appropriate financial instruments (Convergence 2021).

Such projects can integrate revenue from commercial forestry with restoration and conservation activities, build community forestry and out-grower schemes, and use income from environmental markets to generate new investment in landscape scale restoration. For example, the Tropical Landscapes Finance Facility (ADM Capital 2023) combines short- and medium-term loans, in the form of notes, with guarantees and grants for investment in commercial land-based activities, forest conservation, and restoration. These finance structures have great potential to integrate conservation and restoration at a landscape scale (Shames and Sherr 2020) and have increased six-fold between 2007 and 2018 (from USD 20 billion to USD 122 billion). However, only 3% of blended finance was focused on the sustainable development goal (SDG) 15 (life on land) in 2021 (Convergence 2021).

Project scale is important for investors. Intermediaries can help achieve scale by aggregating local reforestation projects, paying specific attention to the differing needs and aspirations of smallholder farmers, forest communities, and indigenous people. For example, in Kenya and Ghana private companies directly prefinance tree planting and ensure markets for the timber (Besacier et al. 2021). For example, in Uganda a nonprofit intermediary organization used development funds in a mechanism to issue carbon payments to restore trees and forests on many small, private

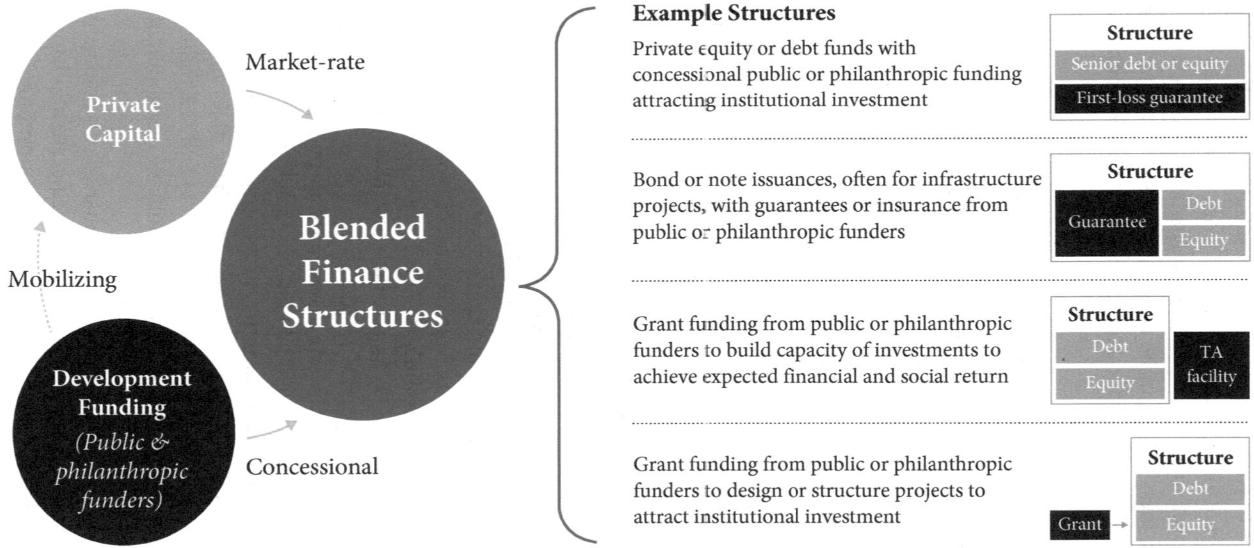

Figure 10.1 Typical blended finance mechanisms and structures (Source: Convergence 2021; reprinted with permission from Convergence©)

farms (Byakagaba et al. 2021). More funding for these types of project development activities is needed (Convergence 2021).

Other forms of innovative finance that could support restoration at a large scale are *green, climate, and sustainability bonds.* In these structures, a group of funders will pledge payments if certain social, environmental, or climate goals are reached. Based on these pledges, investors receive a bond or paper pledge of repayment after a predetermined period. If goals are achieved, pledged funds will be paid to coordinators, who then pay back the money borrowed from the investors, with a loss or profit proportional to the success of the initiative in meeting its goals. An example relevant for restoration are carbon bonds, where investors provide finance for forest restoration, which generates a certain level of carbon sequestration that can be translated into tradable carbon credits. At the designated future time, investors receive their initial investment plus a certain level of return, depending on the quantity and price of the carbon credits sold. While institutional investors are showing more interest in green bonds, in 2017 only 3% of the USD 161 billion in climate bonds was invested in the land sector (Climate Policy Initiative 2022).

Considering these developments and looking ahead, the investable forest asset class for private sector investors could grow from the 65 Mha currently focused on timber production (valued between USD 200–400 billion) to a far larger area with a value of USD 1 trillion, if carbon value and other benefits of restoration are included (Figure 10.2; Brand 2019). While significant, this is a relatively small part of the total current global capital pool of USD 220 trillion (Gadzinski et al. 2018).

In summary, while finance is potentially available to meet global restoration targets, the challenge is to rapidly scale new investment. Blended finance models are clearly not without risk. Bond structures are complex and usually issued by governments, large corporations, or large fund managers. These investors require significant scale, perhaps USD 500 million, to justify developing a product (Burrows 2014). Matching this scale of investment with many smaller restoration projects involving households or communities is a major barrier. Social and institutional arrangements will need to address these scale challenges.

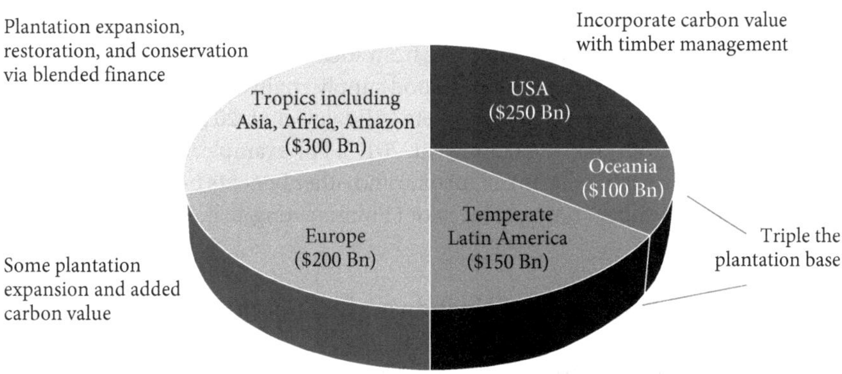

Figure 10.2 Proposed regional strategies to achieve USD 1 trillion worth of forests as an asset class (Source: Brand 2019)

10.8 Social and institutional factors: Who decides?

If it is assumed sufficient, suitable, and degraded forest land exists for large-scale forest restoration and sufficient technical capacity and finance are available to achieve global restoration targets, the main challenge is overcoming social and political barriers to landscape change. Many restoration efforts target degraded forest landscapes or farming lands being used by marginalized communities with limited alternative livelihood options. For example, Erbaugh et al. (2020) found that 294.5 million people in developing countries live on land identified to have tropical forest restoration opportunities; this includes 12% of the total population in low-income countries. Low-income countries have poor natural resource governance capacity and unclear land tenure, and state land administration institutions are poorly accountable, under-resourced, and understaffed. Without due consideration to issues of power, control, and distribution of benefits, local people who depend on degraded land for their livelihoods may be severely impacted. Setting restoration targets and policies without supporting strategies, financial mechanisms, or programs to grant ownership or control over land to local people are likely to fail. This requires time and resources but, if done appropriately, has been found to greatly incentivize participation in forest restoration efforts (Mansourian et al. 2022).

Forest restoration efforts driven by central authorities and state investment, for example, in the post-war "transition" economics of Japan and South Korea (de Jong et al. 2021) have had strong social support. In South Korea, comprehensive reforestation plans started in 1973, providing economic incentives to reach clear, quantifiable goals. Policies also promoted inter-agency cooperation and coordination, especially between the energy and forest sectors, to replace firewood with fossil fuels. These activities increased stocked forest area and average growing stock level (Bae et al. 2012).

More recently, China (Box 10.1) and Vietnam initiated nationwide restoration programs, which greatly increased their forest cover. These state-led and -controlled initiatives depended on community labor (although in China this was not always "voluntary"), and commercial monoculture timber plantations in Vietnam have, in some cases, had negative social or environmental impacts (McElwee 2009). In Vietnam, granting forest land tenure to households and giving them resources and capacity was more successful in generating positive social outcomes from forest restoration (Dang et al. 2019). Social impacts from such restoration can be addressed by integrating smallholders into industrial wood supply chains, providing suitable technical options and access to reliable markets (Midgley et al. 2017) and knowledge of alternative land use options (Chazdon et al. 2017). For example, a plantation project in Laos provides equitable sharing of income and other benefits between smallholder tree growers and a timber company investor (Phimmavong et al. 2019).

10.8.1 Community participation and stakeholder engagement

In other settings, state control over forests has been a constraint on forest restoration, and more collaborative approaches with strong community leadership and participation can overcome problems with weak governance and institutions (FAO and

RECOFTC 2016). Under state control of forests in Nepal for example, communities had little incentive to actively participate in restoration (Ojha and Hall 2021). By building strong social networks with international donors and NGO support, Nepal's community forestry groups have played a significant role in shaping consultation for collective action and reversing forest degradation and restoring forests. By 2019, over 22,000 community forestry user groups were registered across the country, with rights granted to manage nearly 2 Mha of forest (about a third of Nepal's total forest area) and improved forest condition and biodiversity. This system has also contributed positively to livelihoods of nearly a third of the country's population and is developing into a self-sustaining system involving a strong base of policy champions, service providers, and critical action researchers (Ojha and Hall 2021). This demonstrates the benefits of devolving forest rights to local communities and building local capacity (Acharya et al. 2008; Ojha et al. 2009). However, this system is not without challenges: most benefits are often captured by elites within the community, participation by women and those in lower castes is limited, and restored forests are often not available to meet certain community needs (Sapkota et al. 2018).

In other examples in the West African Sahel, farmer-managed processes drive tree and forest restoration while intensifying agriculture to support a growing population, and across Burkina Faso and Niger, large-scale soil and water conservation practices have reduced wind erosion and increased agricultural production (Reij et al. 2009). In these settings, international NGOs and supportive government policies have facilitated coalitions of actors to develop mechanisms for innovation and shared learning. These have expanded the scale of the transformation by strengthening human, social, and political capitals (Magrath 2020; Reij et al. 2009). Restoring trees has provided local income and reduced out-migration with women, in particular, benefiting from improved supplies of water, fuelwood, and other tree products.

These examples indicate that, if done with the right support and policies, forest restoration can generate widespread benefits for local people. However, is important to consider potential trade-offs—that is, who benefits, and who pays, in implementing restoration programs. Inclusive design to ensure restoration programs generate benefits for rural poor are needed to achieve forest restoration goals (Brancalion et al. 2017). These approaches need supporting policy arrangements, development resources, and recognition that community engagement and consultation take time and effort. Large-scale, top-down, time-bound area targets, or short-term demands of public or private investors, can create tensions with local communities, who may be supportive of land use transition but are unclear about local benefits.

10.8.2 New governance models and institutional capacity

Forest restoration is often hamstrung by competition and lack of coordination between sectoral government agencies. Shifting from this sectoral resource governance to the landscape as the unit of governance can better integrate people and different resource sectors and scales in governance and increase the feasibility of restoration (Arts et al. 2017). Designing durable landscape-level governance is challenging, Setting the right scale is important (Anderson et al. 2016), as are commitment and support from the government at higher levels (Kusters et al. 2020) and partnerships between public and private actors (Ros-Tonen and Kusters 2011). The design will vary over time with cultural, political, and socioeconomic settings (Kusters et al. 2020). For example, Nepal's community forestry system, emerged through an evolving approach to governance reform over five decades. Such improved legal frameworks need funding and training support to build implementation capacity. In Ethiopia, for example, the government is setting ambitious FLR targets but lacks capacity and financial resources to implement them (Pistorius et al. 2017). Building institutional capacity, addressing sectoral and social conflicts, and increasing transparency in decision-making are vital components of restoration success (Colfer et al. 2022; Schweizer et al. 2019).

10.9 Synthesis and conclusions

Targets for restoring global forests have been in place for nearly a decade, yet rates of forest expansion, through either planting or natural regeneration, have halved over the same period (FAO 2020), and deforestation continues at a significant rate. To achieve goals like the Bonn Challenge, or the longer-term, "One trillion trees" target, the rate of expansion in the area under forest restoration needs to increase from the current 5 Mha to 35 Mha per year.

In this chapter, we explored the feasibility of achieving these targets by considering the extent of degraded land, ecological and technical knowledge and tools, finance, and social and institutional factors in achieving restoration. These factors interact across scales and across situations. Past programs that have achieved forest restoration at scale (see Box 10.1) have had high levels of capital investment, mainly by the state and have involved payments directly to landowners for tree planting, or to compensate for lost opportunities and significant investment in technical capacity and political and social support.

Ambitious global goals cannot be met by public sector funds. The private sector, including superannuation funds, sovereign wealth funds and private equity, will be a major source of funds. These entities are increasingly looking to invest in climate, conservation, and social outcomes from investments, while also generating a financial return. This will require new investment models and brokers and fund managers capable of facilitating investment. To provide expected financial returns, restoration projects need to integrate productive uses (for example, agricultural products or wood fiber) and generate returns from carbon, water, social, and conservation outcomes (Kremen and Merenlender 2018).

Broadly, forest restoration activity will occur in two types of landscapes or regions:

(i) where rural populations are declining due to economic development and rural-to-urban migration; and
(ii) where rural populations are growing and rural property sizes are decreasing.

In type (i) regions, labor shortage and high costs will constrain active forest restoration—forests will be restored primarily through passive, natural regeneration pathways relying on local propagule sources. Managing these areas will focus on preparing sites and using new technologies to facilitate assisted natural regeneration. Where sites are heavily degraded or few propagule sources exist, planting or seeding of local pioneer or exotic tree species can facilitate natural regeneration. Primary financial drivers in these settings are likely to be carbon, water, or biodiversity payments, with the forest ownership being a mix of public, smaller private, or different forms of common property.

In type (ii) landscapes, increased tree cover will mostly occur on privately held land supported by private or a mix of public–private investment. This will include smallholder and family farmers integrating trees into farming systems using traditional planting methods or "farmer-managed" natural regeneration models (Duriaux-Chavarría et al. 2021) and larger-scale corporate ownership integrating intensively managed agricultural land or tree plantations with substantial areas of forest restored for conservation (Silva et al. 2018). Returns from agriculture or timber production can be integrated with carbon offsets or biodiversity payments through direct payments or blended finance models. Collaborative investment models can integrate these two classes of landowner, although these will need to improve on past arrangements such as out-grower schemes. These have often failed to share benefits fairly or build trust and durability in these company--smallholder relationships, and supply chain complexities have limited smallholder knowledge and power and capacity to participate in markets (NYDF Assessment Partners 2019). New models and approaches can better integrate small landowners with industry supply chains and share value (Phimmavong et al. 2019).

In either type (i) or (ii) landscapes, forest restoration involves operating in complex socio-ecological systems. These systems are, by definition, nested within global policies and capital allocation decisions occurring at larger scales that have downward effects on smaller-scale nested systems. In the past, it has been difficult for small, isolated communities to influence global or national policy and capital investment decisions. But discourse is changing, with the needs and interests of local people increasingly being recognized in both policy and in financial investment decisions. This development can support financially, socially, and politically feasible restoration that links top-down and local commitments (Mansourian et al. 2022).

Design of forest restoration will therefore vary between countries and localities to accommodate different economic circumstances, cultures, and local land uses. Limiting factors will vary between these different settings. In developed countries with secure property rights, financial returns and community and political support are likely to be most limiting. In developing countries with weaker governance and uncertain land tenure, land costs may be cheaper but resolving institutional arrangements and social factors will be key factors in supporting locally appropriate restoration (Mansourian et al. 2022). At the farm or local landscape level, clear land tenure, local regulations, and personal motivations and values; and financial, ecological, and technical issues; are generally the most important factors determining feasibility. While win-win situations between forests, trees, and people are possible, trade-offs are inevitable; these need to be anticipated and incorporated into planning and compensation arrangements.

Framing global targets is also important in building social and political support. Simply indicating hectares or number of trees planted does not describe the desired flow of benefits from increasing tree cover (Brancalion and Chazdon 2017) and does not create wide support for restoration because people may see it as a straightforward competition for land. Targets are better framed in terms of Sustainable Development Goals: meeting growing demands for energy, food and water security; mitigating and adapting climate change; reducing species extinctions; and improving local incomes and supporting new industries (Brancalion et al. 2019; Ghazoul et al. 2019; Guariguata et al. 2019; Louman et al. 2019). Poverty alleviation is also important but our past efforts to reduce deforestation and also address poverty, or other development goals, have proven difficult to sustain and to scale up.

In summary, large-scale, global targets to restore forests are potentially feasible if restoration activities encompass multiple approaches, are based on social and political support, and sufficient finance can be mobilized. Approaches need to involve integrating trees with agriculture and a mix of commercial and environmental objectives. Involving local people in setting goals and providing clear tenure and ownership of land and restored trees and forests can build social and political support. Improving institutional arrangements and coordination across government agencies and private stakeholders can facilitate project finance at the scale needed to achieve global restoration goals.

Note

1. Institutional investors as a class control approximately USD 70 trillion in assets and include pension funds, superannuation funds, insurance and reinsurance companies, sovereign wealth funds, and foundations and endowments.

References

Acharya, K.P., J. Adhikari, and D.R. Khanal. 2008. "Forest Tenure Regimes and Their Impact on Livelihoods in Nepal." *Journal of Forest and Livelihood* 7: pp. 6–18.

ADM Capital. 2023. Tropical Landscapes Finance Facility: Long-term financing for green growth. https://www.admcapital.com/tropical-landscapes-finance-facility/.

African Forestry and Wildlife Commission. 2020. Forests and wildlife: Africa's diversity for shared prosperity and security. Twenty-second session. African Forestry and Wildlife Commission, African Forest Landscape Restoration Initiative – AFR100, Skukuza-Mpumalanga, South Africa.

Anderegg, W.R., A.T. Trugman, G. Badgley, et al. 2020. "Climate-Driven Risks to the Climate Mitigation Potential of Forests." *Science* 368: eaaz7005.

Anderson, Z.R., K. Kusters, J. McCarthy, and K. Obidzinski. 2016. "Green Growth Rhetoric Versus Reality: Insights from Indonesia." *Global Environmental Change* 38: pp. 30–40.

Arts, B., M. Buizer, L. Horlings, V. Ingram, C. Van Oosten, and P. Opdam. 2017. "Landscape Approaches: A State-of-the-Art Review." *Annual Review of Environment and Resources* 42: pp. 439–463.

Bae, J.S., R.W. Joo, and Y.-S. Kim. 2012. "Forest Transition in South Korea: Reality, Path and Drivers." *Land Use Policy* 29: pp. 198–207.

Baral, H., M.R. Guariguata, and R.J. Keenan. 2016. "A Proposed Framework for Assessing Ecosystem Goods and Services from Planted Forests." *Ecosystem Services* 22: pp. 260–268.

Bastin, J.-F., Y. Finegold, C. Garcia, et al. 2019. "The Global Tree Restoration Potential." *Science* 365: pp. 76–79.

Besacier, C., L. Garrett, M. Iweins, and S. Shames. 2021. Local Financing Mechanisms for Forest and Landscape Restoration. Forestry Working Paper 21. Rome: FAO. DOI:10.4060/cb3760en.

Bond, W.J., N. Stevens, G.F. Midgley, and C.E.R. Lehmann. 2019. "The Trouble with Trees: Afforestation Plans for Africa." *Trends in Ecology & Evolution* 34: pp. 963–965.

Borda-Niño, M., D. Hernández-Muciño, and E. Ceccon. 2017. "Planning Restoration in Human-Modified Landscapes: New Insights Linking Different Scales." *Applied Geography* 83: pp. 118–129.

Borda-Niño, M., P. Meli, and P.H.S. Brancalion. 2020. "Drivers of Tropical Forest Cover Increase: A Systematic Review." *Land Degradation & Development* 31: pp. 1366–1379.

Börner, J., K. Baylis, E. Corbera, et al. 2017. "The Effectiveness of Payments for Environmental Services." *World Development* 96: pp. 359–374.

Brancalion, P.H., D. Lamb, E. Ceccon, D. Boucher, et al. 2017. "Using Markets to Leverage Investment in Forest and Landscape Restoration in the Tropics." *Forest Policy and Economics* 85: pp. 103–113.

Brancalion, P.H.S., and R.L. Chazdon. 2017. "Beyond Hectares: Four Principles to Guide Reforestation in the Context of Tropical Forest and Landscape Restoration." *Restoration Ecology* 25: pp. 491–496.

Brancalion, P.H.S., A. Niamir, E. Broadbent, R.Crouzeilles, et al. 2019. "Global Restoration Opportunities in Tropical Rainforest Landscapes." *Science Advances* 5: eaav3223.

Brand, D. 2019. *Transforming the Forestry Asset Class.* Sydney: New Forests.

Brand, D., and D. Singh. 2014. "What Is Needed to Make Markets for Forest Ecosystem Services a Reality?" In *Forests and Globalization: Challenges and Opportunities for Sustainable Development*, edited by W. Nikolakis, and J. Innes, pp. 185–201. London: Routledge.

Broadhurst, L., S. Prober, F. Dickson, and D. Bush. 2017a. "Using Restoration As an Experimental Framework to Test Provenancing Strategies and Climate Adaptability." *Ecological Management & Restoration* 18: pp. 205–208.

Broadhurst, L., C. Waters, and D. Coates. 2017b. "Native Seed for Restoration: A Discussion of Key Issues Using Examples from the Flora of Southern Australia." *The Rangeland Journal* 39: pp. 487–498.

Brockerhoff, E.G., H. Jactel, J.A. Parrotta, and S.F.B Ferraz. 2013. "Role of Eucalypt and Other Planted Forests in Biodiversity Conservation and the Provision of Biodiversity-Related Ecosystem Services." *Forest Ecology and Management* 301: pp. 43–50.

Bryan, B.A., L. Gao, Y. Ye, et al. 2018. "China's Response to a National Land-System Sustainability Emergency." *Nature* 559: pp. 193–204.

Burrows, M. 2014. "Financing Sustainable Forest Landscapes." *Conservation Letters* 7: pp. 499–500.

Byakagaba, P., P.K. Nantongo, and F. Kalibwani. 2021. Finance for Integrated Landscape Management. De-Risking Smallholder Farmer Investments in Integrated

Landscape Management: ECOTRUST's Trees for Global Benefit (TGB) in Uganda. Ede: Environmental Conservation Trust of Uganda and Tropenbos International.

Byerlee, D. 2014. "The Fall and Rise Again of Plantations in Tropical Asia: History Repeated?" *Land* 3 (3): pp. 574–597. DOI:10.3390/land3030574.

Carvalho, W.D., K. Mustin, R.R. Hilário, I.M. Vasconcelos, V. Eilers, and P.M. Fearnside. 2019. "Deforestation Control in the Brazilian Amazon: A Conservation Struggle Being Lost As Agreements and Regulations Are Subverted and Bypassed." *Perspectives in Ecology and Conservation* 17: pp. 122–130.

Cassells, D., X. Luo, and X. Chen. 2020. "Taking Forest and Landscape Restoration to Scale: Lessons from China. Restoring the Earth: The Next Decade." *Unasylva* 71 (252): pp. 9–17.

CBD. 2012. *Resourcing the Aichi Biodiversity Targets: A First Assessment of the Resources Required for Implementing the Strategic Plan for Biodiversity 2011–2020.* Montreal: CBD.

Chazdon, R., and P. Brancalion. 2019. "Restoring Forests as a Means to Many Ends." *Science* 365: pp. 24–25.

Chazdon, R.L., P.H. Brancalion, D. Lamb, L. Laestadius, M. Calmon, and C. Kumar 2017. "A Policy-Driven Knowledge Agenda for Global Forest and Landscape Restoration." *Conservation Letters* 10: pp. 125–132.

Chazdon, R.L., D. Lindenmayer, M.R. Guariguata, R. Crouzeilles, J.M.R. Benayas, and E.L. Chavero. 2020. "Fostering Natural Forest Regeneration on Former Agricultural Land through Economic and Policy Interventions." *Environmental Research Letters* 15: p. 043002.

Chinese Academy of Science. 2018. *Assessment of the Forty Years of Three-North Shelterbelt Programme.* Beijing: Chinese Academy of Science.

Climate Policy Initiative. 2022. "Landscape of Climate Finance for Agriculture, Forestry, other Land Use and Fisheries: Preliminary Findings." Climate Policy Initiative, London. https://www.climatepolicyinitiative.org/wp-content/uploads/2022/11/Landscape-of-Climate-Finance-for-Agriculture-Forestry-Other-Land-Uses-and-Fisheries.pdf

Colfer, C.J.P., R. Prabhu, and A.M. Larson (eds.). 2022. *Adaptive Collaborative Management in Forest Landscapes: Villagers, Bureaucrats and Civil Society.* New York: Routledge.

Convergence. 2021. The State of Blended Finance 2021. Convergence: Blending Global Finance. https://www.convergence.finance/resource/the-state-of-blended-finance-2021/view.

Cook, C.N., E.A. Beever, L.L. Thurman, et al. 2021. "Supporting the Adaptive Capacity of Species through More Effective Knowledge Exchange with Conservation Practitioners." *Evolutionary Applications* 14 (8): pp. 1969–1979.

Costanza, R., R. de Groot, L. Braat, et al. 2017. "Twenty Years of Ecosystem Services: How Far Have We Come and How Far Do We Still Need to Go?" *Ecosystem Services* 28 (Part A): pp. 1–16.

CPIC. 2023. Coalition for Private Investment in Conservation. http://cpicfinance.com/.

Dade, M.C., M.G. Mitchell, C.A. McAlpine, and J.R. Rhodes. 2019. "Assessing Ecosystem Service Trade-Offs and Synergies: The Need for a More Mechanistic Approach." *Ambio* 48: pp. 1116–1128.

Dang, T.K.P., M. Van Der Zouwen, and B. Arts. 2019. "Challenges of Forest Governance: The Case of Forest Rehabilitation in Vietnam." *Public Organization Review* 19: pp. 425–452.

De Groot, R.S., J. Blignaut, S. Van Der Ploeg, J. Aronson, T. Elmqvist, and J. Farley. 2013. "Benefits of Investing in Ecosystem Restoration." *Conservation Biology* 27: pp. 1286–1293.

De Jong, W., J. Liu, and H. Long. 2021. "The Forest Restoration Frontier." *Ambio* 50: pp. 2224–2237.

Dressler, W.H., D. Wilson, J. Clendenning, et al. 2016. "The Impact of Swidden Decline on Livelihoods and Ecosystem Services in Southeast Asia: A Review of the Evidence from 1990 to 2015." *Ambio* 46: pp. 291–310.

Duriaux-Chavarría, J.Y., F. Baudron, S.E. Gergel, et al. 2021. "More People, More Trees: A Reversal of Deforestation Trends in Southern Ethiopia." *Land Degradation & Development* 32: pp. 1440–1451.

Elias, M., M. Kandel, S. Mansourian, et al. 2022. "Ten People-Centered Rules for Socially Sustainable Ecosystem Restoration." *Restoration Ecology* 30 (4): p. e13574.

Enters, T., P.B. Durst, and C.L. Brown. 2006. "Stimulating Forest Plantation Development through Incentives–in Search of the Elusive Blueprint for Success." In *Strategic and Financial Mechanisms for Sustainable Use and Conservation of Forests: Experiences from Latin America and Asia*, edited by S. Appanah, E. Mansur, and R. Krezdorn, pp. 102–119. Chiang Mai: FAO.

Erbaugh, J., N. Pradhan, J. Adams, et al. 2020. "Global Forest Restoration and the Importance of Prioritizing Local Communities." *Nature Ecology & Evolution* 4: pp. 1472–1476.

Fagan, M. 2020. "A Lesson Unlearned? Underestimating Tree Cover in Drylands Biases Global Restoration Maps." *Global Change Biology* 26: pp. 4679–4690.

FAO. 2020. *Global Forest Resources Assessment 2020: Main Report*. Rome: FAO.

FAO and Global Mechanism of the UNCCD. 2015. "Sustainable financing for forest and landscape restoration: Opportunities, challenges and the way forward." Discussion paper. Rome: FAO and UNCCD.

FAO and RECOFTC. 2016. *Forest Landscape Restoration in Asia-Pacific Forests*, edited by S. Appanah. Bangkok: FAO/RECOFT.

Feng, Y., C. Yang, and W. Zhang. 2010. *Study on the Critical Issues of the Fifth Phase of Three-North Shelterbelt Programme*. China Forestry Publishing House.

Ferguson, I. 2014. "Australian Plantations: Mixed Signals Ahead." *International Forestry Review* 16: pp. 160–171.

Fletcher, M.-S., R. Hamilton, W. Dressler, and L. Palmer. 2021. "Indigenous Knowledge and the Shackles of Wilderness." *PNAS* 118 (40): p. e2022218118.

Frelich, L.E., K. Jõgiste, J. Stanturf, A. Jansons, and F. Vodde. 2020. "Are Secondary Forests Ready for Climate Change? It Depends on Magnitude of Climate Change, Landscape Diversity and Ecosystem Legacies." *Forests* 11 (9): p. 965.

Friedlingstein, P., M. Allen, J.G. Canadell, G.P. Peters, and S.I. Seneviratne. 2019. "Comment on 'The Global Tree Restoration Potential.'" *Science* 366: p. eaay8060.

Gadzinski, G., M. Schuller, and A. Vacchino. 2018. "The Global Capital Stock: Finding a Proxy for the Unobservable Global Market Portfolio." *The Journal of Portfolio Management* 44: pp. 12–23.

Ghazoul, J., M. Bugalho, and R. Keenan. 2019. "Forests: Economic Perks of Plantations." *Nature* 570: pp. 307–307.

Gitz, V., F. Place, I. Koziell, et al. 2020. A Joint Stocktaking of CGIAR Work on Forest and Landscape Restoration. FTA Working Paper 4. Bogor: CGIAR Research Program on Forests, Trees and Agroforestry.

Griscom, B.W., J. Adams, P.W. Ellis, et al. 2017. "Natural Climate Solutions." *PNAS* 114: pp. 11645–11650.

Guariguata, M., S. Atmadja, H. Baral, et al. 2021. Forest and Landscape Restoration. Highlight No.4. Bogor: CGIAR Research Program on Forests, Trees and Agroforestry.

Guariguata, M.R., R.L. Chazdon, P.H.S. Brancalion, and L. David. 2019. "Forests: When Natural Regeneration is Unrealistic." *Nature* 570: p. 164.

Gui, L. 2015. "Analysis on the Preserving Status and Related Problems of Afforestation in China." *Forest Resources Management* 3: pp. 15–19.

Hanson, C., K. Buckingham, S. DeWitt, and L. Laestadius. 2015. The Restoration Diagnostic. Washington: World Resources Institute. https://files.wri.org/d8/s3fs-public/WRI_Restoration_Diagnostic_1.pdf.

Holl, K.D. 2017. "Restoring Tropical Forests from the Bottom Up." *Science* 355: pp. 455–456.

Holl, K.D., and P.H. Brancalion. 2020. "Tree Planting Is Not a Simple Solution." *Science* 368: pp. 580–581.

Hua, F., L.A. Bruijnzeel, P. Meli, et al. 2022. "The Biodiversity and Ecosystem Service Contributions and Trade-Offs of Forest Restoration Approaches." *Science* 376: pp. 839–844.

Hyde, W.F. 2012. *The Global Economics of Forestry*. London: Routledge.

IETA University of Maryland and CPLC. 2019. *The Economic Potential of Article 6 of the Paris Agreement and Implementation Challenges*. Washington: International Emissions Trading Association.

Imbach, P.A., B. Locatelli, L.G. Molina, P. Ciais, and P.W. Leadley. 2013. "Climate Change and Plant Dispersal along Corridors in Fragmented Landscapes of Mesoamerica." *Ecology and Evolution* 3: pp. 2917–2932.

IPCC. 2018. "Summary for Policymakers." In *Global Warming of 1.5°C. An IPCC Special Report on the Impacts of Global Warming of 1.5°C above Pre-Industrial Levels and Related Global Greenhouse Gas Emission Pathways, in the Context of Strengthening the Global Response to the Threat of Climate Change, Sustainable Development, and Efforts to Eradicate Poverty*, edited by V. Masson-Delmotte, P. Zhai, H.-O. Pörtner, et al., pp. 3–24. Cambridge: Cambridge University Press.

ITTO. 2020. Guidelines for Forest Landscape Restoration in the Tropics. ITTO Policy Development Series No. 24. Yokohama: ITTO.

Jackson, R.B., E.G. Jobbágy, R. Avissar, et al. 2005. "Trading Water for Carbon with Biological Carbon Sequestration." *Science* 310: pp. 1944–1947.

Kalamandeen, M., E. Gloor, E. Mitchard, et al. 2018. "Pervasive Rise of Small-Scale Deforestation in Amazonia." *Scientific Reports* 8: pp. 1–10.

Kareiva, P.M., B.W. McNally, S. McCormick, T. Miller, and M. Ruckelshaus. 2015. "Improving Global Environmental Management with Standard Corporate Reporting." *PNAS* 112: pp. 7375–7382.

Keenan, R., D. Lamb, O. Woldring, T. Irvine, and R. Jensen. 1997. "Restoration of Plant Biodiversity Beneath Tropical Tree Plantations in Northern Australia." *Forest Ecology and Management* 99: pp. 117–131.

Keenan, R.J. 2015. "Climate Change Impacts and Adaptation in Forest Management: A Review." *Annals of Forest Science* 72: pp. 145–167.

Keenan, R.J., G.A. Reams, F. Achard, J.V. de Freitas, A. Grainger, and E. Lindquist. 2015. "Dynamics of Global Forest Area: Results from the FAO Global Forest Resources Assessment 2015." *Forest Ecology and Management* 352: pp. 9–20.

Kimiti, D.W., A.-M.C. Hodge, J.E. Herrick, A.W. Beh, and L.E. Abbott. 2017. "Rehabilitation of Community-Owned, Mixed-Use Rangelands: Lessons from the Ewaso Ecosystem in Kenya." *Plant Ecology* 218: pp. 23–37.

Koch, A., and J.O. Kaplan. 2022. "Tropical Forest Restoration under Future Climate Change." *Nature Climate Change* 12: pp. 279–283.

Koech, G., L.A. Winowiecki, O. Westermann, et al. 2020. "Regreening Africa: A Bottom-Up Transformation of Degraded Lands." In *Restoring African Drylands*, edited by N. Pasiecznik, and C. Reij, pp. 219–226. Wageningen: Tropenbos International.

Kremen, C., and A.M. Merenlender. 2018. "Landscapes That Work for Biodiversity and People." *Science* 362: p. eaau6020.

Krosby, M., J. Tewksbury, N.M. Haddad, and J. Hoekstra. 2010. "Ecological Connectivity for a Changing Climate." *Conservation Biology* 24: pp. 1686–1689.

Kusters, K., M. De Graaf, L. Buck, et al. 2020. "Inclusive Landscape Governance for Sustainable Development: Assessment Methodology and Lessons for Civil Society Organizations." *Land* 9 (4): p. 128.

Kumar, P. 2012. *The Economics of Ecosystems and Biodiversity: Ecological and Economic Foundations*. Oxon: Routledge.

Lamb, D. 2014. *Large-Scale Forest Restoration*. London: Routledge.

Lamb, D., J. Stanturf, and P. Madsen. 2012. "What is Forest Landscape Restoration?" In *Forest Landscape Restoration*, edited by J. Stanturf, D. Lamb, and P. Madsen, pp. 3–23. Dordrecht: Springer.

Latawiec, A.E., B.B. Strassburg, P.H. Brancalion, R.R. Rodrigues, and T. Gardner. 2015. "Creating Space for Large-Scale Restoration in Tropical Agricultural Landscapes." *Frontiers in Ecology and the Environment* 13: pp. 211–218.

Lewis, S.L., C.E. Wheeler, E.T. Mitchard, and A. Koch. 2019. "Restoring Natural Forests Is the Best Way to Remove Atmospheric Carbon." *Nature* 568: pp. 25–28.

Leys, A.J., and J.K. Vanclay. 2010. "Land-Use Change Conflict Arising from Plantation Forestry Expansion: Views across Australian Fence-Lines." *International Forestry Review* 12: pp. 256–269.

Li, J., J. Merten, G. Burke, and E. Mumford (eds.). 2018. "Application of Restoration Opportunities Assessment Methodology (ROAM) in Asia." Summary of Findings from the First Asia Regional ROAM Learning Exchange. Bangkok: IUCN Asia Regional Office.

Liu, S., S. Wu, and H. Wang. 2014. "Managing Planted Forests for Multiple Uses under a Changing Environment in China." *New Zealand Journal of Forestry Science* 44 (Suppl 1): p. S3.

Long, J.W., and F.K. Lake. 2018. "Escaping Social-Ecological Traps through Tribal Stewardship on National Forest Lands in the Pacific Northwest, United States of America." *Ecology and Society* 23: p. 10.

Louman, B., I. Gutiérrez, J.-F. Le Coq, et al. 2016. "Avances en la comprensión de la transición forestal en fincas costarricenses." *Revibec: revista iberoamericana de economía ecológica* 26: pp. 191–206.

Louman, B., R.J. Keenan, D. Kleinschmit, et al. 2019. "SDG 13: Climate Action – Impacts on Forests and People." In *Sustainable Development Goals: Their Impacts on Forests and People*, edited by P. Katila, C.J. P. Colfer, G. Galloway, W. de Jong, P. Pacheco, and G. Winkel, pp. 419–444. Cambridge: Cambridge University Press.

Louman, B., A. Meybeck, G. Mulder, et al. 2020. Innovative Finance for Sustainable Landscapes. FTA Working Paper 7. Bogor: CIFOR.

Luedeling, E., J. Börner, W. Amelung, K. Schiffers, K. Shepherd, and T. Rosenstock. 2019. "Forest Restoration: Overlooked Constraints." *Science* 366: pp. 315–315.

Maginnis, S., L. Laestadius, M. Verdone, et al. 2014. *A Guide to the Restoration Opportunities Assessment Methodology (ROAM): Assessing Forest Landscape Restoration Opportunities at the National or Sub-National Level.* Gland: IUCN.

Maginnis, S., J. Rietbergen-McCracken, and A. Sarre. 2012. *The Forest Landscape Restoration Handbook.* London: Routledge.

Magrath, J. 2020. "Regreening the Sahel: A Quiet Agroecological Evolution." Oxfam Case Study. Oxford: Oxfam.

Malhi, Y., T.A. Gardner, G. Goldsmith, M.R. Silman, and P. Zelazowski. 2014. "Tropical Forests in the Anthropocene." *Annual Review of Environment and Resources* 39: pp. 125–159.

Malkamäki, A., D. D'Amato, N.J. Hogarth, et al. 2018. "A Systematic Review of the Socio-Economic Impacts of Large-Scale Tree Plantations, Worldwide." *Global Environmental Change* 53: pp. 90–103.

Mansourian, S., H. Kleymann, V. Passardi, et al. 2022. "Governments Commit to Forest Restoration, but What Does It Take to Restore Forests?" *Environmental Conservation* 49 (4): pp. 206–214. DOI:10.1017/S0376892922000340.

Mansourian, S., J.A. Stanturf, M.A.A. Derkyi, and V.L. Engel. 2017. "Forest Landscape Restoration: Increasing the Positive Impacts of Forest Restoration or Simply the Area Under Tree Cover?" *Restoration Ecology* 25: pp. 178–183.

Mather, A.S. 1992. "The Forest Transition." *Area* 24 (4): pp. 367–379.

Matricardi, E.A.T., D.L. Skole, O.B. Costa, et al. 2020. "Long-Term Forest Degradation Surpasses Deforestation in the Brazilian Amazon." *Science* 369: pp. 1378–1382.

McDonald, G., M.W. Appleby, H. Sime, J. Radford, and A.A. Hoffmann. 2021. "Establishing a Climate-Ready Revegetation Trial in Central Victoria–A Case Study." *Ecological Management & Restoration* 22: pp. 256–265.

McElwee, P. 2009. "Reforesting 'Bare Hills' in Vietnam: Social and Environmental Consequences of the 5 Million Hectare Reforestation Program." *Ambio* 38: pp. 325–333.

McLain, R., S. Lawry, M.R. Guariguata, and J. Reed. 2018. "Toward a Tenure-Responsive Approach to Forest Landscape Restoration: A Proposed Tenure Diagnostic for Assessing restoration opportunities." *Land Use Policy* 104: p. 103748.

McLeod, S. 2021. "Feasibility Studies for Novel and Complex Projects: Principles Synthesised through an Integrative Review." *Project Leadership and Society* 2: p. 100022.

Mei, H. 2019. "Dynamic Comprehensive Evaluation and Influencing Factors of Afforestation Efficiency at Province Scale in China." *Chinese Journal of Ecology* 38 (12): pp. 3577–3584.

Meyfroidt, P., and E.F. Lambin. 2011. "Global Forest Transition: Prospects for an End to Deforestation." *Annual Review of Environment and Resources* 36: pp. 343–371.

Midgley, S.J., P.R. Stevens, and R.J. Arnold. 2017. "Hidden Assets: Asia's Smallholder Wood Resources and Their Contribution to Supply Chains of Commercial Wood." *Australian Forestry* 80: pp. 10–25.

Nájera, A., and J.A. Simonetti. 2010. "Enhancing Avifauna in Commercial Plantations." *Conservation Biology* 24: pp. 319–324.

National Forestry and Grassland Administration. 2019. Response to the "Proposal on Strengthening the Rehabilitation of Degraded Forests of the Three North Programme" (No. 1884 of 2019 (Resources and Environment No. 103)). https://www.forestry.gov.cn/main/4862/20191107/160826834623415.html.

NYDF Assessment Partners. 2019. "Protecting and Restoring Forests: A Story of Large Commitments yet Limited Progress." New York Declaration on Forests Five-Year Assessment Report. https://climatefocus.com/wp-content/uploads/2022/06/2019NYDFReport.pdf.

Ojha, H., and A. Hall. 2021. "Transformation as System Innovation: Insights from Nepal's Five Decades of Community Forestry Development." *Innovation and Development* 13 (1): pp. 109–131. DOI:10.1080/2157930X.2021.1917112.

Ojha, H., L. Persha, and A. Chhatre. 2009. "Community Forestry in Nepal: A Policy Innovation for Local Livelihoods." IFPRI Discussion Paper 00913. Washington: International Food Policy Research Institute. http://re.indiaenvironmentportal.org.in/files/Community%20Forestry%20in%20Nepal.pdf.

Olander, L.P., R.J. Johnston, and H. Tallis. 2018. "Benefit Relevant Indicators: Ecosystem Services Measures that Link Ecological and Social Outcomes." *Ecological Indicators* 85: pp. 1262–1272.

Palma, A.C., and S.G. Laurance. 2015. "A Review Of The Use Of Direct Seeding And Seedling Plantings In Restoration: What Do We Know And Where Should We Go?" *Applied Vegetation Science* 18: pp. 561–568.

Pannell, D.J. 2008. "Public Benefits, Private Benefits, and Policy Mechanism Choice for Land-Use Change for Environmental Benefits." *Land Economics* 84: pp. 225–240.

Paudyal, K., H. Baral, and R.J. Keenan. 2018. "Assessing Social Values of Ecosystem Services in the Phewa Lake Watershed, Nepal." *Forest Policy and Economics* 90: pp. 67–81.

Phimmavong, S., T.N. Maraseni, R.J. Keenan, and G. Cockfield. 2019. "Financial Returns from Collaborative Investment Models of Eucalyptus Agroforestry Plantations in Lao PDR." *Land Use Policy* 87: p. 104060.

Pichler, M., M. Schmid, and S. Gingrich. 2022. "Mechanisms to Exclude Local People from Forests: Shifting Power Relations in Forest Transitions." *Ambio* 51 (4): pp. 849–862.

Pistorius, T., S. Carodenuto, and G. Wathum. 2017. "Implementing Forest Landscape Restoration in Ethiopia." *Forests* 8 (3): p. 61.

Pistorius, T., and D. Gromko. 2019. "Economics of Forest Landscape Restoration." Policy brief. Freiburg: Unique Forestry and Land Use GmbH. https://www.unique-landuse.de/en/wp-content/uploads/sites/2/2023/10/Policy-brief-Economics-of-Fo.pdf.

Pugh, T.A., A. Arneth, M. Kautz, B. Poulter, and B. Smith. 2019. "Important Role of Forest Disturbances in the Global Biomass Turnover and Carbon Sinks." *Nature Geoscience* 12: pp. 730–735.

Reij, C., G. Tappan, and M. Smale. 2009. "Agroenvironmental Transformation in the Sahel: Another Kind Of 'Green Revolution.'" IFPRI Discussion Paper r 00914. Washington: International Food Policy Research Institute.

Rinaudo, T., S. McKenzie, T.-B. Huynh, and C.L Sterrett. 2020. "Farmer Managed Natural Regeneration: Community Driven, Low Cost and Scalable Reforestation Approach for Climate Change Mitigation and Adaptation." In *Handbook of Climate Change Management: Research, Leadership, Transformation*, edited by W. Leal Filho, J. Luetz, and D. Ayal, pp. 1–34. Cham: Springer International Publishing.

Ros-Tonen, M.A., and K. Kusters. 2011. "Pro-Poor Governance of Non-Timber Forest Products: The Need for Secure Tenure, the Rule of Law, Market Access and Partnerships." In *Non-Timber Forest Products in the Global Context*, edited by S. Shackleton, C. Shackleton, and P. Shanley, pp. 189–207. Berlin, Heidelberg: Springer.

Russell-Smith, J., C. Monagle, M. Jacobsohn, et al. 2017. "Can Savanna Burning Projects Deliver Measurable Greenhouse Emissions Reductions and Sustainable Livelihood Opportunities in Fire-Prone Settings?" *Climatic Change* 140: pp. 47–61.

Sacande, M., M. Parfondry, and C. Cicatiello. 2020. *Restoration in Action Against Desertification. A Manual for Large-Scale Restoration to Support Rural Communities' Resilience in Africa's Great Green Wall.* Rome: FAO.

Sanders, A.J., R.M. Ford, R.J. Keenan, and A.M. Larson. 2020. "Learning through Practice? Learning from the REDD Plus Demonstration project, Kalimantan Forests and Climate Partnership (KFCP) in Indonesia." *Land Use Policy* 91: p. 104285. DOI:10.1016/j.landusepol.2019.104285.

Sapkota, P., R.J. Keenan, and H.R. Ojha. 2018. "Community Institutions, Social Marginalization and the Adaptive Capacity: A Case Study of a Community Forestry User Group in the Nepal Himalayas." *Forest Policy and Economics* 92: pp. 55–64.

Schweizer, D., P. Meli, P.H. Brancalion, and M.R. Guariguata. 2019. "Implementing Forest Landscape Restoration in Latin America: Stakeholder Perceptions on Legal Frameworks." *Land Use Policy* 104: p. 104244.

Seymour, F., and J. Busch. 2016. *Why Forests? Why Now? The Science, Economics, and Politics of Tropical Forests and Climate Change.* Washington: Center for Global Development.

Shames, S., and S. Scherr. 2020. *Mobilizing Finance across Sectors and Projects to Achieve Sustainable Landscapes: Emerging Models.* Washington: EcoAgriculture Partners.

Sheil, D., and A. Bargués-Tobella. 2020. "More Trees for More Water in Drylands: Myths and Opportunities." In *Restoring African Drylands*, edited by N. Pasiecznik, and C. Reij, pp. 2019–2216. Wageningen: Tropenbos International.

Siles, J., and M. Prebble. 2018." Gender-Responsive Restoration Opportunities Assessment Methodology (ROAM): Engendering National Forest Landscape Restoration Assessments." GLF Brief No. 5. Bogor: CIFOR.

Silva, L.N., P. Freer-Smith, and P. Madsen. 2018. "Production, Restoration, Mitigation: A New Generation of Plantations." *New Forests* 50: pp. 153–168.

Society for Ecological Restoration. 2022. What is Ecological Restoration? https://www.ser-rrc.org/what-is-ecological-restoration/.

Smith, H., P. Kanowski, R.J. Keenan, and S. Phimmavong. 2021. "Lao Plantation Policy: Prospects for Change." *Forests* 12: p. 1132.

Stanturf, J.A., M. Kleine, and S. Mansourian. 2019. "Implementing Forest Landscape Restoration under the Bonn Challenge: A Systematic Approach." *Annals of Forest Science* 76: p. 50.

Stanturf, J.A., and S. Mansourian. 2020. "Forest Landscape Restoration: State of Play." *Royal Society Open Science* 7: p. 201218.

Stanturf, J.A., B.J. Palik, and R.K. Dumroese 2014. "Contemporary Forest Restoration: A Review Emphasizing Function." *Forest Ecology and Management* 331: pp. 292–323.

State Forestry Administration. 2008. *Report of the Thirty Years of Three-North Shelterbelt Programme.* Beijing: China State Forestry Administration.

State Forestry Administration. 2013. *Report of the Eighth National Forest Inventory.* Beijing: China State Forestry Administration.

Sunderland, T.C.H., A. O'Connor, G. Muir, et al. 2019. "SDG 2: Zero Hunger – Challenging the Hegemony of Monoculture Agriculture for Forests and People." In *Sustainable Development Goals: Their Impacts on Forests and People*, edited by P. Katila, C.J.P. Colfer, G. Galloway, W. de Jong, P. Pacheco, and G. Winkel, pp. 48–71. Cambridge: Cambridge University Press.

Svejcar, L.N., and O.A. Kildisheva. 2017. "The Age of Restoration: Challenges Presented by Dryland Systems." *Plant Ecology* 218: pp. 1–6.

TFND. 2022. Developing and Delivering a Risk Management and Disclosure Framework for Organisations to Report and Act On Evolving Nature-Related Risks. Task Force on Nature-related Disclosures. https://kpmg.com/sg/en/home/insights/2022/11/developing-and-delivering-a-risk-management-and-disclosure-frame.html.

The State Council of the People's Republic of China. 2014. Circular of the Ministry of Finance and the State Forestry Administration on Printing and Issuing the Measures for Administration of Central Subsidies for Forestry. In State Council Gazette, Issue 23. http://www.gov.cn/gongbao/content/2014/content_2736117.htm.

Tomaselli, I. 2012. "Forest Financing–Latin America and the Caribbean Region." Final Report. United Nations Forum on Forests (UNFF). https://www.un.org/esa/forests/wp-content/uploads/2014/12/Forest_Financing_LAC_Tomaselli.pdf.

UN Climate Summit. 2021. *Glasgow Leaders Declaration on Forests and Land Use.* Glasgow: UNFCCC. https://ukcop26.org/glasgow-leaders-declaration-on-forests-and-land-use/.

UNEP. 2022. Call on Policymakers to Support Scaling Blended Finance. United Nations Environment Program Finance Initiative. https://www.unepfi.org/wordpress/wp-content/uploads/2022/09/NZAOA_Scaling-Blended-Finance.pdf.

USDA. 2021. *Conservation Reserve Program.* Washington: United States Department of Agriculture.

Veldman, J.W., J.C. Aleman, S.T. Alvarado, et al. 2019. "Comment on 'The Global Tree Restoration Potential.'" *Science* 366: eaay7976.

Viani, R.A., K.D. Holl, A. Padovezi, et al. 2017. "Protocol for Monitoring Tropical Forest Restoration: Perspectives from the Atlantic Forest Restoration Pact in Brazil." *Tropical Conservation Science* 10: pp. 1–8.

Vivid Economics. 2020. *An Investor Guide to Negative Emission Technologies and the Importance of Land Use.* London: Vivid Economics.

Williams, M. 2008. "A New Look at Global Forest Histories of Land Clearing." *Annual Review of Environment and Resources* 33: pp. 345–367.

WRI. 2014. *Atlas of Forest and Landscape Restoration Opportunities.* Washington: World Resources Institute.

Xu, C. 2014. "Forest Management in China from Data of Eight Forest Resources Inventories." *Forest Econ* 4: pp. 8–11.

Zamora Cristales, R., D. Herrador, N. Cuellar, O. Díaz, S. Kandel, and J. Quezada. 2020. *Sustainability Index for Landscape Restoration*. Washington: World Resources Institute.

Zhang, K., C. Song, Y. Zhang, and Q. Zhang. 2017. "Natural Disasters and Economic Development Drive Forest Dynamics and Transition in China." *Forest Policy and Economics* 76: pp. 56–64.

Zhang, L., P. Sun, F. Huettmann, and S. Liu. 2021. "Where Should China Practice Forestry in a Warming World?" *Global Change Biology* 28 (7): pp. 2461–2475.

Zhu, J.-J., and X. Zheng. 2019. "The Prospects of Development of the Three-North Afforestation Program (TNAP): On the Basis of the Results of the 40-Year Construction General Assessment of the TNAP." *Chinese Journal of Ecology* 38: pp: 1600–1610.

Zomer, R.J., A. Trabucco, R. Coe, F. Place, M. van Noordwijk, and J. Xu. 2014. "Trees on Farms: An Update and Reanalysis of Agroforestry's Global Extent and Socio-Ecological Characteristics." Working Paper 179. Bogor: CIFOR.

11
Advances in forest restoration management and technology

John A. Stanturf, R. Kasten Dumroese, Stephen Elliott, Vladan Ivetic, Watit Khokthong, Michael Kleine, Mait Lang, Magnus Löf, Palle Madsen, Cindy Prescott, and Timothy Young

11.1 Introduction

The restoration of global forests is urgently needed to offset climate change, biodiversity loss, and the emergence of viral pandemics (Anderegg et al. 2020; Everard et al. 2020; IPBES 2018). Deforestation continues unabated (Cruz et al. 2020; Reid et al. 2017), and the projected effects of a warmer, drier climate (Allen et al. 2010; Fagan et al. 2020) will increase the area of forests lost or degraded (Stanturf 2020). Calls continue for massive tree planting programs (Mader 2019; Trees that Count 2020), often without regard for ecological, historical, social, or economic context (Fagan et al. 2020; Holl and Brancalion 2020).

Restoration has multiple objectives, and these should be explicitly stated (Stanturf et al. 2019). Forest landscape restoration (FLR) mandates a participatory process to set objectives that recognize the needs of local communities (Mansourian and Parrotta 2018). The choice of restoration technology should be appropriate to site and landscape conditions as well as socioeconomic and cultural contexts. Too often, evidence is insufficient to stipulate what technique is "best"; an adaptive approach is thus critical, one that includes monitoring, feedback, and mechanisms for further intervention (Stanturf et al. 2017). Good practice is context specific; what works "here" may be inappropriate "there" because of different stakeholder objectives or socioecological conditions (Burton and Macdonald 2011).

We recognize that forest restoration is multifaceted and more than simply planting trees. Nevertheless, trees are fundamental to forest restoration, and this chapter focuses on establishing forests where they are lacking. The techniques discussed can establish trees also outside of forests or restore degraded forests. Other chapters address the social (Chapters 4 and 8), ecological (Chapter 7), and governance (Chapters 3 and 4) aspects. The scope of this chapter is limited to the technical aspects of implementation (Stanturf et al. 2017; 2019).

Significant financial, logistical, and socioecological challenges in restoration practice require that advances in technology and management are appropriately deployed to successfully establish new trees in the landscape (Stanturf et al. 2017, 2019). We survey innovative techniques useful across wide ecological and socioeconomic settings that add value in the short term (Rametsteiner and Weiss 2006), making restoration easier or less expensive, especially for inaccessible areas. In the longer term,

John A. Stanturf, R. Kasten Dumroese, Stephen Elliott, Vladan Ivetic, Watit Khokthong, Michael Kleine, Mait Lang, Magnus Löf, Palle Madsen, Cindy Prescott, and Timothy Young, *Advances in forest restoration management and technology*
In: *Restoring Forests and Trees for Sustainable Development.* Edited by: Pia Katila, Carol J. Pierce Colfer, Wil de Jong, Glenn Galloway, Pablo Pacheco, and Georg Winkel, Oxford University Press.
DOI: 10.1093/9780197683958.003.0011

innovations must be robust under a changing climate and reduce the risk of failure if management attention wavers or funding declines.

The ambition to restore large areas, the frequent focus on native species, and the changing social and physical environment exacerbate technical challenges. Limited resources mandate targeting the most responsive sites for the specified objectives with sufficient understanding of social conditions and landscape context. On one hand, treating large areas in the timeframe needed to affect change might favor labor-replacing technologies, hence the current fascination with remotely controlled robotic vehicles. On the other hand, labor-intensive methods might be preferred to provide employment for local communities. Achieving adequate seedling survival requires quality plant material, including genetic composition (Davis and Pinto 2021), and effective planting and tending techniques to overcome low survival rates (Haase et al. 2021). What follows is technology-focused and applicable to stand initiation. Methods for sustainably managing restored forests are described in detail in Stanturf et al. (2014, 2017, 2019).

11.2 Planning and baseline conditions

Because restoration's scope is broad (Stanturf et al. 2017), opportunities to innovate occur at different spatial and temporal scales and multiple entry points. Implementation requires information to decide where to do what, when, by whom, and at what cost (Stanturf et al. 2019). Thus, planning projects (Brancalion and Holl 2020) and justifying investments require understanding and documenting initial conditions and monitoring outcomes (Andries et al. 2021; de Almeida et al. 2020). National and regional assessments of restoration needs, such as the Restoration Opportunities Assessment Methodology (IUCN and WRI 2018), need baseline data, available from existing sources, remote sensing, or ground-based surveys. Increasingly, drones can deliver the same resolution as ground measurements with less effort and expense.

11.2.1 Available information

Surveillance monitoring (Hutto and Belote 2013), such as periodic forest inventories (Chirici et al. 2012; Nesha et al. 2021), can provide baseline data useful for stratifying landscapes into somewhat homogeneous units that will respond similarly to interventions (Stanturf et al. 2017). Soil surveys and maps (Arnold 2016) supply interpreted information about broad differences in soil water regime, ground water level, root impeding layers, and fertility; these conditions can vary over small distances. Land use maps can suggest critical areas of high ecological value requiring protection or degraded areas needing restoration (Hansen et al. 2013).

Remote sensing data are acquired from different platforms: satellites, airplanes, or drones. Some data (USGS GloVis, Google Earth, EU Copernicus) are freely available and provide interpretive maps (Gorelick et al. 2017). Higher resolution products can be purchased. Publicly accessible global land cover maps (Buchhorn et al. 2019; Strahler et al. 1999) and networks such as Global Forest Change and Global

Forest Watch[1] supply analyses of regional and global forest cover changes and wildfires. Change estimates are available at 30 m spatial resolution starting from 2001 (Goldewijk 2001; Hansen et al. 2013).

11.2.2 Remote sensing data

Combinations of remote sensing platforms and sensors are available at the spatial and spectral resolutions needed to meet objectives. Available sensors operate in the visible, near infrared, passive thermal, and microwave spectra. Multispectral sensors and imaging spectrometers supply details across the reflectance spectrum (Liang 2005). Increasingly, drones provide a low-cost sensor platform (Paneque-Gálvez et al. 2014).

Predictive modeling from airborne laser scanning (ALS) uses LiDAR to map canopy height, cover, and topography (Large and Heritage 2009). Flight altitude and maximum scan angle perpendicular to flight direction (usually set to 25–30 degrees) determine the size of the mapped area. Point densities from 0.1 m^{-2} to 4.0 m^{-2} are sufficient for forest inventories. Higher point densities can be obtained at lower altitudes, substantially increasing costs. Predictive models of structure can be constructed if sample plots are available (Kotivuori et al. 2016). Foliage affects ALS, so seasonality must be considered (Arumäe and Lang 2018).

11.2.3 Ground-based data

Pre-restoration site assessments typically involve ground-based sampling. Line transects and "smart sampling" (probability proportional to predictions (PPP) or probability proportional to size (PPS) combined with aerial and terrestrial data) can assess pre-restoration site conditions (Hsu et al. 2021; Latpate et al. 2021). Remnant trees indicate the stage of degradation and help determine numbers and species of planted trees needed versus natural regeneration (Derroire et al. 2016). The ratio of crown cover to weeds (including lianas) is important for assessing the effectiveness of assisted natural regeneration (ANR).

The rapidly expanding GPS capabilities of smartphones are now as accurate as regular handheld GPS (horizontal error is about 1–10 m), putting mapping capabilities into the hands of rural communities (Andrachuk et al. 2019; Cronin et al. 2021). Planning for data entry, analysis, interpretation, and adequate training for users are critical steps for introducing such technological innovations. Handheld devices (dataloggers, GPS, laser range finders) are useful in planning and monitoring. For example, the Regreening Africa App from the World Agroforestry Centre (ICRAF) collects information on how farmers are managing and protecting trees.[2]

11.2.4 Monitoring regeneration outcomes

Monitoring at various times has different aims (Guariguata and Evans 2019; Hutto and Belote 2013). Initial regeneration outcomes, competition from other vegetation, and impacts of damage such as browsing, can trigger remedial action. Satellite imagery can produce vegetation indices to assess regeneration effectiveness over large areas (Liu et al. 2019). Remote sensing of regeneration survival is difficult, however, because seedlings are small (Fernández-Guisuraga et al. 2018). Ground-based sampling is effective over small areas. Sampling artificial regeneration straightforwardly compares seeding or planting density to survival, whereas assessing natural regeneration requires two steps: determining adequate sampling intensity (i.e., number and size of plots) and how many sample plots need to be stocked with desirable species to ensure that the future stand will meet restoration objectives (Matney and Hodges 1991). Local context and experience will influence sampling decisions.

11.2.5 Decision support systems

Web-based and smart-device decision-support tools with Cloud data storage are available for restoration activities, from species/site matching to monitoring (Table 11.1). Limited geographic and species coverage currently constrain the utility of these tools; sufficiently large databases including expert knowledge and advances in artificial intelligence should make these tools widely applicable.

11.3 Implementing forest restoration

Once the landscape has been stratified into homogeneous units, decisions must be taken as to the site-level actions to undertake, based on initial conditions. Four possible starting points are: areas with no or little existing forest, where a forest must be initiated (this can be further differentiated into five stages of degradation as shown in Figure 11.1); areas with some degraded forests requiring rehabilitation; forest remnants that require protection; and areas for farming and other uses. Agroforestry and silvopastoral methods can add trees to the landscape outside of forests (Erdmann 2005).

11.3.1 Stand initiation

The stand initiation phase (Oliver and Larson 1996), or regeneration niche (Grubb 1977), encompasses a sequence of events from seed production to dispersal, germination, establishment, and recruitment into the overstory. Successful regeneration depends upon suitable site and environmental conditions occurring at the appropriate time for each stage. Knowledge about weather seasonality and short and medium-range weather predictions (two weeks ahead) are important for planning stand initiation tasks. Advances in and greater accessibility to weather forecasting are improving operational planning. Although commonly characterized as passive or active (Chazdon et al. 2021; Stanturf et al. 2009), restoration interventions occur over a gradient of intensities (Table 11.2), each with its own benefits and costs.

Table 11.1 On-line decision support tools for Forest Landscape Restoration. Decision categories based on Stanturf et al. (2019)a

Name	Decision
Agroforestry Species Switchboard—Data on 26,000 plant species across a wide range of information sources to support research on trees and tree-based development activities such as agroforestry and wider restoration initiatives.	A
Capfitogen—Software for eco-geographical characterization of the germplasm collecting sites; identifies seed zones.	A
Collect Earth—Augmented Visual Interpretation for Land Monitoring using Google Earth.	S
Diversity for Restoration—Decision-making on the use of appropriate tree species and seed sources for tree-based restoration.	C, A
FAO Forest Restoration Monitoring Tool—Operational guidelines for the restoration of degraded areas, particularly in dryland forests.	S
Regreening Africa App—Phone app to collect data on tree planting and management.	S
Restoration Diagnostic—Method for developing FLR strategies by rapidly assessing the status of key success factors.	V, C
Restoration Opportunities Assessment Methodology (ROAM)—Evaluation tool for restoration readiness, restoration site prioritization, and decisions on intervention types, collaborative costing, and benefit-sharing.	V, C, A
Restor—Open-data platform to access ecological insights at the site level, showing current and potential tree cover, possible flora species, and carbon storage capacity.	C, A
SeedIT—Phone app to track, manage, and diversify seed collections.	A
Seedlot Selection Tool—GIS mapping program designed to help forest managers match seed lots with planting sites based on climate information; can be used to map current or future climates based on selected climate change scenarios.	C, A
SER 5-Star Recovery System—Tool for assessing and ranking a site's degree of recovery over time, compared with a reference ecosystem.	S
Vegetationmap4Africa—Web-based vegetation map with complementary species/site matching tool for Africa.	C
WePlan-Forests—Advanced forest ecosystem restoration planning and spatial modelling to support decision-makers and policy development.	A, C

[a] V = Visioning sets out the aspirational goals for Forest Landscape Restoration (FLR), often at a national or regional level. C = Conceptualizing turns goals into clear and measurable objectives that can be acted upon. A = Acting plans activities to turn objectives into accomplishments by developing a sequenced list of what will be done, where, when, by whom, and at what cost. S = Sustaining FLR over the long term combines management planning with monitoring and evaluation to provide feedback into earlier phases for potential corrective actions.

11.3.2 Passive techniques

Passive techniques rely on natural processes to introduce trees, although active intervention is usually needed. Passive approaches are advertised as low-cost, but this

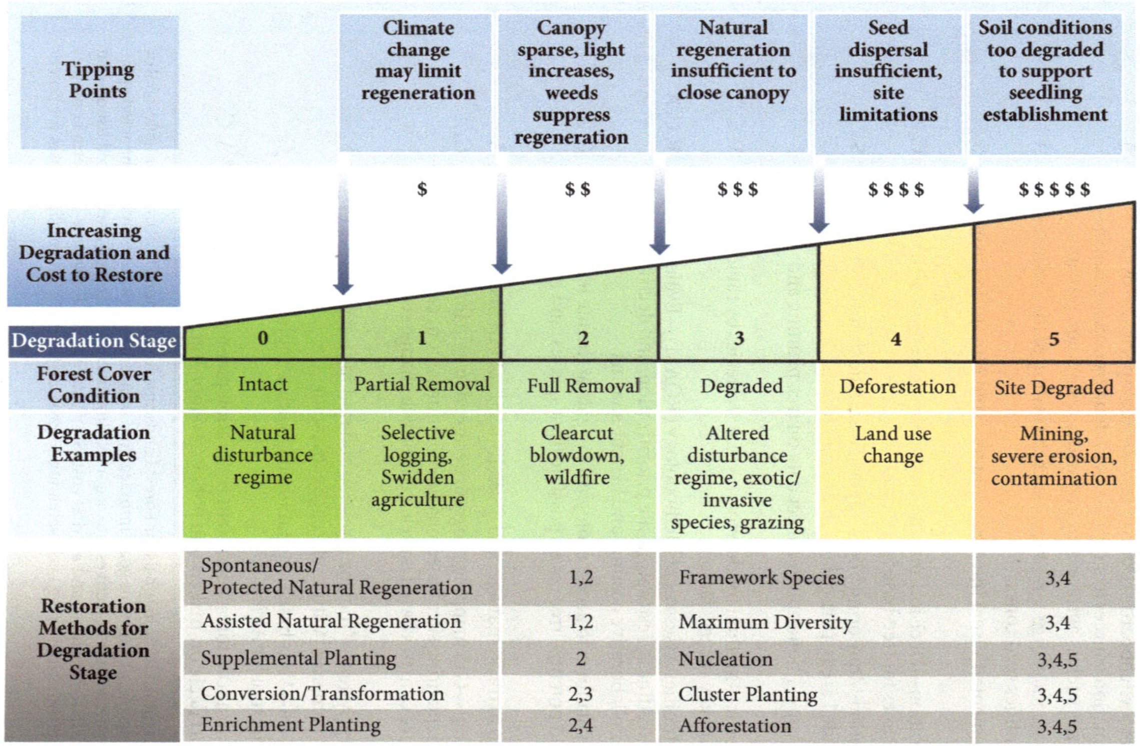

Figure 11.1 Stages of degradation and restoration methods. Tipping points are loss of tree cover that allows competing vegetation, sometimes invasive species, to dominate the regeneration niche, loss of seed sources within dispersal distance, and soil degradation. Distinct degradation agents predominate at different stages, causing progressive loss of tree cover. Tree cover is correlated with ecosystem functioning. As the degradation level increases (1 to 5), so does the cost to recover ecosystem functioning, as well as the intensity needed for restoration methods, from low intensity passive methods to higher intensity active methods (Design: Jim Marin Graphics)

Table 11.2 Comparison of regeneration methods by number of species, age, and spacing. Cost, resulting diversity, and the time to develop beyond the initiation phase are ranked qualitatively (based on Stanturf et al. 2014).

Regeneration Method	Number of Species	Number of Cohorts	Spacing	Cost	Diversity	Development Pace
Natural regeneration	X[a]	2+	Random or clustered	Low	Low to medium	Slow
Assisted Natural Regeneration (ANR)	X[a]	2+	Random or clustered	Low	Low to medium	Slow to medium
ANR with enrichment planting	X[b]	2+	Random, clustered, or dispersed	Low to medium	Medium	Medium
Monoculture	1	1	Uniform or random	High	Low	Fast
Interplanting/ Nurse crop	2	1	Uniform	High[c]	Medium	Fast
Cluster	1	1	Uniform or dispersed	Medium	Medium	Slow to medium
Cluster	2+	1	Uniform or dispersed	Medium	Medium	Slow to medium
Applied nucleation	1	1	Uniform or dispersed	Medium	Medium	Slow to medium
Applied nucleation	2+	1	Uniform or dispersed	Medium	Medium	Slow to medium
Simple mixture	2+	1	Uniform, species rows, or blocks	High	Medium	Fast
High density[d]	6+	1	Dispersed or uniform	Medium to high	Medium to high	Medium to fast
Underplanting or enrichment planting[e]	2+	2+	Uniform or random, with or without overstory gaps or row thinning	Low	Medium	Slow to medium

[a] X number of species depends on the available propagules on the site (e.g., buried seeds, root or stump sprouts, advance regeneration seedlings) or readily dispersed from surrounding forest fragments. Thus, the number of species that will successfully establish is difficult to predict.

[b] Assisted Natural Regeneration with enrichment planting intentionally adds species to those that regenerate naturally.

[c] Overall costs can be reduced if the nurse crop is harvested and sold.

[d] High density planting or direct seeding such as Framework or Miyawaki methods.

[e] Underplanting or enrichment planting is done in an established stand to increase diversity and/or eventually convert the stand to another assemblage pf species. Planting can be done in gaps created by partial harvesting including row thinning in plantations.

oversimplifies the situation (Zahawi et al. 2014). Natural regeneration is appropriate only where the density of available regenerants exceeds that needed to develop a canopy within the desired timeframe, given other constraints, such as grazing.

Natural regeneration

Natural regeneration requires sufficient sources of desirable species within effective dispersal distance (Caughlin et al. 2016). Sources may be patches of native forest or remnant trees within a degraded stand, including root or stump sprouts, seed banks, or advance regeneration (i.e., seedlings or saplings). Wind dispersed seeds, often of pioneer or early successional species, might be the main initial source. Remnant trees provide perches, attracting animals that drop seeds (Charles et al. 2017). In wetland forests, water and even fish dispersal is important. Species such as *Populus* and mangroves disperse vegetative material.

Much literature is available on naturally regenerated, managed forests that predominate globally (Dey et al. 2019; Holger et al. 2015; Scheper et al. 2021). Key factors are distances of effective dispersal, numbers of advance regeneration, and levels of limiting factors to survival and establishment, such as herbivores, fires, flooding, and competing vegetation. Germination and establishment are favored on shaded sites, which have milder environments and moister soil than open sites. Open areas, however, may favor established seedling growth.

Different targets for the number of natural regenerants to secure forest cover of tropical forests have been proposed, from a low of 500 to a high of 3,100 regenerants ha^{-1} taller than 50 cm by the end of the second rainy season for tropical areas with 1,000–2,000 mm annual rainfall (Shono et al. 2007). In dry tropical forests, a relatively high number of tree species have small, dry, wind-dispersed seeds (Vieira and Scariot 2006). In wet tropical forests, animals are important seed dispersers (Palma et al. 2021). Resprouting ability is critical in forests adapted to frequent fires (Vieira and Scariot 2006).

Assisted natural regeneration (ANR)/farmer managed natural regeneration (FMNR)

Competing vegetation hampers the growth of natural regenerants. Suppressing competition around single seedlings or clusters of regenerants can cost-effectively accelerate forest recovery. This is often done by hand, using a machete or a hoe. Alternatively, a wood board (15–30 cm wide, 1–1.5 m long) pushed down by foot onto the ground by an operator without having to bend down, can suppress grasses and weeds around seedling clusters (FAO 2019).

The cost of ANR includes labor and reflects the size of the treated area, transportation costs that depend on remoteness, and maintenance frequency that depends on the time required for the tree canopy to close and suppress weeds (Table 11.3). Added costs are incurred for enrichment planting and protection, such as fencing. Dugan (2011) estimated ANR implementation costs about USD 600 ha^{-1} in the Philippines with firebreak maintenance and weed-pressing eight times over three years. Ong (2011) estimated costs for vine cutting and liberation of potential crop trees in a degraded Malaysian forest at USD 70–100 ha^{-1}. In both cases, planting would have cost USD 1,000 ha^{-1} (not adjusted for inflation).

Table 11.3 Comparison of common restoration tasks accomplished by conventional versus automated methods

	Conventional		**Automated**
Task	Methods	Pros (+) and Cons (-)	Potential Methods
Pre-Restoration Site Assessment	Field team samples vegetation, counts regenerants, estimates weed cover, records barriers to regeneration.	+Eyes at ground level more likely to detect obscured regenerants; -Small sample plots cover low percentage of the site.	• Digital photography by drone. • Interpretation by eye or artificial intelligence.
Weeding	Slashing with blades, digging out roots with hoes, manual application of herbicides.	-Health hazards; -Often lacking available local labor.	• Smart herbicides and/or smart spraying by drones, combined with plant recognition systems.
Locating Seed Trees	Scanning underside of forest canopy with binoculars by walking along trails.	-Scans small areas; -Seeds tend to be collected from few well-known trees; -Narrow genetic diversity.	• Recognition of fruiting trees from above, by imagery from drones. • Seeds collected by drone-mounted tools.
Enrichment Planting	Conventional tree planting.	+Seedlings survive better than direct seeding; -Practical only near vehicular access; -Often lacking available local labor.	• Aerial seeding by drone.
Canopy Closure	Hemispherical photography from the ground.	-Expensive; -Equipment must be moved manually from point to point.	• Digital photography by drone from above canopy. • Interpretation by eye or artificial intelligence.
Forest Structure and Carbon	Basal area measured by hand. Allometry.	-Allometric equations to estimate carbon storage have high error margins.	• Drone-mounted LiDAR. Below canopy measures.
Tree Species Richness/ Diversity	Ground survey.	-Sample plots cover small part of area; -Needs field botanist or herbarium access.	• Digital photograph by drone with pattern-matching algorithms to identify species.
Wildlife Recovery	Direct observation or recording tracks and signs.	+Small animals more likely to be recorded; -High transportation costs; -Wildlife disturbed; -Health and safety hazards for field workers.	• Advanced camera traps and microphone arrays. • Data transfer by phone network or drone. • Thermal imagery from drones.

11.3.3 Active approaches

Active measures can increase stem density and accelerate canopy closure, increase species richness, and/or reintroduce rare species. Conventionally this is done through direct seeding or planting of nursery-raised seedlings propagated from locally collected seeds or vegetatively from cuttings, or a combination of planting and seeding, sometimes augmented by natural regeneration. Tree planting is a multi-year process that requires more attention and infrastructure than simply digging a hole and planting a seedling (Figure 11.2).

Direct seeding

Direct seeding is an alternative to planting that avoids some of the risks of relying on natural regeneration. Seeds must be collected, stored, sown at the optimal time for germination, and treated to maximize germination (Waiboonya and Elliott 2020). Direct seeding is cheaper than planting (Camargo et al. 2002; Engel and Parrotta 2001), but only if successful (Bullard et al. 1992). Cost savings range from 30–84% (Baumhauer et al. 2005; Summers et al. 2015), but are mostly >50% (Grossnickle and Ivetić 2017; Pérez et al. 2019). Directly seeded trees might have better root development and grow faster than poor quality nursery trees (Tunjai 2012), but direct seeding only works with some species and certain site conditions. Sowing a mixture of many species can be effective if seeds are readily available (Miyawaki 2004), although the high densities mandated (20,000 ha^{-1}) increase costs and logistical challenges.

Few seeds germinate and fewer seedlings survive to maturity (Bonilla-Moheno and Holl 2010; Cole et al. 2011). Sometimes multiple seeds are placed in a seeding spot to increase the chance that one survives; this leads to a three-fold or higher increase in the amount of seeds needed. Even so, sowing many seeds might not achieve targets; mean establishment rates (seedlings alive at least one year) are low (17% for tropical trees, 28% for temperate hardwoods, and 16% for temperate conifers) and nearly half of direct-seeding operations are unsuccessful, failing to meet stocking objectives (Grossnickle and Ivetić 2017).

Species that regenerate well through direct seeding tend to have spherical seeds larger than 0.1 g dry mass, with 36–70% moisture content (Tunjai and Elliott 2012). Large seeds have more food reserves, persist longer, and produce more robust seedlings compared to smaller seeds. Many legumes fix nitrogen, giving them a competitive advantage against non-leguminous weeds.

New germinants are smaller than nursery-raised seedlings and might require careful weeding to persist and additional costs (Doust et al. 2006; Tunjai 2012). Newly developed technologies have shown to increase establishment success by overcoming barriers to seed and germinant desiccation, seed and seedling predation, browsing, and competition from herbaceous weeds. Desiccation and predation can be overcome by selecting tree species with thick seed coats, burying seeds, or mulching seeding spots (Doust et al. 2006; Hau 1997; Woods and Elliott 2004). Pre-planting seed treatments to break dormancy and initiate germination reduces the time seeds are vulnerable to predators (Woods and Elliott 2004). Chemical repellents and protective structures to deter rodent seed predators have yielded mixed results (Castro

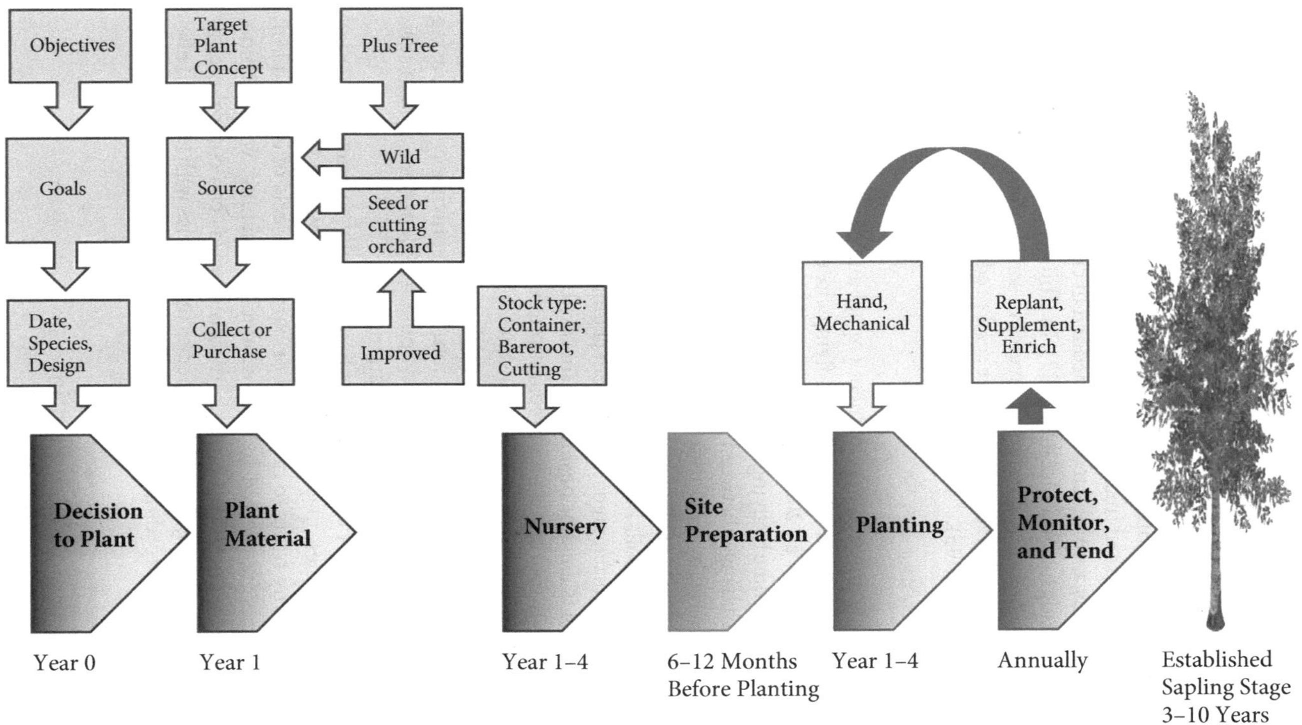

Figure 11.2 Simplified planting timeline: details are in the text. Planting is an active method of FLR and should be seen as an integral component of long-term forest ecosystem management. Participatory processes will have set goals and objectives; the aim of planting is to successfully establish saplings of the target species. Material can be purchased or produced from collected seeds or vegetative material, including wildings that might require a hardening period in a nursery. The nursery phase can last from a few months to multiple years, depending on the target plant material. If no nursery exists, it might take several years to start up a new facility. Site preparation can include fencing or wildfire protection. Tending most frequently includes weeding and fertilizer application. If initial planting produces lower survival rates than desired, additional plants can be added (Design: Jim Marin Graphics)

et al. 2015). Carnivores that prey on small seed-eating mammals should be encouraged and protected around direct seeded sites.

Aerial seeding

Aerial seeding has been used to establish pines or eucalypts, particularly for post-fire restoration. Aerial seeding is less reliable than direct seeding because precise placement, seedbed preparation, and burial are not options. Seldom used for restoring mosaic landscapes, it could be an option for large remote areas, using many of the preferred species and pre-sowing seed treatments developed for direct seeding. In North America and Australia, seeds, unprotected or embedded in clay pellets, were dropped by plane or helicopter (Hodgson and McGhee 1992). In China, the method has established millions of hectares of mostly conifers to reverse desertification (Sannai 2003). Chemical repellents have been developed to deter seed predators and prevent desiccation (Nuyun and Jingchun 1995).

Drone seeding

Drone seeding is a relatively untested development (Aghai and Manteuffel-Ross 2020; Amorós and Ledesma 2020; Elliott 2016). Drones flying close to the ground can drop seeds with precision (Mohan et al. 2021). Disadvantages of low payload capacity and short battery life can be overcome by flying multiple sorties of multiple drones with rapid seed reloading and battery changes between flights. Short battery life can also be overcome by using drones powered by combustion engines. An effective delivery system for species mixtures consists of seeds in biodegradable projectiles, capable of penetrating the weed cover and lodging in the soil. Projectiles might contain polymer gel (to prevent desiccation), slow-release fertilizer, predator-repellent chemicals, and microbial inocula.

Ineffective weed control is a major obstacle to successful drone seeding. Although aerial herbicide spraying is routine in site preparation, few herbicides are available that selectively kill weeds but not tree seedlings. Besides nontarget damage to desired tree species, drawbacks to herbicides include drift to distant areas, high costs, impacts on human health, and environmental contamination. Smart spraying systems, using computer vision to distinguish between target and nontarget plants, are used in agriculture and might be transferrable to forest restoration.

Exploiting the allelopathic properties of plants that naturally colonize deforested sites, particularly pioneer species, may be an alternative to herbicides (Intanon and Sangsupan 2020). Allelopathy refers to beneficial or harmful effects of one plant on another via allelochemicals transferred by root exudation, leaching, volatilization, or decomposition. Allelochemicals that target herbaceous weeds can control pre- or post-emergence weeds. Identifying desired tree species or varieties with elevated allelopathic effects on herbaceous weeds is a promising area for research.

11.3.4 Planting

Planting is done by hand or is mechanized; tree planting crews use hand tools (shovels, hoes, dibbles) depending upon the stock type (bare root, container), or sometimes mechanical augurs for large seedlings. Mechanized systems usually include a platform pulled by a prime mover with a system for feeding seedlings and closing the planting slit. Feeding can be done automatically or by one or more planting personnel (Ersson et al. 2018).

Increased computing power and artificial intelligence have stimulated the design of robots and semi-automated machines for agriculture and horticulture, with potential for site preparation, planting, cleaning, and thinning (Aguiar et al. 2020). Robots need task maps with a main axis orientation with respect to north; robotic arms then work relative to the main axis. This requires precise data about stones, debris, stumps, local microrelief, advance regeneration, and retention trees. Fully automated systems detect and interpret close objects using onboard data collection via LiDAR, radar, or photographic devices (Orumaa et al. 2021).

Planting designs

The simplest planting design is a single species planted in equidistant rows and space between plants, resulting in a specified density. While this single-cohort, uniform design might not meet restoration objectives calling for diverse stands, it can act as an interim stage to recover severely degraded soils (Stanturf et al. 2014). Dispersed and random planting designs appear more natural and random mixtures result in stands diverse in composition and structure (Lamb 2011). Designed mixtures require considerable knowledge of species traits and growth patterns to avoid stagnation or crowding out of slower growing species (Lockhart et al. 2008).

Planting only a portion of an area and relying on natural regeneration to fill openings increases diversity while lowering costs. This can effectively restore large-seeded species that do not readily disperse. Applied nucleation, where planted islands are dispersed across the landscape, can include one or more species (Holl et al. 2020; Wilson et al. 2021). Examples of this and other methods are described in Stanturf et al. (2014).

11.3.5 Regeneration methods

Each planting method has its own advantages and disadvantages for particular situations and objectives; choosing the "best" approach depends on context, including local knowledge and capacity as well as available material (Table 11.2). Natural regeneration costs the least while planting is the least risky; direct seeding is in between. Natural regeneration guarantees locally adapted genotypes, but planting can control species composition and introduce genetically improved or climatically adapted material. Seed availability limits direct seeding. Direct comparisons of regeneration methods are few (see Stanturf et al. 2009), and failures are seldom reported, thus recommendations might risk implicit bias (Reid et al. 2018).

11.4 Plant materials and genetic sources

Ensuring biodiversity is critical during the collection, development, and deployment of plant materials. For dioecious species, ensuring sexual diversity is necessary for sustainable reproduction, especially when vegetative propagation is involved (Hultine et al. 2016; Landis et al. 2003). Maintaining sexual and genetic diversity within species, as well as species diversity across the landscape (Thompson et al. 2009), is key to enhancing ecosystem resilience (Liang et al. 2016; Seidl et al. 2016).

Plant material supply can be a bottleneck in restoration (Acevedo et al. 2021; Bannister et al. 2018; Fargione et al. 2021). It takes time and investment to scale up the supply of seeds and plants to be available at the right time of year. The capacities, equipment, and facilities to collect, store, pretreat, and sow the seeds in nurseries must be developed and implemented to achieve planting success. Appropriate genetic material might be limited or not yet identified. Biotechnology provides advanced reproduction techniques to restore rare or extirpated species as well as identifying species or provenances better adapted to the future climate.

Biotechnology allows land managers to identify and use the "target plant concept"—materials with (morpho-physiological and genetic) characteristics that support tree survival and growth when confronted with (biogeophysical and societal) site-limiting factors (Davis and Pinto 2021; Dumroese et al. 2016). Biotechnology runs the gamut from defining plant transfer guidelines to traditional tree breeding programs to molecular genetics/genomics to genetic engineering. Materials can include seeds or plants, and plants can originate from seeds or vegetative propagation, including macro-cuttings (traditional hard- or softwood cuttings), or micro-cuttings (tissue culture; embryogenesis). Seeds and cuttings can be collected from the wild, seed orchards, or hedging/stooling beds. Employing biotechnology can be a complex, controversial challenge (Dumroese et al. 2015). Ignoring the biological, technological, and sociological aspects of plant material procurement and delivery, however, will impede successful forest restoration (Acevedo et al. 2021; Bannister et al. 2018; Haase and Davis 2017).

11.4.1 Transferring plant materials across the landscape

Because trees might be unable to migrate as rapidly as climate changes (Aitken et al. 2008), assisted migration (Dumroese et al. 2015) has been advanced as a tool to facilitate the continuation of adapted populations across landscapes (Pedlar et al. 2012; Williams and Dumroese 2013), albeit not without opposition (Aubin et al. 2011). Temperature and precipitation are primary drivers of adaptation for many species (Aitken and Bemmels 2016; Leites et al. 2019). This relationship is the basis for adaptive plant material transfers: the traditional static zone is replaced with dynamic approaches that rely on climate models predicting future conditions (Fremout et al. 2021).

Fixed and flexible guidelines

Foundational to restoration is that plant materials are adapted to their intended sites; a basic tenet is that local sources are best adapted to local conditions (Broadhurst et al. 2008). Specific zones enclosing similar conditions of climate, soils, etc. delineate the "local" boundaries within which plant materials can be moved ostensibly without maladaptation; these provisional zones guide transfers until data prove otherwise

(Bower et al. 2014; Pike et al. 2020a). Fixed or continuous zones for transferring plant materials provide recommendations for how far materials can be transferred latitudinally, longitudinally, and altitudinally from their point of origin (Parker and Niejenhuis 1996).

The risk of extreme or anomalous weather events such as drought need to be considered. Provenance trials and common garden studies have identified sources with desired traits for many commercially important tree species that may or may not originate locally. These data have allowed the development of formal guidelines for collecting and transferring plant materials beyond "local" distances and for developing seed orchards. Nevertheless, local adaptation can be highly variable among species (Ennos et al. 1998).

Understanding patterns of local adaptation or provenance data can further refine these approaches (Aitken et al. 2008). Operational tools, such as the Seedlot Selection Tool[3] and British Columbia's climate-based seed transfer system (O'Neill et al. 2017), can aid decision-making and potentially allowing wider seed deployment, thereby helping mitigate limited seed supplies. Perceived risks of assisted migration can be managed by "provenancing": restoration sites receive mixed sources, augmenting the local one with one(s) matched to future climates (Aitken et al. 2008; Broadhurst et al. 2008).

11.4.2 Seed orchards

Seed orchards provide substantial amounts of diverse plant materials but require significant long-term investments (Funda and El-Kassaby 2013). Retaining a broad genetic base and concurrently combining desired characteristics into the progeny is a fundamental goal for orchards (Ruņģis et al. 2019; Sønstebø et al. 2018; White 1987). Although climate is considered during the initial placement of seed orchards (Enescu 1987; Germishuizen and Gardner 2015), changes in climate can threaten the long-term viability of these resources (Ivetić and Aleksić 2016).

11.4.3 Genomics, molecular genetics, genetic engineering, and genome editing

Rapid advancements in genomics, molecular genetics, and bioengineering are transforming plant material sourcing, development, and deployment. Genomics looks at genes and their interrelationships to understand their combined influence on growth. This can help assess climate adaptation within plant species (Hamilton et al. 2013), enabling guideline adjustments for material transfers (O'Neill et al. 2017); it can also identify suites of genes for breeding desired traits, especially resistance to invasive species (Bauman et al. 2014; Pike et al. 2020b). Molecular genetics can inform breeding strategies; rapidly improving technology is reducing the resources required to obtain genomes, facilitating plant transfers, and responding to climate change (Neale and Kremer 2011).

Genetic engineering and genome editing modify DNA to include a new trait (e.g., pest resistance), offering options beyond traditional breeding (Dumroese et al. 2015).

Genetic engineering uses a viral or bacterial carrier to insert DNA, whereas genome editing removes or replaces targeted DNA (Mackelprang and Lemaux 2020). One editing tool, CRISPR-Cas, is increasingly used across multiple disciplines (Jaganathan et al. 2018). In natural forests, where many tree species are at risk from invasive species, genetic editing to rescue at-risk species is quicker than traditional breeding (Novak et al. 2018). CRISPR can "stack" multiple favorable genes to increase the durability of introduced traits, especially for pest resistance (Vincelli 2016), or to obtain multiple desired outcomes (Chanoca et al. 2019).

The rapid evolution of genetic modification techniques continues to focus global attention on biosafety; the debate is shifting from "what is possible" to "what is acceptable" (Gamborg and Sandoe 2010; Kuzma 2019; Mackelprang and Lemaux 2020). There may be less regulatory constraints on CRISPR-derived plants than genetically engineered ones (Waltz 2018).

Large-scale commercial use of somatic embryogenesis provides a pathway for deploying genetically engineered trees (Denchev and Grossnickle 2019; Grossnickle et al. 1996). The higher costs of these materials relative to nursery-produced plants (Bonga 2015) are justified when viewed through an ecological, economic, or societal lens (Jacobs et al. 2013). Addressing multiple bottlenecks throughout the entire somatic embryogenesis production chain (Egertsdotter et al. 2019) and the use of robotics (Hata et al. 2006) can upscale delivery through gained efficiencies. Using somatic embryogenesis to develop artificial seeds, in combination with conventional nursery practices, could reduce costs and foster implementation (Guan et al. 2016; Sharma et al. 2013).

11.4.4 Nursery practices

Given the large number of plants needed to meet current tree planting goals (Haase and Davis 2017), maximizing the efficiency of nurseries is critical to producing quality materials. Only recently has attention been directed toward restoration objectives, explicitly linking quality plant materials, appropriately outplanted, to restoration success (Bannister et al. 2018; Höhl et al. 2020; Löf et al. 2019; Stanturf et al. 2014;). Implementation has demonstrated that the target plant concept can help achieve restoration goals (Dumroese et al. 2016; Kildisheva et al. 2017).

11.4.5 Upscaling seed collection

Locating seed sources is a critical step in propagating planting stock and in verifying that nearby sources provide a sufficiently diverse and dense seed rain to naturally regenerate species richness (Caughlin et al. 2016; Elliott 2016). Locating enough quality seed trees is a major bottleneck (Jalonen et al. 2018; León-Lobos et al. 2020). Broad adoption of drone or direct seeding will dramatically increase demand for seeds; a single drone can deliver hundreds of thousands of seeds per day and the conversion rate (from seed to established tree) is low, about 5% or less (Aghai and Manteuffel-Ross 2020). The seed-to-seedling ratio for direct seeding can be as high as 2:1 for temperate conifers (Grossnickle and Ivetić 2017), but restoration projects are lower, from 10:1 to complete failure (Commander et al. 2013; Merritt and Dixon 2011).

Conventional seed-collection methods in natural stands are slow, inefficient, and expensive. Collectors must search for target species bearing ripe seeds, then climb trees or collect fallen seeds. Identifying and mapping seed trees from above could reduce search times. Seed collectors, using handheld GPS devices, could then plan efficient collection routes, minimizing search and travel times. Initial efforts to recognize seed trees from aircraft (Garzon-Lopez et al. 2013) have been replaced with more sophisticated approaches, including hyperspectral remote sensing in combination with LiDAR (Baldeck et al. 2015; Singh et al. 2015). Spectral signatures, however, vary among individual trees within the same species (Miyoshi et al. 2020), so reliability and accuracy might be questionable and small 4-band (RGB + NIR) cameras on drones can be more cost-effective than hyperspectral remote sensing.

Imaging from airplanes registers gross crown morphology and cannot determine the presence or ripeness of fruits (Frame and Garzon-Lopez 2020). Consumer drones are less expensive than airplanes, require no airport, and provide practitioners with miniaturized hyperspectral and LiDAR sensors. Disadvantages include limited battery life, small area covered per flight, and high costs (Table 11.3). Drones fly a few meters above the forest canopy, providing close-up views of taxonomic characteristics. A drone equipped with a 20-megapixel RBG camera and 4K video can locate potential seed sources of distinctive tree species, such as palms or cycads in a tropical forest. More species can be easily located when they flower.

Flight-planning software enables waypoint flying (such as Litchi[4]), allowing operators to position drones repeatedly above the same trees, enabling phenological observations to determine favorable seed-collection times. Species-identification tools using artificial intelligence, such as smartphone apps (Pl@ntNet and Leafsnap[5]), may soon be deployed on drones (Bonnet and Frame 2020). Image databases must be expanded with tree crown samples for recognizing species and the presence/absence of ripe fruit.

Currently, collecting tree seeds requires skilled tree climbers, bucket trucks in seed orchards, or helicopters with an aerial raking device.[6] In the horticulture industry, robots equipped with computer vision to recognize fruits, and devices to remove and stow fruits on-board, are now commonplace (Hua et al. 2019). Such technology could be applied to drones for the aerial harvesting of fruits; consumer drones already have accessories to pick up objects, but none have the dexterity or stability needed to pull fruits from trees. The autonomous, ground-based, commercial fruit-harvest systems available are bulky and slow; but if miniaturized and mounted on drones—with visual feedback able to control the arm-joint positions and flight-positioning systems—such technology could collect fruits from a forest canopy. Additionally, each fruit type requires a separate recognition algorithm. Ground-based robot arms harvest apples using a stereo camera to triangulate the position of each apple (Onishi et al. 2019). Systems are available for other fruits and vegetables (Defterli 2016; Hua et al. 2019) and some can detect ripeness (Hayashi et al. 2014).

11.5 Site modification

Technology for manipulating sites can improve seedling survival and establishment but are underutilized in restoration. Site modification methods include altering the physical, chemical, or biological environment. Some methods treat an entire area, others only the planting spot.

11.5.1 Physical soil manipulation

Mechanical site preparation (MSP) often relies on heavy machinery, but alternative machines, such as mini excavators are being developed (Collet et al. 2021). MSP methods include scarification, mounding, and subsoiling to improve soil conditions and reduce competing vegetation, facilitate planting, and improve seed germination and seedling survival and growth (Prévost 1992; Sutton 1993). Species respond differently due to multiple interactions among soil properties, weather, and biotic agents (Bergquist et al. 2009). MSP should be carefully implemented to protect soil health and biodiversity (Löf et al. 2015). As most research and development have concentrated on commercial species (Löf et al. 2012; Stanturf et al. 2003), the scientific basis for MSP during restoration is weak and research is needed to match MSP methods to tree species and sites.

Water harvesting and conservation in semi-arid and arid regions

Aridity and degradation often interact; low soil fertility decreases plant water availability and erosion on steep slopes reduces water infiltration (Löf et al. 2015). Water harvesting and conservation during seedling establishment have been practiced for many years in some areas (Tenbergen et al. 1995). Examples include semicircular water harvesting bunds in Kenya[7] and trenches, fanya juu terraces, and gabion check dams in Ethiopia.

Micro-catchments and fog collecting systems improve water availability by changing the soil structure (deeper planting holes, constructed channels and reversed slopes, stone wells) to direct runoff toward seedlings, facilitate infiltration, and reduce surface evaporation. Micro-catchments are low-cost alternatives to other watering systems, but results vary (Vallejo et al. 2012). Some techniques ignore cost or impact (Löf et al. 2015). For example, fog collecting systems require regular maintenance and might have a negative visual impact on the landscape.

Mulching and water-absorbing gels

Surface mulches can improve plant survival and growth in water-limited environments by reducing evaporation and temperature at the soil surface, thereby increasing water infiltration and storage (Coello et al. 2018; Prosdocimi et al. 2016). Even a thin layer of wood-chips on the soil surface can reduce erosion and increase plant survival on steep slopes (Breton et al. 2016). Biodegradable mulches, such as wood chips, paper residues, or jute, are preferable to plastic.

Soil conditioners (hydrogels) absorb moisture during rain events and release it slowly (Hüttermann et al. 2009). Hydrogels can enhance early survival and growth

(Chirino et al. 2011; Hueso-González et al. 2016; Hüttermann et al. 1999) particularly in coarse-textured soils (Coello et al. 2018) but might be ineffective in clay soils (Werden et al. 2018). While most hydrogels are non-biodegradable synthetic polymers (Hüttermann et al. 2009), current research focuses on biodegradable hydrogels (Kumar et al. 2019).

Extracellular polymeric substances (EPS) secreted by soil microorganisms are sources of stable soil organic matter (SOM) and offer promise to restore degraded soils by enabling microorganisms to "stick" to pore surfaces and soil particles, promoting soil structure and stability (Chang et al. 2015; Costa et al. 2018; Wu et al. 2014). Applying EPS extracted from cyanobacteria increased water-holding capacity of a dryland soil (Adessi et al. 2018) and increased seed germination, ion uptake, and photosynthetic activity of a shrub in a desert soil (Xu et al. 2013). Inoculating soil with EPS-producing organisms, especially cyanobacteria that exude exopolysaccharides, is effective in gluing soil particles together, thereby improving water infiltration and reducing soil erosion (Mager and Thomas 2011) and hastening the accumulation of SOM and nutrients (Adessi et al. 2018; Chamizo et al. 2019).

11.5.2 Fertility amendments

Degraded soils are often low in SOM, which limits water storage, nutrients, and habitat for soil organisms. Mulches, composts, and manure can increase SOM and facilitate plant establishment (Hueso-González et al. 2016) and significantly increase productivity on infertile or coarse-textured soils (Chen et al. 2018; Fuentes et al. 2010). Nutrient-embedded hydrogels increase both water availability and nutrient supply (Coello et al. 2018; Kumar et al. 2019); clays incorporated into the hydrogel matrix form (nano)composites that improve water absorption and ion-exchange capacity while reducing production costs (Kuśtrowski et al. 2018).

11.5.3 Biostimulants

Natural and synthetic biostimulants can improve nutrient use efficiency as well as adding nutrients. Commercially available biostimulants include seaweed extracts, bacteria and fungi-based biofertilizers, protein hydrolysates, synthetic formulae containing antioxidant properties, plant growth regulators, plant growth promoting bacteria (PGPB), fermentation metabolite-based products, and humic acids. Transplanting native soils from remnant forests may have similar effects to biostimulants (Rivas et al. 2021).

Biostimulants hold potential for restoration use by enhancing seed germination (Domínguez-Castillo et al. 2021), increasing seedling drought tolerance (Tiepo et al. 2020), and as biofertilizers (García-Fraile et al. 2015; Vessey 2003). Biostimulants have improved seedling development in laboratory and field situations (Alves et al. 2019; Karličić et al. 2016; Khan et al. 2015; Ozyhar et al. 2019; Padda et al. 2021; Puri et al. 2020; Radhapriya et al. 2018). In field experiments, PGPB increased wood yield of *Schizolobium parahyba* var. *amazonicum* compared to fertilizer alone (Cely et al.

2016), and increased *Platanus × acerifolia* seedling height, root collar diameter, and total dry biomass on a reclamation site (Karličić et al. 2017). When combined with other bioactive substances, PGPB improved survival and root size of *Abies gracilis* (Trofimuk et al. 2020).

Inoculating seedlings with PGPB before outplanting is inexpensive, environmentally benign, and easily applied in the nursery; this process enhances the infusion of (desired) ectomycorrhizal fungi into the root systems (Chanway 1997). Seeds can be coated before sowing, roots dipped in liquid before transplanting, and/or leaves sprayed. Liquid biostimulants can be applied in irrigation systems, or as powdered or encapsulated products in centrifugal spreaders.

Despite the promise PGPB shows, more research is needed to better understand the underlying mechanisms governing its interactions with plants. Responses to PGPB inoculation are specific to plant–bacteria and genotype–genotype combinations (de Souza et al. 2020). Evidence for PGPB efficacy is inconclusive; comparing efficacy and functionality of commercial products is hampered by a lack of standardization and quantitative characterization.

11.6 Protection and tending

Browsing and trampling by livestock and wildlife can cause regeneration failure. Individual plants can be protected by physical barriers such as tree shelters (Padilla et al. 2011; Reis et al. 2021) or fencing entire areas (Burns et al. 2012; Löf et al. 2021; Miller et al. 2017). Temporary, non-lethal high-voltage electric fencing is used in Africa and South Asia to deter elephants. Solar-electric fencing is used in Malawi (Thuma and Dedza-Salima Forest Reserves) to keep elephants within wildlife reserves and away from agricultural fields and settlements.[8]

Lower-cost physical barriers include placing coarse woody debris around young plants to reduce mammalian browsing to acceptable levels. Combined with adequate regulation of wildlife density, this can be an effective restoration method (Stapleton et al. 2017). In India, perimeter trenches excavated along the forest boundary excluded cattle and helped resolve boundary disputes. Planting a live fence of *Caesalpinia bonducella* (fever nut), a tree with medicinal properties, provided additional protection and livelihood benefits (Stanturf et al. 2020).

Nurse plants might protect desired regeneration (Urretavizcaya and Defossé 2013). Cluster plantings, where nurse seedlings are planted around a desired species (Saha et al. 2016), utilize facilitation effects whereby species interact to the benefit of at least one with harm to none (Filazzola and Lortie 2014; Ren et al. 2008; Soliveres et al. 2012). Clusters facilitate establishment by altering microclimates and providing physical protection. Dense regeneration—natural, artificial, or a combination—allows high browser densities before impacting regeneration success (Baumhauer et al. 2005; Rooney et al. 2015). Browsing-tolerant regeneration emulates dense natural regeneration; direct seeding to produce densely stocked regeneration (50,000–100,000 seedlings ha^{-1}) offsets ungulate browsing. This was done operationally for *Fagus sylvatica* and *Abies alba* in Germany (Baumhauer et al. 2005). In Denmark, mixtures of ten species including pioneers (*Betula, Larix, Pinus*) were sown as sacrificial plants to provide protection against deer for the main species (*Fagus, Acer, Quercus, Pseudotsuga, Abies*) (Madsen and Madsen 2014). Smaller tree and shrubs (*Sorbus, Malus, Crataegus, Rosa*) were used as sacrificial nurse trees in the same spot/hole as the main species.

Two methods that might be useful under limited conditions deserve mention. Chemical repellents can be applied to seeds to deter seed predators, or to seedlings to deter browsers. This method may be too expensive for large-scale use as it requires reapplication after rain. (Villalobos et al. 2019; 2020). Large plants, taller than typical browsing height, can impede ungulate herbivory (Stanturf et al. 2004; Stanturf and van Oosten 2014). This might require temporarily excluding livestock to allow saplings to establish. For example, in South America, planting poles of *Populus* is used in silvopastoral systems; this technique could be adapted to restoration.

11.7 Conclusion and other considerations

Conventional methods for many restoration activities are slow, labor intensive, and expensive (Table 11.3). With high transportation costs, they are suitable only for sites near a road. Nevertheless, these methods can be implemented quickly with simple skill sets and minimal training. Standardized methods require little research and development. Automated systems are rapid and enable restoration on inaccessible sites with lower transportation costs than conventional methods. While development costs are high, running costs are low. Automated systems require highly skilled researchers to develop mechanisms but could be implemented with simple skill sets (Table 11.3). Research is required to improve existing technologies, develop new ones, and reduce costs.

Rapidly changing climate and disturbance regimes increase the need for techniques to improve resilience of existing stands, as well as to manage or create novel forests (Stanturf et al. 2015; 2018). Specifically, techniques are needed to convert non-native plantations to native species, transform structures from simple to complex or single species to multi-species stands, and to reintroduce natural disturbance regimes (Stanturf et al. 2014; Stanturf and Mansourian 2020). Increasing trees outside of forests is another tool (Djanibekov et al. 2016; Miccolis et al. 2019). Farmer-assisted natural regeneration in drylands (Chirwa and Mahamane 2017), taungya in West Africa (Foli et al. 2018), and shade-grown coffee and cacao (Badari et al. 2020) are examples of how forest restoration can enhance livelihoods while increasing biodiversity.

Fast-growing species can be used in value-added processing (Harrison et al. 2020). Incorporating biofuel plantings into restoration could have triple benefits—a local energy source, carbon sequestration, and reduced impact on native forests (Harvey and Guariguata 2021). Utilizing invasive plants is another restoration approach (Bandara and Udadini 2018; Stafford et al. 2018).

Enhanced technical and social capacity is needed to successfully implement FLR projects and to take full advantage of emerging technical innovations. Private financing promises to expand restoration efforts beyond the resources of public entities. High-profile tree planting initiatives might have limited impact (Holl and Brancalion 2020), but private investors seeking a return on investment could go beyond simplistic solutions by requiring an enabling environment, and evidence-based estimates of costs and benefits to reduce risk (Gitz et al. 2020).

Acknowledgments

This article is a collaboration with IUFRO Task Force Transforming Forest Landscapes for Future Climates and Human Well-being. We acknowledge the useful discussions and shared ideas of the Task Force members. The research was supported in part by the USDA Forest Service, Rocky Mountain Research Station. Our thanks to Jim Marin Graphics for assistance with visualizations. The findings and conclusions in this publication are those of the authors and should not be construed to represent any official USDA or US Government determination or policy, or indeed, any government or institution position.

Notes

1. https://www.globalforestwatch.org
2. https://regreeningafrica.org/wp-content/uploads/2020/01/Regreening_Africa_App_User_Guide_English-1.pdf
3. https://www.fs.usda.gov/research/pnw/products/dataandtools/tools/seedlot-selection-tool
4. https://flylitchi.com/
5. Pl@ntNet (m.plantnet-project.org) and Leafsnap (leafsnap.com)
6. http://www.coneharvesters.com/ib20.pdf
7. Water harvesting bunds dug by the Masai in Kenya; https://justdiggit.org/project/kenya-kuku-2/
8. https://www.elefence.org/

References

Acevedo, M., C. Alvarez, R.K. Dumroese, et al. 2021. "Native Plant Production in Chile. Is It Possible to Achieve Restoration Goals by 2035?" *Land* 10 (1): p. 71. DOI:10.3390/land10010071.

Adessi, A., R.C. de Carvalho, R. De Philippis, C. Branquinho, C., and J.M. da Silva. 2018. "Microbial Extracellular Polymeric Substances Improve Water Retention in Dryland Biological Soil Crusts." *Soil Biology and Biochemistry* 116: pp. 67–69. DOI:10.1016/j.soilbio.2017.10.002.

Aghai, M., and T. Manteuffel-Ross. 2020. "Enhancing Direct Seeding Efforts with Unmanned Aerial Vehicle (UAV) 'Swarms' and Seed Technology." *Tree Planters' Notes* 63: pp. 32–48.

Aguiar, A.S., F.N. dos Santos, J.B. Cunha, H. Sobreira, H., and A.J. Sousa. 2020. "Localization and Mapping for Robots in Agriculture and Forestry: A Survey." *Robotics* 9: p. 97. DOI:10.3390/robotics9040097.

Aitken, S.N., and J.B. Bemmels. 2016. "Time to Get Moving: Assisted Gene Flow of Forest Trees." *Evolutionary Applications* 9: pp. 271–290. DOI:10.1111/eva.12293.

Aitken, S.N., S. Yeaman, J.A. Holliday, T. Wang, and S. Curtis-McLane. 2008. "Adaptation, Migration or Extirpation: Climate Change Outcomes for Tree Populations." *Evolutionary Applications* 1: pp. 95–111. DOI:10.1111/j.1752-4571.2007.00013.x.

Allen, C.D., A.K. Macalady, H. Chenchouni, et al. 2010. "A Global Overview of Drought and Heat-Induced Tree Mortality Reveals Emerging Climate Change Risks for Forests." *Forest Ecology and Management* 259: pp. 660–684. DOI:10.1016/j.foreco.2009.09.001.

Alves, N.F., S.F. de Lima, A.P.L. de Lima, C.A. Simon, and P.P.V. Damis. 2019. "Biostimulant and Micronutrient Applications in the Production of *Acacia mangium* Seedlings." *Journal of Experimental Agriculture International* 38 (5): pp. 1–11. DOI:10.9734/jeai/2019/v38i530313.

Amorós, L., and J. Ledesma. 2020. "Aerial Robotics and Forest Management and Seeding." In *Automated Forest Restoration: Could Robots Revive Rain Forests*? Proceedings of a Brain-Storming Workshop, edited by S. Elliott, G. Gale, and M. Robertson, pp. 102–111. Chiang Mai: Chiang Mai University.

Anderegg, W.R.L., A.T. Trugman, G. Badgley, et al. 2020. "Climate-Driven Risks to the Climate Mitigation Potential of Forests." *Science* 368: p. eaaz7005. DOI:10.1126/science.aaz7005.

Andrachuk, M., M. Marschke, C. Hings, and D. Armitage. 2019. "Smartphone Technologies Supporting Community-Based Environmental Monitoring and Implementation: A Systematic Scoping Review." *Biological Conservation* 237: pp. 430–442. DOI:10.1016/j.biocon.2019.07.026.

Andries, A., R.J. Murphy, S. Morse, and J. Lynch. 2021. "Earth Observation for Monitoring, Reporting, and Verification within Environmental Land Management Policy." *Sustainability* 13 (6): p. 9105. DOI:10.3390/su13169105.

Arnold, R.W. 2016. "Soil Survey and Soil Classification." In *Environmental Soil-Landscape Modeling*, edited by S. Grunwald, pp. 50–72. Boca Raton: CRC Press.

Arumäe, T., and M. Lang. 2018. "Estimation of Canopy Cover in Dense Mixed-Species Forests Using Airborne Lidar Data." *European Journal of Remote Sensing* 51: pp. 132–141. DOI:10.1080/22797254.2017.1411169.

Aubin, I., C. Garbe, S. Colombo, et al. 2011. "Why We Disagree about Assisted Migration: Ethical Implications of a Key Debate Regarding the Future of Canada's Forests." *Forestry Chronicle* 87: pp. 755–765. DOI:10.5558/tfc2011-092.

Badari, C., L. Bernardini, D. Almeida, et al. 2020. "Ecological Outcomes of Agroforests and Restoration 15 Years after Planting." *Restoration Ecology* 28: pp. 1135–1144. DOI:10.1111/rec.13171.

Baldeck, C.A., G.P. Asner, R.E. Martin, et al. 2015. "Operational Tree Species Mapping in a Diverse Tropical Forest with Airborne Imaging Spectroscopy." *PloS One* 10: p. e0118403. DOI:10.1371/journal.pone.0118403.

Bandara, W., and K. Udadini. 2018. "Community and Industry Attitude towards Using Invasive *Prosopis juliflora* as a Dry Matter Energy Source for Small Scale Industries: A Case Study in Hambantota, Sri Lanka." *Journal of Tropical Forestry and Environment* 8: pp. 64–72. DOI:10.31357/jtfe.v8i1.3484.

Bannister, J.R., R. Vargas-Gaete, J.F. Ovalle, et al. 2018. "Major bottlenecks for the Restoration of Natural Forests in Chile." *Restoration Ecology* 26: pp. 1039–1044. DOI:10.1111/rec.12880.

Bauman, J.M., C.H. Keiffer, and B.C. McCarthy. 2014. "Growth Performance and Chestnut Blight Incidence (*Cryphonectria parasitica*) of Backcrossed Chestnut Seedlings in Surface Mine Restoration." *New Forests* 45: pp. 813–828. DOI:10.1007/s11056-014-9439-3.

Baumhauer, H., P. Madsen, J. Stanturf. 2005. "Regeneration by Direct Seeding—A Way to Reduce Costs of Conversion." In *Restoration of Boreal and Temperate Forests*, edited by J.A. Stanturf, and P. Madsen, pp. 349–354. Boca Raton: CRC Press.

Bergquist, J., M. Löf, and G. Örlander. 2009. "Effects of Roe Deer Browsing and Site Preparation on Performance of Planted Broadleaved and Conifer Seedlings When Using Temporary Fences." *Scandinavian Journal of Forest Research* 24: pp. 308–317. DOI:10.1080/02827580903117420.

Bonga, J. 2015. "A Comparative Evaluation of the Application of Somatic Embryogenesis, Rooting of Cuttings, and Organogenesis of Conifers." *Canadian Journal of Forest Research* 45: pp. 379–383. DOI:10.1139/cjfr-2014-0360.

Bonilla-Moheno, M., and K.D. Holl. 2010. "Direct Seeding to Restore Tropical Mature-Forest Species in Areas of Slash-and-Burn Agriculture." *Restoration Ecology* 18: pp. 438–445. DOI:10.1111/j.1526-100X.2009.00580.x

Bonnet, P., and D. Frame. 2020. "Basics of Automated Plant Identification." In *Automated Forest Restoration: Could Robots Revive Rain Forests? Proceedings of a Brain-Storming Workshop*, edited by S. Elliott, G. Gale, and M. Robertson, pp. 158–167. Chiang Mai: Chiang Mai University.

Bower, A.D., J.B.S. Clair, and V. Erickson. 2014. "Generalized Provisional Seed Zones for Native Plants." *Ecological Applications* 24: pp. 913–919. DOI:10.1890/13-0285.1.

Brancalion, P.H., and K.D. Holl. 2020. "Guidance for Successful Tree Planting Initiatives." *Journal of Applied Ecology* 57: pp. 2349–2361. DOI:10.1111/1365-2664.13725.

Breton, V., Y. Crosaz, and F. Rey. 2016. "Effects of Wood Chip Amendments on the Revegetation Performance of Plant Species on Eroded Marly Terrains in a Mediterranean Mountainous Climate (Southern Alps, France)." *Solid Earth* 7: pp. 599–610. DOI:10.5194/se-7-599-2016.

Broadhurst, L.M., A. Lowe, D.J. Coates, et al. 2008. "Seed Supply for Broadscale Restoration: Maximizing Evolutionary Potential." *Evolutionary Applications* 1: pp. 587–597. DOI:10.1111/j.1752-4571.2008.00045.x.

Buchhorn, M., B. Smets, L. Bertels, et al. 2019. Copernicus Global Land Service: Land Cover 100m: Epoch 2015. Globe, Version V2. 0.2.

Bullard, S., J.D. Hodges, R.L. Johnson, and T.J. Straka. 1992. "Economics of Direct Seeding and Planting for Establishing Oak Stands on Old-Field Sites in the South." *Southern Journal of Applied Forestry* 16: pp. 34–40. DOI:10.1093/sjaf/16.1.34.

Burns, B., J. Innes, and T. Day. 2012. "The Use and Potential of Pest-Proof Fencing for Ecosystem Restoration and Fauna Conservation in New Zealand." In *Fencing for Conservation*, edited by M.J. Somers and M.W. Hayward, pp. 65–90. Heidelberg: Springer.

Burton, P.J., and S.E. Macdonald. 2011. "The Restorative Imperative: Challenges, Objectives and Approaches to Restoring Naturalness in Forests." *Silva Fennica*. 45: pp. 843–863. DOI:10.7939/R34746R3T.

Camargo, J.L.C., I.D.K. Ferraz, and A.M. Imakawa. 2002. "Rehabilitation of Degraded Areas of Central Amazonia Using Direct Sowing of Forest Tree Seeds." *Restoration Ecology* 10: pp. 636–644. DOI:10.1046/j.1526-100X.2002.01044.x.

Castro, J., A.B. Leverkus, and F. Fuster. 2015. "A New Device to Foster Oak Forest Restoration via Seed Sowing." *New Forests* 46: pp. 919–929. DOI:10.1007/s11056-015-9478-4.

Caughlin, T.T., S. Elliott, and J.W. Lichstein. 2016. "When Does Seed Limitation Matter for Scaling Up Reforestation from Patches to Landscapes?" *Ecological Applications* 26: pp. 2439–2450. DOI:10.1002/eap.1410.

Cely, M.V., M.A. Siviero, J. Emiliano, et al. 2016. "Inoculation of *Schizolobium parahyba* with Mycorrhizal Fungi and Plant Growth-Promoting Rhizobacteria Increases Wood Yield under Field Conditions." *Frontiers in Plant Science* 7: p. 1708. DOI:10.3389/fpls.2016.01708.

Chaín, J.M., E. Tubert, C. Graciano, et al. 2020. "Growth Promotion and Protection from Drought in *Eucalyptus grandis* Seedlings Inoculated with Beneficial Bacteria Embedded in a Superabsorbent Polymer." *Scientific Reports* 10: pp. 1–17. DOI:10.1038/s41598-020-75212-4.

Chamizo, S., A. Adessi, G. Mugnai, A. Simiani, and R. De Philippis. 2019. "Soil Type and Cyanobacteria Species Influence the Macromolecular and Chemical Characteristics of the Polysaccharidic Matrix in Induced Biocrusts." *Microbial Ecology* 78: pp. 482–493. DOI:10.1007/s00248-018-1305-y.

Chang, I., J. Im, A.K. Prasidhi, and G.-C. Cho. 2015. "Effects of Xanthan Gum Biopolymer on Soil Strengthening." *Construction and Building Materials* 74: pp. 65–72. DOI:10.1016/j.conbuildmat.2014.10.026.

Chanoca, A., L. De Vries, and W. Boerjan. 2019. "Lignin Engineering in Forest Trees." *Frontiers in Plant Science* 10: p. 912. DOI:10.3389/fpls.2019.00912.

Chanway, C. 1997. "Inoculation of Tree Roots with Plant Growth Promoting Soil Bacteria: An Emerging Technology for Reforestation." *Forest Science* 43: pp. 99–112. DOI:10.1093/forestscience/43.1.99.

Charles, L.S., J.M. Dwyer, and M.M. Mayfield. 2017. "Rainforest Seed Rain into Abandoned Tropical Australian Pasture is Dependent on Adjacent Rainforest Structure and Extent." *Austral Ecology* 42: pp. 238–249. DOI:10.1111/aec.12426.

Chazdon, R.L., D.A. Falk, L.F. Banin, et al. 2021. "The Intervention Continuum in Restoration Ecology: Rethinking the Active–Passive Dichotomy." *Restoration Ecology*: p. e13535. DOI:10.1111/rec.13535.

Chen, Y., M. Camps-Arbestain, Q. Shen, B. Singh, and M.L. Cayuela. 2018. "The Long-Term Role of Organic Amendments in Building Soil Nutrient Fertility: A Meta-Analysis and Review." *Nutrient Cycling in Agroecosystems* 111: pp. 103–125. DOI:10.1007/s10705-017-9903-5.

Chirici, G., R.E. McRoberts, S. Winter, et al. 2012. "National Forest Inventory Contributions to Forest Biodiversity Monitoring." *Forest Science* 58: pp. 257–268. DOI:10.5849/forsci.12-003.

Chirino, E., A. Vilagrosa, and V.R. Vallejo. 2011. "Using Hydrogel and Clay to Improve the Water Status of Seedlings for Dryland Restoration." *Plant and Soil* 344: pp. 99–110. DOI:10.1007/s11104-011-0730-1.

Chirwa, P.W., and L. Mahamane. 2017. "Overview of Restoration and Management Practices in the Degraded Landscapes of the Sahelian and Dryland Forests and Woodlands of East and Southern Africa." *Southern Forests* 79: pp. 87–94. DOI:10.2989/20702620.2016.1255419.

Coello, J., M. Piqué, P. Rovira, C. Fuentes, and A. Améztegui. 2018. "Combining Innovative Mulches and Soil Conditioners in Mountain Afforestation with Ash (*Fraxinus excelsior* L.) in the Pyrenees (NE Spain)." *Forest Systems* 27(3): p. e017. DOI:10.5424/fs/2018273-13540.

Cole, R.J., K.D. Holl, C. Keene, R.A. Zahawi. 2011. "Direct Seeding of Late-Successional Trees to Restore Tropical Montane Forest." *Forest Ecology and Management* 261: pp. 1590–1597. DOI:10.1016/j.foreco.2010.06.038.

Collet, C., F. Vast, C. Richter, and R. Koller. 2021. "Cultivation Profile: A Visual Evaluation Method of Soil Structure Adapted to the Analysis of the Impacts of Mechanical Site Preparation in Forest Plantations." *European Journal of Forest Research* 140: pp. 65–76. DOI:10.1007/s10342-020-01315-2.

Commander, L., D. Rokich, M. Renton, K. Dixon, and D. Merritt. 2013. "Optimising Seed Broadcasting and Greenstock Planting for Restoration in the Australian Arid Zone." *Journal of Arid Environments* 88: pp. 226–235. DOI:10.1016/j.jaridenv.2012.08.012.

Costa, O.Y., J.M. Raaijmakers, and E.E. Kuramae. 2018. "Microbial Extracellular Polymeric Substances: Ecological Function and Impact on Soil Aggregation." *Frontiers in Microbiology* 9: p. 1636. DOI:10.3389/fmicb.2018.01636.

Cronin, D.T., S. Benbow, R.A. Bergl, et al. 2021. "Empowering Rangers through Technology and Innovation." *Parks Stewardship Forum* 37(1): p. 196. DOI:10.5070/P537151750.

Cruz, D., J. Benayas, G. Ferreira, S. Santos, and G. Schwartz. 2020. "An Overview of Forest Loss and Restoration in the Brazilian Amazon." *New Forests* 52: pp. 1–16. DOI:10.1007/s11056-020-09777-3.

Davis, A.S., and J.R. Pinto. 2021. "The Scientific Basis of the Target Plant Concept: An Overview." *Forests* 12: p. 1293. DOI:10.3390/f12091293.

de Almeida, D.R., S.C. Stark, R. Valbuena, et al. 2020. "A New Era in Forest Restoration Monitoring." *Restoration Ecology* 28: pp. 8–11. DOI:10.1111/rec.13067.

Defterli, S.G. 2016. "Review of Robotic Technology for Strawberry Production." *Applied Engineering in Agriculture* 32: pp. 301–318. DOI:10.13031/aea.32.11318.

Denchev, P., and S.C. Grossnickle. 2019. "Somatic Embryogenesis for Conifer Seedling Production: The Biology of Scaling." *Reforesta* 9: pp.109–137. DOI:10.21750/REFOR.7.08.70.

Derroire, G., R. Coe, and J.R. Healey. 2016. "Isolated Trees as Nuclei of Regeneration in Tropical Pastures: Testing the Importance of Niche-Based and Landscape Factors." *Journal of Vegetation Science* 27: pp. 679–691. DOI:10.1111/jvs.12404.

de Souza, N.L., S.S. Rocha, N.T. Narezzi, et al. 2020. "Differential Impacts of Plant Growth-Promoting Bacteria (PGPB) on Seeds of Neotropical Tree Species with Contrasting Tolerance to Shade." *Trees* 34: pp. 121–132. DOI:10.1007/s00468-019-01902-w.

Dey, D.C., B.O. Knapp, M.A. Battaglia, et al. 2019. "Barriers to Natural Regeneration in Temperate Forests across the USA." *New Forests* 50: pp. 11–40. DOI:10.1007/s11056-018-09694-6.

Djanibekov, U., K. Dzhakypbekova, J. Chamberlain, et al. 2016. "Agroforestry for Landscape Restoration and Livelihood Development in Central Asia." *ICRAF Working Paper* 186: pp. 1–31. Kunming, China: World Agroforestry Centre East and Central Asia.

Domínguez-Castillo, C., J.M. Alatorre-Cruz, D. Castañeda-Antonio, et al. 2021. "Potential Seed Germination-Enhancing Plant Growth-Promoting Rhizobacteria for Restoration of *Pinus chiapensis* ecosystems." *Journal of Forestry Research* 32: pp. 2143–2153.

Doust, S.J., E.D. Erskine, and D. Lamb. 2006. "Direct Seeding to Restore Rainforest Species: Microsite Effects on the Early Establishment and Growth of Rainforest Tree Seedlings on Degraded Land in the Wet Tropics of Australia." *Forest Ecology and Management* 234: pp. 333–343. DOI:10.1016/j.foreco.2006.07.014.

Dugan, P. 2011. "Cost Comparison Analysis of ANR Compared to Conventional Reforestation." In *Forests Beneath the Grass: Proceedings of the Regional Workshop on Advancing the Application of Assisted Natural Regeneration for Effective Low-Cost Forest Restoration*, edited by P.B. Durst, P. Sajise, and R.N. Leslie, pp: 9–13. Bohol: FAO.

Dumroese, R.K., T.D. Landis, J.R. Pinto, et al. 2016. "Meeting Forest Restoration Challenges: Using the Target Plant Concept." *Reforesta* 1: pp. 37–52. DOI:10.21750/REFOR.1.03.3.

Dumroese, R.K., M.I. Williams, J.A. Stanturf, and J.B. St. Clair. 2015. "Considerations for Restoring Temperate Forests of Tomorrow: Forest Restoration, Assisted Migration, and Bioengineering." *New Forests* 46: pp. 947–964. DOI:10.1007/s11056-015-9504-6.

Egertsdotter, U., I. Ahmad, and D. Clapham. 2019. "Automation and Scale Up of Somatic Embryogenesis for Commercial Plant Production, with Emphasis on Conifers." *Frontiers in Plant Science* 10: p. 109. DOI:10.3389/fpls.2019.00109.

Elliott, S. 2016. "The Potential for Automating Assisted Natural Regeneration of Tropical Forest Ecosystems." *Biotropica* 48: pp. 825–833. DOI:10.1111/btp.12387.

Enescu, V. 1987. "Climate and the Choice of Seed Orchard Sites." *Forest Ecology and Management* 19: pp. 257–265. DOI:10.1016/0378-1127(87)90035-1.

Engel, V.L., and J.A. Parrotta. 2001. "An Evaluation of Direct Seeding for Reforestation of Degraded Lands in Central Sao Paulo State, Brazil." *Forest Ecology and Management* 152: pp. 169–181. DOI:10.1016/S0378-1127(00)00600-9.

Ennos, R., R. Worrell, and D. Malcolm. 1998. "The Genetic Management of Native Species in Scotland." *Forestry* 71: pp. 1–23. DOI:10.1093/forestry/71.1.1-a

Erdmann, T.K. 2005. "Agroforestry as a Tool for Restoring Forest Landscapes." In *Forest Restoration in Landscapes*, edited by S. Mansourian, and D. Vallauri, pp. 274–284. New York: Springer.

Ersson, B.T., T. Laine, and T. Saksa. 2018. "Mechanized Tree Planting in Sweden and Finland: Current State and Key Factors for Future Growth." *Forests* 9: p. 370. DOI:10.3390/f9070370.

Everard, M., P. Johnston, D. Santillo, and C. Staddon. 2020. "The Role of Ecosystems in Mitigation and Management of Covid-19 and Other Zoonoses." *Environmental Science & Policy* 111: pp. 7–17. DOI:10.1016/j.envsci.2020.05.017.

Fagan, M.E. 2020. "A Lesson Unlearned? Underestimating Tree Cover in Dryland Biases Global Restoration Maps." *Global Change Biology* 26: pp. 4679–4690. DOI:10.1111/gcb.15187.

Fagan, M.E., J.L. Reid, M.B. Holland, J.G. Drew, and R.A. Zahawi. 2020. "How Feasible are Global Forest Restoration Commitments?" *Conservation Letters* 1: p. e12700. DOI:10.1111/conl.12700.

FAO. 2019. Restoring Forest Landscapes through Assisted Natural Regeneration (ANR) – A Practical Manual. Bangkok: FAO.

Fargione, J., D. Haase, O. Burney, et al. 2021. "Challenges to the Reforestation Pipeline in the United States." *Frontiers in Forests and Global Change* 4: p. 629198 DOI:10.3389/ffgc.2021.629198.

Fernández-Guisuraga, J.M., E. Sanz-Ablanedo, S. Suárez-Seoane, and L. Calvo. 2018. "Using Unmanned Aerial Vehicles in Postfire Vegetation Survey Campaigns through Large and Heterogeneous Areas: Opportunities and Challenges." *Sensors* 18: p. 586. DOI:10.3390/s18020586.

Filazzola, A., and C.J. Lortie. 2014. "A Systematic Review and Conceptual Framework for the Mechanistic Pathways of Nurse Plants." *Global Ecology and Biogeography* 23: pp. 1335–1345. DOI:10.1111/geb.12202.

Foli, S., M.A.F. Ros-Tonen, J. Reed, and T. Sunderland. 2018. "Natural Resource Management Schemes as Entry Points for Integrated Landscape Approaches: Evidence from Ghana and Burkina Faso." *Environmental Management* 62: pp. 82–97. DOI:10.1007/s00267-017-0866-8.

Frame, D., and C. Garzon-Lopez. 2020. "Applications of Remote Sensing for Tropical Forest Restoration: Challenges and Opportunities." In *Automated Forest Restoration: Could Robots Revive Rain Forests? Proceedings of a Brain-Storming Workshop*, edited by S. Elliott, G. Gale, and M. Robertson, pp. 46–53. Chiang Mai: Chiang Mai University.

Fremout, T., E. Thomas, K.T. Bocanegra-González, et al. 2021. "Dynamic Seed Zones to Guide Climate-Smart Seed Sourcing for Tropical Dry Forest Restoration in Colombia." *Forest Ecology and Management* 490: p. 119127. DOI:10.1016/j.foreco.2021.119127.

Fuentes, D., A. Valdecantos, J. Llovet, J. Cortina, and V.R. Vallejo. 2010. "Fine-Tuning of Sewage Sludge Application to Promote the Establishment of *Pinus halepensis* Seedlings." *Ecological Engineering* 36: pp. 1213–1221. DOI:10.1016/j.ecoleng.2010.04.012.

Funda, T., and Y.A. El-Kassaby. 2013. "Seed Orchard Genetics." *CABI Reviews 2012*: pp. 21–43. DOI:10.1079/PAVSNNR201270.

Gamborg, C., and P. Sandøe. 2010. "Ethical Considerations of Genetically Modified Trees." In *Forests and Genetically Modified Trees*, pp. 163–176. Rome: FAO.

García-Fraile, P., E. Menéndez, and R. Rivas. 2015. "Role of Bacterial Biofertilizers in Agriculture and Forestry." *AIMS Bioengineering* 2: pp. 183–205. DOI:10.3934/bioeng.2015.3.183.

Garzon-Lopez, C.X., S.A. Bohlman, H. Olff, and P.A. Jansen. 2013. "Mapping Tropical Forest Trees Using High-Resolution Aerial Digital Photographs." *Biotropica* 45: pp. 308–316. DOI:10.1111/btp.12009.

Germishuizen, I., and R.A. Gardner. 2015. "A Tool for Identifying Potential *Eucalyptus nitens* Seed Orchard Sites Based on Climate and Topography." *Southern Forests* 77: pp. 123–130. DOI:10.2989/20702620.2014.984554.

Gitz, V., A. Meybeck, V. Garavaglia, and B. Louman. 2020. "Upscaling Restoration: How to Unlock Finance." *Unasylva* 252: pp. 109–118.

Goldewijk, K.K. 2001. "Estimating Global Land Use Change over the Past 300 Years: The HYDE Database." *Global Biogeochemical Cycles* 15: pp. 417–433. DOI:10.1029/1999GB001232.

Gorelick, N., M. Hancher, M. Dixon, S. Ilyushchenko, D. Thau, and R. Moore. 2017. "Google Earth Engine: Planetary-Scale Geospatial Analysis for Everyone." *Remote Sensing of Environment* 202: pp. 18–27. DOI:10.1016/j.rse.2017.06.031.

Grossnickle, S.C., D. Cyr, and D.R. Polonenko. 1996. "Somatic Embryogenesis Tissue Culture for the Propagation of Conifer Seedlings: A Technology Comes of Age." *Tree Planters Notes* 47: pp. 48–57.

Grossnickle, S.C., and V. Ivetić. 2017. "Direct Seeding in Reforestation–A Field Performance Review." *Reforesta* 4: pp. 94–142. DOI:10.21750/REFOR.4.07.46.

Grubb, P.J. 1977. "The Maintenance of Species-Richness in Plant Communities: The Importance of the Regeneration Niche." *Biological Reviews* 52: pp. 107–145.

Guan, Y., S.-G. Li, X.-F. Fan, and Z.-H. Su. 2016. "Application of Somatic Embryogenesis in Woody Plants." *Frontiers in Plant Science* 7: p. 938. DOI:10.3389/fpls.2016.00938.

Guariguata, M., and K. Evans. 2019. "A Diagnostic for Collaborative Monitoring in Forest Landscape Restoration." *Restoration Ecology* 28: pp. 742–749. DOI:10.1111/rec.13076.

Haase, D.L., K. Bouzza, L. Emerton, et al. 2021. "The High Cost of the Low-Cost Polybag System: A Review of Nursery Seedling Production Systems." *Land* 10: p. 826. DOI:10.3390/land10080826.

Haase, D.L., and A.S. Davis. 2017. "Developing and Supporting Quality Nursery Facilities and Staff Are Necessary to Meet Global Forest and Landscape Restoration Needs." *Reforesta* 4: pp. 69–93. DOI:10.21750/REFOR.4.06.45.

Hamilton, J.A., C. Lexer, and S.N. Aitken. 2013. "Differential Introgression Reveals Candidate Genes for Selection Across a Spruce (*Picea sitchensis* × *P. glauca*) Hybrid Zone." *New Phytologist* 197: pp. 927–938. DOI:10.1111/nph.12055.

Hansen, M.C., P.V. Potapov, R. Moore, et al. 2013. "High-Resolution Global Maps of 21st-Century Forest Cover Change." *Science* 342: pp. 850–853. DOI:10.1126/science.1244693.

Harrison, R.D., T. Swinfield, A. Ayat, et al. 2020. "Restoration Concessions: A Second Lease on Life for Beleaguered Tropical Forests?" *Frontiers in Ecology and the Environment* 18: pp. 567–575. DOI:10.1002/fee.2265.

Harvey, C.A., and M.R. Guariguata. 2021. "Raising the Profile of Woodfuels in the Forest Landscape Restoration Agenda." *Conservation Science and Practice* 3: p. e342. DOI:10.1111/csp2.342.

Hata, S., T. Hiroyasu, J. Hayashi, S. Takahashi, and H. Hojo. 2006. "Clone Nursery Plant Trans-Planting Robot System with 3-D Vision System." In *2006 International Conference on Mechatronics and Automation. IEEE*: pp. 916–921. DOI:10.1109/ICMA.2006.257747.

Hau, C.H. 1997. "Tree Seed Predation on Degraded Hillsides in Hong Kong." *Forest Ecology and Management* 99: pp. 215–221. DOI:10.1016/S0378-1127(97)00207-7.

Hayashi, S., S. Yamamoto, S. Saito, et al. 2014. "Field Operation of a Movable Strawberry-Harvesting Robot Using a Travel Platform." *Japan Agricultural Research Quarterly* 48: pp. 307–316. DOI:10.6090/jarq.48.307.

Hodgson, B., and P. McGhee. 1992. "Development of Aerial Seeding for the Regeneration of Tasmanian Eucalypt Forests." *Tasforests* 4: pp. 77–85.

Höhl, M., V. Ahimbisibwe, J.A. Stanturf, et al. 2020. "Forest Landscape Restoration—What Generates Failure and Success?" *Forests* 1: p. 938. DOI:10.3390/f11090938.

Holger, F., H. Franka, H. Ulrike, H., and S. Wagner. 2015. "Developing Restoration Strategies for Temperate Forests Using Natural Regeneration Processes." In *Restoration of Boreal and Temperate Forests*, edited by J.A. Stanturf, pp. 103–164. 2nd ed. Boca Raton: CRC Press.

Holl, K., and P. Brancalion. 2020. "Tree Planting Is Not a Simple Solution." *Science* 368: pp. 580–581. DOI:10.1126/science.aba8232.

Holl, K., J.L. Reid, R.J. Cole, et al. 2020. "Applied Nucleation Facilitates Tropical Forest Recovery: Lessons Learned from a 15-Year Study." *Journal of Applied Ecology* 57: pp. 2316–2328. DOI:10.1111/1365-2664.13684.

Hsu, Y.-H., J.A. Kershaw Jr, M.J. Ducey, T.-R. Yang, and H. Wang. 2021. "Sampling with Probability Proportional to Prediction (3P Sampling) Using Covariates Derived from Spherical Images." *Canadian Journal of Forest Research* 51: pp. 1140–1147. DOI:10.1139/cjfr-2020-0498.

Hua, Y., N. Zhang, X. Yuan, et al. 2019. "Recent Advances in Intelligent Automated Fruit Harvesting Robots." *The Open Agriculture Journal* 13: pp. 101–106. DOI:10.2174/1874331501913010101.

Hueso-González, P., J.F. Martínez -Murillo, and J.D. Ruiz-Sinoga. 2016. "Effects of Topsoil Treatments on Afforestation in a Dry Mediterranean Climate (Southern Spain)." *Solid Earth* 7: pp. 1479–1489. DOI:10.5194/se-7-1479-2016.

Hultine, K.R., K.C. Grady, T.E. Wood, et al. 2016. "Climate Change Perils for Dioecious Plant Species." *Nature Plants* 2: p. 16109. DOI:10.1038/nplants.2016.109.

Hüttermann, A., L.J. Orikiriza, and H. Agaba. 2009. "Application of Superabsorbent Polymers for Improving the Ecological Chemistry of Degraded or Polluted Lands." *CLEAN–Soil, Air, Water* 37: pp. 517-526. DOI:10.1002/clen.200900048.

Hüttermann, A., M. Zommorodi, and K. Reise. 1999. "Addition of Hydrogels to Soil for Prolonging the Survival of *Pinus halepensis* Seedlings Subjected to Drought." *Soil and Tillage Research* 50: pp. 295–304. DOI:10.1016/S0167-1987(99)00023-9.

Hutto, R.L., and R. Belote. 2013. "Distinguishing Four Types of Monitoring Based on the Questions They Address." *Forest Ecology and Management* 289: pp. 183–189. DOI:10.1016/j.foreco.2012.10.005.

Intanon, S., and H. Sangsupan. 2020. "Allelopathy for Weed Management in Forest Restoration." In *Automated Forest Restoration: Could Robots Revive Rain Forests? Proceedings of a Brain-Storming Workshop*, edited by S. Elliott, G. Gale, and M. Robertson, pp. 148–157. Chiang Mai: Chiang Mai University.

IPBES. 2018. *Summary for Policymakers of the Thematic Assessment Report on Land Degradation and Restoration of the Intergovernmental Science-Policy Platform on Biodiversity and Ecosystem Services* edited by R. Scholes, L. Montanarella, A. Brainich, et al. Bonn: IPBES Secretariat.

IUCN and WRI. 2018. A guide to the Restoration Opportunities Assessment Methodology (ROAM) Assessing Forest Landscape Restoration Opportunities at the National or Sub-National Level. Gland: IUCN.

Ivetić, V., and J.M. Aleksić. 2016. "Response of Rare and Endangered Species *Picea omorika* to Climate Change: The Need for Speed." *Reforesta* 1: pp. 81–89. DOI:10.21750/REFOR.2.09.24.

Jacobs, D.F., H.J. Dalgleish, and C.D. Nelson. 2013. "A Conceptual Framework for Restoration of Threatened Plants: The Effective Model of American Chestnut (*Castanea dentata*) Reintroduction." *New Phytologist* 197: pp. 378–393. DOI:10.1111/nph.12020.

Jaganathan, D., K. Ramasamy, G. Sellamuthu, S. Jayabalan, and G. Venkataraman. 2018. "CRISPR for Crop Improvement: An Update Review." *Frontiers in Plant Science* 9: p. 985. DOI:10.3389/fpls.2018.00985.

Jalonen, R., M. Valette, D. Boshier, J. Duminil, and E. Thomas. 2018. "Forest and Landscape Restoration Severely Constrained by a Lack of Attention to the Quantity and Quality of Tree Seed: Insights from a Global Survey." *Conservation Letters* 11: p. e12424. DOI:10.1111/conl.12424.

Karličić, V., V. Golubović-Ćurguz, and V. Raičević. 2016. "The Alleviation of Reforestation Challenges by Beneficial Soil Microorganisms." *Reforesta* 1: pp. 238–260. DOI:10.21750/REFOR.1.12.12.

Karličić, V., D. Radić, J. Jovičić-Petrović, et al. 2017. "Use of Overburden Waste for London Plant (*Platanus* × *acerifolia*) Growth: The Role of Plant Growth Promoting Microbial Consortia." *iForest-Biogeosciences and Forestry* 10: pp. 692–699. DOI:10.3832/ifor2135-010.

Khan, Z., S.L. Kandel, D.N. Ramos, et al. 2015. "Increased Biomass of Nursery-Grown Douglas-Fir Seedlings upon Inoculation with Diazotrophic Endophytic Consortia." *Forests* 6: pp. 3582–3593. DOI:10.3390/f6103582.

Kildisheva, O.A., M.M. Aghai, K. Bouazza, and A.S. Davis. 2017. "Improving Restoration Success through Research-Driven Initiatives: Case Studies Targeting *Pinus pinea* Reforestation Stock Development in Lebanon." *Plant Ecology* 218: pp. 39–53. DOI:10.1007/s11258-016-0632-7.

Kotivuori, E., L. Korhonen, and P. Packalen. 2016. "Nationwide Airborne Laser Scanning Based Models for Volume, Biomass and Dominant Height in Finland." *Silva Fennica* 50: pp. 1–28. DOI:10.14214/sf.1567.

Kumar, V., H. Mittal, and S.M. Alhassan. 2019. "Biodegradable Hydrogels of Tragacanth Gum Polysaccharide to Improve Water Retention Capacity of Soil and Environment-Friendly Controlled Release of Agrochemicals." *International Journal of Biological Macromolecules* 132: pp. 1252–1261.

Kuśtrowski, P., P. Natkański, A. Rokicińska, and E. Witek. 2018. "Polymer Hydrogel-Clay (Nano) Composites." In *Polymer Gels. Gels Horizons: From Science to Smart Materials*, edited by V. Thakur, and M. Thakur, pp. 1–62. Singapore: Springer. DOI:10.1007/978-981-10-6086-1_1.

Kuzma, J. 2019. "Procedurally Robust Risk Assessment Framework for Novel Genetically Engineered Organisms and Gene Drives." *Regulation and Governance* 15: pp. 1144–1165. DOI:10.1111/rego.12245.

Lamb, D. 2011. *Regreening the Bare Hills: Tropical Forest Restoration in the Asia-Pacific Region*. Dordrecht: Springer Science & Business Media.

Landis, T.D., D.R. Dreesen, and R.K. Dumroese. 2003. "Sex and the Single *Salix*: Considerations for Riparian Restoration." *Native Plants Journal* 4: pp. 110–117. DOI:10.3368/npj.4.2.110.

Landis, T.D., and R.K. Dumroese. 2007. "Applying the Target Plant Concept to Nursery Stock Quality." In *Plant Quality: A Key to Success in Forest Establishment*, Proceedings of the COFORD Conference, p. 1–10. Dublin: National Council for Forest Research and Development.

Large, A.R., and G.L. Heritage. 2009. "Laser Scanning–Evolution of the Discipline." In *Laser Scanning for the Environmental Sciences*, edited by A.R. Largem and G.L. Heritage, pp. 1–20. Chichester: Wiley-Blackwell.

Latpate, R., J. Kshirsagar, V.K. Gupta, and G. Chandra. 2021. "Probability Proportional to Size Sampling." In *Advanced Sampling Methods*, pp. 85–98. Singapore: Springer. DOI:10.1007/978-981-16-0622-9_7.

Leites, L.P., G.E. Rehfeldt, and K.C. Steiner. 2019. "Adaptation to Climate in Five Eastern North America Broadleaf Deciduous Species: Growth Clines and Evidence of the Growth-Cold Tolerance Trade-Off." *Perspectives in Plant Ecology, Evolution and Systematics* 37: pp. 64–72. DOI:10.1016/j.ppees.2019.02.002.

León-Lobos, P., M.A. Bustamante-Sánchez, C.R. Nelson, et al. 2020. "Lack of Adequate Seed Supply Is a Major Bottleneck for Effective Ecosystem Restoration in Chile: Friendly Amendment to Bannister *et al.* (2018)". *Restoration Ecology* 28: pp. 277–281. DOI:10.1111/rec.13113.

Liang, J., T.W. Crowther, N. Picard, et al. 2016. "Positive Biodiversity-Productivity Relationship Predominant in Global Forests." *Science* 354: pp. 192–194. DOI:10.1126/science.aaf8957.

Liang, S. 2005. *Quantitative Remote Sensing of Land Surfaces*. Hoboken, New Jersey: John Wiley & Sons.

Liu, C.-C., Y.-H. Chen, M.-H.M. Wu, C. Wei, and M.-H. Ko. 2019. "Assessment of Forest Restoration with Multitemporal Remote Sensing Imagery." *Scientific Reports* 9, 7279. DOI:10.1038/s41598-019-43544-5.

Lockhart, B.R., E. Gardiner, T. Leininger, and J. Stanturf. 2008. "A Stand-Development Approach to Oak Afforestation in the Lower Mississippi Alluvial Valley." *Southern Journal of Applied Forestry* 32: pp. 120–129. DOI:10.1093/sjaf/32.3.120.

Löf, M., J. Barrere, M. Engman, L.K. Petersson, and A. Villalobos. 2021. "The Influence of Fencing on Seedling Establishment during Reforestation of Oak Stands: A Comparison of Artificial and Natural Regeneration Techniques Including Costs." *European Journal of Forest Research* 140: pp. 807–817. DOI:10.1007/s10342-021-01369-w.

Löf, M., D.C. Dey, R.M. Navarro, and D.F., Jacobs. 2012. "Mechanical Site Preparation for Forest Restoration." *New Forests* 43: pp. 825–848. DOI:10.1007/s11056-012-9332-x.

Löf, M., B.T. Ersson, J. Hjältén, et al. 2015. "Site Preparation Techniques for Forest Restoration." In *Restoration of Boreal and Temperate Forests*, edited by J.A. Stanturf, pp. 85–102. 2nd ed. Boca Raton: CRC Press.

Löf, M., P. Madsen, M. Metslaid, J. Witzell, and D.F. Jacobs. 2019. "Restoring Forests: Regeneration and Ecosystem Function for the Future." *New Forests* 50: pp. 139–151. DOI:10.1007/s11056-019-09713-0.

Mackelprang, R., and P.G. Lemaux. 2020. "Genetic Engineering and Editing of Plants: An Analysis of New and Persisting Questions." *Annual Review of Plant Biology* 71: pp. 659–687. DOI:10.1146/annurev-arplant-081519-035916.

Mader, S. 2019. "Plant Trees for the Planet: The Potential of Forests for Climate Change Mitigation and the Major Drivers of National Forest Area." *Mitigation and Adaptation Strategies for Global Change* 25: pp. 519–536. DOI:10.1007/s11027-019-09875-4.

Madsen, P., and T. Madsen. 2014. "Skovbrug og kronvildt." In *Kronvildt på Sjælland. Resultater af 3 års praksisorienteret forskning og forvaltning*, edited by N. Kanstrup, P. Madsen, K. Stenkjær, R. Buttenschøn, and A. Jensen, p. 156. København: Institut for Geovidenskab og Naturforvaltning, Københavns Universitet.

Mager, D., and A. Thomas. 2011. "Extracellular Polysaccharides from Cyanobacterial Soil Crusts: A Review of Their Role in Dryland Soil Processes." *Journal of Arid Environments* 75: pp. 91–97. DOI:10.1016/j.jaridenv.2010.10.001.

Mansourian, S., and J. Parrotta (eds.). 2018. *Forest Landscape Restoration: Integrated Approaches to Support Effective Implementation.* Abingdon: Routledge.

Matney T.G., and J.D. Hodges. 1991. "Evaluating Regeneration Success." In *Forest Regeneration Manual*, edited by M.L. Duryea, and P.M. Dougherty, pp. 321-331. Dordrecht: Kluwer Academic Publishers.

Merritt, D.J., and K.W. Dixon. 2011. "Restoration Seed Banks–A Matter of Scale." *Science* 332: pp. 424–425. DOI:10.1126/science.1203083.

Miccolis, A., F.M. Peneireiro, D.L.M. Vieira, H.R. Marques, and M.R.M. Hoffmann. 2019. "Restoration Through Agroforestry: Options for Reconciling Livelihoods with Conservation in the Cerrado and Caatinga Biomes in Brazil." *Experimental Agriculture* 55: pp. 208–225. DOI:10.1017/S0014479717000138.

Miller, G.W., P.H. Brose, and K.W. Gottschalk. 2017. "Advanced Oak Seedling Development as Influenced by Shelterwood Treatments, Competition Control, Deer Fencing, and Prescribed Fire." *Journal of Forestry* 115: pp. 179–189. DOI:10.5849/jof.16-002.

Miyawaki, A. 2004. "Restoration of Living Environment Based on Vegetation Ecology: Theory and Practice." *Ecological Research* 19: pp. 83–90. DOI:10.1111/j.1440-1703.2003.00606.x.

Miyoshi, T., N.N. Imai, A.M. Garcia Tommaselli, M.V. Antunes de Moraes, and E. Honkavaara. 2020. "Evaluation of Hyperspectral Multitemporal Information to Improve Tree Species Identification in the Highly Diverse Atlantic Forest." *Remote Sensing* 12: p. 244. DOI:10.3390/rs12020244.

Mohan, M., G. Richardson, G. Gopan, et al. 2021. "UAV-Supported Forest Regeneration: Current Trends, Challenges and Implications." *Remote Sensing* 13, 2596. DOI:10.3390/rs13132596.

Neale, D.B., and A. Kremer. 2011. "Forest Tree Genomics: Growing Resources and Applications." *Nature Reviews Genetics* 12: pp.111–122. DOI:10.1038/nrg2931.

Nesha, M.K., M. Herold, V. De Sy, et al. 2021. "An Assessment of Data Sources, Data Quality and Changes in National Forest Monitoring Capacities in the Global Forest Resources Assessment 2005-2020." *Environmental Research Letters* 16: p. 054029. DOI:10.1088/1748-9326/abd81b.

Novak, B.J., T. Maloney, and R. Phelan. 2018. "Advancing a New Toolkit for Conservation: From Science to Policy." *The CRISPR Journal* 1: pp. 11–15. DOI:10.1089/crispr.2017.0019.

Nuijten, R., N. Coops, C. Watson, and D. Theberge. 2021. "Monitoring the Structure of Regenerating Vegetation Using Drone-Based Digital Aerial Photogrammetry." *Remote Sensing* 13: p. 1942. DOI:10.3390/rs13101942.

Nuyun, L., and Z. Jingchun. 1995. "China Aerial Seeding Achievement and Development." *Forestry and Society Newsletter* 3: pp. 9–11.

Oliver, C.D., and B.C. Larson. 1996. *Forest Stand Dynamics: Updated Edition.* New York: John Wiley and Sons.

O'Neill, G., T. Wang, N. Ukrainetz, et al. 2017. A Proposed Climate-Based Seed Transfer System for British Columbia. Technical Report. British Columbia: Ministry of Forests, Lands and Natural Resource Operations.

Ong, R. 2011. "Recent Forest Restoration Initiatives in Sabah, Malaysia." In: *Forests Beneath the Grass: Proceedings of the Regional Workshop on Advancing the Application of Assisted Natural Regeneration for Effective Low-Cost Forest Restoration*, edited by P.B. Durst, P. Sajise, and R.N. Leslie, pp. 119–124. Bohol: FAO.

Onishi, Y., T. Yoshida, H. Kurita, et al. 2019. "An Automated Fruit Harvesting Robot by Using Deep Learning." *ROBOMECH Journal* 6 (13). DOI:10.1186/s40648-019-0141-2.

Orumaa, A., P. Vellak, M. Lang, et al. 2021. "How Can Remote Sensing Reduce Required Human Intervention in Robotic Forest Regeneration." *Forests* 12 (12): 1802. DOI:10.3390/f12121802.

Ozyhar, T., G. Mughini, and M. Marchi. 2019. "Influence of Biostimulant Application in Containerized *Eucalyptus globulus* Labill. Seedlings after Transplanting." *Dendrobiology* 82: pp. 17–23. DOI:10.12657/denbio.082.003.

Padda, K.P., A. Puri, and C. Chanway. 2021. "Long-Term Effect of Inoculating Lodgepole Pine Seedlings with Plant Growth-Promoting Bacteria Originating from Disturbed Gravel Mining Ecosystem." *Canadian Journal of Forest Research* 51: pp. 533–545. DOI:10.1139/cjfr-2020-0333.

Padilla, F.M., J.d.D. Miranda, R. Ortega, et al. 2011. "Does Shelter Enhance Early Seedling Survival in Dry Environments? A Test with Eight Mediterranean Species." *Applied Vegetation Science* 14: pp. 31–39. DOI:10.1111/j.1654-109X.2010.01094.x.

Palma, A.C., M. Goosem, R.J. Fensham, et al. 2021. "Dispersal and Recruitment Limitations in Secondary Forests." *Journal of Vegetation Science* 32: e12975. DOI:10.1111/jvs.12975.

Paneque-Gálvez, J., M.K. McCall, B.M. Napoletano, S.A. Wich, and L.P. Koh. 2014. "Small Drones for Community-Based Forest Monitoring: An Assessment of Their Feasibility and Potential in Tropical Areas." *Forests* 5: pp. 1481–1507. DOI:10.3390/f5061481.

Parker, W.H., and A.V. Niejenhuis. 1996. "Regression-Based Focal Point Seed Zones for *Picea mariana* from Northwestern Ontario." *Canadian Journal of Botany* 74: pp. 1227–1235. DOI:10.1139/b96-148.

Pedlar, J.H., D.W. McKenney, I. Aubin, et al. 2012. "Placing Forestry in the Assisted Migration Debate." *BioScience* 62: pp. 835–842. DOI:10.1525/bio.2012.62.9.10.

Pérez, D.R., F. González, C. Ceballos, M.E. Oneto, and J. Aronson. 2019. "Direct Seeding and Outplantings in Drylands of Argentinean Patagonia: Estimated Costs, and Prospects for Large-Scale Restoration and Rehabilitation." *Restoration Ecology* 27: pp. 1105–1116. DOI:10.1111/rec.12961.

Pike, C., K.M. Potter, P. Berrang, et al. 2020a. "New Seed-Collection Zones for the Eastern United States: The Eastern Seed Zone Forum." *Journal of Forestry* 118: pp. 444–451. DOI:10.1093/jofore/fvaa013.

Pike, C.C., J. Koch, and C.D. Nelson. 2020b. "Breeding for Resistance to Tree Pests: Successes, Challenges, and a Guide to the Future." *Journal of Forestry* 119: pp. 96–105. DOI:10.1093/jofore/fvaa049.

Prévost, M. 1992. "Effets du Scarifiage sur les Propriétés du Sol, la Croissance des Semis et la Compétition: Revue des Connaissances Actuelles et Perspectives de Recherches au Québec." *Annales Des Sciences Forestières* 49: pp. 277–296. DOI:10.1051/forest:19920306.

Prosdocimi, M., A. Jordán, P. Tarolli, et al. 2016. "The Immediate Effectiveness of Barley Straw Mulch in Reducing Soil Erodibility and Surface Runoff Generation in Mediterranean Vineyards." *Science of the Total Environment* 547: pp. 323–330. DOI:10.1016/j.scitotenv.2015.12.076.

Puri, A., K.P. Padda, and C.P. Chanway. 2020. "Sustaining the Growth of Pinaceae Trees under Nutrient-Limited Edaphic Conditions via Plant-Beneficial Bacteria." *Plos One* 15: e0238055. DOI:10.1371/journal.pone.0238055.

Radhapriya, P., A. Ramachandran, and P. Palani. 2018. "Indigenous Plant Growth-Promoting Bacteria Enhance Plant Growth, Biomass, and Nutrient Uptake in Degraded Forest Plants." *3 Biotech* 8: 154. DOI:10.1007/s13205-018-1179-1.

Rametsteiner, E., and G. Weiss. 2006. "Innovation and Innovation Policy in Forestry: Linking Innovation Process with Systems Models." *Forest Policy and Economics* 8: pp. 691–703. DOI:10.1016/j.forpol.2005.06.009.

Reid, J.L., M.E. Fagan, and R.A. Zahawi. 2018. "Positive Site Selection Bias in Meta-Analyses Comparing Natural Regeneration to Active Forest Restoration." *Science Advances* 4: eaas9143. DOI:10.1126/sciadv.aas9143.

Reid, J.L., S.J. Wilson, G.S. Bloomfield, et al. 2017. "How Long Do Restored Ecosystems Persist?" *Annals of the Missouri Botanical Garden* 102: p. 258. DOI:10.3417/2017002.

Reis, L.K., G.A. Damasceno Junior, L.L. Battaglia, and L.C. Garcia. 2021. "Can Transplanting Seedlings with Protection against Herbivory Be a Cost-Effective Restoration Strategy for Seasonally Flooded Environments?" *Forest Ecology and Management* 483: 118742. DOI:10.1016/j.foreco.2020.118742.

Ren, H., L. Yang, and N. Liu. 2008. "Nurse Plant Theory and Its Application in Ecological Restoration in Lower Subtropics of China." *Progress in Natural Science* 18: pp. 137–142. DOI:10.1016/j.pnsc.2007.07.008.

Rivas, M.B.R., D. Douterlungne, L.G. Aparicio, E.I. Badano, and J.A.F. Cano. 2021. "Restoration Interventions Produce Opposite and Non-Additive Benefits on Tree Establishment in Degraded Forest Clearings." *Restoration Ecology* 2021: e13539. DOI:10.1111/rec.13539.

Rooney, T.P., R. Buttenschøn, P. Madsen, et al. 2015. "Integrating Ungulate Herbivory into Forest Landscape Restoration." In *Restoration of Boreal and Temperate Forests*, edited by J.A. Stanturf, pp. 69–84. 2nd ed. Boca Raton: CRC Press.

Ruņģis, D., S. Luguza, E. Bāders, V. Šķipars, and A. Jansons. 2019. "Comparison of Genetic Diversity in Naturally Regenerated Norway Spruce Stands and Seed Orchard Progeny Trials." *Forests* 10: p. 926. DOI:10.3390/f10100926.

Saha, S., C. Kuehne, and J. Bauhus. 2016. "Lessons Learned from Oak Cluster Planting Trials in Central Europe." *Canadian Journal of Forest Research* 47: pp. 139–148. DOI:10.1139/cjfr-2016-0265.

Sannai, J. 2003. "Assisted Natural Regeneration in China." In *Advancing Assisted Natural Regeneration (ANR) in Asia and the Pacific*, edited by P.C. Dugan, P.B. Durst, D.J. Ganz, and P.J. McKenzie, pp. 29–31. Bangkok: FAO Regional Office for Asia and the Pacific. DOI:10.1016/j.ijbiomac.2019.04.023.

Scheper, A.C., P.A. Verweij, and M. van Kuijk. 2021. "Post-Fire Forest Restoration in the Humid Tropics: A Synthesis of Available Strategies and Knowledge Gaps for Effective Restoration." *Science of the Total Environment* 771: 144647. DOI:10.1016/j.scitotenv.2020.144647.

Seidl, R., T.A. Spies, D.L. Peterson, S.L. Stephens, and J.A. Hicke. 2016. "Searching for Resilience: Addressing the Impacts of Changing Disturbance Regimes on Forest Ecosystem Services." *Journal of Applied Ecology* 53: pp. 120–129. DOI:10.1111/1365-2664.12511.

Sharma, S., A. Shahzad, and J.A.T. da Silva. 2013. "Synseed Technology—A Complete Synthesis." *Biotechnology Advances* 31: pp.186–207. DOI:10.1016/j.biotechadv.2012.09.007.

Shono, K., E.A. Cadaweng, and P.B., Durst. 2007. "Application of Assisted Natural Regeneration to Restore Degraded Tropical Forestlands." *Restoration Ecology* 15: pp. 620–626. DOI:10.1111/j.1526-100X.2007.00274.x.

Singh, M., D. Evans, B.S. Tan, and C.S. Nin. 2015. "Mapping and Characterizing Selected Canopy Tree Species at the Angkor World Heritage Site in Cambodia Using Aerial Data." *PloS One* 10: e0121558. DOI:10.1371/journal.pone.0121558.

Soliveres, S., D.J. Eldridge, F. Hemmings, and F.T. Maestre. 2012. "Nurse Plant Effects on Plant Species Richness in Drylands: The Role of Grazing, Rainfall and Species Specificity." *Perspectives in Plant Ecology, Evolution and Systematics* 14: pp. 402–410. DOI:10.1016/j.ppees.2012.09.003.

Sønstebø, J., M. Tollefsrud, T. Myking, et al. 2018. "Genetic Diversity of Norway Spruce (*Picea abies* (L.) Karst.) Seed Orchard Crops: Effects of Number of Parents, Seed Year, and Pollen Contamination." *Forest Ecology and Management* 411: pp. 132–141. DOI:10.1016/j.foreco.2018.01.009.

Stafford, W.H., G.P. Von Maltitz, and H.K. Watson. 2018. "Reducing the Costs of Landscape Restoration by Using Invasive Alien Plant Biomass for Bioenergy." *Wiley Interdisciplinary Reviews: Energy and Environment* 7: e272. DOI:10.1002/wene.272.

Stanturf, J.A. 2020. "Landscape Degradation and Restoration." In *Soils and Landscape Restoration,* edited by J.A. Stanturf, and M.A. Callaham, pp. 125–159. New York: Academic Press. DOI:10.1016/B978-0-12-813193-0.00005-9.

Stanturf, J.A., W.H. Conner, E.S. Gardiner, C.J. Schweitzer, A.W. Ezell. 2004. "Recognizing and Overcoming Difficult Site Conditions for Afforestation of Bottomland Hardwoods." *Ecological Restoration* 22: pp. 183–193. https://www.jstor.org/stable/43442898.

Stanturf, J.A., E.S. Gardiner, J.P. Shepard, et al. 2009. "Restoration of Bottomland Hardwood Forests across a Treatment Intensity Gradient." *Forest Ecology and Management* 257: pp. 1803–1814. DOI:10.1016/j.foreco.2009.01.052.

Stanturf, J.A., P. Kant, J.-P.B. Lillesø, et al. 2015. *Forest Landscape Restoration as a Key Component of Climate Change Mitigation and Adaptation.* Vienna: International Union of Forest Research Organizations.

Stanturf, J.A., R.C. Kellison, F. Broerman, and S.B. Jones. 2003. "Innovation and Forest Industry: Domesticating the Pine Forests of the Southern United States, 1920–1999." *Forest Policy and Economics* 5: pp. 407–419. DOI:10.1016/S1389-9341(03)00039-X.

Stanturf, J.A., M. Kleine, S. Mansourian, et al. 2019. "Implementing Forest Landscape Restoration under the Bonn Challenge: A Systematic Approach." *Annals of Forest Science* 76: 50. DOI:10.1007/s13595-019-0833-z.

Stanturf, J.A., P. Madsen, K. Sagheb-Talebi, and O.K. Hansen. 2018. "Transformational Restoration: Novel Ecosystems in Denmark." *Plant Biosystems* 152: pp. 536–546. DOI:10.1080/11263504.2018.1435586.

Stanturf, J.A., and S. Mansourian. 2020. "Forest Landscape Restoration: State of Play." Royal Society Open Science 7: 201218. https://doi.org/10.1098/rsos.201218.

Stanturf, J.A., S. Mansourian, A. Darabant, et al. 2020. "Forest Landscape Restoration Implementation: Lessons Learned from Selected Landscapes in Africa, Asia and Latin America. Occasional Paper No. 33. Vienna: IUFRO.

Stanturf, J., S. Mansourian, and M. Kleine. 2017. Implementing Forest Landscape Restoration: A Practitioner's Guide. Vienna: IUFRO.

Stanturf, J., B. Palik, and R.K. Dumroese. 2014. "Contemporary Forest Restoration: A Review Emphasizing Function." *Forest Ecology and Management* 331: pp. 292–323. DOI:10.1016/j.foreco.2014.07.029.

Stanturf, J.A., and C. van Oosten. 2014. "Operational Poplar and Willow Culture. In *Poplars and Willows: Trees for Society and the Environment*, edited by J.C. Isebrands, and J.K. Richardson, pp. 200–257. Wallingford: CABI.

Stapleton, J.P., K. Ikin, and D. Freudenberger. 2017. "Coarse Woody Debris Can Reduce Mammalian Browsing Damage of Woody Plant Saplings in Box-Gum Grassy Woodlands." *Ecological Management and Restoration* 18: pp. 223–230. DOI:10.1111/emr.12270.

Strahler, A., D. Muchoney, J. Borak, et al. 1999. MODIS Land Cover Product Algorithm Theoretical Basis Document (ATBD). Version 5.0" MODIS Land Cover and Land-Cover Change. Boston MA: Center for Remote Sensing, Department of Geography, Boston University.

Summers, D.M., B.A. Bryan, M. Nolan, and T.J. Hobbs. 2015. "The Costs of Reforestation: A Spatial Model of the Costs of Establishing Environmental and Carbon Plantings." *Land Use Policy* 44: pp. 110–121. DOI:10.1016/j.landusepol.2014.12.002.

Sutton, R. 1993. "Mounding Site Preparation: A Review of European and North American Experience." *New Forests* 7: pp. 151–192. DOI:10.1007/BF00034198.

Tenbergen, B., A. Günster, and K.-F. Schreiber. 1995. "Harvesting Runoff: The Minicatchment Technique--An Alternative to Irrigated Tree Plantations in Semiarid Regions." *Ambio* 24: pp. 72–76. https://www.jstor.org/stable/4314299.

Thompson, I., B. Mackey, S. McNulty, and A. Mosseler. 2009. "Forest Resilience, Biodiversity, and Climate Change." *Technical Series* 43: pp. 1–67. Montreal: Secretariat of the Convention on Biological Diversity.

Tiepo, A.N., L.V. Constantino, T.B. Madeira, et al. 2020. "Plant Growth-Promoting Bacteria Improve Leaf Antioxidant Metabolism of Drought-Stressed Neotropical Trees." *Planta* 251: pp. 1–11. DOI:10.1007/s00425-020-03373-7.

Trees that Count. 2020. New Zealanders Are Helping to Make a Difference by Planting Native Trees. Wellington, New Zealand: Project Crimson Trust. https://www.treesthatcount.co.nz/.

Trofimuk, L.P., P.S. Kirillov, and A.A. Egorov. 2020. "Application of Biostimulants for Vegetative Propagation of Endangered *Abies gracilis*." *Journal of Forestry Research* 31: pp. 1195–1199. DOI:10.1007/s11676-019-00903-2.

Tunjai, P. 2012. "Direct Seeding for Restoring Tropical Lowland Forest Ecosystems in Southern Thailand." Thesis. Thailand: The Graduate School. Walailak University.

Tunjai, P., and S. Elliott. 2012. "Effects of Seed Traits on the Success of Direct Seeding for Restoring Southern Thailand's Lowland Evergreen Forest Ecosystem." *New Forests* 43: pp. 319–333. DOI:10.1007/s11056-011-9283-7.

Urretavizcaya, M.F., and G.E. Defossé. 2013. "Effects of Nurse Shrubs and Tree Shelters on the Survival and Growth of two *Austrocedrus chilensis* Seedling Types in a Forest Restoration Trial in Semiarid Patagonia, Argentina." *Annals of Forest Science* 70: pp. 21–30. DOI:10.1007/s13595-012-0234-z.

Vallejo, V.R., A. Smanis, E. Chirino, et al. 2012. "Perspectives in Dryland Restoration: Approaches for Climate Change Adaptation." *New Forests* 43: pp. 561–579. DOI:10.1007/s11056-012-9325-9.

Vessey, J.K. 2003. "Plant Growth Promoting Rhizobacteria as Biofertilizers." *Plant and Soil* 255: pp. 571–586. DOI:10.1023/A:1026037216893.

Vieira, D.L., and A. Scariot. 2006. "Principles of Natural Regeneration of Tropical Dry Forests for Restoration." *Restoration Ecology* 14: pp. 11–20. DOI:10.1111/j.1526-100X.2006.00100.x.

Villalobos, A., G. Olsson, M. Birkedal, and M. Löf. 2019. "The Effects of Four Repellents on Bank Vole Consumption and Germination of Beech Nuts and Acorns." *New Forests* 50: pp. 241–254. DOI:10.1007/s11056-018-9660-6.

Villalobos, A., F. Schlyter, G. Olsson, J. Witzell, and M. Löf. 2020. "Direct Seeding for Restoration of Mixed Oak Forests: Influence of Distance to Forest Edge, Predator-Derived Repellent and Acorn Size on Seed Removal by Granivorous Rodents." *Forest Ecology and Management* 477: p. 118484. DOI:10.1007/s11056-018-9660-6.

Vincelli, P. 2016. "Genetic Engineering and Sustainable Crop Disease Management: Opportunities for Case-by-Case Decision-Making." *Sustainability* 8 (5): p. 495. DOI:10.3390/su8050495.

Waiboonya, P., and S. Elliott. 2020. "Sowing Time and Direct Seeding Success of Native Tree Species for Restoring Tropical Forest Ecosystems in Northern Thailand." *New Forests* 51: pp. 81–99. DOI:10.1007/s11056-019-09720-1.

Waltz, E. 2018. "With a Free Pass, CRISPR-Edited Plants Reach Market in Record Time." *Nature Biotechnology* 36: pp. 6–7. DOI:10.1038/nbt0118-6b.

Werden, L.K., J.P. Alvarado, S. Zarges, et al. 2018. "Using Soil Amendments and Plant Functional Traits to Select Native Tropical Dry Forest Species for the Restoration of Degraded Vertisols." *Journal of Applied Ecology* 55: pp. 1019–1028. DOI:10.1111/1365-2664.12998.

White, T.L. 1987. "A Conceptual Framework for Tree Improvement Programs." *New Forests* 1: pp. 325–342. DOI:10.1007/BF00031742.

Williams, M.I., and R.K. Dumroese. 2013. "Preparing for Climate Change: Forestry and Assisted Migration." *Journal of Forestry* 114: pp. 287–297. DOI:10.5849/jof.13-016.

Wilson, S.J., N.S. Alexandre, K.D. Holl, et al. 2021. Applied Nucleation Restoration Guide for Tropical Forests. Arlington: Conservation International.

Woods, K., and S. Elliott. 2004. "Direct Seeding for Forest Restoration on Abandoned Agricultural Land in Northern Thailand." *Journal of Tropical Forest Science* 16: pp. 248–259. https://www.jstor.org/stable/23616517.

Wu, N., H.X. Pan, D. Qiu, and Y.M. Zhang. 2014. "Feasibility of EPS-Producing Bacterial Inoculation to Speed Up the Sand Aggregation in the Gurbantunggut Desert, Northwestern China." *Journal of Basic Microbiology* 54: pp. 1378–1386. DOI:10.1002/jobm.201400355.

Xu, Y., F. Rossi, G. Colica, et al. 2013. "Use of Cyanobacterial Polysaccharides to Promote Shrub Performances in Desert Soils: A Potential Approach for the Restoration of Desertified Areas." *Biology and Fertility of Soils* 49: pp. 143–152. DOI:10.1007/s00374-012-0707-0.

Zahawi, R.A., J.L. Reid, and K.D. Holl. 2014. "Hidden Costs of Passive Restoration." *Restoration Ecology* 22: pp. 284–287. DOI:10.1111/rec.12098.

12

Sustainable and fair forest and land restoration

Balancing goals, interests, and trade-offs

Wil de Jong, Pia Katila, Georg Winkel, Pablo Pacheco, Glenn Galloway, and Carol J. Pierce Colfer

12.1 The ascent of the restoration era

Restoration of nature, ranging from forests and coastal areas to larger land- and seascapes, has coalesced into a major global effort, one that is endorsed and supported by a broad array of governments, civil society organizations, and social movements that are concerned about the diminishing benefits that nature provides to humans. Restoration has become an imperative not only to sustain or enhance the contributions of nature to people but also—due to interactions between nature loss and climate change—is increasingly promoted as a critical contribution to climate change mitigation and adaptation. Next to environmental and social benefits, restoration is assumed to also generate significant economic contributions that would increase over longer time horizons.

Restoration of forests, forest ecosystems, and forest landscapes, the focus of this book, are prominent manifestations of this contemporary societal pursuit of global nature restoration. This chapter uses the term forest restoration, recognizing that alternative terms are used in the literature and in other chapters of this book, each with their own nuances and context of use. Section 12.2.1 reviews the terminology used by authors of the chapters, as well as among forest restoration protagonists over the last five decades or so.

The various benefits expected from restored forests are commonly referred to as ecosystem services and are classified as provisioning, regulating and maintenance, and cultural services (Haines-Young and Potschin 2018). Forests are currently being restored for various reasons: for instance, to produce timber, protect watersheds, capture atmospheric carbon, conserve biodiversity, improve scenic beauty, and provide outdoor recreation. While forest restoration projects often target environmental benefits, there is increasing recognition of the importance of the well-being of people who depend on forests in multiple ways and who are negatively affected by deforestation and forest degradation. Restored forests can maintain and improve livelihoods, as forests play an important part in local knowledge and contribute to key cultural and symbolic values.

Forest restoration as a practice has historical roots worldwide. Some modern-day forest restoration practices and ideas can be traced to past restoration experiences and

Wil de Jong, Pia Katila, Georg Winkel, Pablo Pacheco, Glenn Galloway, and Carol J. Pierce Colfer, *Sustainable and fair forest and land restoration* In: *Restoring Forests and Trees for Sustainable Development.* Edited by: Pia Katila, Carol J. Pierce Colfer, Wil de Jong, Glenn Galloway, Pablo Pacheco, and Georg Winkel, Oxford University Press.
DOI: 10.1093/9780197683958.003.0012

outcomes (de Jong et al. 2021; Chapter 2). There is increased attention on, interest in, and call for forest restoration; this manifests in multiple ways. A broad array of international initiatives have sprung up that address forest restoration (Chapters 1, 3). The number of academic publications that appear when searching for "forest restoration" in the Scopus databases has increased from 21 in 1990 to over 2000 in 2021; similar trends can be seen for "forest rehabilitation" and "reforestation." This rapid increase in publications exerts an influence on international and national policies (Reinecke and Blum 2018), while increasing political focus on forest restoration has invigorated academic interest. Since the 1960s, the area of restored forest globally has consistently expanded and the rate of expansion has increased over the last 30 years (de Jong et al. 2021). Concurrently, capacity enhancement networks, tree-planting technology, and private forest restoration service providers have emerged in response to the growing demand (Urzedo et al. 2023).

The rise of forest restoration on the policy agenda is a response to increasing global environmental challenges. These challenges threaten populations the world over, especially at-risk populations and localities, who are thereby rendered more vulnerable. They were flagged in high-profile publications: The Limits of Growth (Meadows et al. 1972), Our Common Future (Bruntland 1987), the Millennium Ecosystem Assessment (MEA 2005), Planetary Boundaries (Rockström et al. 2009.), the Intergovernmental Panel on Climate Change reports,[1] and the latest Intergovernmental Science-Policy Platform on Biodiversity and Ecosystem Services Global Assessment Report on Biodiversity and Ecosystem Services (IPBES 2019). These all share a grim outlook of how overexploitation and pollution are negatively affecting terrestrial and marine ecosystems, including forest biomes.

Forests yield diverse ecosystem services (Constanza et al. 2014), leading to high expectations placed on forest restoration in the myriad of proposals to overcome global and local environmental challenges, allowing restored forests to contribute to the Sustainable Development Goals (Katila et al. 2019). If the global community manages to restore forests on a grand scale, the ecosystem services that forests provide will increase concurrently. This would, to an important degree, mitigate many environmental challenges that the world now faces, contributing in particular to reversing the concurrent crises of climate change and biodiversity loss (Verkerk et al. 2020).

These high expectations have resulted in ambitious plans and international programs to restore vast areas of land that are considered degraded or have been deforested. Staggering figures of restoration targets have been provided by various sources: 350 Mha of restored forests by 2030 as proposed by the Bonn Challenge; a 3% increase in the total global forest area, or about 120 Mha, during the same period by the UN Strategic Plan for Forests, or planting 3 billion trees (as promoted by the EU biodiversity and forest strategy) or even 1 trillion trees as advocated by the World Economic Forum. All these, however, still represent only a fraction of the more than 2 billion ha that are estimated to "offer opportunities for restoration" (Minnemeyer 2011).

The present volume, initiated under the auspices of IUFRO's Special Project World Forests, Society and Environment investigates multiple dimensions of what can be referred to as "the global forest restoration project." There are a suite of questions that can be asked related to these efforts. What are the key objectives and types of restoration? How feasible are they? How can or should they be organized and governed? Who are behind the global initiatives? Do these initiatives coincide or compete? Do we have the know-how and technology to implement forest restoration at a larger scale on sites of variable quality and productivity? How can this be financed? What are the benefits of different forest restoration approaches across different contexts; and more importantly, who will benefit, who will lose out, and who will bear the costs? Furthermore, who will track the emerging outcomes of these efforts to extract and disseminate important lessons learned to improve the performance of the global forest restoration project over time? These, and other questions formulated in Chapter 1, are addressed throughout the book. This chapter synthesizes findings, addressing such questions, and commenting on the implications for ongoing debates on forest restoration.

The chapter is divided into six sections. Section 12.2 brings together evidence on drivers of forest restoration, explores forest restoration history and trends, and summarizes some projections of what can be expected of future forest restoration. Are the ambitious goals of the global forest restoration project realistic and achievable? What are a range of forest restoration and restored forests scenarios that may play out? These are questions that draw on Chapters 2, 3, 4, 5, 6, and 10 of this volume, as well as on related discussions in the literature.

Section 12.3 focuses on the goals, costs, and actual and potential social, economic, and environmental benefits of forest restoration, and how likely it is that such benefits will be achieved in practice. It is acknowledged there may be trade-offs between social, economic and environmental benefits, and these will play out across short and long-time horizons. Forest restoration is recognized to result not only in benefits but, in certain contexts, also yields "ecosystem disservices," i.e., negative outcomes. This section draws on information from Chapters 3 and 6–11 and related discussions in the literature.

Section 12.4 focuses on restoration governance at international, national, and subnational levels. While some emphasis has been placed on understanding the enabling conditions facilitating forest restoration processes at national and subnational scales, there has been limited analysis carried out on how restoration governance at the international level takes shape, how it relates to other international processes, or how it links to national and subnational forest restoration governance. These nested, multi-scale and multi-actor governance processes are explored drawing on Chapters 2, 3, and 4 and related discussions in the literature. A synthesis of local-level governance, based on Chapters 2, 3, 4, and 6, addresses the questions: Who decides whether forest restoration is necessary? How is it carried out? How are costs and benefits distributed? Section 12.5 extracts major insights from the discussions of Sections 12.2 to 12.4, and from all the chapters of this book, to discuss the potential of forest restoration and how to achieve it.

12.2 Forest restoration status, drivers, and future projections

12.2.1 Drivers of forest restoration

Great expectations associated with forest restoration have led to ambitious forest restoration goals and plans. It is opportune to consider the likelihood that these plans will come to fruition. While accurate predictions of the scale and outcomes of future forest restoration are hard to make, it is possible to gain informed insights into probable future trends. In this section, drivers of forest restoration and past developments provide clues into how the global forest restoration project may play out.

In the past, reforestation and afforestation were commonly used terms for processes leading to forest area increase. Reforestation is generally understood as bringing back forest cover where it had existed in the recent past, for example, less than 50 years ago, while afforestation is understood as establishing tree cover where it has not been a recent component of the landscape (UNFCCC 2013). During the 1990s and into the 2000s, the term forest rehabilitation gained traction among forest professionals, defined as: "efforts to bring back tree cover on formerly forested grasslands, brushlands, scrublands, or barren areas through planting, seeding and assisted natural regeneration for the purposes of producing industrial timber, sustaining livelihoods, or restoring forest environmental functions" (Chokkalingam et al. 2005, 405).

Since the first decade of this century, the term forest rehabilitation was gradually replaced with the term forest restoration, defined as: "returning trees to former forest land and improving the condition of degraded forests" (Gann et al. 2019, S6). De Jong et al. (2021) argue that this represents an ideological shift from reestablishing tree cover to produce wood and some compatible ecosystem services (e.g., reducing erosion) toward "actions to restore ecological processes, which accelerate recovery of forest structure, ecological functioning and biodiversity levels towards those typical of climax forest" (Elliot et al. 2013, 12). The latest iteration of related terms is forest landscape restoration, also referred to as forest and landscape restoration. Chapter 5 defines it as "efforts to address the drivers of deforestation and land degradation, to restore ecological functions, and to promote the sustainable use of land and forest resources with positive outcomes at the landscape scale for livelihoods, ecosystem services, and biodiversity conservation."

Considering the variable definitions of forest restoration, estimating how much forest has been "restored" in the past (Chapter 2) or in recent years (Chapters 5, 10) is not easy. The increase in the extent of planted forests can be taken as a relevant indicator (de Jong et al. 2021). The Food and Agriculture's (FAO) quinquennial Global Forest Resource Assessment (FAO 2020) provides data on the increase of planted forest for each country since 1990. Other sources give estimates of its expansion since the mid-1960s (FAO 1988). In 2020, 290 Mha, or 7% of the global forest cover was planted forests, while the rest was "naturally regenerating forests" according to the FAO (2020) classification (Chapter 10). Important variations in recent restoration progress are observable across individual countries; a proposed grouping of countries in early, intermittent, or advanced forest restoration stages is presented in Chapter 5. The latter group includes countries that have made national engagement with forest restoration and have been or are currently recipients of overseas development cooperation for forest restoration (Chapter 5).

A series of indirect drivers influence the demand for forest ecosystem services. These include such drivers as population growth and density, changing and inequitable wealth, and consumption patterns and preferences, as well as the increase in atmospheric carbon combined with the expectation that restoring tree cover in large areas of land can mitigate climate change through carbon sequestration (Chapters 3,

6). Some drivers are linked to specific restoration modalities, e.g., rural outmigration has resulted in natural forest regrowth in some areas (Frei et al. 2020); fuelwood demand in other areas has resulted in efforts to establish short-rotation plantations.

The actual and projected future demand for ecosystem services from restored forests provides a confounding picture. The forest restoration discourse emphasizes addressing the unfolding climate crisis, widely predicted to worsen in coming years (IPCC 2022; Chapter 4). Among the multiple forums that advocate large-scale forest restoration, the primary service demanded from forest restoration, including agroforestry, is capturing CO_2 from the atmosphere and storing it long term in tree vegetation and forest soils. Predictions are, however, that about 45% of forest restoration under the Bonn Challenge will be undertaken as tree plantations to produce timber, and smaller proportions as natural regeneration and agroforestry (FAO 2022; NYDF Assessment Partners 2019; Chapters 5, 10). While each of those contributes to terrestrial carbon storage, wood production responding to market demands appears likely to continue to be the main driver for planting trees, assumedly because it is more profitable than profits obtained from capturing atmospheric carbon. This is especially evident in tropical South America, Southeast Asia, South Asia, and Eastern Africa (Chapter 5). However, wood production and its use in, for example, construction, substituting for fossil fuel-based materials, is also advocated as an important contribution to climate change mitigation (Zhao et al. 2022).

One key challenge to achieving any of the ambitious forest restoration goals is financing; not only to establish restored forests but also to ensure their long-term conservation and management. Private sector financing of forest restoration is expected to increase more than public investments (FAO 2022; Chapter 10). Sustainable financing, investments that generate profits but are also environmentally, economically, and socially sustainable, is advocated in forestry. Future trends of private sustainable financing, however, remain to be seen. A plethora of challenges, including risks—especially those related to predicted climate change and its impact on planted forests—and the perceived public goods status of forests, which are expected to generate a broad range of benefits, including carbon capture and biodiversity conservation, are perceived as limiting private investment options in forests and forestry (Begemann et al. 2023). While this is the case, commercial tree plantations are appreciated for ecosystem services beyond producing wood. National forest restoration commitments of any type are included by many countries under their Nationally Determined Contributions (NDC) to meet their Paris Agreement commitments (NYDF Assessment Partners 2019; Chapters 5, 10).

A similar argument can be made for agroforestry. Agroforestry is recognized as an important option with significant potential to yield sociocultural and economic benefits. Agroforestry production has been part of local land-use history in many locations where they are often established and managed to increase food security and resilience of agricultural production, but also to generate products for markets. The profitability of agroforestry production is widely hailed (e.g., Waldén et al. 2020) but also questioned (e.g., Ajayi et al. 2009). Increasing the number of trees in agricultural landscapes is considered an important future carbon sink (Zomer et al. 2022). A further motivation for forest restoration is generating bioenergy (Chapter 3), a sector that is expected to double in size within the next decade (Gade 2024). Bioenergy's negative environmental impacts, however, are questioned by OECD country consumers. Concerns are over contribution to air pollution and the conversion of forest landscapes to wood production landscapes. Bioenergy forest plantations are also not as effective a carbon sink as permanent restored forests (Reid et al. 2020).

Land availability is a fundamental factor influencing forest restoration opportunities. Global estimates postulate the availability of large areas of land for forest restoration (e.g., Bastin et al. 2019; WRI 2014), but such estimates have been questioned by others (e.g., Lewis et al. 2019; Veldman et al. 2019). To assess land availability for forest restoration, land-use dynamics within national territories are more relevant than those at the global scale, as the former must take into consideration important contextual factors. Contextualized estimates suggest sufficient land area available for restoration (e.g., Brancalion et al. 2019; Chapter 10): Brazil, Indonesia, India, Madagascar, and Colombia, for instance, are reported to have more than 100 Mha of land with a "high restoration opportunity" (Chapter 10). Veldman et al. (2019) and Lewis et al. (2019) argue that the estimates of the amount of "degraded" land available for forest restoration are incorrect, not least because many of these lands are claimed as territory by resident people, or they constitute natural non-forest landscapes. One important factor that may constrain land availability for restoration is the projected global expansion of agricultural lands. Food demand is predicted to increase by 35–56% by 2050 (van Dijk et al. 2021), requiring an area of similar magnitude as that targeted for forest restoration by the various international initiatives between now and 2030 (van der Esch et al. 2022). Tenure issues, complex and unclear throughout the tropics, represent another potential challenge, as do other competing land uses and security problems that characterize many parts of the world. The topic of land availability for forest restoration remains a contentious one that requires more focused research.

There are multiple factors negatively influencing forest restoration potential, referred to as barriers. For example, Meli et al. (2023) assess deficient legislative and policy environments, lacking political will, limited financial and technical capacity, gaps in scientific research, and little public awareness as important barriers. Many of these factors are reported in earlier assessments on forest restoration (e.g., Mansourian et al. 2005), and are explored in Chapters 5 and 10. Chapter 5 identifies political and legal frameworks, institutional capacities and effectiveness, economic incentives and investment, partnerships, and access to technology and technological innovations as being deficient in many countries with commitments to international forest restoration programs. However, considerable variation in such deficiencies exists across these countries. Contrasting with this view is the conclusion by Keenan et al. (Chapter 10), who argue that technological capacity, including traditional ecological knowledge to implement forest restoration, is available.

12.2.2 Scenarios of restored forests

The expansion of planted forests since 1990, as evidenced from FAO data, shows a declining annual expansion of the area of planted forests. Over the last 30 years, the average annual forest expansion increased from 8 Mha/year during the 1990–2000 period to 11 Mha/year during the 2000–2010 period. Since then, the increase in area of planted forest declined to 7 Mha/year 2010–2015 and to 5 Mha/year 2015–2020. During these years, net forest loss declined from 7.8 Mha per year in the decade 1990–2000 to 5.2 Mha per year in 2000–2010 and to 4.7 Mha per year in 2010–2020. Extrapolating the trend of planted forest suggested by these numbers results in an increase of the area of planted forests between 2020 and 2030 by about 83.4 Mha or less, falling far short of the areas that are targeted to be restored under the various international initiatives.

Assessments of the possible expansion of planted forests over the next decades suggest that demand drivers, especially for timber, will influence the expansion of planted forests (FAO 2022; Nepal et al. 2019). The estimates for future timber demand, however, vary greatly. Past assessments, like Carle and Holmgren (2008), predicted that until 2030 annual roundwood production from planted forests would increase to between 1.2 and 1.9 billion m^3. Korhonen et al. (2021) estimate an increase in commercial forest plantations of 20–40 Mha by 2050, but this estimate considers only high-production plantations (FAO 2022). Mishra et al. (2021), on the other hand, estimate the need for an additional 141 Mha in plantations before 2100 to meet future timber and wood demands.

The earlier estimates have been followed up by scenarios for restored forest expansion. Korhonen et al. (2014) use a range of social and economic factors to model the expansion of planted forests. Later reports (e.g., Korhonen et al. 2021; Nepal et al. 2019) link planted forest expansion to five different shared socioeconomic pathways (SSP), alternative scenarios of expected global socioeconomic change up to 2100 identified with the following key terms: sustainability, middle of the road, regional rivalry, inequality, and fossil fuel development. They anticipate that SSP 5, fossil fuel development, will result in the highest area of planted forests, reaching 470 Mha by 2070. The scenario with the lowest expansion is SSP 3, regional rivalry, anticipated to result in 380 Mha of planted forest by 2070.

In addition to planted forest dynamics, naturally regenerated forests constitute an important contribution to forest restoration (FAO 2020). Figure 12.1 indicates countries that have experienced natural forest increase in the decade between 2010 and 2020, further distinguishing those that already had an increase during the 1990–2020 period or 2000–2020 period. About half of those countries are in Europe and one quarter in Asia. Except for a few countries that have a stable area of natural forests, all other countries in the FAO's 2020 Global Forest Resource Assessment experienced natural forest decline during the 2010–2020 period. Data from FAO (2020) show a very limited number of countries that have experienced a net forest increase over recent years, while still experiencing a decline in naturally regenerated forest area. In these countries, the increase in planted forest area exceeds the decline of naturally regenerated natural forests.

12.3 Goals, costs, benefits, and disservices of restored forests

12.3.1 Restoration goals and modalities

Forest restoration goals can be summarized as generating or regenerating forest ecosystem services. These include carbon sequestration, water regulation, biodiversity

Figure 12.1 Countries with net natural forest increase during 1990–2020, 2000–2020, or 2010–2020. Author's elaboration, based on FAO (2020). Only countries that had no net forest decrease during the 2010–2020 period are included.

restoration and conservation, and timber and non-timber forest product production, among others. There is a suite of complementary goals that can be pursued through restoration, such as improving livelihoods, preserving cultural integrity, and sustaining economic activities that rely on nature (Chapters 3, 5, 8, 9, 10). Complementary forest restoration goals have been reported since the 1970s, but especially since the 1990s. For instance, restoring forest ecological functions and improving rural livelihoods are emphasized together in academic treatises on forest restoration (de Jong et al. 2021).

Achieving these goals depends on which type of restoration is pursued. To begin, forest restoration can be achieved through the application of active or passive forest restoration measures (Morrison and Lindell 2011). The former implies direct seeding or planting trees, while the latter refers to allowing tree regeneration to occur naturally through the germination of existing or naturally distributed seeds or tree remnants. Both active and passive restoration have multiple variations, including, for instance, natural regeneration and assisted natural regeneration, but also single-species tree plantations of different densities, or planting trees as part of agroforestry (Table 12.1). Often a distinction is made between using native species or non-native species in tree planting projects (Chapter 11). Under the forest and landscape restoration concept, agroforestry and silvopastoral production are included as restoration modalities (Chapters 5, 8, 9, 10).

Table 12.1 Examples of restoration modalities and their potential contributions towards different goals (Chapters: 7–10)

Restoration Modality	Main Goal			
	Biodiversity Improvement	**Carbon Sequestration**	**Livelihood Improvement**	**Timber Production**
Natural regeneration	●●	●●	●	●
Assisted natural regeneration/enrichment planting/secondary forest management	●●	●●	●	●
Large scale plantations with native species	●	●●	○○	●●
Large scale plantations with exotic species	○○	●●	○○	●●
Agroforestry/silvopastoral systems	●	●	●●	○
Woodlots, trees on farm, small scale plantations	●	●	●●	●

● Potential towards the goal; ●● Strong potential towards the goal

○ Possible adverse effects towards the goal; ○○ Possible strong adverse effects towards the goal

Which restoration modalities are used and what techniques are applied depend on who is undertaking or promoting forest restoration, on biophysical and climatic site-conditions, on the degree of degradation of the land to be restored, on past land use, as well as the social and cultural context. Passive restoration will lead to more favorable biodiversity outcomes, as well as the generation of some regulating ecosystem services. Active restoration is more commonly pursued if provisioning ecosystem services are preferred, that is, when production of timber or other wood products is the major interest. Even when regulating or cultural ecosystem services are the primary objectives, single-species tree plantations may be adopted because they offer options for profitable uses, which can be justified from a financial standpoint. In addition, active restoration is undertaken when goals of carbon sequestration or soil erosion mitigation are pursued. Table 12.1 provides a simplified overview of forest restoration modalities and how they contribute to four key goals.

12.3.2 Restoration costs, benefits, and trade-offs

As already discussed, forest restoration generates multiple benefits and costs, as well as potential trade-offs between both. The costs borne by actors undertaking restoration include land, labor, and other inputs; and the costs of financing, planning, and contracting. Costs often go far beyond the legal owner of the land assigned to be restored (see Chapters 4, 8, and Section 12.3.3). Costs also include opportunity costs, that is, the lost income from the land use that restoration replaces, whether the actor is the owner or a previous user. The costs of restoration measures increase with the degree of degradation of the land and the complexity of the restoration modality being pursued. In situations with similar production costs, the costs of forest regeneration increase along a continuum from natural regeneration to assisted natural regeneration, seeding, and tree planting. Agroforestry systems often have the highest establishment and management costs, particularly when they involve the establishment of high-value tree crops such as cacao or coffee (Chapters 8, 9). The silvicultural management of plantations is often quite costly, depending on the scale, requiring considerable investments at different stages of development.

Social costs of restoration include, inter alia, the negative impacts that restoration measures can inflict on previous land or forest users, others not engaged in the restoration itself, and those not enjoying the benefits generated. However, society at large may enjoy forest restoration benefits, even in cases where private actors undertake restoration to generate a limited set of benefits or services. Many of the social costs and benefits are difficult to evaluate, such as those related to biodiversity and the enhancement or reduction in social and cultural resilience (Chapters 4, 8, 9). Costs and benefits also vary according to the spatial scale over which restoration is implemented. Some ecosystem services require large areas, for example, positive impacts on water provision and water quality. If biodiversity restoration is the primary goal, connectivity between restored areas is often very beneficial, indeed essential. Some ecosystem services, such as timber, become more cost effective when larger areas are restored. Furthermore, forest restoration outcomes and benefits change over time. Reaching biodiversity goals or producing timber, for example, takes years, even decades. The permanence and management of restored areas are crucial for realizing benefits and their long-term provision.

Biodiversity is associated with the provision of ecosystem services in complex and multifaceted ways. More biodiversity is seen as being positively correlated to ecosystem resilience, which again contributes to the provisioning of many ecosystem services (Aerts and Honnay 2011; Chapter 7). How biodiversity recovery is achieved through restoration, however, depends crucially on the restoration modalities used, subsequent management, and time span. Biodiversity restoration and conservation are likely to decline along the natural regeneration, assisted natural regeneration, seeding, tree-planting continuum, if other conditions are equal. When considering biodiversity conservation or recovery, natural forest regeneration tends to yield better outcomes, but tree plantations are also not without merits. However, the impacts of plantations on biodiversity and forest ecosystem functions depend on the type of primary vegetation cover and the land cover composition of the landscapes where plantations are established (Chapter 7). For instance, if forests are restored in previously homogenous and intensively used agricultural landscapes with little structural diversity, single-species plantations may provide substantial biodiversity benefits. The opposite may be true if plantations are planted on extensively used marginal sites that harbor substantial biodiversity, especially if planted trees are left to largely occupy the site.

In recent years forest restoration has become prioritized as an important means to address the ongoing global climate crisis (Busch et al. 2019; Daigneault et al. 2022; Lewis et al. 2019). While forest restoration is hailed as a prospective solution (Bastin et al. 2019), there is little consensus over the mitigation potential of different restoration modalities. Existing studies use different research methods, timeframes, areas, and management regimes, and consider different carbon pools (above ground/below ground) including or excluding wood harvesting and subsequent use (Bernal et al. 2018; Waring et al. 2020; Chapters 6, 7).

As a point of consensus, to achieve significant amounts of atmospheric carbon removal through forest restoration, significant areas need to be restored (Bastin et al. 2019) and managed as forests for long time periods—periods during which deforestation and forest degradation need to be halted. To a significant degree, the challenges in meeting the latter conditions pose a serious threat to meeting the Bonn Challenge and other international forest restoration programs. Many countries with significant commitments to any of the international restoration programs continue to suffer uncontrolled deforestation (Fagan et al. 2020). In addition, as explained in Section 12.2, future demand for land for food production, not to mention struggles over land access and ownership, create doubts regarding the presumed widespread availability of land suitable for restoration. Indeed, these factors require greater attention in the forest-based climate change mitigation debate (Van der Esch et al. 2022; Chapter 6).

In some circumstances, forest restoration can lead to ecosystem disservices negatively affecting human well-being, mainly when restoration takes place through tree plantations, but also in connection with natural regeneration. Tree plantations can negatively affect freshwater supply (Creed and van Noordwijk 2018). Exotic tree species may provide poor habitats for native fauna and may even invade native forests (Campagnaro et al. 2018), or suppress native understory plant species. Restored forests can be the origin of pests and pathogens for humans, crops, or livestock and increase human–wildlife conflicts. In Japan, for instance, forests restored since the Pacific War are increasingly causing pollinosis, wildlife damage, and river flooding when landslides result in trees blocking rivers (Takahashi et al. 2021).

12.3.3 Forest restoration and local livelihoods

When social benefits are discussed in current forest restoration debates, a key reference group are rural residents in restoration locations. The impact of forest restoration on local livelihoods varies greatly across different contexts. As discussed above, since the 1990s sweeping claims have been made of win-win opportunities of successful forest restoration, that is, that by restoring forests it is possible to improve the livelihoods of local populations (e.g., Erbaugh and Oldekop 2018). The evidence of the actual impact of forest restoration on local people, however, indicates that outcomes vary greatly, depending on the specific contexts considered. Various meta studies and edited volumes on forest restoration have been published since 2000. At least for the earlier ones, impact on local communities was not a major focus. A six-country study undertaken by the Center for International Forestry Research (CIFOR) observes positive reception among communities of restoration support, but little evidence is provided of how communities benefitted from these initiatives (Chokkalingam et al. 2005). Subsequent review studies have tried to capture socioeconomic impacts of forest restoration more concretely (e.g., Adams et al. 2016; Aronson et al. 2010; FAO 2021; Malkamäki et al. 2018). Aronson et al. (2010, 150) point out that "linking ecosystem services to beneficiaries of ecosystem restoration, and demonstrating their values to society, has only recently begun to enter the mainstream academic literature."

The review of Adams et al. (2016) on the consequences of restoration on local livelihoods shows that the impact on income is assessed most frequently, followed by livelihood diversification, employment, equity, and fuelwood and timber production for personal use. Serious gaps include assessments of impacts on local cultures and equity. The impact on income is influenced by household characteristics, land productivity, and markets, among others (Adams et al. 2016). Malkamäki et al.'s (2018) review of large-scale commercial plantations finds many negative impacts on employment, access to land, people's livelihoods, and other social indicators. Similar observations are made by Péroches et al. (2022) and Scheidel and Work (2018). For their part, D'Amato et al. (2017) discover a mix of negative and positive experiences with commercial tree plantations among resident communities. Alternative arrangements in the case of commercial plantations, like contract farming or production by independent smallholders, are reported to have yielded economic benefits (Nigussie et al. 2021; van der Meer Simo 2020). An assessment focusing on non-concession tree plantations in the Greater Mekong Region suggests that while planted forests contribute to income and livelihood improvements, they reduce grazing land and associated livestock production (FAO 2021).

Case studies on forest restoration involving communities and smallholders report how forest restoration not only results in direct benefits and costs but also impacts local production factors and livelihood capitals, resulting in changes in livelihood strategies. Reforestation projects implemented under the Clean Development Mechanism (CDM) in China directly affect the livelihood strategies of residents, sometimes negatively. For example, when previous land -use options such as goat raising were prohibited, migration to regional towns was the result (Ba Feng et al. 2020). Studies assessing longer time periods indicate that land cover changes brought on by land-use changes have complex drivers. External forest restoration support, which operates for a limited duration, is only one of these drivers (Hernández-Aguilar et al. 2021; Truong et al. 2017). Other reports show communities' enduring ability and desire to be involved and to act in their own and others' best interests (Mukasa et al. 2022; Yuliani et al. 2023). An in-depth study in communities affected by Indonesian oil palm plantations show worrying equity impacts. These cases are cause for concern should similar patterns prove to be the norm in forest restoration efforts (Elmhirst et al. 2017; Li and Semedi 2021).

The chapters of this volume report similar mixed impacts of forest restoration on communities. Historical coercive forest restoration has resulted in the displacement of forest communities and the curtailing of environmentally benign or even favorable customary land use (Chapter 2). The review of historical restoration suggests similarities of past and contemporary forest restoration discourses and practices and the role assigned to local communities by national or international restoration protagonists, while ignoring local ontologies, epistemologies, and related values and preferences, as well as impacting local livelihood strategies (Chapters 2, 4, 8). Other chapters of this volume signal opportunities for creating employment, producing goods or services that can be traded in markets such as carbon sequestration, protecting watersheds, and preserving cultural services (Chapters 7, 8, 9, 10). If agroforestry is included as one broad and diverse forest restoration modality, the number of restoration options that link tree production to markets and opportunities to improved livelihoods increases. It may also contribute to improving food security, increasing the resilience of production systems (Chapters 8, 9, 10), and strengthening or restoring valued cultural practices (e.g., complex swidden agriculture practices). Many indigenous production systems traditionally include trees, offering a way to build collaboratively on existing traditional knowledge when further developing these practices (Chapter 8).

12.4 Evolving forest restoration governance

12.4.1 International restoration governance

Forest restoration paradigms have evolved dramatically over the past half century, and different priorities have gained prominence in international and national forums. Chapter 2 summarizes past restoration efforts taking place in many locations across the globe during the 19th and early 20th centuries, promoted by governments in national and international territories. During the 1970s, FAO and partners considered communal tree planting a top priority to avoid a looming fuelwood scarcity that never materialized (Dewees 1989). Since the 1990s, forest restoration discourses have advocated reversing the loss of forest-regulating ecosystem services and forest biodiversity, while improving local livelihoods, a narrative adopted by nearly all modern day international and national forest restoration initiatives, including those that primarily aim to address the global climate crisis.

The shifts in forest restoration paradigms are reflected in how forest restoration has become institutionalized in nonlegally binding international forest restoration arrangements and legally binding treaties. An international forest restoration governance architecture has emerged that is driven by such initiatives as the Bonn Challenge and the One Billion Tree Initiative. Leading to various regional initiatives, they all draw on voluntary commitments. These are promoted by civil society actors, or networks thereof, like the International Union for Conservation of Nature (IUCN), the Collaborative Partnership on Forests, and the World Economic Forum. Forest restoration, or at least goals and targets that can partly be met through forest restoration, are included in international treaties like the Convention on Biological Diversity (CBD), the United Nations Framework Convention on Climate Change (UNFCCC), and the United Nations Convention to Combat Desertification and protocols for their implementation. None of these specifically demand signatories to restore forests, but forest restoration can contribute to meeting obligations.

The three key logics identified in Chapter 3 as driving international forest and forest landscape restoration efforts have focused on addressing 1) global environmental challenges, especially climate change and biodiversity loss; 2) plantation establishment to fulfill the increasing demand for wood; and 3) restoration for local development. These key logics reflect the different priorities and goals that originate from the international level and are translated to national and, often, subnational goals. Following these logics, all international forest restoration initiatives, whether they constitute nonlegally binding governance arrangements, like the Bonn Challenge, or are part of international law, like UNFCCC or CBD, have as their primary mandate either biodiversity conservation (IUCN, CBD), economic development (World Economic Forum), or climate change mitigation (UNFCCC). While many current forest restoration initiatives prioritize a global environmental challenges logic, they have also adopted a prominent integrated restoration and development discourse, thus combining their primary logic with a community development one. However, this is neither enshrined in hard obligations (see Chapter 4) nor implemented routinely. Commercial production, as argued in Section 12.2, is largely driven by market demands and assuring profits for investors; however, these are often attached to environmental objectives, specifically climate mitigation (Begemann et al. 2023). In sum, the argument among many forest restoration advocates is that forest restoration can achieve multiple goals, that is, generating benefits to local and nonlocal residents alike. Looking deeper into the different forest restoration paradigms, initiatives, and practices, it appears that trade-offs are necessary, but they need to be assessed and governed on a case-by-case basis.

Recognizing the need to strive toward more equitable, just, and inclusive forest restoration that is adapted to local ecological, social, and cultural contexts, the UN Decade on Ecosystem Restoration initiative has developed 10 restoration principles to guide restoration efforts (FAO, SER, and IUCN CEM 2023). There is a widely held moral imperative, although not a legally imposed obligation, to uphold these principles and respect the Free Prior Informed Consent (FPIC), or similar principles. At least to some extent, companies and other actors engaged in forest restoration will be driven by corporate social responsibility principles or seek social licenses to operate, which may well be good news for resident communities.

12.4.2 Local governance and justice in forest restoration

Forest restoration takes place in diverse social, economic, and institutional contexts. It is pursued by different actors, where land is used for different purposes and is in different biophysical states of degradation. Forest governance is about authority, power, and capacities; how these are used to define or influence who has access to resources; how resources are managed and used; and how the related costs and benefits are distributed. Forest governance configurations, especially at the local level where restoration is implemented, determine processes and influence outcomes. This emphasizes the importance of local level governance (Chapter 4).

Close to 300 M indigenous and other rural peoples are estimated to live on land in the so-called Global South that can be targeted for restoration (Erbaugh et al. 2020). At least 74% of this potential land is on areas with variable population densities, where forests are combined with communal or smallholder agriculture, settlements, and other land uses (Minnemeyer 2011), suggesting great variation in local forest restoration governance. Little is known about local governance of forest restoration (for exceptions, see Butler and Schultz 2019; Mansourian and Parrotta 2018). In contrast, there are many studies on local governance of communal forests (e.g., Arts and Visseren-Hamakers 2012; Pagdee et al. 2006). Communal forest transition and CDM project implementation studies signal intra-communally coordinated actions (e.g., Hernández-Aguilar et al. 2021), but also coercion by communal authority (Ba Feng et al. 2020; Cano Cardona et al. 2014).

A typical forest restoration scenario, evident in many of the chapters of this volume, is that of an external agency engaging residents of locations where restoration is implemented. The need for inclusive and effective participation of all stakeholders in restoration and equitable distribution of restoration benefits and costs has been argued many times over (e.g., Lund et al. 2009; Chapter 5). Standards have been adopted in international law and initiatives, identified with terms like participation, empowerment, co-governance, co-management, and the FPIC principles.

The FPIC principles theoretically grant resident communities the right to decide freely after being fully informed on matters pertaining to any activity initiated by outsiders. The right is recognized internationally, in the UN Declaration on the Rights of Indigenous Peoples among others; it is also included in national legislations worldwide. FPIC principles are advocated in formal international restoration governance (Anderson 2011) but are considered insufficient according to the environmental justice framework (Chapter 4). The latter differentiates procedural, distribution, and recognition justice principles, which are more demanding than the FPIC principles. The decolonial environmental justice principles go one step further and call for wide-reaching autonomy of indigenous communities or other forest residents over territories targeted for forest restoration (Chapter 4). Governments worldwide are making commitments to restore forests over large areas. If, as predicted, the private sector's prioritization of tree plantations or planted forests dominates future forest restoration (NYDF Assessment Partners 2019), one can anticipate tension between these developments and ecosystem restoration principles (FAO, SER and IUCN CEM 2023) and FPIC principles.

It is likely that government agencies in countries with low performance on governance indicators will continue to favor actors who do not consider such participation or justice principles their priority. There is evidence that forests customarily claimed by indigenous and non-indigenous local people without legal tenure have often been allocated for tree planting, sometimes resulting in the displacement of local residents (Fairhead et al. 2012; Leach and Fairhead 2015). In many cases FPIC principles have been respected only in name; examples of where they have been outright ignored abound. Given the biodiversity restoration or conservation, climate change mitigation, and timber production goals dominating the forest restoration agendas (Chapter 3), the economic and environmental benefits of much future forest restoration will often be enjoyed by distant beneficiaries, with the costs borne by indigenous and other local communities and smallholders (Chapters 3, 4, 8, 10).

The issues of FPIC or environmental justice are furthermore complicated because of intra-communal diversity and power imbalances (Ba Feng et al. 2020; Cano Cardona et al. 2014; Chapters 4, 8). Within a landscape and a single community, different groups and individuals may value restoration benefits differently, and the costs of restoration can have distributional impacts that disproportionately affect the poorest and most marginalized (Chapters 4, 8, 10). Equal participation of different social groups across gender, ethnicity, caste, religion, etc. with different priorities and interests, is therefore important for just restoration outcomes. The criticism against importing Western conceptualizations and ideas, which includes gender equality, sustainable resource management, participatory governance, and institutional models, should be considered against existing injustices, the roots of which can be traced to colonial and postcolonial eras, postcolonial national policies, and/or to customary resource management and allocation, which in many cases is dominated by and favors local men and elites (Arora-Jonsson et al. 2019).

Natural resource management and restoration have also been criticized for importing and imposing Western worldviews, knowledge systems, scientific knowledge, and management models and ignoring traditional/indigenous knowledge, cultural practices, values, and resource management practices. This is reflected in the selection of restoration modalities and tree species, even in projects with explicit objectives to combine restoring forests and generating positive outcomes for the rural communities targeted by these initiatives (Chapter 4). These attempts suffer from assumptions about the passivity of local communities. Many projects try to do things for local people rather than with local people. Gaining the understanding that is truly needed for just restoration requires closer long-term involvement with local communities than has been the usual practice to date.

12.5 Overcoming challenges and nurturing potentials

12.5.1 Need to balance forest restoration modalities and pathways

The multiple calls and international initiatives for forest restoration anticipate different forest restoration modalities (Table 12.1). Evidence from this chapter and others in the book suggests that the expansion of commercial forest plantations, which includes the typical fast-growing species like acacia and eucalyptus but also a suite of other slower-growing coniferous species, has accelerated since the mid-20th century.

At the same time, commercial forest plantations are included in forest landscape restoration targets (Chapters 5, 8–10).

For instance, governments and industry might promote the substitution of fossil fuel-based construction materials for wood-based materials. This could in turn lead to an increase in plantations to meet the additional demand. This would generate additional, though limited, forest ecosystem service benefits beyond wood and timber, depending on which areas are reforested and as long as no existing forests are replaced with forest plantations. An important use of wood obtained from commercial tree plantations is fuelwood. It is predicted that fuelwood production will double in value between 2020 and 2028 (Gade 2024). Increased wood pellet production in some regions will likely result in the replacement of long-rotation production forest with short-rotation ones (Aguilar et al. 2020), resulting in a loss of related extra benefits. Such dominance of production-oriented forest restoration would hence have mixed environmental consequences and could limit opportunities to achieve livelihood or well-being improvements for local residents.

The remainder of modalities in Table 12.1 all relate to forests and trees on community or smallholder lands. Their extent is not systematically reported. The proportion of the total of around 2.2 billion ha of global agricultural land on which trees are present is already significant, and increasing the presence of trees can yield 4–19 GtC according to estimates (Zomer et al. 2022). An increase of 10% tree cover in agricultural lands is estimated to yield 18 GtC stored carbon. It is difficult to estimate how much communities, smallholders, and their land will be affected by such an increase, nor what socioeconomic impacts are likely to occur. Existing studies of rural communities in the Americas (e.g., Hernández-Aguilar et al. 2021), Africa (e.g., Leach and Scoones 2015), and Asia (e.g., Ba Feng et al. 2020; Truong et al. 2017), commonly considered as possible targets for integrated forest restoration development support, suggest that an increase of tree and forest cover is a long-term process, that results from multiple drivers, the most important of which are progressive transformations toward more intensive land use and outmigration; external tree-planting assistance projects appear but one determining driver. This suggests that the realization of such forest landscape restoration approaches—while looking promising with regard to their ability to provide multiple ecosystem services—will depend greatly on the context, including the interplay of local communities and external investors or projects.

12.5.2 Needs and options for good forest restoration governance

Within the current global forest restoration regime, forest restoration policies, regulations, and administration are largely a national affair. Countries voluntarily make pledges to such initiatives as the Bonn Challenge, the New York Declaration on Forests, or any of the other initiatives; countries include forest restoration plans as part of their NDCs to meet their Paris Agreement commitments and restoration efforts for forest and other natural habitats under the CBD or UN Decade on Ecosystem Restoration. International organizations, bilateral actors, and internationally operating civil society actors can and do provide incentives for forest restoration.

For instance, a mix of incentives encourages meeting good governance standards (Addink 2019) for national governments, and their agencies, as well as for actors with the financial and political clout to influence choices and decisions on when, where, and how to pursue and implement forest restoration. The need to incorporate community actors more meaningfully in such decision-making is one of the challenges.

A suite of forest restoration governance challenges can be envisioned in years to come. Where will commercial forest plantations be located? What sort of arrangements will be implemented, not only with resident populations but also with national and subnational governments, among others? How will these commercial plantations be linked to climate change mitigation efforts? In recent years, much attention has been given to corporate social responsibility, social licenses to operate, and an environmental, social, and governance approach to doing business that is more responsive to stakeholder concerns (de Jong and Humphreys 2016; Hajdúchová et al. 2019; Han and Hansen 2012; Stuart et al. 2022). In addition, an increasing number of countries are implementing FPIC-like regulations and legislations. FPIC principles are advocated by most actors who engage in forest restoration collaboration or assistance with local residents. While this is the case, how different groups, possibly with conflicting interests, are identified, and what their interests are, the degree to which they understand the relevant issues, and how they are accommodated in forest restoration assistance remain contentious (Chapter 4). Key questions in this regard are what interests or rights are to be considered, including rights of self-determination, recognition, distribution of benefits, between different parties and among single sociocultural groups. Even where FPIC principles are fully embraced by parties in a single forest restoration effort, how true participation in choices and decisions is to be achieved remains an area of attention and debate.

A sensible approach, given the diversity of contexts emphasized throughout this book, is a dual strategy of strengthening formal governance mechanisms and simultaneously developing working relationships with a broad array of communities in diverse areas. The purpose of such work in each community would include gaining shared understandings of goals and developing strategies to move forward with applicable restoration, as well as identifying insights that could likely apply to nearby similar communities. The forest-based rural development ODA has long developed tools to facilitate collaboration between local and nonlocal parties, some dealing specifically with restoration (Colfer and Prabhu 2023). However, some collaboration approaches have been criticized as favoring nonlocal agendas (Chapter 4).

Future efforts may foster communities of practice that allow stakeholders to share technical, economic, and social governance experiences to foster collaborative learning. Additional opportunities to increase adherence to environmental justice principles include establishing instruments similar to the safeguards now adopted as part of the REDD+ process (Duchelle et al. 2017). Once these are available, secretaries of international initiatives can undertake sensitization efforts with countries who have made pledges under any international initiative. Parties engaged in promoting or supporting forest restoration overseas usually advocate for FPIC and related ways of interacting with local residents. National government agencies and civil society actors may be part of national democratic and participation ethics that are different from those embraced by internationally accepted good governance and FPIC principles. The best way forward to achieve agreement and consensus on these principles is open, transparent, and fair negotiations between parties with forest restoration interests.

Concurrently, efforts need to be expanded at the grass-roots level (see Chapters 4, 8, 10), working directly with communities in areas where restoration is being planned. Without such direct involvement, there is little likelihood that local priorities and capabilities can be incorporated. Moving forward effectively will require recognition of the uniqueness of ecological and human communities and the necessity for longer term involvement in each.

12.5.3 Need to harmonize goals, benefits, costs, and threats

The prevailing discourses on forest restoration emphasize opportunities to pursue multiple goals simultaneously within single restoration interventions, or concurrently through multiple interventions within specific geographic regions. Table 12.1 illustrates that the services achieved depend on the restoration modality utilized. For instance, carbon sequestration can be achieved using all six modalities, but not at the same speed and scale. While timber and wood production is best obtained with large-scale plantations and woodlots, livelihood improvements are best obtained when smallholder plantations or agroforestry are adopted as the main modalities.

Forest biodiversity has played a rather subordinate role in the forest restoration debates so far, as has the local cultural relevance and meaning of forests. There is growing recognition that this needs to change. Biodiversity conservation and the maintenance of key cultural elements (sacred sites, well-loved species, unique and valued skills, and important rituals) may be best obtained when natural regeneration and its variants are chosen. Choices for one modality as opposed to others will lead to different benefits. The reality is that achieving priority outcomes of forest restoration implies choices on whose interests are to prevail, as well as how the different modalities are managed; it also depends on the productivity of the sites where forests are established.

Long-time horizons are often required for restoration benefits to materialize, and restoration modalities need to be adaptable to changes in the ecological circumstances due to climate change as well as changes in the social and economic context, such as outmigration, and changes in land management and local actors, priorities, and interests. In some contexts, security threats and corruption pose serious risks to forest restoration initiatives that must be considered. Restoration efforts will lead over time to changes in local actors' relationships with and dependency on land and forest resources. The long-term perspective needed for restoration emphasizes the need for flexibility and adaptation in the face of changing circumstances (Chapters 5, 6, 8, 9).

Forest restoration is often seen as a positive development with only positive outcomes, and the prevailing forest restoration discourses promise benefits to low-income residents of locations where forest restoration is undertaken (Erbaugh and Oldekop 2018; Chapters 3, 5, 8–10). Risks and possible undesired outcomes, that is, ecosystem and sociocultural disservices, are less recognized. The literature suggests more nuanced and partially context-dependent restoration outcomes, depending on which forest restoration modality is pursued. Factors like productivity, management, market demand, governance, and public administration greatly influence impacts on people and their livelihoods. It is crucially important to take up a nuanced perspective when engaging in forest restoration and to thoroughly consider possible consequences for the environment, economy, and (local) people.

Forests and their multiple benefits have been hailed in recent years as an important solution to major global challenges and win-win or win-win-win opportunities, that is, beneficial for climate and biodiversity, the economy, and resident communities (Chapter 3). This is the justification used to promote restoring forests on a grand scale. The implementation of the current ambitious plans, however, will take place through a large number of initiatives, each with its own particular circumstances, challenges, and conflicting interests. There is a great need to better understand the opportunities and constraints of the economic, social, cultural, political, and technical dimensions of forest restoration in markedly different contexts. Ongoing efforts could benefit well from a much better understanding of the rich forest restoration experiences that the world has known to date. Investing time and resources in increasing the knowledge of these dimensions of forest restoration should be of the highest priority. This will require both better understanding of the impacts of existing projects and policies, as well as more in-depth attention to the needs, goals, and desires of the communities affected.

Note

1. https://www.ipcc.ch/reports/

References

Adams, C., S.T. Rodrigues, M. Calmon, and C. Kumar. 2016. "Impacts of Large-Scale Forest Restoration on Socioeconomic Status and Local Livelihoods: What We Know and Do Not Know." *Biotropica* 48 (6): pp. 731–744.

Addink, H. 2019. *Good governance: Concepts and context.* Oxford: Oxford University Press.

Aerts, R., and O. Honnay. 2011. "Forest Restoration, Biodiversity and Ecosystem Functioning." *BMC Ecol* 11, 29. DOI:10.1186/1472-6785-11-29.

Aguilar, F.X., A. Mirzaee, R.G. McGarvey, et al. 2020. Expansion of US Wood Pellet Industry Points to Positive Trends but the Need for Continued Monitoring. *Scientific Reports* 10: p. 18607. DOI:10.1038/s41598-020-75403-z.

Ajayi, O.O., F.K. Akinnifesi, G.W. Sileshi, and W. Kanjipite. 2009. "Labour Inputs and Financial Profitability of Conventional and Agroforestry-Based Soil Fertility Management Practices in Zambia." *Agrekon* 48 (3): pp. 246–292. DOI:10.1080/03031853.2009.9523827.

Anderson, P. 2011. *Free, Prior, and Informed Consent: Principles and Approaches for Policy and Project Development.* Bangkok: RECOFTC and GIZ.

Aronson, J., J.N. Blignaut, S.J. Milton, et al. 2010. "Are Socioeconomic Benefits of Restoration Adequately Quantified? A Meta-Analysis of Recent Papers (2000–2008) in Restoration Ecology and 12 other Scientific Journals." *Restoration Ecology* 18: pp.143–154.

Arora-Jonsson, S., S. Agarwal, C.J.P. Colfer, et al. 2019. "SDG 5: Gender Equality – A Precondition for Sustainable Forestry." In *Sustainable Development Goals: Their Impacts on Forests and People*, edited by P. Katila, C.P. Colfer, W. de Jong, et al., pp. 146–177. Cambridge: Cambridge University Press.

Arts, B.J.M., and I.J. Visseren-Hamakers. 2012. "Forest Governance: A State of the Art Review." In *Forest People Interfaces: Understanding Community Forestry and Biocultural Diversity*, edited by B.J.M. Arts, S. van BommJel, M.A.F. Ros-Tonen, and G.M. Verschoor, pp. 241–257. Wageningen Academic Publishers.

Ba Feng, J. Liu, T. Zhu, Y. Liu, J. Zhao. 2020. "CDM Forest Carbon Sequestration Projects in Western China: An Analysis Using Actor-Centered Power Theory." *Forest Policy and Economics* 115: p. 102139.

Bastin, J.-F., Y. Finegold, C. Garcia, et al. 2019. "The Global Tree Restoration Potential." *Science* 365: pp. 76–79.

Begemann, A., C. Dolriis, and G. Winkel. 2023. "Rich Forests, Rich People? Sustainable Finance and Its Links to Forests." *Journal of Environmental Management* 326, 116808.

Bernal, B., L.T. Murray, and T.R.H. Pearson. 2018. "Global Carbon Dioxide Removal Rates from Forest Landscape Restoration Activities." *Carbon Balance Management* 13, 22. DOI:10.1186/s13021-018-0110-8.

Besseau, P., S. Graham, and T. Christophersen (eds.). 2018. *Restoring Forests and Landscapes: The Key to a Sustainable Future.* Global Partnership on Forest and Landscape Restoration. Vienna: IUFRO. https://afr100.org/sites/default/files/2023-08/GPFLR_FINAL%2027Aug_0.pdf

Brancalion, P., Aidin Niamir, Eben Broadbent, et al. 2019. "Global Restoration Opportunities in Tropical Rainforest Landscapes." *Science Advances* 5 (7). DOI:10.1126/sciadv.aav3223.

Brand, D., and D. Singh. 2014. "What is Needed to Make Markets for Forest Ecosystem Services a Reality?" In *Forests and Globalization: Challenges and Opportunities for Sustainable Development*, edited by W. Nikolakis and J. Innes, pp. 185–201. London: Routledge.

Bruntland, G.H. 1987. *Our Common Future: Report of the World Commission on Environment and Development.* Geneva: UN-Dokument A/42/427.

Busch, J., J. Engelmann, S. C. Cook-Patton, et al. 2019. "Potential for Low-Cost Carbon Dioxide Removal through Tropical Reforestation." *Nature Climate Change* 9: pp. 463–466. DOI:10.1038/s41558-019-0485-x.

Butler, W.H., and C.A. Schultz (eds.). 2019. *A New Era for Collaborative Forest Management: Policy and Practice Insights from the Collaborative Forest Landscape Restoration Program.* London: Earthscan/Routledge.

Campagnaro, T., G. Brundu, and T. Sitzia. 2018. "Five Major Invasive Alien Tree Species in European Union Forest Habitat Types of the Alpine and Continental Biogeographical Regions." *Journal for Nature Conservation* 43: pp. 227–238.

Cano Cardona, W., W. de Jong, P. Zuidema, and R. Boot. 2014. "Diverse Local Regulatory Responses to a New Forestry Regime in Forest Communities in the Bolivian Amazon." *Land Use Policy* 39: pp. 224–232; DOI:10.1016/j.landusepol.2014.02.013.

Carle, J., and P. Holmgren. 2008. "Wood from Planted Forests: A Global Outlook 2005–2030." *Forest Products Journal* 58 (12): pp. 6–18.

Chokkalingam, U., C. Sabogal, E. Almeida, E., et al. 2005. "Local Participation, Livelihood Needs, and Institutional Arrangements: Three Keys to Sustainable Rehabilitation of Degraded Tropical Forest Lands." In *Forest Restoration in Landscapes: Beyond Planting Trees*, edited by S. Mansourian, D. Vallauri, and N. Dudley, pp. 405–414. New York: Springer.

Colfer, C.J.P., and R. Prabhu (eds.). 2023. *Responding to Environmental Issues through Adaptive Collaborative Management: From Forest Communities to Global Actors.* London: Earthscan/Routledge.

Costanza, R., R. de Groot, P. Sutton, et al. 2014. "Changes in the Global Value of Ecosystem Services." *Global Environmental Change* 26: pp. 152–158.

Creed, I. F., and M. van Noordwijk (eds.). 2018. *Forest and Water on a Changing Planet: Vulnerability, Adaptation and Governance Opportunities.* A Global Assessment Report, IUFRO World Series Volume 38. Vienna: IUFRO.

Daigneault, A.J.S. Baker, J. Guo, et al. 2022. "How the Future of the Global Forest Sink Depends on Timber Demand, Forest Management, and Carbon Policies." *Global Environmental Change* 76: 102582. DOI:10.1016/j.gloenvcha.2022.102582.

D'Amato D., A. Malkamäki, N.J. Hogarth, and H. Baral. 2017. "A Descriptive Plantation Typology and Coding System to Aid the Analysis of Ecological and Socio-Economic Outcomes." *Current Forestry Reports* 3: pp. 296–307. DOI:10.1007/s40725-017-0066-z.

D'Annunzio, R., M. Sandker, Y. Finegold, and Z. Min. 2015. "Projecting Global Forest Area Towards 2030." *Forest Ecology and Management* 352: pp. 124–133.

Dave, R., C. Saint-Laurent, L. Murray, et al. 2019. *Second Bonn Challenge progress report. Application of the Barometer in 2018.* Gland: IUCN.

De Jong, W., and D. Humphreys. 2016. "A Failed Social License to Operate for the Neoliberal Modernization of Amazonian Resource Use: The Underlying Causes of the Bagua Tragedy of Peru." *Forestry* 89 (5): pp. 552–564. DOI:10.1093/forestry/cpw033.

De Jong, W., J. Liu, and H. Long, 2021. "The Forest Restoration Frontier." *Ambio* 50: pp. 2224–2237. DOI:10.1007/s13280-021-01614-x.

Dewees, P.A. 1989. "The Woodfuel Crisis Reconsidered: Observations on the Dynamics of Abundance and Scarcity." *World Development* 17 (8): pp. 1159–1172.

Dobson, A.P., S.L. Pimm, L. Hannah, et al. 2020. "Ecology and Economics for Pandemic Prevention." *Science* 369 (6502): pp. 379–381. DOI:10.1126/science.abc3189.

Dove, M. 1994. "A Revisionist View of Tropical Deforestation and Development." *Environmental Conservation* 20 (1): pp. 17–24.

Du, Y., and K. Takeuchi. 2019. "Can Climate Mitigation Help the Poor? Measuring Impacts of the CDM in Rural China." *Journal of Environmental Economics and Management* 95: pp. 178–197.

Duchelle, A.E., C. de Sassi, P. Jagger, et al. 2017. "Balancing Carrots and Sticks in REDD+ : Implications for Social Safeguards." *Ecology and Society* 22 (3): p.2. DOI:10.5751/ES-09334-220302.

Elliott, S., D. Blakesley, and K. Hardwick 2013. *Restoring Tropical Forests: a Practical Guide.* Kew: Royal Botanical Garden.

Elmhirst, R., M. Siscawati, B. Sijapati Basnett, and D. Ekowati. 2017. "Gender and Generation in Engagements with Oil Palm in East Kalimantan, Indonesia: Insights from Feminist Political Ecology." *The Journal of Peasant Studies* 44 (6): pp. 1135–1157. DOI: 10.1080/03066150.2017.1337002.

Erbaugh, J.T., and A. Oldekop. 2018. "Forest Landscape Restoration for Livelihoods and Well-being." *Current Opinion in Environmental Sustainability* 32: pp. 76–83. DOI: 10.1016/j.cosust.2018.05.007.

Erbaugh, J.T., N. Pradhan, J. Adams, et al. 2020. "Global Forest Restoration and the Importance of Prioritizing Local Communities." *Nature Ecology & Evolution* 4: pp. 1472–1476.

European Union. 2023. *Guidelines on Biodiversity-Friendly Afforestation, Reforestation and Tree Planting*. Commission Staff Working Document. Brussels: European Union.

Fagan, M.E., J.L. Reid, M.B. Holland, J.G. Drew, and R.A. Zahawi. 2020. "How Feasible Are Global Forest Restoration Commitments?" *Conservation Letters* 13: p. e12700. DOI:10.1111/conl.12700.

Fairhead, J., M. Leach, and I. Scoones. 2012. "Green Grabbing: A New Appropriation of Nature?" *The Journal of Peasant Studies* 39 (2): pp. 237–261. DOI: 10.1080/03066150.2012.671770.

FAO. 1988. *An Interim Report on the State of Forest Resources in the Developing Countries*. Rome: Food and Agricultural Organization.

FAO. 2020. *Global Forest Resources Assessment 2020 – Main Report*. Rome: FAO.

FAO. 2021. *Understanding the Impact of Planted Forest on Smallholder Livestock Farmers and Their Livelihoods in the Greater Mekong Subregion*. Bangkok: FAO.

FAO. 2022. *Global Forest Sector Outlook 2050*. Rome: FAO.

FAO, SER, and IUCN CEM. 2023. *Standards of Practice to Guide Ecosystem Restoration. A Contribution to the United Nations Decade on Ecosystem Restoration*. Summary report. Rome: FAO.

Frei, T., J. Derks, C.R. Fernández-Blanco, and G. Winkel. 2020. "Narrating Abandoned Land: Perceptions of Natural Forest Regrowth in Southwestern Europe." *Land Use Policy* (99): p.105034.

FSC. 2012. *Strategic Review on the Future of Forest Plantations*. Helsinki: Forest Stewardship Council.

Gade, S. 2024. *Wood Fuel Market Latest Trends and Analysis, Future Growth Study by 2032*. https://www.linkedin.com/pulse/wood-fuel-pellets-market-analysis-present-future-7spie.

Gann, G.D., T. McDonald, B. Walder, et al. 2019. "International Principles and Standards for the Practice of Ecological Restoration. Second edition." *Restoration Ecology* 27: pp. S1–S46.

Gibb, R., D.W. Redding, K.Q. Chin, et al. 2020. "Zoonotic Host Diversity Increases in Human-Dominated Ecosystems." *Nature* 584: pp. 398–402. DOI:10.1038/s41586-020-2562-8.

Giessen, L. 2014. "Reviewing the Main Characteristics of the International Forest Regime Complex and Partial Explanations for Its Fragmentation." *International Forestry Review* 15 (1): pp. 60–70.

Haines-Young, R., and M. Potschin. 2018. *Common International Classification of Ecosystem Services (CICES) V5.1. Guidance on the Application of the Revised Structure*. Nottingham: Fabis Consulting.

Hajdúchová, I., C.H. Mikler, and B. Giertliová. 2019. "Corporate Social Responsibility in Forestry." *Journal of Forest Science* 65: pp. 423–427.

Han, X., and E. Hansen. 2012. "Corporate Social Responsibility Implementation in the Global Forest Sector." *The Journal of Corporate Citizenship* 47: pp. 101–118.

Hernández-Aguilar, J.A., E. Durán, A. Velázquez, W. de Jong, and G. Pérez-Verdín. 2021. "Understanding Direct and Indirect Factors of Forest Transition in the Mixteca Alta Landscape of Oaxaca, Mexico." *Forest Policy and Economics* 131: p. 102542.

Hoch, L., B. Pokorny, and W. de Jong. 2012. "Financial Attractiveness of Smallholder Tree Plantations in the Amazon: Bridging External Expectations and Local Realities." *Agroforestry Systems* 84: pp. 361–375.

Hua, F., L.A. Bruijnzeel, P. Meli, et al. 2022. "The Biodiversity and Ecosystem Service Contributions and Trade-Offs of Forest Restoration Approaches." *Science* 376 (6595): pp. 839–844. DOI:10.1126/science.abl4649.10.1126/science.abl4649(2022).

Humphreys, D. 2017. "Rights of Pachamama: The Emergence of an Earth Jurisprudence in the Americas." *Journal of International Relations and Development* 20 (3): pp. 459–484.

IPBES. 2019. *Global Assessment Report on Biodiversity and Ecosystem Services of the Intergovernmental Science-Policy Platform on Biodiversity and Ecosystem Services.* Edited by E.S. Brondizio, J. Settele, S. Díaz, and H. T. Ngo. Bonn: IPBES secretariat. DOI:10.5281/zenodo.3831673.

IPCC. 2022: *Climate Change 2022: Impacts, Adaptation, and Vulnerability.* Contribution of Working Group II to the Sixth Assessment Report of the Intergovernmental Panel on Climate Change. Cambridge: Cambridge University Press DOI: 10.1017/9781009325844.

Katila, P., C.J.P. Colfer, W. de Jong, et al. (eds.). 2019. *Sustainable Development Goals: Their Impacts on Forests and People.* Cambridge: Cambridge University Press.

Koo, J.-C. 2021. *Korean Forest Policy and Current Issues.* FFTC Agricultural Policy Platform. https://ap.fftc.org.tw/article/2859.

Korhonen, J., J.P. Prestemon, F. Cubbage, and P. Nepal. 2021. "Projecting Global and Regional Outlooks for Planted Forests under the Shared Socioeconomic Pathways." *New Forests* 52 (4): pp. 1–20.

Korhonen, J., A. Toppinen, F. Cubbage, and J. Kuuluvainen. 2014. "Factors Driving Investment in Planted Forests: A Comparison Between OECD and non-OECD Countries." *International Forestry Review* l6 (1): pp. 67–77.

Le, H. D., C. Smith, J. Herbohn, and S. Harrison. 2012. "More Than Just Trees: Assessing Reforestation Success in Tropical Developing Countries." *Journal of Rural Studies* 28: pp. 5–19.

Leach, M., and J. Fairhead. 2015. *Misreading the African Landscape; Society and Ecology in a Forest-Savanna Mosaic.* Cambridge: Cambridge University Press.

Leach, M., and I. Scoones. 2015. *Carbon Conflicts and Forest Landscapes in Africa.* London: Routledge.

Lewis, S.L., C.E. Wheeler, E.T.A. Mitchard, and A. Koch. 2019. "Restoring Natural Forests Is the Best Way to Remove Atmospheric Carbon." *Nature* 568 (7750): pp. 25–28. DOI: 10.1038/d41586-019-01026-8.

Li, T.M., and P. Semedi. 2021. *Plantation Life: Corporate Occupation in Indonesia's Oil Palm Zone.* Durham: Duke University Press.

Lund, J.F., K. Balooni, and T. Casse. 2009. "Change We Can Believe In? Reviewing Studies on the Conservation Impact of Popular Participation in Forest Management." *Conservation & Society* 7 (2): pp. 71–82.

Malkamäki A., D. D'Amato, N.J. Hogarth, et al. 2018. "A Systematic Review of the Socio-Economic Impacts of Large-Scale Tree Plantations, Worldwide." *Global Environmental Change* 53: pp. 90–103. DOI:10.1016/j.gloenvcha.2018.09.001.

Mansourian, S., and J. Parrotta (eds.). 2018. *Forest Landscape Restoration: Integrated Approaches to Support Effective Implementation.* New York: Routledge/Earthscan.

Mansourian, S., D. Vallauri, and N. Dudley. 2005. *Forest Restoration in Landscapes. Beyond Planting Trees.* New York: Springer. DOI:10.1007/0-387-29112-1.

Meadows, D.H., D.L. Meadows, J. Randers, and W.W. Behrens III. 1972. *The Limits to Growth; A Report for the Club of Rome's Project on the Predicament of Mankind.* New York: Universe Books.

Meli, P., D. Schweizer, L. A. Winowiecki, et al. 2023. "Mapping the Information Landscape of the United Nations Decade on Ecosystem Restoration Strategy." *Restoration Ecology* 31 (1): p. e13810.

Millennium Ecosystem Assessment. 2005. *Ecosystems and Human Well-Being: Synthesis.* Washington: Island Press.

Minnemeyer, S. 2011. *Atlas of Forest and Landscape Restoration Opportunities.* Washington: World Resources Institute.

Mishra, A., F. Humpenöder, J.P. Dietrich, et al. 2021. "Estimating Global Land System Impacts of Timber Plantations Using MAgPIE 4.3.5." *Geoscientific Model Development* 14: pp. 6467–6494. DOI:10.5194/gmd-14-6467-2021.

Morrison, E.B., and C.A. Lindell. 2011. "Active or Passive Forest Restoration? Assessing Restoration Alternatives with Avian Foraging Behavior." *Restoration Ecology* 201: pp. 170–177.

Mukasa, C., A. Tibazalika, A. Banana, E. Mwangi, and T. Mutimukuru-Maravanyika. 2022. "Strengthening Women's Tenure Rights and Participation in Community Forestry in Central Uganda." In *Adaptive Collaborative Management of Forest Landscapes: Villagers, Bureaucrats and Civil Society*, edited by C. J. P. Colfer, R. Prabhu, and A. Larson, pp. 106–132. London: Earthscan/Routledge.

Nepal, P., J. Korhonen, J.P. Prestemon, F. W. Cubbage. 2019. "Projecting Global Planted Forest Area Developments and the Associated Impacts on Global Forest Product Markets." *Journal of Environmental Management* 240 (15): pp. 421–430.

Nigussie, Z., A. Tsunekawa, N. Haregeweyn, et al. 2021. "The Impacts of Acacia Decurrens Plantations on Livelihoods in Rural Ethiopia." *Land Use Policy* 100: p. 104928.

NYDF Assessment Partners. 2019. "Protecting and Restoring Forests: A Story of Large Commitments Yet Limited Progress." New York Declaration on Forests Five-Year Assessment Report. https://climatefocus.com/wp-content/uploads/2022/06/2019NYDFReport.pdf.

Ota, L., J. Herbohn, N. Gregorio, and S. Harrison. 2020. "Reforestation and Smallholder Livelihoods in the Humid Tropics." *Land Use Policy* 92 (2): p. 104455. DOI: 10.1016/j.landusepol.2019.104455.

Pagdee, A., Y.-S. Kim, P.J. Daugherty. 2006. "What Makes Community Forest Management Successful: A Meta-Study from Community Forests Throughout the World." *Society and Natural Resources* 19: pp. 33–52.

Péroches A., H. Baral, M. Chesnes, A. Lopez-Sampson, and G. Lescuyer. 2022. "Suitability of Large-Scale Tree Plantation Models in Africa, Asia and Latin America for Forest Landscape Restoration Objectives." *Bois et Forêts des Tropiques* 351: pp. 29–44. DOI: 10.19182/bft2022.351.a36870.

Pokorny, B., and W. de Jong. 2015. "Smallholders and Forest Landscape Transitions: Locally Devised Development Strategies of the Tropical Americas." *International Forestry Review* 17 (S1): pp. 1–19. http://www.ingentaconnect.com/content/cfa/ifr/2015/00000017/a00101s1.

Reid, W.V., M.K. Ali, C.B. Field. 2020. "The Future of Bioenergy." *Global Change Biology* 26: pp. 274–286.

Reinecke, S., M. Blum. 2018. "Discourses across Scales on Forest Landscape Restoration." *Sustainability* 10 (3): p. 613. DOI:10.3390/su10030613.

Rockström, J., W. Steffen, K. Noone, et.al. 2009. "Planetary Boundaries: Exploring the Safe Operating Space for Humanity". *Ecology and Society* 14 (2): p. 32.

Scheidel, A., and C. Work. 2018. "Forest Plantations and Climate Change Discourses: New Powers of 'Green' Grabbing in Cambodia." *Land Use Policy* 77: pp. 9–18.

Sears, R.R., P. Cronkleton, F.P. Villanueva, M.M. Ruiz, M.P.-O. del Arco. 2018. "Farm-Forestry in the Peruvian Amazon and the Feasibility of Its Regulation through Forest Policy Reform." *Forest Policy and Economics* 87: pp. 49–58.

Stuart, A., A.J. Bond, A.M.A. Franco, et al. 2022. "Conceptualising Social Licence to Operate." *Resources Policy* 85, Part A, 103962, DOI:10.2139/ssrn.4191438.

Takahashi, T., W. de Jong, H. Kakizawa, et al. 2021. "New Frontiers in Japanese Forest Policy: Addressing Ecosystem Disservices in the 21st Century." *Ambio* 50: pp. 2272–2285.

Truong, D.M., M. Yanagisawa, Y. Kono. 2017. "Forest Transition in Vietnam: A Case Study of Northern Mountain Region." *Forest Policy and Economics* 76: pp. 72–80.

UNFCCC. 2013. *Afforestation and Reforestation Projects under the Clean Development Mechanism: A Reference Manual.* https://cdm.unfccc.int/public_inputs/2013/arcdm_01/AR_CDM_Manual_Draft_01.pdf.

Urzedo, D., M. Westerlaken, and J. Gabrys. 2023. "Digitalizing Forest Landscape Restoration: A Social and Political Analysis of Emerging Technological Practices." *Environmental Politics* 32 (3): pp. 485–510. DOI:10.1080/09644016.2022.2091417.

Van der Esch, S., A. Sewell, M. Bakkenes, et al. 2022. The Global Potential for Land Restoration: Scenarios for the Global Land Outlook 2. The Hague: PBL Netherlands Environmental Assessment Agency PBL publication number 4816.

Van der Meer Simo, A. 2020. "Livelihood Impacts of Plantation Forests on Farmers in the Greater Mekong Subregion: A Systematic Review of Plantation Forest Models." *Forests* 11 (11): p. 1162. DOI:10.3390/f11111162.

Van Dijk, M., T. Morley, M.L. Rau, et al. 2021. "A Meta-Analysis of Projected Global Food Demand and Population at Risk of Hunger for the Period 2010–2050." *Nature Food* 2: pp. 494–501. DOI:10.1038/s43016-021-00322-9.

Veldman, J.W., J.C. Aleman, S.T. Alvarado, et al. 2019. Comment on "The Global Tree Restoration Potential". *Science* 366(6463): pp. eaay7976. DOI:10.1126/science.aay7976.

Verkerk, P.J., R. Costanza, L. Hetemäki, et al. 2020. "Climate-Smart Forestry: The Missing Link." *Forest Policy and Economics* 115: p. 102164.

Waldén, P., M. Ollikainen, and H. Kahiluoto. 2020. "Carbon revenue in the Profitability of Agroforestry Relative to Monocultures." *Agroforestry Systems* 94: pp. 15–28 (2020). DOI:10.1007/s10457-019-00355-x.

Waring, B., M. Neumann, I. C. Prentice, et al. 2020. "Forests and Decarbonization – Roles of Natural and Planted Forests." *Frontiers for Global Change* 3: 58. DOI:10.3389/ffgc.2020.00058.

World Resource Institute. 2014. *Atlas of Forest Restoration Opportunities.* https://www.wri.org/data/atlas-forest-and-landscape-restoration-opportunities.

Yuliani, L., M. Moeliono, Y. Kusumanto, et al. 2023. "Revisiting Baru Pelepat: Life after ACM (Indonesia)." In *Responding to Environmental Issues through Adaptive Collaborative Management: From Forest Communities to Global Actors,* edited by C. P. Colfer and R. Prabhu, pp. 25–55. London: Earthscan/Routledge.

Zhao, J., X. Wei, L. Li. 2022. "The Potential for Storing Carbon by Harvested Wood Products." Frontiers in Forests and Global Change 5. DOI:10.3389/ffgc.2022.1055410.

Zomer R.J., D.A. Bossio, A. Trabucco, M. Noordwijk, and J. Xu. 2022. "Global Carbon Sequestration Potential of Agroforestry and Increased Tree Cover on Agricultural Land." *Circular Agricultural Systems* 2: 3. DOI:10.48130/CAS-2022-0003.